Geobotany Studies

Basics, Methods and Case Studies

The series includes outstanding monographs and collections of papers on a range of topics in the following fields: Phytogeography, Phytosociology, Plant Community Ecology, Biocoenology, Vegetation Science, Eco-informatics, Landscape Ecology, Vegetation Mapping, Plant Conservation Biology, and Plant Diversity. Contributions should reflect the latest theoretical and methodological developments or present new applications on large spatial or temporal scales that will reinforce our understanding of ecological processes acting at the phytocoenosis and vegetation landscape level. Case studies based on large data sets are also considered, provided they support habitat classification refinement, plant diversity conservation or vegetation change prediction. ***Geobotany Studies: Basics, Methods and Case Studies*** is the successor to *Braun-Blanquetia*, a journal published by the University of Camerino from 1984 to 2011 in cooperation with the *Station Internationale de Phytosociologie* (Bailleul, France) and the *Dipartimento di Botanica ed Ecologia (*Università di Camerino, Italy) and under the aegis of the *Société Amicale Francophone de Phytosociologie*, the *Société Française de Phytosociologie*, the *Rheinold-Tüxen-Gesellschaft* and the *Eastern Alpine and Dinaric Society for Vegetation Ecology*. This series promotes the expansion, evolution, and application of the invaluable scientific legacy of the Braun-Blanquet school.

More information about this series at http://www.springer.com/series/10526

Franco Pedrotti · Elgene Owen Box
Editors

Tools for Landscape-Scale Geobotany and Conservation

Springer

Editors
Franco Pedrotti
Department of Botany and Ecology
University of Camerino
Camerino, Italy

Elgene Owen Box
Department of Geography
University of Georgia
Athens, GA, USA

ISSN 2198-2562 ISSN 2198-2570 (electronic)
Geobotany Studies
ISBN 978-3-030-74949-1 ISBN 978-3-030-74950-7 (eBook)
https://doi.org/10.1007/978-3-030-74950-7

This Springer imprint is published by the registered company Springer Nature Switzerland AG
The registered company address is: Gewerbestrasse 11, 6330 Cham, Switzerland

Preface

Tools for Landscape-Scale Geobotany and Conservation

This book presents a variety of methodologies, tools (s.l.), and other approaches for ecological analysis and evaluation of sites and landscapes, especially for geobotanical study and conservation purposes. Most of the chapters represent talks from special sessions at two annual meetings of the International Association for Vegetation Science (IAVS):

1) Pirenopolis (Brazil, 2016), on "Applied Mapping for Conservation and Management: from Plant to Landscape"
2) Palermo (Sicily, 2017), on "Vegetation Patterns in relation to multi-scale levels of ecological complexity: from associations to geoseries"

The individual papers are grouped by methodology: mapping, phytosociology, basic ecological analysis, and conservation strategy.

Mapping

Mapping was a main theme of both special sessions and is an often indispensable tool in vegetation study at all scales. Javier Loidi presents two chapters on various aspects of vegetation mapping. The first represents a good introduction to major concepts of biogeography in general and then considers how to map biogeographical territories at different ranks, from kingdom to down to district, using floristic, vegetation, and landscape criteria. The second chapter explains and develops further the concepts of vegetation catenas, series, and geosigmeta. A catalog of different types of sigmeta is presented, along with a methodology for mapping dynamic vegetation catenas, as a tool for ecological restoration and conservation.

The science of symbols and symbolism is called semiology, and some basic principles of semiology are explored by Perrin and Bioret, especially the use of hues and luminosity as well as colors *per se* to represent different phytosociological units. Another mapping consideration involves bias due to differential sampling effort. How to map this bias explicitly, in Europe, is treated by Garzon-Lopez and Rocchini. The concept of ecosystem services is rather anthropocentric but is also well established in various aspects of environmental policy. A pilot project using detailed data of many kinds from many sources is described in detail by Mesquita et al. to assess and map ecosystem services in southern Portugal. In addition to ecosystem services, this project defines criteria for assessing the value, threat level, rarity, naturalness, phytodiversity, and replaceability of the ecosystems treated.

Mountains pose particular problems for both field study of vegetation and mapping. In the Bolivian Andes, at elevations over 3000m, wetland mapping for sustainable management is described by Liberman Cruz, who also explains in detail the local traditional system for managing the domesticated native camelids that depend on the wetlands. Also in Bolivia, Pedrotti applies the concepts of sigmeta and geosigmeta to geobotanical mapping in three quite different environments, presenting maps for evaluation of the conservation status of the quite different vegetation of each area.

The last two chapters in this section treat the mapping of two quite different aspects at of vegetation mapping at local scales. Raimondo presents some of the first attempts to map plant diversity in Italy, with three examples from areas near Palermo in Sicily. The mapping work involved photographs, field sampling, and GIS software to assess biodiversity, degrees of endemism and naturalness. In cultural landscapes, economic and environmental values must be combined with social and cultural values for effective sustainable management. As a result, people's mental representations of remaining patches of more or less natural vegetation may become important, especially fragments of forest. Roulier et al. consider how to map people's cognition of forest fragments in France.

Phytosociology

Traditional phytosociology still has its place, especially in more local and regional vegetation studies. A good example is the updated phytosociological treatment of the ultramafic vegetation of California by Sánchez Mata, which proposes some new contributions to the taxonomic nomenclature for California. In the eastern Carpathian Mountains of Romania, Oroian and co-authors use traditional phytosociology to identify associations of *Petasites* and related tall forbs along montane riverbanks. Ramirez and co-authors use traditional phytosociology to identify halophytic plant communities, especially those involving *Arthrocaulon (Arthrocnemum) macrostachyum* along coastlines and in inland saline areas in the Iberian Peninsula, Balearic and Canary Islands. In recent years, traditional phytosociology has expanded to treat more integrative, synthetic aspects of vegetation patterns in the field. This sub-field, known as symphytosociology, studies landscape mosaics and provides the underpinning for the concepts of sigmeta and geosigmeta already introduced in the preceding section on mapping. Bioret provides a good introduction to the concept and uses of geosymphytosociology, as a tool for bioevaluation. The use of symphytosociology as a tool for landscape monitoring over time is explained by Theurillat et al., using a case study from the Swiss Alps. This study, involving data from 1982 to 2014, combined a detailed list of phytosociological vegetation units, aspects of physiognomic vegetation structure, physical environmental factors, land use, and other human activities.

Ecology

Ecological analysis of vegetation may involove various methodologies, some applied simultaneously, to understand the effects of climate, substrate, topography, and other physical factors. Climatic envelope modeling is used by Box to suggest a climatic basis for the apparent high diversity of plant life forms and vegetation structures in Mediterranean regions. Also in the Mediterranean region, Acosta considers habitat types and environmental constraints along gradients in Italian coastal dune vegetation. In the warm-temperate climatic region of the southeastern USA, there are only small areas of zonal evergreen broad-leaved forest. Climatic, edaphic, and other constraints on this forest type, now and under global warming, are considered by Box and Fujiwara, using relevé as well as physical data. Finally, the role of riparian buffers in reducing pollution and ameliorating other consequences of intensive agriculture in Romania is treated by Gheorghe and colleagues. This detailed study considers hydrology, different agricultural crops, and nutrient loads in different river basins.

Conservation

Conservation is often the ultimate goal of applied vegetation studies. This final section presents three papers on management strategies and landscape evaluation for conservation purposes. Especially in Asia but also elsewhere, sacred sites often preserve natural vegetation. Sacred sites in northern Morocco are described by Deil and co-authors, who ask critical questions concerning the effectiveness of such sites for conserving the native flora, vegetation in general, and forest patches in particular. Cultural preservation can often be a useful strategy for conservation of natural ecosystems. Fujiwara describes how this is done along the Kaga coast of central Japan, where preservation of a historically significant coastal dune pinewood landscape and of habitat for a unique, tree-climbing crab are combined for conservation of natural ecosystems. Finally, small islands differ from larger islands in various characteristics, some affecting conservation strategies and biodiversity. Different methods for measuring biodiversity, at different fine scales, are considered by Domina and co-authors on small Mediterranean islands. They conclude that, on small islands, distribution mapping may be the most effective method for assessing biodiversity.

Elgene Owen Box
Franco Pedrotti

Contents

Conservation

Mapping

Basic Semiological Principles to Define Coherent Colour Schemes for Vegetation Mapping

Gwenhael Perrin and Frédéric Bioret

Abstract Colour is the most used visual element in cartography for symbolising vegetation because, more than many other signs, it translates information in an implicit and condensed way. While the diversity of vegetation and thematic mapping preclude the possibility of a universal colorimetric norm, certain pioneer works have tended to establish a methodological framework allowing the rendering of ecological and bioclimatic conditions. Thus, the evolution of colorimetric models, linked to the progression of infographic tools, make it possible to consider novel approaches.

This article proposes an intuitive method for using colorimetric parameters to represent phytosociological and synphytosociological units. It allows, through colour only, without any overlay, the representation of three to four ecological dimensions. Hue and luminosity are used for representing vegetation types or vegetation series and their physiognomy. Saturation, an overlooked parameter in mapping, represents the level of anthropization. Used in GIS applications, the system permits easy establishment of numerically calibrated palettes and contributes to improving map understanding.

Keywords Semiology · Cartography · Vegetation series mapping · Colours · Ecological and bioclimatic factors

1 Introduction

Vegetation maps are good tools for scientific knowledge and research but also represent tools of communication for applications to land management, as they can be presented to stakeholders and agencies dealing with different types of projects related to land management or elaboration of nature conservation strategies. Maps have to be easily understandable if their elaboration integrates basic rules.

G. Perrin (✉) · F. Bioret
Univ Brest, EA 7462 Géoarchitecture, 29200 Brest, France
e-mail: gwenhael.perrin@univ-brest.fr

F. Pedrotti and E. O. Box (eds.), *Tools for Landscape-Scale Geobotany and Conservation*, Geobotany Studies, https://doi.org/10.1007/978-3-030-74950-7_1

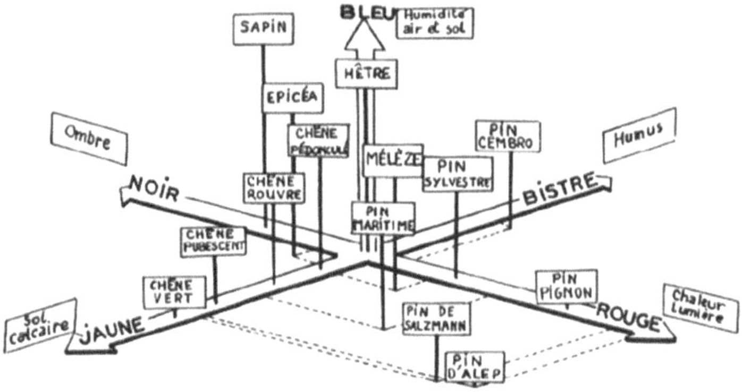

Fig. 1 Colorimetric space of Gaussen published in Ozenda [9]. The different colours of the series are located in a 3D volume, with a red-black axis, a yellow-brown axis and a blue vertical axis

A few authors have developed methodological approaches to harmonise and make more obvious to the reader the symbology of vegetation maps. Gaussen [4, 5] proposed the use of shades of hues in mapping vegetation series. The CNRS 1:200 000 map of France was used this principle to represent different dynamic stages by the same hue as the climax community but with colours ranging from light to dark in order to show the gap between the climax and the actual vegetation. Ozenda [9] introduced the notion of chromatic synthesis and preliminary colorimetric principles (Fig. 1), which, however, presents some incoherencies. For example, bistre, *i.e.* blackish brown, symbolises vegetation types on organic soils when the balance between black for shadow and red for light generates a colour close to bistre. The Ozenda system also presents counter-intuitive choices for the way colours are organised in more recent systems. More recently, Pedrotti [11] proposes two other ways to represent vegetation: use of monochrome palettes, from white to black, with superposition of hatching; use of colours without pre-established rules.

The French national CarHAB program aims to map habitats and vegetation series at the 1:25,000 scale all over the metropolitan territory up to 2025 [8]. For this, a preliminary theoretical reflection on semiology and vegetation mapping has been developed [12] and integrated into a methodological guide.

In his historical reference book, Bertin [1] proposed basic rules for colour use in representing qualitative or quantitative data, taking our perceptions into account. He explains that, according to its terminology, the colour—or hue—has a semiological value only if we can distinguish its colour (in other words, its hue) from its value (amount of black in the hue). The two graphical variables do not have the same symbolic properties and cannot be used for the same type of semiological components. In other words, the hue is not ordered because of its selective force that allows the eye to distinguish entities as belonging to different categories. It allows representing quantitative components, as in the case of vegetation types defined according to a categorisation based on ordinal to quantitative data analysis. Always following Bertin, an ordinal scale can nevertheless be foreseen within a colour temperature range. In this case, the eye is able to establish an order relation, contrarily to schemes based on the whole chromatic spectrum, from red to purple. On the contrary, the value suits ordinal components for which it is possible to establish a hierarchical relation between categories. Numerous semi-quantitative ecological variables expressed along gradients (trophy, humidity, bioclimatic belts…) can be linked to this component type.

Thus, after having reviewed some aspects of the colour symbolism applied to vegetation and having presented the colorimetric models commonly used in computer graphics, we will make some proposals useful for the development of semiological charts.

2 The Colour Metric in Infographics

Among the numerous available colorimetric spaces, three models are widely used for colour coding by computers. Each model has three parameters: RGB, HSL and HSV. Different proportions of the three base colours allow reproduction of all the possible colour impressions.

The RGB system reconstitutes a colour by mixing in additive synthesis the three basic colours, red (R), green (G), and blue (B). If the RGB model represents a main advantage in computer coding and storage, it adopts a purely physical conception and seems less adapted to the human colour vision that we do not perceive as an addition of primary colours but rather as a complex mixing of sensations associating brightness as perception of luminous flux intensity and chromaticness as perception of chromacity factors. Chromaticness results from two visual sensations: the hue that sensitively renders the dominant wavelength and saturation that renders purity of the colour.

The other two coding systems, HSL and HSV, developed by Smith [14], are based precisely on this colour perception, as they are said to be perceptual. They use two common variables, the hue of colour (H), in which all the shades are included inside a circle, and its saturation (S), varying from 0 to 100%. The result is defined according to a sensibly different method for each of the two models. The third parameter differs in the two models: HSV uses the value (V) of the colour, i.e. the brightness (the distance of one hue from black), while HSL uses luminosity (L), i.e. the degree of darkness or lightness. The main distinction between the two models corresponds to the way totally saturated hues are treated. The variables V and L range between 0 and 1 (Fig. 2).

Fort each hue, the variation of luminosity and purity range can result in four main colour types: bright (light and saturated), pale (light and unsaturated) deep (dark and saturated) and dull (dark and unsaturated).

If the HSV and HSL model conceptions are closed and offer the possibility of varying the saturation and intensity without affecting hue, nevertheless some differences make HSL preferable for mapping applications. If pure black ($L = 1$ and $V = 1$) is located at the base of the cylinder for both models, white is located at the top of the cylinder for HSL ($L = 1$) and at $V = 1$ and $S = 0$ for HSV. The brightest colours for HSV ($V = 1$) are equivalent to intermediate luminosity colours in HSL

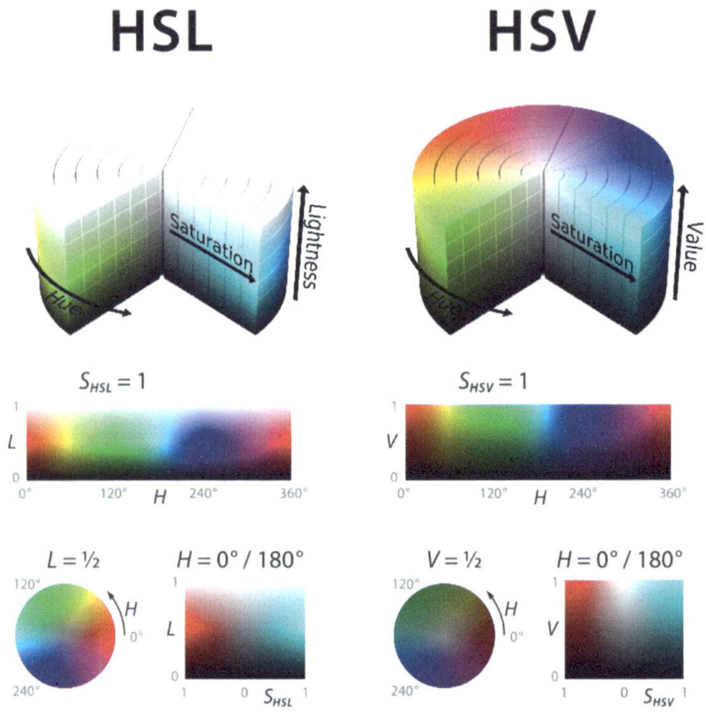

Fig. 2 HSL and HSV models (top: radial representation in 3 dimensions; middle: HL and HV spaces; bottom: colour wheel and SL spaces for two different colours)

(L = 0.5). HSL appears to be a symmetric system offering a wider range along the L axis, from black to the selected hue to white. It seems to be more intuitive for varying colour intensity (form dark to light). At constant hue, saturation gradients are dependent on the luminosity in HSL, which limits intensity variations from one extreme to the other, reducing the differences of perception.

The most commonly used GIS (ArcGIS, MapInfo, QGIS…) integrate these models in a more or less complete way, allowing precise numerical parameterization of colours. RGB and HSV are usually available, while HSL is less common. Its use requires a conversion with easily available algorithms. The following equations do the HSL-to-HSV conversion:

$$H_{HSV} = H_{HSL}$$

$$V = \frac{2L + S_{HSL}(1 - |2L - 1|)}{2}$$

$$S_{HSV} = \frac{2(V - L)}{V}$$

3 Colour Symbolism and Its Potential Use in Vegetation Mapping

It appears to be possible to translate the impression given by an object (natural, artificial, feeling, state…) by varying the three parameters according to a colorimetric language. This is the most common confusion in vegetation mapping: lack of explicitly established rules, at least based on a feeling or a shared symbolism, even if perceptions can differ inside the occidental world and over historical time [3, 10]. This is why we often recognise invariants in colour choice (red for warm and dry, blue for cold and wet) but also a broad range of choices from one cartographer to the other, based on the phenomena to be represented (edaphic gradients *vs* bioclimatic gradients) or representation modes. For all these reasons, we propose a flexible system to harmonise vegetation mapping semiology.

Table 1 proposes a correspondence between colorimetric parameters, symbolism, and vegetation or vegetation series for each opposed value of a colorimetric parameter (warm hue *vs* cold, saturated *vs* unsaturated colours…). The ecological significance is considered in the light of symbolic interpretation, which is generally realised according to what the colours seem to indicate. The proposed ecological and bioclimatic terms can be adapted easily to the cartographer's set of themes and to local context.

Table 1 Symbolic and ecological interpretation of the three colorimetric parameters

Variable	Value 1	⇔		Value 2
	Symbolic and ecological / bioclimatic equivalence	Colorimetric value		*Symbolic and ecological / bioclimatic equivalence*
Hue	*Sky, air, atmosphere, cold, sea, water, humidity, reduction* Cryophile, (edapho-) hygrophile, hydrophile, flooded, pergelid, ultrahyperhumid, boreal and polar, oceanic	Cyan (H=180°)	Red (H=0°)	*Fire, heat, drought, oxidation* Thermophile, (edapho-) xerophile, drained, infrathermic, ultrahyperarid, tropical and mediterranean, desert
	Planet Earth, darkness, night, humidity, reduction Sciaphilous, (edapho-) hygrophile, hydrophile, flooded, ultrahyperhumid, polar, series	Blue (H=240°)	Yellow (H=60°)	*Sun, shine, light, clarity, drought, oxidation* Heliophilous, (edapho-) xerophile, drained, ultrahyperarid, tropical and mediterranean, permaseries
	Unidentified natural symbolism In opposition to green: unproductive, stress tolerant, oligotrophic, acidophilic, basiphile, mor, euhaline, azonal, intrazonal	Magenta (H=300°)	Green (H=120°)	*Vegetation, vitality, fertility, growth* Productive, competitive, eutrophic, neutrophilic, mull, limnic, zonal, extrazonal
Luminosity	*Dark, shady, cloudy, heavy, deep, carbon* Closed, high, dense, evergreen, sciaphilous, series, organic	Dark (L=0)	Light (L=1)	*Bright, sunny, light, surficial* Open, low, sparse, deciduous, heliophilous, permaseries, mineral
Saturation	*Dull, bland, extinguished, monotonous, sad, impure, dusty, poor* Oligotrophic, ruderal, secondary, artificial	Unsaturated (S=0)	Saturated (S=1)	*Bright, vibrant, vibrant, varied, cheerful, pure, rich* Eutrophic, undisturbed, primary, natural

4 Setting of Colorimetric Variables

4.1 General Principle

Hue appears to be the main way to represent particular ecological factors. A colorimetric gradient can be produced by the opposition of divergent hues for elaborating a colorimetric spectrum (*e.g.* gradient from red to cyan). The temperature contrast between warm and cold colours is mainly used [6]. The polar chromatic of the range should be adapted for mapping themes and the cartographer's objectives (Table 1).

According to needs, two other colorimetric variables, L and S can be added as secondary ecological dimensions in order to represent, respectively, physiognomy —in the way Itten considers the progressive light-dark contrast—and the level of vegetation or vegetation-series degradation, by means of quality contrast (see horizontal and vertical axes of Fig. 3). This method permits one, in a certain way, to reconstitute the state of a dynamic system. Other sets of themes can also be developed, always by following the elements listed in Table 1.

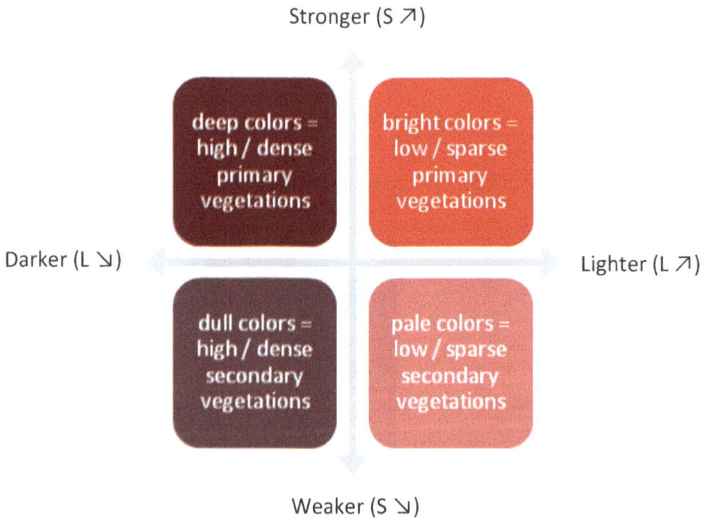

Fig. 3 The four divergent colours of a same hue

5 Symbolisation of First-Order Ecological Dimensions

The construction of a hue range can follow some simple rules, making it possible to use the entire available spectrum and to obtain the best balance possible.

In the case of a single ecological variable, the cartographer will define two hues located at the extremes of the spectrum in order to define the palette.

In the case of two ecological variables and in order to exploit one half of the chromatic circle, a nominal variable is defined first, in order to subdivide the circle into two parts (*e.g.* acidophilous in magenta *vs* neutrophilous in green, oligotrophic *vs* eutrophic, azonal *vs* zonal…) (Fig. 4).

A second ecological axis can be defined between two other chromatic poles, in order to symbolise an ordinal variable (*e.g.* from xerophilous in red to hydrophilous in cyan). Each vegetation type or vegetation series can then be located on the circle according to its values for the two variables.

In certain cases it is necessary to divide the circle in other ways. One triad could be used to symbolise zonality, according to a humidity gradient and the following scheme: from yellow to cyan for zonal vegetation or vegetation series (around green), from cyan to magenta for azonal tending to hygrophilous vegetation (around blue), and from magenta to yellow for azonal tending to xerophilous vegetation (around red).

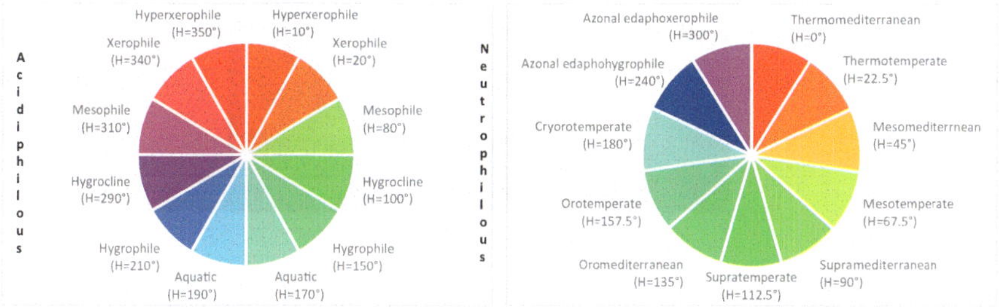

Fig. 4 Examples of chromatic symbolizations (diagram on the left: symbology adapted to vegetations according to a humidity gradient: left hemisphere coding for acidiphilic vegetations, right for neutrophilic ones; right diagram: complex symbology adapted to zonal series following a thermal gradient and to azonal series following a moisture series gradient: green-tending colours code for climatophilic series, magenta-tending colours correspond to azonal series)

6 Symbolisation of Physiognomic, Dynamic and Anthropogenic Dimensions

Once the hues are defined, luminosity and saturation levels are fixed. These two parameters are usually applied to ordinal type variables, and ecological factors to be described are categorised above all according to a quantitative scale. Thus, colours range from pale for more open vegetation, to intermediate luminosity for herbaceous and shrubby formations, to dark for closed forests (Table 2). It is similar for the three main vegetation series types of Rivas-Martínez (Table 3).

By the colour language described above, primary vegetation types and low levels of disturbance are represented by saturated colours, and secondary vegetation and high levels of disturbance by unsaturated colours; achromatic colours (*i.e.* grey) correspond to artificial environments (Table 4). The openness or density of artificial environments is represented in the same way, by a colour gradient from light to dark (Table 5). Addition of the saturation variable eventually provides a great variety of palettes, as illustrated by Table 6 and Table 7.

Table 2 Colorimetry of formations types

S=1

	H													
	0	30	60	90	120	150	180	210	240	270	300	330	360	L
Sparse annual herbaceous														0.8
Open perennial herbaceous														0.7
Closed perennial herbaceous														0.6
Dwarf shrubby														0.5
Shrubby														0.4
Open or pioneer woody														0.3
Closed woody														0.2

Table 3 Colorimetry of series types

S=1

	H													
	0	30	60	90	120	150	180	210	240	270	300	330	360	L
Permaseries														0.75
Minoriseries														0.50
Series														0.25

Table 4 Colorimetry of the degree of artificialization of a vegetation type

L=0.5	H													
	0	30	60	90	120	150	180	210	240	270	300	330	360	S
Primary														1.00
Replacement secondary														0.75
Regressive secondary														0.50
Anthropogenic														0.25
Artificial														0

Table 5 Colorimetry for artificial units

S=0		L
Annual crops		0.8
Perennial herbaceous crops, temporary meadows		0.6
Shrub plantations		0.4
Tree plantations		0.2

Table 6 Colorimetry of the formation type according to the degree of artificialization

	H1=0°				S				H2=180°	
	1.00	0.75	0.50	0.25	0.00	0.25	0.50	0.75	1.00	L
Sparse annual herbaceous										0.8
Open perennial herbaceous										0.7
Closed perennial herbaceous										0.6
Dwarf shrubby										0.5
Shrubby										0.4
Open or pioneer woody										0.3
Closed woody										0.2

Table 7 Colorimetry of the degree of artificialization according to the formation type

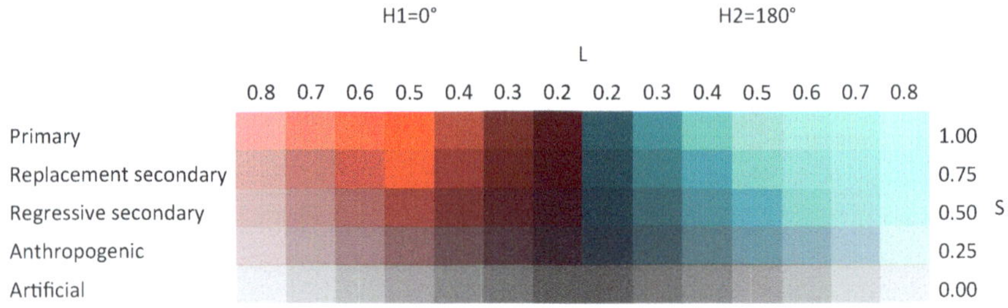

7 Additional Symbologies: Outlines, Overlays and Transparencies

Maps are not only constituted of solid colours. They may also integrate other symbolic elements to provide complementary thematic information that is useful to the cartographer's discourse, or more simply to the spatial localisation that can guide the reader's eye. Use of these other symbolic elements requires some rules.

An area outline represents a limit or a boundary between two polygons and is shown by a line. The nature of this limit could potentially be represented by one or several graphic variables specific to linear forms (thickness, form, grain…). Nevertheless, since polygons are generally contiguous to other polygons, setting up such a methodology seems complicated. In order not to multiply superfluous signs, it appears preferable to suppress less visible outlines when surfaces become smaller or when the scale increases. Their use could be reserved for particular cases such as insufficient contrast between adjacent colours or for thematic information useful for locations, such as administrative entities.

The use of fill symbols overlying flat tints allows restitution of additional information. These are composed of various visual variables: form (dots, lines, symbols), orientation of the form, grain, or colorimetric parameters [1]. In order to make the message less confused, overlays should be used with moderation, for two reasons. The first risk is to overcharge the senses by multiplying information from different themes. The second risk is to create an overcharge of signs. In general, overlays should be complementary and not redundant with the main colour raster. In principle, overlays should not be used to describe vegetation types or vegetation series, but rather should be used for additional information useful for interpreting the map. Different topics may be involved:

- for disturbance: grazing in dots, mowing in hatching, fire…; grain can express the intensity of the phenomenon;
- dynamic facies: colonisation by shrubs or trees can be represented by dots or symbols of different grains according to the level of colonisation;

- pedologic or geologic information: turf, sand, lithology, etc. can be represented by filling polygons with dots or other more suitable symbols.

It also seems important to ensure that the map integrates enough landmarks to offer the best reading conditions for the reader. Geographic maps tend from the outset to erase remarkable landscape elements, by generalising information. This aspect is often neglected, with visual elements poorly designed (*e.g.* topographic raster maps overprinted on thematic map). Weighted addition of vector data can represent a solution in this case. Another approach consists in using transparencies by merging thematic vegetation layers with a raster background (*e.g.* shaded relief generated from DEM, aerial photographs), thanks to layer-rendering tools of the mapping software, which improves locations [2, 13].

8 Conclusion

Despite the lack of bibliographic references and of any uniformly recognised semiological chart for vegetation mapping, it seems interesting to propose some principles for increasing the readability of maps, especially for integrated mapping of vegetation series and dynamic maps. In any case, the colour choice linked to an unequivocal typology appearing in the legend is a preliminary principle in vegetation mapping [7].

Also, the main purely semiological question is: can palettes be harmonious? If a landscape is heterogeneous, due to its ecological heterogeneity and often to its high level of human impact, is it preferable *a priori* that mapping appears disharmonious. Potentials of the colorimetric spaces of Munsell or the Natural Colour System elaborated for colour harmony more adapted to the human eye should be explored in order to check whether their adaptation to represent vegetation is possible.

References

1. Bertin J (2005) Sémiologie graphique. Les diagrammes, les réseaux, les cartes, 4th edn. Editions EHESS, Paris
2. Bioret F, Caillon N, Glemarec E (2014) Cap d'Erquy et Cap Fréhel. Doc Phytosociologiques, s.3, 1:25–65
3. Caivano JL (1998) Color and semiotics: a two-way street. Color Res Appl 23:390–401. https://doi.org/10.1002/(SICI)1520-6378(199812)23:6%3c390::AID-COL7%3e3.0.CO;2-%23
4. Gaussen H (1936) Le choix des couleurs dans les cartes botaniques. Bull De La Société Botanique De France 83:474–480. https://doi.org/10.1080/00378941.1936.10837343
5. Gaussen H (1947) L'emploi des couleurs en cartographie. Comptes Rendus Hebdomadaires des Séances de l'Académie des Sciences 224:450–452
6. Itten J (1961) Kunst der Farbe. Otto Maier, Ravensburg
7. Küchler AW (1967) Vegetation Mapping. Ronald Press Company, New York

8. Lalanne A (2016) Le programme de cartographie nationale des végétations naturelles et semi-naturelles de France: CarHAB. Doc Phytosociologiques 6:236–244
9. Ozenda P (1986) La cartographie écologique et ses applications, p 159
10. Pastoureau M, Simonnet D (2005) Le petit livre des couleurs. Panama, Paris
11. Pedrotti F (2013) Plant and Vegetation Mapping. Springer, Heidelberg
12. Perera S (2016) Compte-rendu du groupe de travail sur la sémiologie CarHab. Représentations graphique et cartographique de la végétation
13. Perrin G (2014) Etang du Moulin Neuf. Doc phytosociologiques, s.3, 1:123–154
14. Smith AR (1978) Color gamut transform pairs. ACM SIGGRAPH Comput Graph 12:12–19. https://doi.org/10.1145/965139.807361

Plant Species Distributions and Ecological Complexity: Mapping Sampling-Effort Bias Explicitly

Carol X. Garzon-Lopez and Duccio Rocchini

Abstract In geobotany, mapping plant species distributions properly is crucial to guarantee a proper estimate of their dispersal variability in space and time, also considering habitat suitability. In most cases, uncertainty in the modelling procedures has been disregarded. However, hidden uncertainty or bias may hamper robust estimates of the distribution of plant species or species assemblages. In this paper, we propose an approach to mapping uncertainty properly, mainly deriving from sampling effort bias, when mapping plant species distributions.

Keywords Biodiversity · GBIF · Sampling schemes · Uncertainty

1 Introduction

In ecology, a number of studies have dealt with the prediction of species distribution and diversity over space and its changes over time, based on a set of environmental predictors related to environmental variability, productivity, spatial constraints and climate drivers [10].

C. X. Garzon-Lopez
Ecology and Vegetation Physiology Group (EcoFiv), Universidad de los Andes, Cr. 1E No 18A, Bogotá, Colombia

D. Rocchini (✉)
Center Agriculture Food Environment, University of Trento, Via E. Mach 1, 38010 S. Michele All'Adige, TN, Italy

D. Rocchini
Centre for Integrative Biology, University of Trento, Via Sommarive, 14, 38123 Povo, TN, Italy

D. Rocchini
Research and Innovation Centre, Department of Biodiversity and Molecular Ecology, Fondazione Edmund Mach, Via E. Mach 1, 38010 S. Michele All'Adige, TN, Italy

F. Pedrotti and E. O. Box (eds.), *Tools for Landscape-Scale Geobotany and Conservation*, Geobotany Studies, https://doi.org/10.1007/978-3-030-74950-7_2

Species distribution models have been acknowledged as the most powerful methods to map the spread of plant (and animal) species. The basic approach used to create maps based on predictors is to rely on linear models to create gridded landscapes of potential distribution of species based on point or local polygon data [5].

In most cases, the output is a density function in two dimensions representing the distribution S_x of species x. In general, boundaries are sharply defined based on thresholds of predictors/factors (e.g. when mainly based on land cover) or continuous, if based on the continuous variability of predictors (e.g. the continuous variation of temperature).

Uncertainty in such models derives mainly from pseudo-absences in input data [8] as well as from model bias, i.e. the error deriving from the model being chosen (GAM, GLM, Maximum entropy models, etc.,). Hence, the representation of uncertainty in two dimensions is strongly suggested but is disregarded in most cases. However, its importance is apparent. In fact, areas with a high or low probability of species occurrence might also be related to a high or low error rate. As an example, in case of invasive species, it might be crucial to represent uncertainty spatially, in order to allow better decision making.

As an example, a high (or low) invasion rate might be related to high or low errors in the output model being used by decision makers. The most dangerous situation occurs when a low predicted invasion rate is related to a high error in the modelling procedure. In this case decision makers might underestimate the effort needed against invasion, suspected to be low from the species distribution map.

In this case, misconceived use of a species distribution map might be dangerous, for example in case of a low dispersion probability of an invasive species but with a high error in the model. The prediction of the distribution of an invasive species might be low but with a high error; hence its spread could be underestimated in some parts of an area.

The aim of this study is to provide a straightforward and robust mapping procedure to show explicitly the spatial uncertainty related to sampling problems like sampling effort (relying on commonly used datasets) when dealing with species distribution modelling.

2 Uncertainty Related to Sampling Effort: Representing Sampling Bias by a Diffusion-Density Algorithm

For bias related to sampling effort, we will rely on one of the most used datasets in biodiversity study at large spatial extents, namely the Global Biodiversity Information Facility (GBIF) dataset (https://www.gbif.org/). GBIF data comprise a huge range of species occurrence observations collected with a wide variety of sampling methods, ranging from well-established plot censuses to direct observations collected during field trips. Consequently, some of the data points are at the center of censused grids (each point comprises the species located at a specific-size quadrant) or correspond to single observations of one (or more) individuals of the same species. These differences also depend on the methodologies used to observe/record occurrences per taxon. Plots, and plots within transects, are common designs in vegetation censuses, while transects, point counts and live traps are preferred in the case of animals.

The variation in factors, such as per country biodiversity monitoring schemes, funding schemes, focal ecosystems, accessibility to remote areas, and others, adds another source of variation, especially at multinational scales [1].

Undoubtedly, all those sources of variation result in non-homogeneous sampling that has important consequences not only for the development of accurate species distribution maps but, more importantly, for conservation and management decisions [14]. The aim of this study was to show explicitly the spatial uncertainty in the sampling effort of the GBIF data.

In this study we aim to quantify and map the uncertainty derived from the variation in observations due to differences in sampling efforts. In particular, cartograms are used, in which the shape of objects (countries and grid cells) is directly related to a certain property, in our case to uncertainty. Cartograms build on the standard treatment of diffusion, in which the current density is given by:

$$J = v(r,t)p(r,t) \tag{1}$$

where v(r, t) and p(r, t) are the velocity and density at position r and time t. See Gastner and Newman [9] for additional information.

Cartograms facilitate the visualization of spatial uncertainty in the results by changing the size of the polygons based on the density of information contained (number of observations, variation, etc.). In this paper, the cartograms were developed using the free, open-source software ScapeToad (https://scapetoad.choros.ch/).

After having superimposed a grid on the GBIF set over Europe, we modified the shape of different grid cells based on their relative oversampling, quantified as the number of dates in which a cell was sampled. In this, i) the colour of the map cells represents the occurrence possibility of a certain species; while ii) cell size is related to the sampling effort. The greater the distortion, the higher will be the number of visits in a certain area, hence the sampling effort bias.

By this strategy, the maps generated show the differences in species observations per area in Europe for three different plant species: *Quercus coccifera*, *Fagus sylvatica* and *Betula pendula*. Their differences in relative sampling bias are shown in Fig. 1.

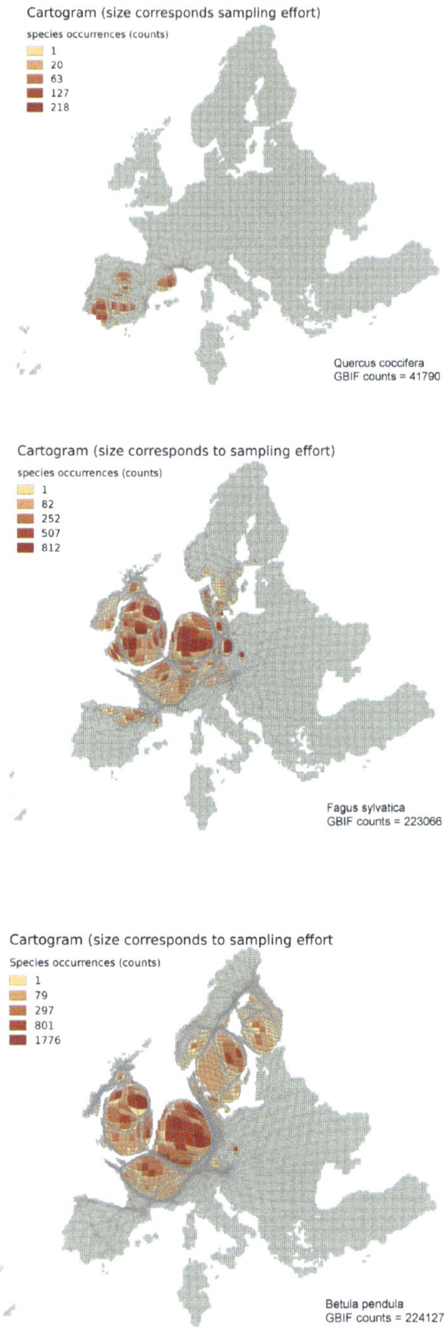

Fig. 1 Examples of cartograms for different plant species, from those sampled with a lower sampling bias to those based on cells sampled several times, hence with high sampling bias on occurrences of the species

3 Discussion and Open Challenges

Mapping sampling bias properly is a crucial step when aiming to develop robust species distribution models or to represent species variability over space (and time).

Various methods have been proposed to account for uncertainty in such models. One of the most important is certainly related to Bayesian statistics, in which the model can be'informed' with previous knowledge related to the distribution of a certain species. Uncertainty, in our case based on sampling effort bias, might then be directly part of the species distribution model, as additional information about spatial bias. This could be particularly important when forecasting species distribution in time, directly taking into account potential oversampling and avoiding hidden uncertainty [13].

On the other hand, tests of oversampling could be performed by relying on virtual species in order to estimate synthetically the amount of effort due to properly sampling a certain species and to avoid oversampling, and hence a higher possibility of occurrence. This is true considering both single species distributions and species assemblages, in which complexity is expected to increase exponentially. In such cases, correlative distribution models could benefit for complete control during the experimental phase of input and response (e.g. species or assemblage distribution) variables [7].

Mapping of different objects in 2-D space has always been a peculiar problem in geobotany and in cartography more generally [2]. This is particularly true when dealing with the distribution over space of plant species and community assemblages [6]. We think that the cartogram-based approach proposed in this paper has an explicitly high robustness and an implicitly high reproducibility, since it is based only on free, open-source algorithms. We refer to Keil et al. [12] for a dissertation about the development of free, open-source concepts in recent years.

Besides free, open-source methods, free data are crucial for further development of proper species distribution models, including both free environmental data and joint field data archiving procedures [11]. Future developments in SDMs should include the visualization of species distribution model predictions combined with the map of uncertainty presented here. This could lead to a proper spatial cross-correlation investigation to check for potential autocorrelation in the sampling process [3]. This is true when considering mapping of plant species, and more so for vegetation properties at global scales (e.g. [4]).

References

1. Barbosa AM, O'Hara RB (2015) fuzzySim: applying fuzzy logic to binary similarity indices in ecology. Methods Ecol Evol 6:853–858
2. Baselga S (2018) Fibonacci lattices for the evaluation and optimization of map projections. Comput Geosci 117:1–8

3. Chen Y (2015) A new methodology of spatial cross-correlation analysis. PLoS ONE 10: e0126158
4. Crowther TW, Glick HB, Covey KR, Bettigole C, Maynard DS, Thomas SM, Smith JR, Hintler G, Duguid MC, Amatulli G, Tuanmu M-N, Jetz W, Salas C, Stam C, Piotto D, Tavani R, Green S, Bruce G, Williams SJ, Wiser SK, Huber MO, Hengeveld GM, Nabuurs G-J, Tikhonova E, Borchardt P, Li C-F, Powrie LW, Fischer M, Hemp A, Homeier J, Cho P, Vibrans AC, Umunay PM, Piao SL, Rowe CW, Ashton MS, Crane PR, Bradford MA (2015) Nature 525:201
5. Elith J, Graham CH (2009) Do they? How do they? WHY do they differ? On finding reasons for differing performances of species distribution models. Ecography 32:66–77
6. Feilhauer H, Schmidtlein S (2011) On variable relations between vegetation patterns and canopy reflectance. Ecol Inform 6:83–92
7. Fernandes RF, Scherrer D, Guisan A (2018) How much should one sample to accurately predict the distribution of species assemblages? A virtual community approach. Ecol Inform 48:125–134
8. Foody GM (2011) Impacts of imperfect reference data on the apparent accuracy of species presence-absence models and their predictions. Glob Ecol Biogeogr 20:498–508
9. Gastner MT, Newman MEJ (2004) Diffusion-based method for producing density-equalizing maps. PNAS 101:7499–7504
10. Guisan A, Thuiller W (2005) Predicting species distribution: offering more than simple habitat models. Ecol Lett 8:993–1009
11. Huang X, Qiao G (2011) Biodiversity databases should gain support from journals. Trends Ecol Evol 26:377
12. Keil P, MacDonald AA, Ramirez KS, Bennett JM, Garcia-Pena GE, Yguel B, Bourgeois B, Meyer C (2018) Macroecological and macroevolutionary patterns emerge in the universe of GNU/Linux operating systems. Ecography 41:1–13
13. Ramin M, Arhonditsis GB (2013) Bayesian calibration of mathematical models: Optimization of model structure and examination of the role of process error covariance. Ecol Inform 18:107–116
14. Rocchini D, Hortal J, Lengyel S, Lobo JM, Jiménez-Valverde A, Ricotta C, Bacaro G, Chiarucci A (2011) Accounting for uncertainty when mapping species distributions: The need for maps of ignorance. Prog Phys Geogr 35:211–226

Mapping of Biogeographical Territories: Flora, Vegetation and Landscape Criteria

Javier Loidi

Abstract One of the tasks of biogeography has traditionally been the definition of territorial units by means of their biotic content. Five main ranks have been recognized, in a downward scale: kingdom, region, province, sector and district. It is assumed that for this purpose biogeography can be used as a synonym of phytogeography, as plants are particularly indicative of the substrate (soil, bedrock, geomorphology) and atmospheric (climatic) conditions. In this chapter, the way in which traditional plant taxonomical information is enriched with information about vegetation (plant communities) and a certain type of landscape analysis are explained, as far as they reflect the conditions prevailing in each territory. Floristic specificity reaches the level of family at kingdom level, while it decreases as we move towards the lower ranks, being null in the districts. Plant communities have high specificity at the level of formation at kingdom level with all the classes being endemic at this level; at the destrict level of district on the other hand, the specificity is null. Concerning landscape, the analysis of the repeated ordered arrangement of vegetation units across a gradient (vegetation zonation) in each territory is considered. The normal topographic zonation is the crest-slope-valley bottom model (geosigmetum). The rest of the eco-topographic zonations, i.e. those considered exceptional (permageosigmetum), are also considered. They differ minimally between the lower rank units, the districts, i.e. they differ in at least one geopermasigmetum. Combining information from three sources: taxonomy, vegetation and landscape analysis, it is possible to establish efficiently accurate and precise limits between territorial units.

Keywords Biogeography · Provincialism · Floristic element · Phytogeographical element · Geoflora · Geosigmetum · Biogeographic kingdom · Biogeographic region · Biogeographic province · Biogeographic sector · Biogeographic district

---

J. Loidi (✉)
Department of Plant Biology and Ecology, University of the Basque Country (UPV/EHU), P.O. Box 644, 48080 Bilbao, Spain
e-mail: javier.loidi@ehu.eus

F. Pedrotti and E. O. Box (eds.), *Tools for Landscape-Scale Geobotany and Conservation*, Geobotany Studies, https://doi.org/10.1007/978-3-030-74950-7_3

1 Introduction. Concept of Biogeography

The term Biogeography is understood by the scientific world to mean a discipline that deals with the distribution of living beings in a geographical scale; in words of Brown and Lomolino [10], it attempts to document and understand spatial patterns of biodiversity. It could be defined as the study of the geographical distribution of organisms, but this simplicity hides the great complexity of the subject [31] as it attempts to find patterns of distribution in both the present and in the past [23]. Such patterns of species distribution are usually explained by means of a set of historical factors such as speciation, extinction, continental drift and glaciation, in combination with current environmental conditions, such as climate, substrate, geography and orography, thus, the current distribution of species over the earth is the result of the current and the past physical and chemical conditions of the substrate and atmosphere. The inclusion of comprehensive timescale has been part of the definition since the early geobotanists, such as A. Grisebach [24] and Ch. Flauhault [20]. Therefore, **Biogeography** is devoted to the study of the distribution of species and ecosystems across geographical space and through (geological) time.

If we focus on the current distribution of a taxon, we see that it is the result of: a) historical processes, such as the origin and location of the lineage and its evolution, in the frame of climatical and geographical changes; b) the dispersal possibilities, determined by the plant´s abilities (anemochory, endozoochory, barochory, etc.) and the geographical barriers and migratory pathways; and c) the physiological and current ecological conditions enabling its survival [35]. The former two can be considered the core of the genuine “biogeographical” component of the concept while the ecological part deals with the current environmental opportunities and constraints driving species distribution. Both aspects, equally important, need to be clearly separated. Certainly, the distribution of living beings is driven mainly by ecological factors, but the other factors, i.e. the dispersal capacity of the organisms and their evolutionary history in the geologic time scale, are also necessary to explain their distribution at a geographical scale. Not considering the genuinely biogeographical factors has led to a widespread confusion between biogeography and ecology, to the extent that many texts hardly distinguish between them [46], and the boundaries between both disciplines have become blurred [30]. Biogeography, sensu stricto, takes into account phenomena such as migration, geographical distribution, speciation, extinction, etc., which are not uniquely explained by current environmental conditions but by ecological and geographical conditions that existed in the past and evolved with time. Very often we have difficulties to explain the reason of the geographical limit of a species range if a clear ecological or geographical barrier is lacking. In that case, we can guess that such limit has to do with competitiveness with other species or simply because it needs more time to expand beyond its current border until it reaches its natural ecological limit.

In any case, it is clear that biogeography is a synthesis discipline which encompasses a set of phenomena governing the distribution patterns of organisms, from their current or past dispersal abilities to their migrations and ecological adaptations, in the frame of a changing world in a historical perspective: climatic changes, geographical changes (continental drift, orogenic processes, island upliftments and sinkages and many others). All of them have their influence in the present day distribution of living beings. Thus, biogeographical research necessarily combines information and concepts from many fields: the physiological and ecological opportunities and constraints influencing the dispersal of organisms, the dispersal strategies of the different species, the geological, geographical and climatic changes operating at global spatial scales combined with the evolutionary phenomena which take place in the geological-scale time frame.

2 Definition of Biogeographical Territories

One of the products of traditional biogographical surveys has been the establishment of territories on earth or seas characterized by their biodiversity content. Thus, the definition of such territories is an activity which has been called **provincialism** [10]. It has produced a hierarchical system of units ranked, in a descending order, as realm (kingdom), region, province, sector and district. Botanists and zoologists have contributed to this field since the early nineteenth century [32]. We can mention A.P. De Candolle [13, 14] and also Schow [42], who produced the first map of biogeographical regions covering the entire world. Since then several important syntheses covering the globe or large areas of it have been produced, including those by Drude [15], Meusel et al. [29], Good [23], Takhtajan [45] and Schroeder [43].

This definition of territories provides a description of the different areas by means of their current biological content, which is related to their current climatic, geological and geographical conditions, and also to the evolutionary history of their biota, forged by the climatic and geological history. From that point of view, provincialism can be considered a relevant task in biogeography.

Biogeography "sensu stricto" considers organisms of all groups (plants, animals, fungi, algae, bacteria, etc.), however, there is a tendency of describing biogeographical territories by using preferably plants, as they are particularly indicative of the substrate (soil, bedrock, geomorphology) and atmospheric (climatic) conditions. Thus, the term biogeography is often used as a synonym of phytogeography. In the following paragraphs I shall focus on the vascular plants and the vegetation.

3 Floristic Approach

Initially, all the surveys and maps of the biogeographical units were done by using taxa, either high rank such as families or genera or low rank such species and subspecies. In this approach the floristic content of the different territories is considered. On the other hand, the distribution area of the taxa follows certain regular patterns. As Myers and Giller put it "*The existence of non-random patterns in species distribution implies the operation of some general causal processes and from an understanding of them one can work towards a reconstruction of the history of life that can explain the present day distribution of species*" [31]. In that sense, concepts such as floristic **element were coined** [22]: this can refer to a group of plants having similar distribution areas (geographical element), or having originated in the same region (genetic element), or that immigrated at the same time (historical element) or following the same route (migratory element). Initially this concept was introduced by Christ [12] in a geographical sense, but later Braun-Blanquet (1923) used the term **phytogeographical element** with the aim to include the flora and plant communities characteristic of a territory, i.e. "*the floristic and phytosociologic expression of a vast and defined territory*", so that we could speak about the Mediterranean element or the Boreal element as the group of plants and communities living exclusively or preferentially in the mentioned areas (Mediterranean or Boreal) and constituting their living features. The different floristic elements which participate in the flora of a territory are the footprints of the successive migrations arrived in the past, and the endemics are the result of the local biological evolution and witnesses of geographical isolation [41].

Another concept used in biogeography is that of **geoflora**, particularly applied to the Arcto-Tertiary Geoflora and the Paleotropical Geoflora [28, 47]. Those terms were established by Engler [17] and have been used mostly in paleobotany or paleobiogeography [4, 28, 47] referring to the flora constituent of large vegetation zones which persisted unchanged over long periods of time. Those species have thus a common evolutionary history along a relevant period of their existence on earth and reflect certain environmental conditions that existed in the past or at present.

Traditionally, distribution areas of taxa have been studied (chorology), and their distribution patterns have been a fundamental source for the construction of the biogeographical theory and spatial units. Some outstanding examples of this activity are shown in the "Atlas of Flora Europaea, vols. 1–16" [26], the "Atlas of British Flora" [34], the "Atlas partiell de la flore de France" [16] or the "Atlas der Farn- und Blütenpflanzen der BRD" [25].

Such information is used to apply the knowledge of the species pool or floristic content of the different territories, which is considered the main criterion for their biogeographical definition. The species pool concept has already been used in relation to diversification and migratory phenomena in biogeographical analysis [48]. As a result, a number of biogeographical territorial maps (also called chorionomical synthesis) have been drawn, with a considerable degree of accuracy, especially on broad geographical scales. All of them exclusively use the distribution of taxa to define the biogeographical territories [23, 45].

The use of plant taxa is profoundly indicative of both the plant diversity of the concerned territory and also of relevant evolutionary and paleoecological issues affecting it. However, in the definition of lower rank biogeographical categories, and particularly for their geographical delimitation, the flora or the floristic concepts have a poor diagnostic power [41].

4 Vegetational Approach

Since the spatial distribution pattern of plants (or other organisms) is conditioned by the multiple environmental factors converging in each territory of the world, it is possible to figure out that the environmental scenarios have a geographical variability, consequently responsible for the variability, not only of the territorial floras or faunas, but also of the ecological frameworks which determine and organize the existing communities, and that question is central in biogeography.

With the floristic approach we can at most know the plant species (or other higher rank taxa) list of any territory, but no information is provided about the relative abundance, distribution pattern and the assemblage patterns into communities of each of these taxa. Thus, in a step forward, it could be relevant to know how this species pool assemble into communities, i.e. how plant species combine to form associations [19]. Such phenomena depend on local climatic, lithological and geomorphological conditions, as well as on the human land use system, the biogeographical territories are the specific scenario in which those phenomena occur (Fig. 1).

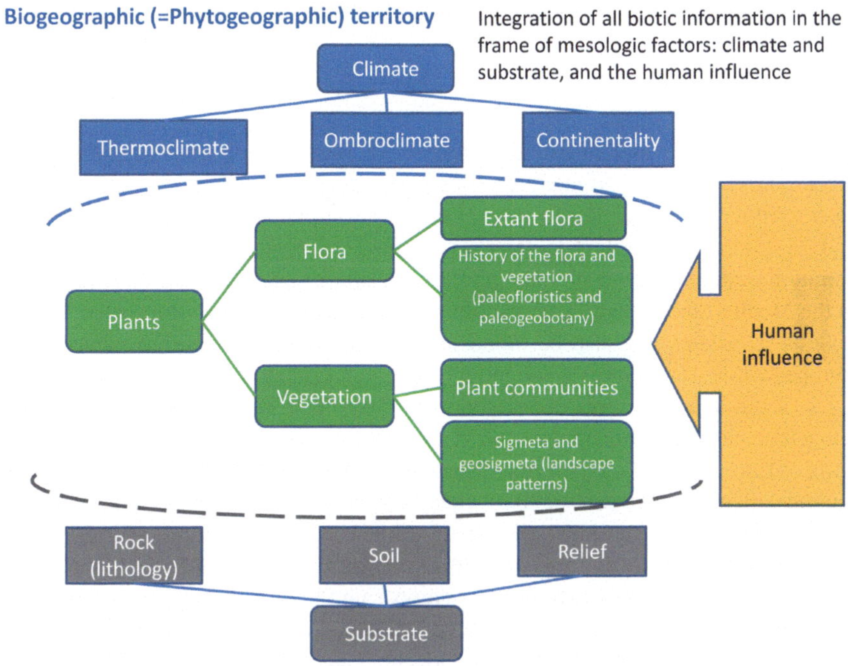

Fig. 1 Schematic representation of the biogeographical synthesis in a given area

Plant taxa

	Kingdom (realm)	Region	Province	Sector	District
Families	some	few	none	none	none
Genera	many	some	few	none	none
Species	all	most	many	few	none
Differences in species pool	Total ca. 100%	very high > 50%	Substantial < 50%	Small < 10%	Irrelevant (circumstantial < 2 %)

Assemble into:

Vegetation units

Formations	several	few	none	none	none
Classes	all	some	none	none	none
Alliances	all	most	few	none	none
Associations	all	all	most	few	none
Subassociatio ns	all	all	all	all	few

Fig. 2 Floristic and vegetational content of the biogeographical categories

The abundance and distribution pattern of plants is of basic importance in biogeography. They are directly determined by the geology, climate and history of each territory. Consider, for instance, two common plants in Europe: *Urtica dioca* and *U. urens,* both linked to strongly nitrified habitats. The first is particularly common in the temperate climatic regions of Europe, while the second is prevalent in the Mediterranean part. Nevertheless, they are both profusely represented in both regions: *U. dioica* can also be found in the hydromorphic soils in Mediterranean Europe, while *U. urens* is frequently found in the relatively dry and continental areas of the temperate part of the continent. Dot maps do not make this situation altogether clear, as both species are represented by dense clouds of dots in both parts of Europe, masking their real distribution pattern and relative abundance. At a more detailed scale we can see that a rocky country will have abundant populations and communities of chasmophytic plants, while a neighbouring one with less rock outcrops will have lower abundance of those particular taxa and vegetation. Such differences will be poorly expressed by comparing floristic catalogs or by observing dot maps. Indeed, the latter can even give a distorted idea of the true abundance and distribution of some plants; even some localities are famous for having rare plants that are often located on the edge of their area and which are over-represented in herbaria and literature. Quantifying the "role" of every plant in a particular territory and considering the assemblage rules among species is a task for a more accurate biogeography, as it reflects geological, climatic and historical phenomena. Taking vegetation into account is one way of advancing in this direction.

The use of vegetation units in the form of phytosociological plant communities in biogeography started early in the twentieth century with Flauhault [20, 21], Braun and Furrer [7] and Braun-Blanquet [8, 9] and has been widely accepted later [39, 41]. They established the first rules for applying knowledge of syntaxa and their distribution to the definition of territorial units. Figure 2 summarizes the taxonomic and vegetational content in the different biogeographical categories.

5 Landscape Approach

For defining higher rank biogeographical categories, kingdom or region, the array of taxonomic criteria are sufficient, as they are large areas bearing profound differences in their flora resulting from very different evolutionary histories; we can easily understand that floristic differences are enough to sustain the independence of Holarctic and Paleotropic kingdoms. When descending in rank, i.e. province or sector, the vegetation analysis in terms of plant communities (syntaxa) becomes important for documenting the units and especially for a definition and a precise drawing of the limits between the categories in a map. Floras can be too similar between two neighbouring provinces to provide criteria for such precise delimitation. However, in the low ranks, limits cannot be precisely defined by using either the flora or the plant communities, we need more precise criteria which are provided by a specific form of landscape analysis called Integrated phytosociology [1] or dynamic-catenal phytosociology [37].

The incorporation of this landscape analysis in the provincialism has been carried out in some territories, mostly in southern Europe, in the last decades of the twentieth century [3, 5, 6, 33, 36, 44]. On a detailed scale, concerning mostly the low-rank units, Alcaraz et al. [2] and Loidi and Fernández Prieto [27] made substantial contributions.

In order to briefly explain the application of a certain type of landscape analysis to the biogeographical definition of territories, Alcaraz [1] attempts to formalize the criteria (under the name of integrated phytosociology). He tries to define the basic unit, i.e. the district, as the absolute homogeneous territory from the biogeographical point of view, resulting from its climatic and geomorphological homogeneity. As indicated above, differentiating closely-related neighbouring districts using only floristic criteria or a bare list of plant communities is almost impossible. Such lists of species and of plant communities share almost all of the elements and would not show any substantial differences between the districts. However, in applying landscape analysis, the rules of distribution of communities into the landscape can change minimally, so that it can result in certain and clearly different distribution patterns across the territory. To detect this, we can focus on the ordered arrangement of vegetation units across a gradient, i.e. the vegetation zonation, which, when repeated in a certain territory, can be typified and named with the prefix geo-. For that, the name [50] **geosigmetum** was established by Rivas-Martínez et al. [37, 38]. These zonations can follow a universal topographic

Table 1 Summary of the defining criteria for the biogeographical categories

	Flora	Plant communities	Integrated phytosociology
Kingdom (Realm)	Endemic families, subfamilies and tribes; very high generic and specific endemism. Paleogenic endemism of high rank	Practically all the classes are endemic, only synatropic or aquatic ones are cosmopolitan or subcosmopolitan	
Region	High generic and specific endemism. Few families endemic or with predominant position in the region. Paleogenic endemism of intermediate rank (genus)	Several exclusive classes and orders. Classes with their optimum within the region involved	
Province	Some endemic genera (usually monotypic or oligotypic). Still relevant specific endemism; paleogenic endemism of low rank (species). Statistically distinct assemblages of species	Some alliances or suballiances endemic. Relevant association endemism	Some exclusive climatophilous sigmeta. Particular altitudinal (mountain) vegetation zonation (cliserial geosigmetum)
Sector	Low or absence of species endemism; long isolated islands increase endemic element	Some associations endemic	Several exclusive climatophilous geosigmeta
District	Usually no endemism in the flora. Slight floristical differences with neighbour districts are due to geomorphology: relief or substrata, or to slight differences in climate	No syntaxonomical endemism; in some cases lower rank syntaxa like subassociation, variant and facies are exclusive. Some associations are lacking in neighbouring districts	Geomorphologically homogeneous territory. At least one climatophilous geosigmetum and some geopermasigmeta exclusive to the territory

pattern crest-slope-valley bottom model. This gradient model is universal and can be studied in any territory of the world, being called the climatophilous geosigmetum. In a broader scale, the zonation of vegetation belts along the altitudinal gradient of a mountain has been called **cliserial geosigmetum**. Cliserial geosigmeta are strongly linked to regional environmental conditions (regional climate and substrate) and thus have a high biogeographical specificity (Fig. 6). Other cases are those gradients occurring in exceptional situations, as in highly stressed areas under extremely limiting conditions such as coastal dunes, salt marshes, cliffs, screes, ponds, mires, fens, etc. In those cases the geosigmetum becomes formed by permanent vegetation types and the term **geopermasigmetum** is used [38]. It has been established that the minimum difference needed to separate two districts is that they differ in at least one permageosigmetum present in one of them and absent in the other,the rest, i.e. all climatophilous geosigmeta, as well as the rest of the permageosigmeta, may be shared. Logically, the greater the differences in geosigmeta

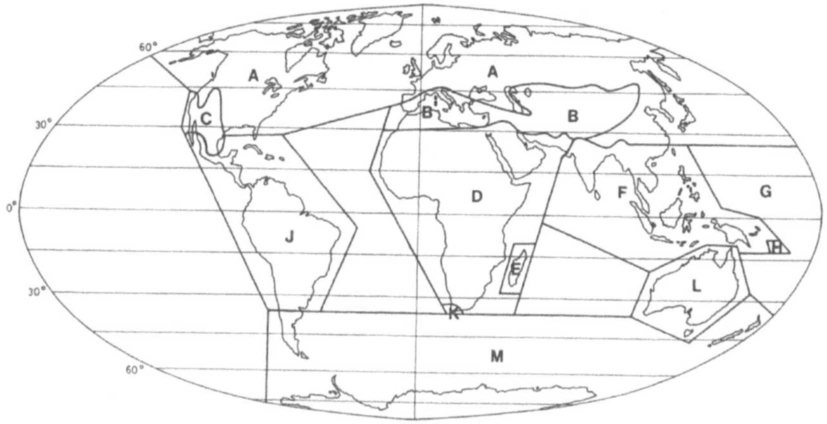

Fig. 3. Floristic kingdoms and subkingdoms of the world, after Takhtajan [45]. A–C Holarctic kingdom (A Boreal subkingdom, B Tethyan subkingdom, C Madrean subkingdom); D–H Paleotropical kingdom (D African subkingdom; E Madagascan subkingdom; F Indo-Malaysian subkingdom; G Polynesian subkingdom; H Neocaledonian subkingdom), J Neotropical kingdom; K Cape kingdom; L Australian kingdom; M Anctarctic kingdom

(climatophilous or permageosigmeta), the greater the biological distance between the territories analyzed. This system helps to provide a very accurate definition of the units and to draw the frontiers between those districts; each piece of land can be submitted to a rigorous and quick analysis leading to its inclusion in one district or the other. In practical work, such frontiers have been drawn at a relatively detailed scale, sometimes at 1:50.000. This is the starting point for efficient work in biogeographical provincialism at the elementary level.

Table 1 shows the floristic, pure phytosociological and integrated phytosociological criteria that can be used to characterise each biogeographical rank. They are taken from Braun-Blanquet [9], Takhtajan [45], Rivas-Martínez [36–38] and Alcaraz [1]. In it, the decreasing importance of floristic criteria when descending in rank parallel to the increasing importance of landscape criteria becomes evident. Vegetation units play an intermediate role.

6 Definition of the Biogeographical Territorial Ranks

6.1 *Kingdom (Realm)*

It is the highest rank unit. It is usually a territory of great extent, encompassing one or several continents and concealing entirely several high rank taxonomic entities, such as families, subfamilies or tribes, as well as most of the genera and practically all the species occurring in it. Often one particular vegetation formation is exclusive of one kingdom. In the world there are six kingdoms traditionally recognized: Holarctic, Paleotropical, Neotropical, Capensis, Australian and Antarctic (Fig. 3). Some authors include the Capensis in the Paleotropical [35] and even split the Antarctic and allocate its fragments to the neighbouring kingdoms [11].

6.2 Region

The kingdoms are divided into regions, which are vast territories characterized by a high specificity in their flora, having often paleogenic endemics such as a small family or some families which dispersal centre lies in the region, so that although not strictly endemic, they can be considered having their optimum in the concerned region (e.g. Cistaceae for the Mediteranean region). The level of endemicity increases with descending taxonomic rank, so that several genera and most of the species are endemic to the region. The regions can be characterized by some vegetation classes exclusive to them and by some others for which the optimum is in the concerned region, eg. Quercetea ilicis has its optimum in the Mediterranean region while Molinio-Arrhenatheretea develops mostly in the Eurosiberian region. At lower levels, such as order and alliance, endemicity increases. There are 35 regions around the world [45] (Fig. 4).

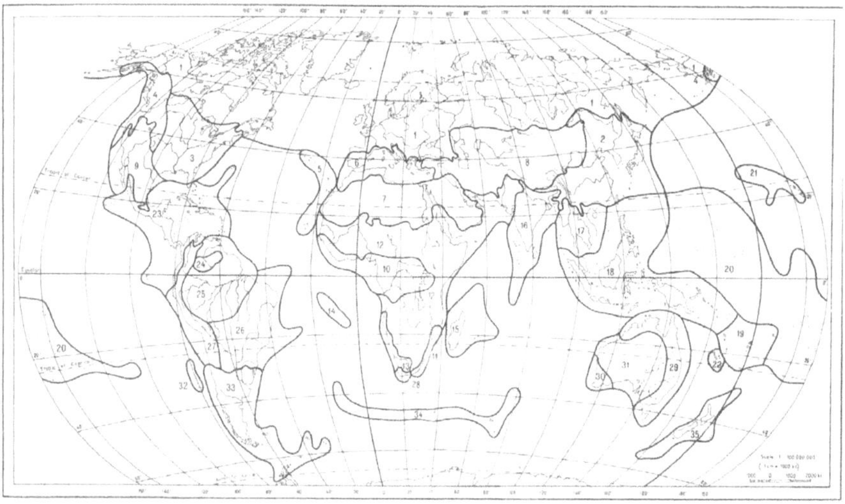

Fig. 4 Biogeographical regions of the world [23]

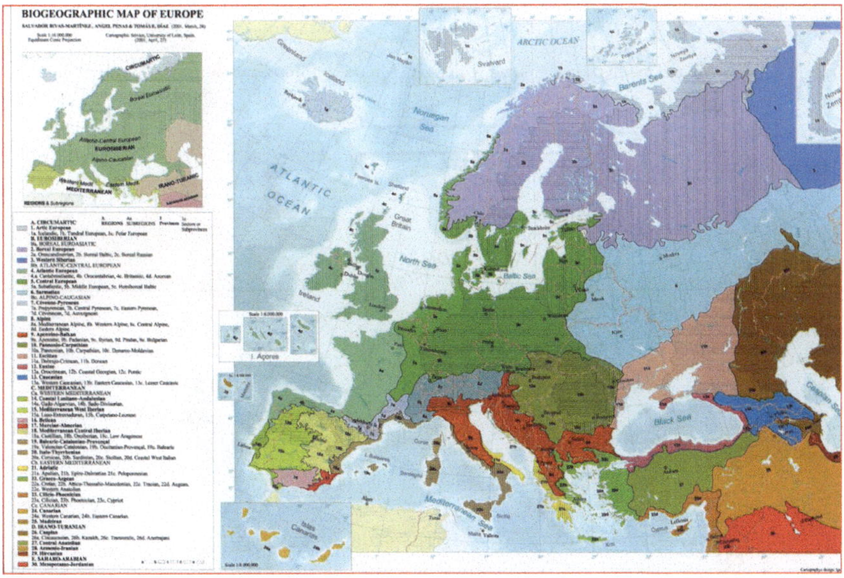

Fig. 5 Biogeographical map of Europe with the regions and provinces [39]

6.3 *Province*

It is a territory of regular extent, highly variable depending on spatial niche diversification. Endemicity hardly reaches the genus level with some mono- or oligospecific genera. The number of endemic species is high. Syntaxonomically, there are endemic alliances but at the association level there are more, even in vegetation types representing mature stages of succession. Vegetation zonation patterns (geosigmeta) are entirely particular to the province (Fig. 5). Vegetation belts zonation in mountains (cliserial geosigmeta) are characteristic of the province (Fig. 6).

6.4 *Sector*

It is a territory of modest extent (Fig. 7) where endemicity concerns mostly subspecies, reaching to the species level in few cases (this is not applicable to mountain areas and to true islands, where endemicity increases). The floristic differences refer to the species pool. It can have some endemic associations but not of mature stages of succession. One important feature to distinguish sectors belonging to the same province is their specific zonation pattern (geosigmetum) in the crest-slope-valley bottom model. Slight variations in cliserial geosigmeta can also be used to separate closely related sectors [2].

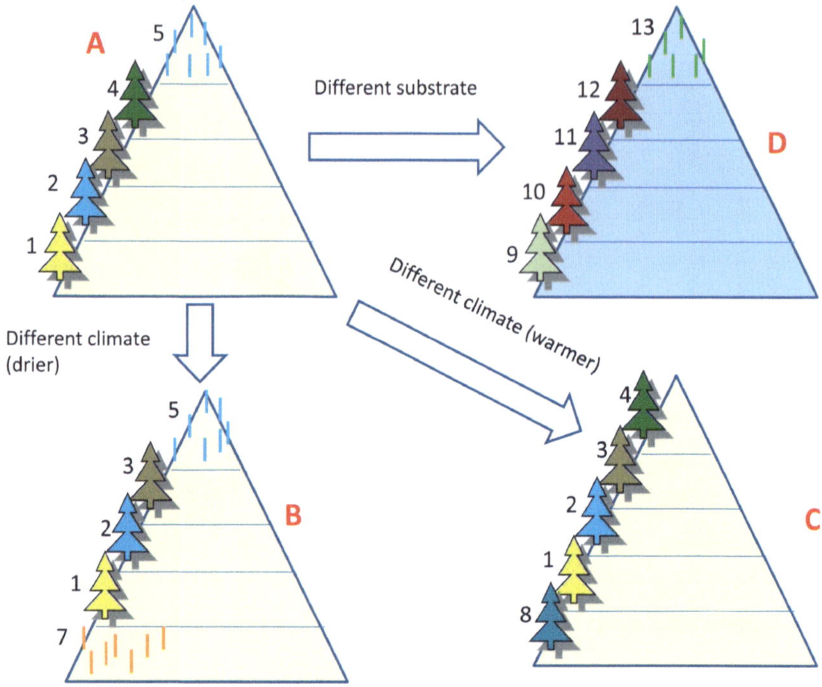

Fig. 6 The differences in the altitudinal zonation pattern in different mountains are the result of profound differences in climate, substrate and/or vegetation history and can thus be used as a criterion for differentiating biogeographics units, such as provinces or even sectors, depending the importance of such differences. Between A and D there is a difference in substrate while between A and B and C there are differences in climate. In both cases, the differences entail a change in the zonation pattern (cliserial geosigmetum) which, if accompanied by differences in historic elements, can be used to substantiate with more or less arguments, the independence between biogeographic territories

6.5 District

It is the lowest unit in the classification system, and also the most homogeneous in terms of mesoclimate, relief, lithology and history of the flora living within it. It is a territory of small extent (Fig. 8) without endemicity at any taxonomic level (perhaps some accidentally). As regards vegetation, there is no endemicity even in the association level. For Braun-Blanquet [9] a district is an area defined only by "*particular communities of lower rank (variants, facies), lacking endemic species*". The differences with a contiguous district in terms of flora and vegetation are due to slight changes in mesologic conditions, not due to facts related to a "phylogenetic

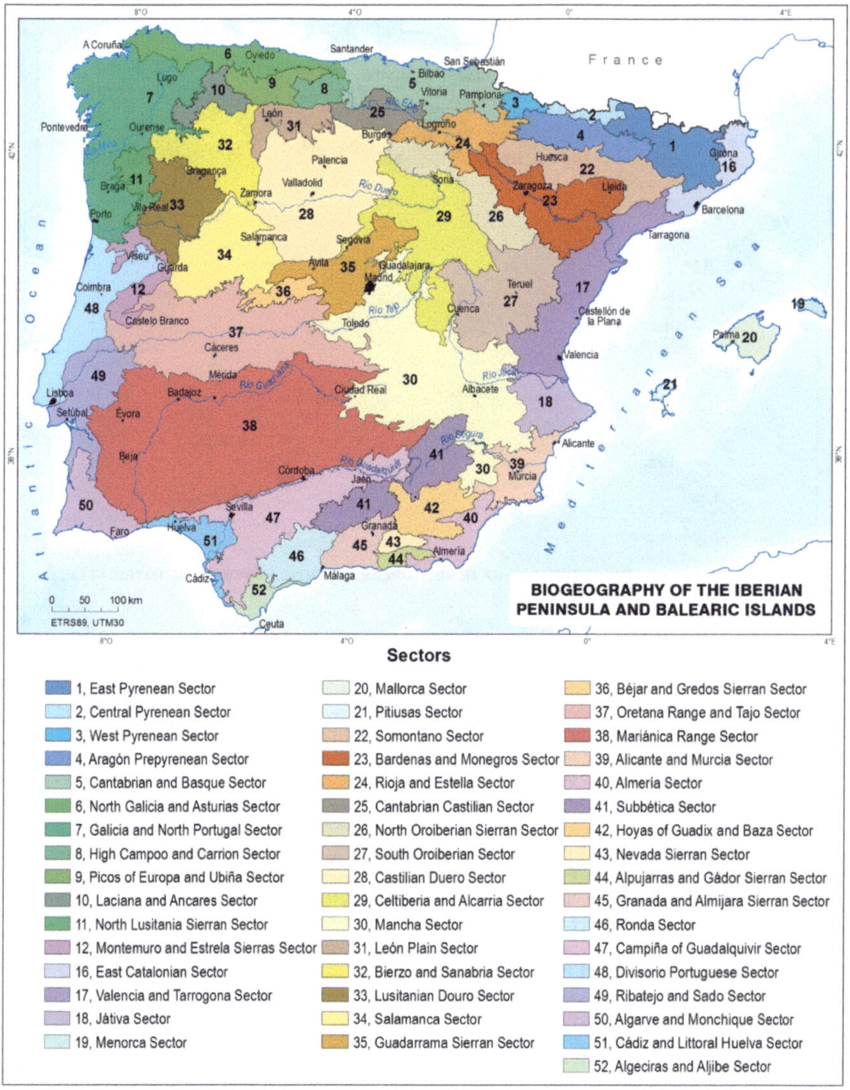

Fig. 7 Biogeography of the Iberian Peninsula at the level of sector (Rivas-Martínez et al. [40])

phytogeography" in the sense of Engler [18]. Regarding landscape, the pattern of distribution and sorting of the plant communities concerning their zonal and successional relationships remains within its limits, i.e. the formal zonation pattern, or geosigmetum, remains unaltered within the district [1]. Any change in these patterns (affecting at least one geopermasigmetum) due to a slight change in climate, lithology ot relief, means a change to another district within the same sector.

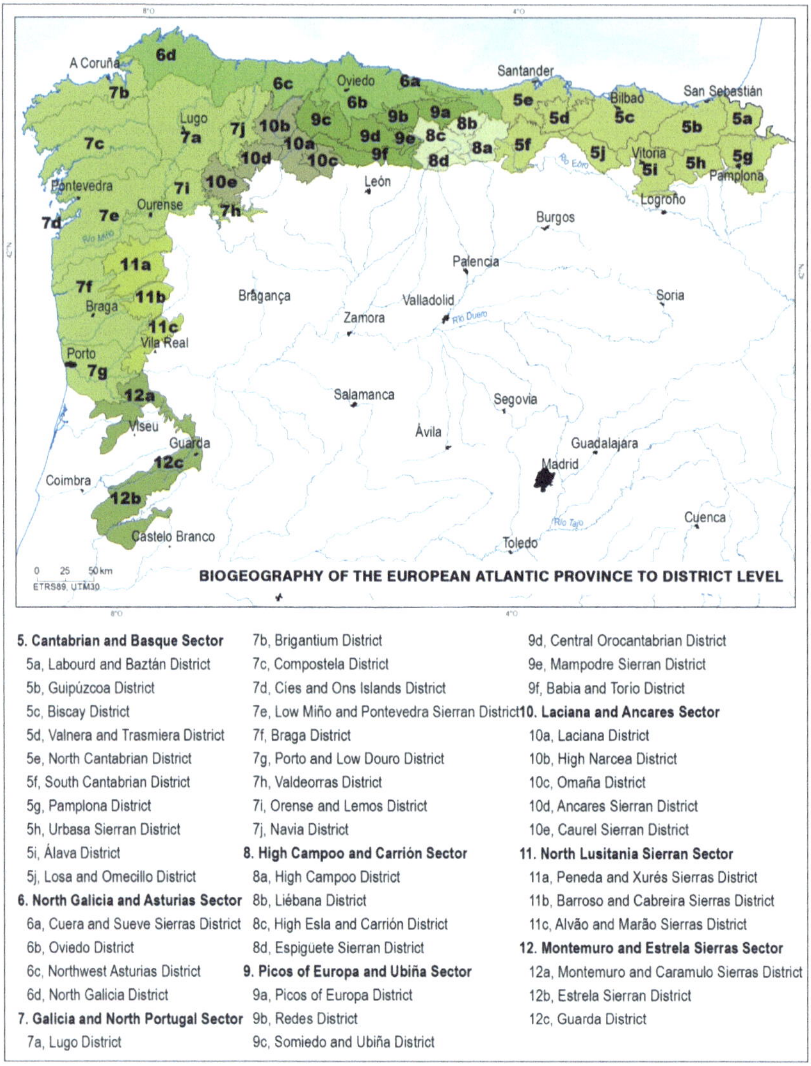

Fig. 8 Biogeography of the North-west of the Iberian Peninsula at the level of district (Rivas-Martínez et al. [40])

References

1. Alcaraz F (1996) Fitosociología integrada, paisaje y biogeografía. In: Loidi J (ed) Advances in phytodociology, pp 59–94. Ed. UPV/EHU
2. Alcaraz F, Sánchez-Gómez P, De La Torre A (1991) Biogeografía de la provincia Murciano-Almeriense hasta el nivel de sub sector. Rivasgodaya 6:77–100

3. Asensi A 1996 Fitosociología y paisaje (Una aproximación histórica). In: Loidi, J. (ed.) Avances en Fitosociología, pp 43–58. Ed. UPV/EHU
4. Axelrod DI (1966) Origin of deciduous and evergreen habits in temperate forests. Evolution 20(1):1–15
5. Biondi E (1996) L'analisi fitosociologica nello studio integrato del paesaggio. In: Loidi J (ed) Avances en fitosociología, pp 13–22. Ed. UPV/EHU
6. Blasi C (ed) (2010) La Vegetazione d'Italia con Carta delle Serie di Vegetazione in scala 1:500000. Palombi Editori, Roma
7. Braun J, Furrer E (1913) Remarques sur l'étude des groupements de plantes. Soc Languedoc Geogr Bull 36:20–41
8. Braun-Blanquet J (1919) Essai sur la notion d'element et de territoire phytogeographyques. Arch Sci Phys Nat Geneva Ser 5, 1:497–512
9. Braun-Blanquet J (1951) Pflanzensoziologie. Grundzüge der Vegetationskunde. Springer, Wien
10. Brown JH, Lomolino MV (1998) Biogeography, 2nd edn. Sinauer, Sunderland
11. Cox B (2001) The biogeographic regions reconsidered. J Biogeogr 28:511–523
12. Christ H (1867) Ueber die Verbreitung der Pflanzen der alpinen Region der europäischen Alpenkette, Basel
13. De Candolle AP (1820) Essai élémentaire de Géographie botanique, vol 18. du Dictionnaire des Sciences Naturelles. F.G. Levrault, Strasbourg
14. De Candolle AP (1855) Géographie Botanique Raisonné, vol 2. V. Masson, Paris
15. Drude O (1884) Die Florenreiche der Erde (The floristic realms of the Earth). Petermanns Geogr. Mitt., Erg änzungsheft, p 74
16. Dupont P (1990) Atlas partiel de la flore de France. Secrét Faune et Flore Mus Nat D'Histoire Naturelle, Paris
17. Engler A (1879–1882) Versuch einer Entwicklungsgeschichte der Pflanzenwelt seit der Tertiärperiode. Leipzig Engelmann, pp 1–2
18. Engler A (1899) Entwickelung der Pflanzengeographie in den letzten hundert Jahren und weitere Aufgaben derselben. Wissenschaftliche Beiträge zum Gedächtnis der hundertjährigen Wiederkehr des Antritts von Alexander von Humboldt's Reise nach Amerika. Gesellschaft für Erdkunde zu Berlin, Berlin
19. Eriksson O (1993) The species pool hypothesis and the plant community diversity. Oikos 68 (2):371–374
20. Flauhault C (1901a) La flore et la Végétation de la France. In: Coste, H. Flore descriptive et illustrée de la France, de la Corse et des contées limitrophes, vol 1. Libr. Sci. Et Techn. Albert Blanchard, Paris, pp 1–52
21. Flauhault C (1901) Premier essai de nomenclature phytogeographique. Bull Spc Langedoc Géogr 24:157–192
22 Font Quer P (1953) Diccionario de Botánica Ed. Labor, Barcelona
23 Good R (1974) The geography of the flowering plants, 4th edn. Longman Group Ltd., London, p 557
24 Grisebach A (1884) Vegetation der Erde, vol 1. Verlag von W. Engelmann, Leipzig, p 230
25 Haeupler H, Schönfelder P (1989) Atlas der Farn- und Blütenpflanzen der Bundesrepublik Deutschland Ed. Eugen Ulmer, Stuttgart
26 Jalas J, Suominen J (1972–2013) Atlas Florae Europaea, vol 1–16. Committee for mapping the Flora of Europe, Helsinki
27 Loidi J, Fernández Prieto JA (1986) Datos sobre la biogeografía y la vegetación del sector castellano-cantábrico (España). Docum. Phytosociologiques N.S. 10:323–362
28 Mai DH (1991) Palaeofloristic changes in Europe and the confirmation of the Arctotertiary-Palaeotropical geofloral concept. Rev Palaeobot Palynol 68:29–36
29 Meusel H, Jäger E, Rauschert S, Weinert E (1965–1992) Vergleichende Chorologie der zentraleuropäischen Flora. Ed Gustav Fischer, Jena
30 Müller P (1986) Biogeography. Harper & Pow, New York, p 377
31 Myers AA, Giller PS (eds) (1988) Analytical biogeography. Chapman & Hall, London, p 578

32. Nelson G (1978) From Candolle to Croizat: comments on the history of biogeography. J Hist Biol 11:269–305
33 Pedrotti F, Venanzoni R, Moretti G, Cianficconi F (1987) Biogeografia e sue prospettive attuali. Aspetti e problem della geografia, pp 557–588. Ed. Marzorati, Milano
34 Perring FH, Walters SM (1962) Atlas of British Flora. Ed. Nelson, London & Edinburgh
35 Pfadenhauer JS, Klötzli FA (2014) Vegetation der Erde. Springer, p 643
36 Rivas-Martínez S (1987) Memoria del mapa de series de vegetación de España, 1:400.000. ICONA, Madrid
37. Rivas-Martínez S, et al (2007) Mapa de series, geoseries y geopermaseries de vegetación de España Memoria del mapa de vegetación potencial de España] Parte I. Itinera Geobotanica 17:5–436
38 Rivas-Martínez S, et al (2011) Mapa de series, geoseries y geopermaseries de vegetación de España [Memoria del mapa de vegetación potencial de España]. Parte II. Itinera Geobotanica 18 (1):5–424
39 Rivas-Martínez S, Penas A, Díaz González TE (2001) Biogeographic map of Europe, León
40 Rivas-Martínez S, Penas A, Díaz González TE, Cantó P, Del Río S, Costa JC, Herrero L, Molero J (2017) Biogeographic units of the Iberian Peninsula and Baelaric Islands to district level. A concise sinopsis. In: Loidi J The Vegetation of the Iberian Peninsula
41 Roisin P (1969) Le domaine phytogéographique atlantique d'Europe. Les presses agronomiques de Gembloux A. S. B. L. Ed. Duculot. Gembloux, p 262
42 Schow JF (1823) Grundzüge einer allgemeinen Pflanzengeographie, Berlin
43 Schroeder F-G (1998) Lehrbuch de Pflanzengeographie. Verlag Quelle & Meyer, Wiesbaden, p 457
44 Schwabe A (1997) Sigmachorology as a subject of phytosociological research. Phytocoenologia 27(4):463–507
45 Takhtajan A (1986) Floristic regions of the world. University of California Press, Berkeley
46 Vincent P (1990) The biogeography of the British Isles. Routledge, New York, p 315
47 Wolfe JA (1981) Paleoclimatic significance of the Oligocene and Neogene floras of Eastern Asia and relation to forests of other regions of the Northern Hemisphere and Australasia. United States Geol Surv Prof Pap 1106:1–37
48 Zobel M, Otto R, Laanisto L, Naranjo-Cigala A, Pärtel M, Fernández-Palacios JM (2011) The formation of species pools:historical habitat abundance affects current local diversity. Glob Ecol Biogeogr 20:251–259
49. Braun-Blanquet J (1923) L'origine et le développement des flores dans le massif central de France avec aperçu sur les migrations des Flores dans l'Europe sud-occidentale. Annales de la Société linnéenne de Lyon et des Société botanique de Lyon, Société d'anthropologie et de biologie de Lyon réunies 69(1):1–16
50. Loidi J (2017) Dynamism in Vegetation. Vegetation changes in a short time scale. In: Loidi J (Ed.) The vegetation of the Iberian Peninsula vol. 1:81–99.

Dynamic-Catenal Vegetation Mapping as a Tool for Ecological Restoration and Conservation Policy

Javier Loidi

Abstract Traditional or standard vegetation mapping represents a static view of the ecosystems living on earth. The units represented in the map are vegetation types that express one or several features of the existing reality, i.e. species composition, size, structure, etc.; they express little, however, about the dynamic context in which they develop and even less about the spatial relationships network of which they form part. This chapter presents an approach that develops a system of complex geobotanical entities expressing the temporal and spatial relationships between simple vegetation units found in the field. The basic relationships are the successional and ecological gradients. Consequently two basic concepts are established: the sigmetum (or vegetation series), in which the vegetation units are related by successional series; and the geosigmetum, which encompasses the zonation associated with an ecological gradient. Within these categories variants such as minorisigmeta and geominorisigmeta as well as permasigmeta and geopermasigmeta are considered. Maps drawn using these units provide valuable synthetic information for ecological restoration and land management.

Keywords Geosigmetum · Potential natural vegetation · Sigmetum · Tessela · Vegetation catena · Vegetation series

1 Introduction

Vegetation Science—or Geobotany—has developed working on the initial intuitive idea that there are patterns or laws which determine the existence and distribution of the different plant communities (vegetation units) observed in the landscape [36]. After two centuries, we have constructed different explanatory scientific models that formalize such organization. A milestone in this history was the formalization and

J. Loidi (✉)
Department of Plant Biology and Ecology, University of the Basque Country (UPV/EHU), P.O. Box 644, 48080 Bilbao, Spain
e-mail: javier.loidi@ehu.eus

F. Pedrotti and E. O. Box (eds.), *Tools for Landscape-Scale Geobotany and Conservation*, Geobotany Studies, https://doi.org/10.1007/978-3-030-74950-7_4

systematic description of plant communities after their floristic composition (phytodiversity content), led by the Zürich-Montpellier school, developed mostly in continental Europe and Japan. This approach has been extensively called phytosociology [13]. A further level of integration in this survey for patterns in vegetation distribution was to address the possibility of recognizing complexes or assemblages of plant communities which can reflect some existing patterns in the distribution, not only of plants, which produce a regularly based plant community typology, but also in the spatial and temporal distribution of the plant communities, this would enable them to be grouped into categories of higher level. This idea of a higher level of integration of vegetation units has been underpinning the thinking of vegetation scientists more or less intuitively since early times of phytosociology and geography [12, 67]. It was fueled by the assumption of the successionist postulates of Clements [17, 18] in the early 20th Century, which permitted the universal acceptance of the dynamic nature of vegetation. However, until the seventies, no serious effort was made to produce a scientific construction that would order the previously observed patterns of relations between different plant communities. Nonetheless, in the long period between 1910 and 1973 there were some fundamental theoretical contributions that made further developments possible. The first was by Braun-Blanquet, who accepted the Clementsian succession theory, in his original "Vocabulary of Phytosociology" [15] there are several concepts which group a number of plant associations in time (série) or space (climax-complexe, complexe des groupements). Tüxen [71] made a crucial contribution by establishing the concept of Potential Natural Vegetation (PNV), largely used by European and Japanese phytosociologists [35]. Later, Bolòs [10, 11] defined some important concepts, such as tessella, and discussed the catenal arrangement of plant communities, accepting the ideas of vegetation series, PNV, primitive vegetation and climax. But the first attempt to construct a formal theoretical model begins with Tüxen [72], and has continued with many later contributions: Géhu [25–31], Beguin et al. [3], Béguin and Hegg [4], Tüxen [73, 75] and Rivas-Martínez [54]. The work of Géhu and Rivas-Martínez [33] on basic conceptions in phytosociology brought this phase to an end. This period was followed by another in which the above-mentioned concepts were developed and tested in the field, often accompanied by different kinds of maps [43, 58]. Some comprehensive and conceptual reviewing papers about this issue of vegetation complexes are worth being mentioned, such as those by Matuskiewicz [46], Theurillat [69], Rivas-Martínez [60], Gillet and Gallandat [34], Schwabe [68], Vigo [76] and Biondi [6].

2 What is Dynamic Catenal Phytosociology?

The term "integrated phytosociology" [32], initially established and later replaced by dynamic-catenal phytosociology [64], was coined by Rivas-Martínez [58] in an attempt to integrate the fields of phytosociology, synphytosociology and geosynphytosociology. Integrated phytosociology could be seen as a sort of phytosociology squared or even cubed [33] or as a kind of high phytosociology (*hohere Pflanzensoziologie*), as Tüxen put it [74]. Recently, the terms integrated phytosociology and landscape phytosociology tend to be replaced by the more explicit dynamic-catenal phytosociology [65]. This discipline studies the associations of associations, or association complexes, distinguishable in the landscape, their organization patterns and the factors determining such organization.

The main issue to be settled when starting a scientific grouping of plant communities is the selection of the criteria to be applied, i.e. to determine the assembly rules governing such groupings. The definition of such criteria and their relative acceptance took decades, from the beginning of phytosociology to the early seventies. When phytosociology was still in its infancy, only a handful of authors actually wrote down ideas on the question of how communities could be combined within the landscape, and occasionally left the limits of the proposed combinations unclear. A good example is the basic work of Braun-Blanquet and Pavillard [15], in which they defined their "complex of communities" (Gesellschaftkomplex) as a combination of plant communities presenting a repetitive pattern in the landscape due to ecological and geomorphological coincidences. This is still a rather loose concept with no clear-cut limits. Later contributions helped to make things somewhat clearer, particularly Tüxen's contribution [72], who decided that the sample, i.e. the sigma-relevé or relevé of the plant communities of a particular area, should be made within the limits of the area of a single association of Potential Natural Vegetation.

In a synthetic way we can establish that there are only two different criteria to be used to distinguish groups of vascular plant communities on the territory and make sensitive groupings out of them: the **successional** and the **zonal**. This is because **succession** and **zonation** are the two main phenomena involving changes in plant communities in time or space respectively and thus encompassing all the variability in plant communities in a given area; succession is related to disturbance, which includes human influence. In other words, any plant community will have a determined position in succession and zonation, we only need to know how to place it in the correct position (Fig. 1). Thus, if we are able to use both criteria for organizing the existing plant communities in any area, we will be able to somehow express such phenomena in the frame of landscape ecology involving the existing phytodiversity. Vigo [76] addressed this question using the term **serie** for the temporal dimension and **arrangement** for the spatial one. Thus the task of **dynamic-catenal phytosociology** is to apply both criteria defining two integration levels, one for each criterion.

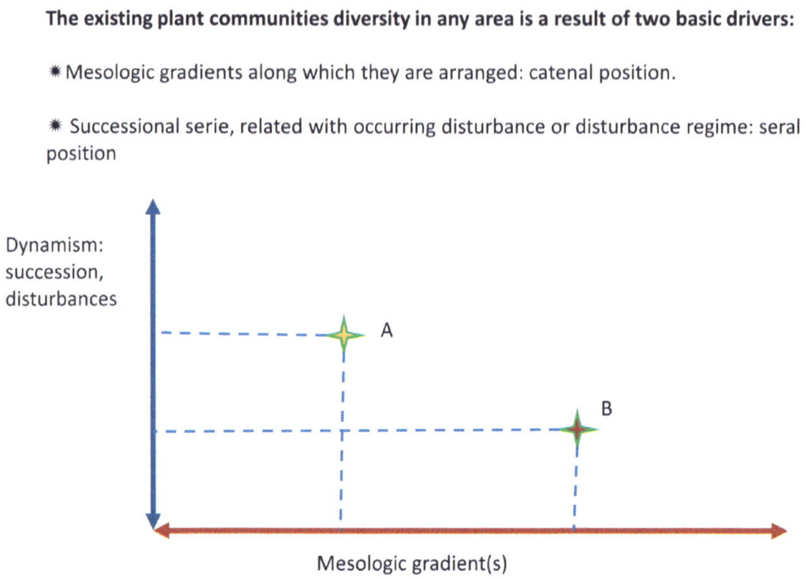

Fig. 1 Plant community diversity in any area is a result of two drivers: 1. **Mesological gradient** and 2. **Successional position,** being the latter narrowly related to disturbance regime. Plant communities A and B are determined by their particular position both in the gradient and in the succession.

3 Main Levels of Integration

The successionist criterion groups communities that succeed each other in the same area (serie); this type of process would end in a stable maximum in equilibrium with the current environmental conditions. This implies, of course, the acceptance of the concept of Potential Natural Vegetation (PNV) as formulated by Tüxen [71], as well as the idea of vegetation series [12] and tessella [10, 11].

The other criterion is based on the zonation of plant communities caused by one or several mesologic gradients: The zonal distribution of different vegetation types is called a catena (= chain) and used to be represented by a schematic section of the terrain. The scale of the zonation varies according to the gradient represented; changes in vegetation due to climatic gradient use very reduced scales involving for example a whole mountain or an entire region, while zonation due to soil moisture gradient in a pond will use a much more detailed one. So it is clear that a catena is a different type of representation to a successionist model. In the latter, the plant communities represented share the same ecological space, while in the former ecological parameters change causing change in the vegetation types. The term catena was initially used in soil science [49, 66] to express the sequences of soil types in a valley or small region [48], and later was formalized by Bolòs [10] for vegetation science.

Table 1 Integration levels in phytosociology.

Level	Studied units	Driving factor	Scientific discipline
1	**plant communities**	Floristic composition	classic phytosociology
2	**sigmeta** (or vegetation series)	succession	synphytosociology
3	**geosigmeta** (or geoseries)	catena	geosynphytosociology

Thus, integrated or dynamic-catenal phytosociology as landscape science [2, 5] is involved in assemblages of communities linked by dynamic relationships on the one hand and by zonal ones on the other. In the words of Rivas-Martínez [59], it develops on three levels: associations (Braun-Blanquet or classic phytosociology), vegetation series, also called sigmeta (synphytosociology), and geoseries, also called geosigmeta (geosynphytosociology). The first refers to assemblages of plants, the second to succession and the third to zonation (Table 1).

4 Main Concepts

Potential Natural Vegetation (PNV). This idea has been widely debated and is one of the key concepts of dynamic-catenal phytosociology. The old idea of Clements [18] that there was an optimal phase in vegetation in balance with regional climate beyond which succession could never achieve a higher degree of complexity and size, called the *climax*, is a concept that resonates with the logic of ecology. Any system has its carrying capacity, which cannot be exceeded, and the vegetation also reaches that carrying capacity or limit in the possibilities of development according the climate, the available resources and the available flora of the region (plants having the possibility to reach a site by their own dispersal means). This initial idea was strongly connected with the regional climate and supposed that different substrata would converge towards the same climax if there was a similar climate. This accepted the presumption that such a climax would be reached after a sufficient length of time, after which no disturbance would happen. The idea was modified by Tüxen [72], who formulated the concept of Potential Natural Vegetation (*Potentielle Natürliche Vegetation*). He wrote. "*... is the vegetation that would finally develop if all human influences on the site and its immediate surroundings would stop at once, and if the terminal stage would be reached at once*". This text apparently suggests that we should have a lot of time at our disposal to test the validity of each type of PNV we hypothesize. But over such a long time, environmental conditions could change, such as the climate for instance, deviating the succession tendency in its course and reaching to an unexpected end. However, it is not the case for the concept, since Tüxen's paper clearly explains that the PNV assumption is according to the current (*heutige*)

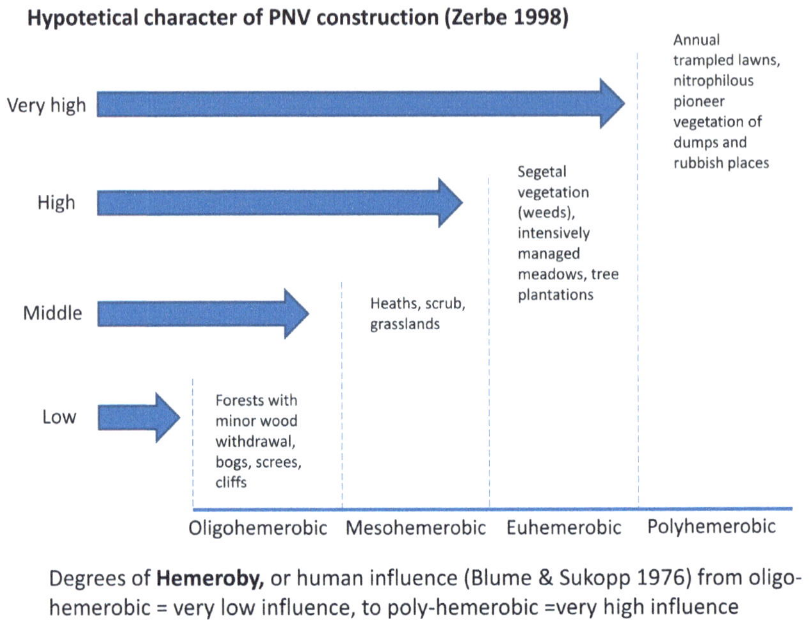

Fig. 2 Hypothetical character in PNV construction [77]. The lower the position of the community in the successional series is, i.e. higher hemerobie [9], the greater the hypothetical character in the PNV is in that site.

conditions. The PNV has to be imagined as if succession would happen instantaneously (*schlagartig*). Then it is not a set of hypothesis for succession processes, it is a set of assumptions of ecological potentialities in a given territory. However, in a landscape covered by a variety of plant communities forming a mosaic, the assumption of its PNV has a variable degree of hypothetical character [77], depending on the distance a concrete plant community found in a site is from the PNV, the more hemerobic a community is [9] the longer the distance to the PNV is (Fig. 2).

Intense discussion of the PVN concept has continued until recently [16, 42, 45, 50] and, although some authors consider it unnecessary [19], a large number of vegetation scientists use it as a descriptor of territory in land planning for conservation and as a target guide in restoration projects.

Tessella. This term is taken from the Latin *tesella* (paving stone)—*Fliese* in German [80], *piastrella* in Italian [53]—and has been used profusely with different meanings. In general, it has been used to refer to each of the surface elements that compose a landscape, like each of the pieces of a mosaic. Each tessella coincides with each of the visible plant communities; their geometrical shape produces a perfect assemblage, like the pieces of a puzzle. In this sense it is purely descriptive and does not define any ecological quality of the element considered.

Fig. 3 Scheme of plant communities in a Sigmetum (Querco rotundifolae-Sigmetum) lying on its tessella. The forest (PNV) and its substitution communities: mantle, scrub and other herb communities whose appearance is dependent on human influence (from [43]). 1- Quercetum rotundifoliae (PNV); 2- Rhamno-Quercetum cocciferae (mantle); 3- Salvio-Ononidetum fruticosae (scrub); 4- Ruto-Brachypodietum retusi (grassland); 5- Trachynion (annual grassland); 6- Lygeo-Stipetum lagascae (steppic grassland); 7- Pinus halepensis plantations; 8- Olive tree plantations; 9- Vineyards; 10- Cereal crop fields.

Another meaning of tessella is the sense in which it was used by Bolòs [10, 11], who gave an inspired ecological and successionist definition of the term: "*any ecologically homogeneous portion of land and thus able to sustain a determined stable community as well as its substitution associations ... The tessella is the elementary component of the mosaic of pieces of ground of different quality which constitute the territorial basis of the landscape. Homogeneous in its inherent properties, each tessella used to be occupied, in a primeval landscape, by a particular community. However, in human-influenced landscapes, the homogeneity of the tessellar areas is often masked because of the different treatment applied to its different parts.*" The tessella is thus a piece of land, a physical territory homogeneous in its intrinsic mesologic conditions: climate, soil, topography, hydric and chemical conditions of substratum, etc. Such homogeneity results in an ecological equipotentiality which supports a unique association of PNV: this occurs only if enough disturbance-free time passes. PNV is usually scarce in the real landscape of most areas, strongly influenced by human activity; most of the surface is covered by seral communities (*Ersatzgesellschaften*) of different kinds, depending on the various treatments given by man. From this circumstance, it is possible to relate all these communities and group them together to build a higher entity (Fig. 3).

This interpretation of tessella, largely accepted by Rivas-Martínez and other authors, implies the acceptance of two main ideas:

1. The assumption of the concept of PNV, which may coincide -or not- with primitive vegetation.
2. The almost reciprocal one-to-one relationship between tessella and PNV association. Each tessella (or approximate pluri-tessellar complex) sustains a unique association of PNV due to the equipotentiality of the entire tessellar area. Inversely, each PNV association occupies its own tessella, each PNV association has to have its defined place.

Sigmetum (also called *Vegetation Series*). Sigmetum is the oldest concept of integrated phytosociology, and was already used, under the name of series, by Braun and Furrer [12] and Braun-Blanquet and Pavillard [13]. They distinguished the initial and transformation stages which terminate in a final stage to bring the dynamic process called *succession* to a close [18]. However, the idea of series was only extensively used much later, especially in France [24, 51, 52], Spain [10, 41, 43, 44, 55–58] and Italy [6–8]. Sigmetum is a geobotanical concept which includes one PNV association as well as all the communities which substitute it as a result of regressive succession (Fig. 3). In Bolòs' words sigmetum "*is the ordinated group of communities which, in the* ***same*** *place, substitute each other in time*" (Fig. 4). Hence, the sigmetum expresses all the vegetal content organized in communities, linked by dynamic relationships, which are living on and being supported by a tessella (Fig. 5). Certainly sigmetum and tessella are as closely related concepts, as PNV and tessella, and are often used as synonyms. However, since sigmetum comprises all the communities living on the tessella, its use is more appropriate when speaking about the tessella's vegetation. Other terms used in the literature, such as synassociation [54], sigmassociation [21, 72, 73–75] or, more repeatedly, vegetation series, express the same concept [55–58, 60, 62, 64, 65].

A sigmetum in a particular area is usually built following chrono-sequences, i.e. space-for-time substitutions. Each sigmetum is built up by detecting the most mature vegetation type within each tesella, attributing to it the PNV role and adding the rest of the plant communities (seral stages) observed in the vicinity within the tessella. The seral stages are usually numerous and diverse, being necessary to take into account the hemeroby [9] degree associated with each of them (Fig. 6). The whole set constitutes the sigmetum.

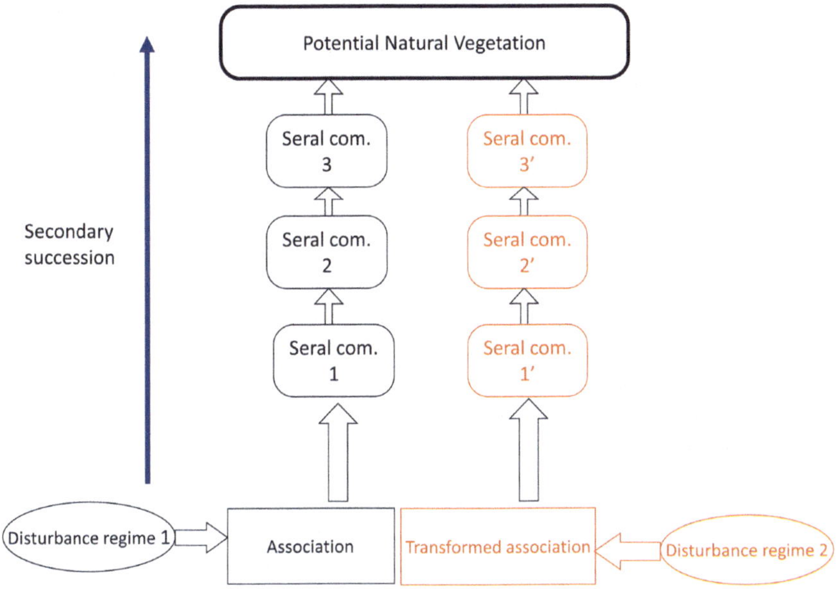

Fig. 4 Schematic representation of the concept of sigmetum and its dynamic significance. In a particular place, a plant community (association) is conditioned by a disturbance regime (1), when disturbance ceases, secondary succession drives the vegetation towards PNV along several stages represented by seral communities 1 to 3. If the disturbance regime is a different one (2), the resulting association is different (transformed association) and if that second disturbance regime ceases, the secondary succession will drive the vegetation to the same endpoint but along a different set of stages, seral communities 1′ to 3′. All the communities of the seral stages in addition to the PNV constitute the Sigmetum or Vegetation Series.

It is important to formalize and systematize sampling procedures, establishing a sampling methodology that renders all collected samples comparable to each other as a condition for building a typology on which new models can be based. A first step in this direction was made by Bolòs [10], who proposed the following scale to assess the abundance of the associations existing within a sigmetum when making a sigma-relevé:

1. Dominant community (dom): covering over 50% of the area
2. Extensive community (ext): >10% <50%
3. Localized community (loc): covers small areas but occurs regularly in favorable sites
4. Sporadic community (spor): only exceptionally observed

Béguin et al. [3] and later Géhu and Rivas-Martínez [33] proposed, with much more success, the use of the Braun-Blanquet scale to assess the abundance of each

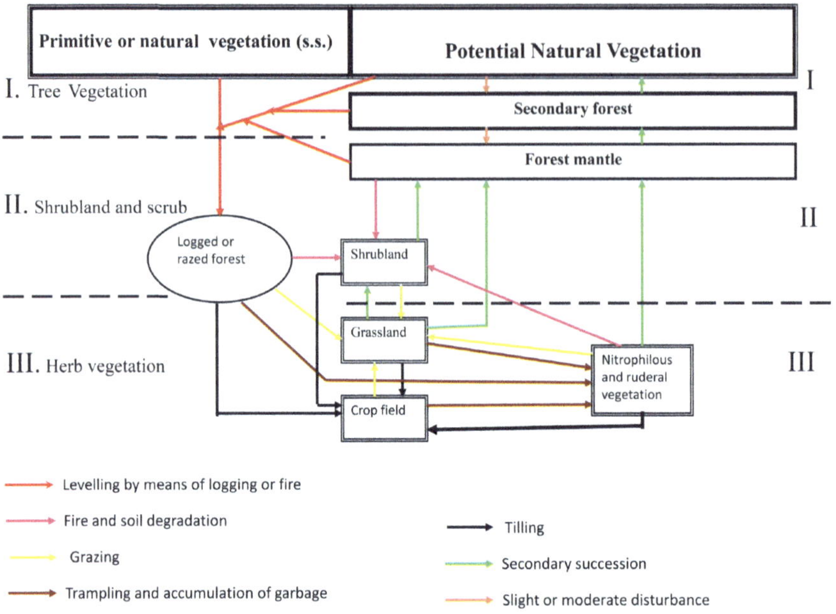

Fig. 5 Schematic representation of possible plant communities existing in a sigmetum, their successional relationships and associated disturbance regimes.

community of the sigma-relevé. Certain symbols can be also added to indicate the type of distribution of the community on the land:

O: Spatial distribution (e.g. *Cynosurion* meadows in a bocage landscape)
/: Linear distribution (e.g. *Prunetalia* hedges in the same landscape)
0: Spatial-linear distribution, as in a broad fringe
•: Point distribution.

This system combines an assessment of the abundance of each association occurring in one tessella with an estimation of their quality or nature expressed through their distribution type on the land. It is possible, then, to collect samples (sigma-relevés) which can be tabulated and treated numerically by using multivariate methods or other statistical analysis.

- *Ecologic Variability*. Despite the homogeneity of the mesological parameters over their area (tessella) as a whole, ecological and geographical factors usually give the sigmetum a certain internal variability [1, 41]. Ecological variations are generally related to minor lithological and trophic changes incapable of promoting substantial changes in the PNV. Such variations may influence changes in some of the substitution communities such as scrub or grasslands. A good example of this is provided by gypsum soils, frequent in vast areas in central and

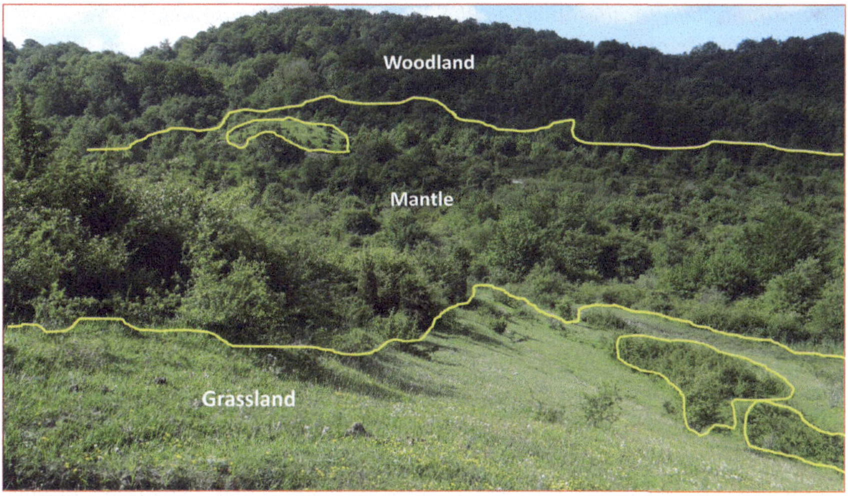

Fig. 6 Landscape representing a fragment of a successional series, with the woodland, the mantle and the grassland, living on a mesologic homogeneous area (tessella). Each of the three plant communities is determined by its position in the successional series and the disturbance regime exerted upon it: the woodland represents a terminal position in the succession, probably the PNV or a near stage; the mantle is a previous stage in the succession and is linked to episodic disturbance due to fire or cutting; the grassland is a more initial stage in the successional trend and is linked to a heavy grazing regime by sheep or cattle.

southern Spain, which determine the existence of specific scrub communities but have no influence on the PNV of those areas. Another example is the neighbouring effect observed in the transitional zones between two vegetation belts in a mountain. In this case, the variability of a sigmetum is due to the border effect and depends on the number of sigmeta in its vicinity which comes into contact with.

- *Variations due to biogeography*. Geographical (or, more accurately, biogeographical) factors give rise to another type of variability within a vegetation series or sigmetum. If the area of a sigmetum (tessella) is large, floristic differences inevitably appear in its different "regions". They often support biogeographically-related phytosociological diversity, usually involving substitution stages of the PNV. Loidi [41] shows how the central Spanish association *Quercetum rotundifoliae* (incl. *Asparago-Quercetum rotundifoliae*), which occupies a vast area, can be replaced by no less than 12 geographical associations of scrub.

  Both types of variability, ecological and biogeographical, are essentially different in kind and should be expressed at different levels. The former is due mainly to lithological changes or to the border effect, while the latter is the result of differences in climate (past or current) and often of differences in vegetation history. This suggests that biogeographical variability is more significant than ecological variability and should therefore prevail when constructing a typology.

- *Variability due to alterations in the successional trend.* Often, environmental conditions are extreme in the sense that they impede the completion of a successional trend. This happens in places where exceptional, often stressing, conditions occur, leading to mosaics of microhabitats occupied by plant communities more dependent on such conditions than on the general climate. These plant communities, apparently arrested in succession, have traditionally been called *permanent communities*. This term indicates that the stressing conditions are more relevant than the general climate, which plays a secondary role in conditioning the life of the community. Under such conditions, the succession is slowed down or even canceled, remaining the community stable throughout time (permanent). Each of these permanent communities usually develops on a small-sized tessella which supports a small-sized sigmetum called a **permasigmetum (perma-series)**. In these situations, the permasigmetum often has a unique plant community; seral stages are lacking or occasionally restricted to annual communities. In this category we can also include the extremely cold climates, i.e. polar or alpine, as far as they impose severe conditions on plant life. The permasigmeta often appear imbricated forming a mosaic or arranged along a gradient [61, 64].

A variant in this is the **minorisigmetum (minori-series)**, a concept initially launched by Lazare and Lanniel [40] under the name of *mesosérie* and later renamed as *curtosérie* or *série tronquée* (truncated) [39]. Finally the last term for this concept was proposed by Rivas-Martínez et al. [65] as minorisigmetum. Similarly to the permasigmetum, the minorisigmetum occurs in situations in which the succession is restricted or blocked by a stressing condition which does not allow it to reach the potential vegetation corresponding with the climate and substrate. The difference with the permasigmetum is that the seral stages can be various and have perennial non-nitrophilous stages, such as grasslands or scrub, completely lacking in the permasigmetum. It is as if the higher potential vegetation stage is prevented by the stressing conditions but not the seral ones. They can be found in coastal areas (dune, cliffs and marshes), river banks, peat bogs, wind exposed crests, etc., usually adjacent to the perma-series subjected to more stressing conditions. The minorisigmetum can be regarded as an intermediate category between the permasigmetum and the standard sigmetum.

Geosigmetum. This concept deals with the phenomenon of zonation. It can also be called *Geosigmassociation* and *Geoseries* and is the basic unit of what is known as geosynphytosociology. It is formed by a group of sigmeta arranged along a gradient: each sigmetum, occupying its own tessella, will span a certain stretch of the gradient. Thus, the geosigmetum has a spatial expression and represents a catena of sigmeta (Fig. 7). Geosigmetum is the most synthetic concept in phytosociology and, owing to the extreme variability and complexity of the situations occurring in nature, some approaches must be made in order to formalize observations and sampling methodology and to develop a system able to organize the resulting knowledge.

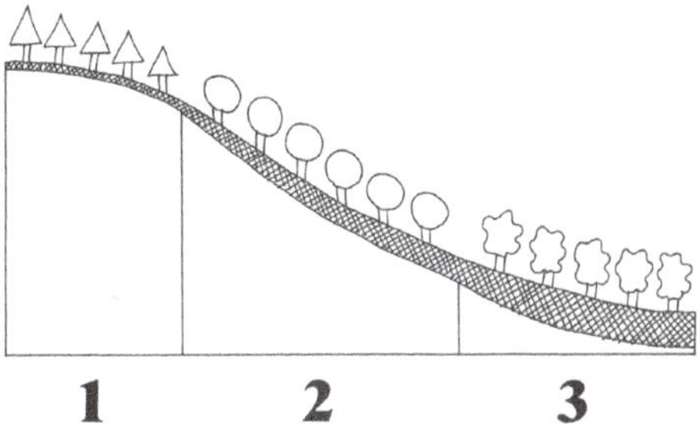

Fig. 7 Schematic representation of a Geosigmetum. The sigmeta are arranged along a slope gradient, a complex gradient which encompasses soil depth, nutrient richness and soil moisture.

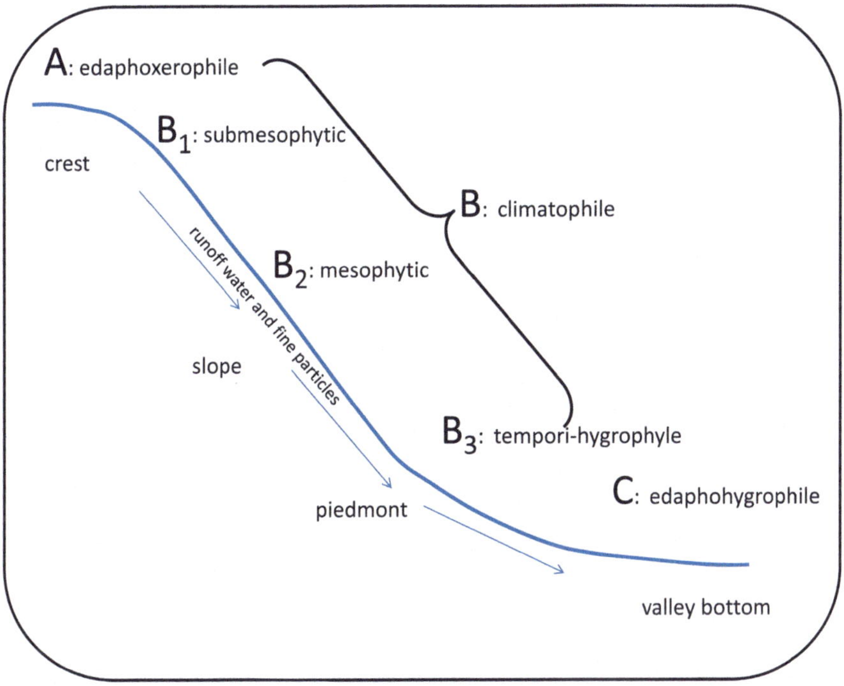

Fig. 8 Schematic representation of the topographic geosigmetum. The crest-slope-piedmont-valley bottom model is of universal application. A: edphoxerophile sigmetum (the most xerophytic), B: climatophile sigmeta (B1: submesophytic, B2: mesophytic, B3: temporihygropile). C: edaphohygrophile (the most humid). They are arranged along a soil-moisture and soil-depth gradient, favored by the gravity-driven oblique runoff of water and fine particles in the soil. Erosion is prevalent on the crest and sedimentation on the valley bottom, where fine particles, nutrients and water accumulate.

Types of Geosigmeta

Several attempts have been made to study the variability of geosigmeta after observation in the field. Theurillat [70] was the first, resulting from extensive field work in the Swiss Alps. He established four major groups of geosigmeta depending on their extent. Another classification of geo-permasigmeta was done by Alcaraz [1], and it partially inspires what we propose here. We can divide the geosigmeta into three main types depending on scale: topographic, altitudinal, and geographic [64].

Geosigmeta reflecting general conditions. The three categories expressed below refer to zonation models at three spatial scales, from that representing the general conditions of local climate (mesoclimate), to the climatic variation patterns across vast areas, in the frame of the lithology and geomorphology existing in the different territories of the world.

A). **Topographic geosigmetum**. Its expression can be outlined in the crest-slope-piedmont-valley model (Fig. 8), and it bears the representation of the fundamental distribution pattern of vegetation units in a territory. It is related to the predominating climatic and geological conditions of that territory, and can therefore be considered as its genuine geobotanical representation. In it, the most xerophytic sigmetum (edaphoxerophilous) will occupy the driest and most eroded crest position, the most hygrophilic (edaphohygrophilous) in the bottom of the valley on the moister and deeper soils, and the intermediate stretch encompassing the middle slopes and piedmonts will be occupied by the climatophilous sigmeta—which can be divided into sub-mesophitic, mesophitic and tempori-hygrophilous sigmeta. As outlined in Table 2, the middle stretch of the slope reflects the ombroclimatic (euclimatope) conditions of the area, while the crest will be under drier and the bottom under moister conditions [64].
This basic model is recognized within a particular vegetation belt and a concrete territory (a biogeographical district). It represents the fundamental pattern of vegetational variability in an area of homogeneous climate, substrate and history. The topographic geosigmetum is the main geobotanical descriptor of any territory as it summarizes the great majority of the variability and relationships of its vegetation. It can also be used as an indicator of paleohistoric information since edaphoxerophilous sigmeta correspond to past periods of dry climate in which xerophilous vegetation could invade the area and, vice versa, the edaphohygrophilous sigmetum to episodes of moister climate. These dry and wet complexes can serve as a bridgehead for colonization in temporary episodes of climatic change. The interpretation of thermophilic and cryophilic vegetation examples in a given territory can similarly correspond to warm and cold periods in the past. A smart interpretation of the geobotanical conditions in a given area can provide us with important information about past migrations driven by climatic changes [62, 64].

B). **Cliserial geosigmetum**. Cliserial refers to altitudinal zonation and is expressed at the scale of an entire mountain or mountain range. The cliserial geosigmetum consists genuinely of the altitudinal vegetation belts on any mountain. Increasing elevation entails decreasing temperatures and, most commonly,

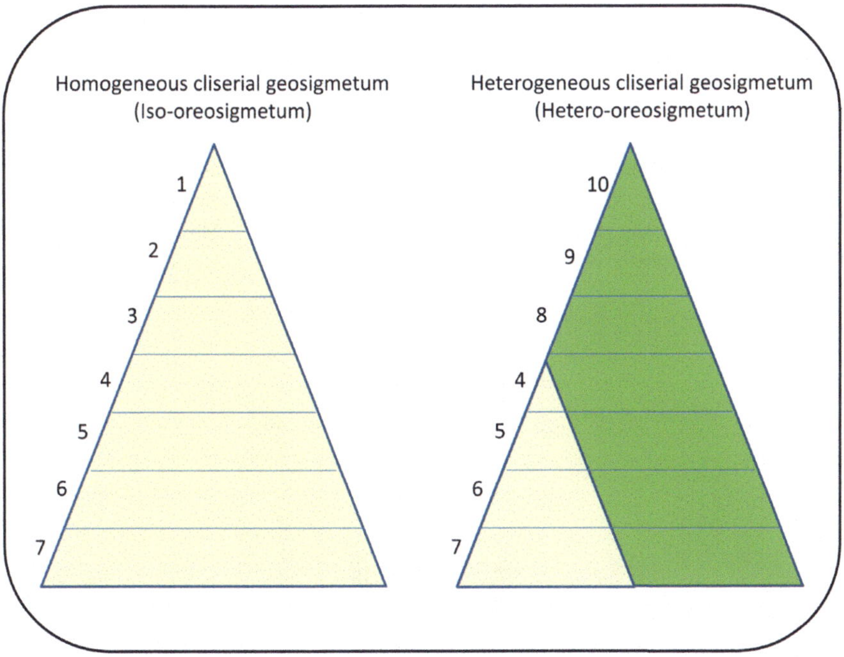

Fig. 9 Schematic representation of the cliserial geosigmetum. There are two variants, homogeneous and heterogeneous depending on whether the rock material of the mountain is homogeneous or heterogeneous.

increasing precipitation. If the mountain is high enough, the summit will bear a tundra-like vegetation, or even perennial ice, while in the lower level a more thermophilous vegetation will develop. This altitudinal zonation resembles the broad-scale geographical latitudinal zonation existing in a continent by which climatic zones arranged in parallel bands are inhabited by their particular biomes, giving way to the rule of thumb that 1 m in altitude is equivalent to 1.3 km in latitude [38]. The zonation will be affected depending on whether the rock material of the mountain is homogeneous or heterogeneous (Fig. 9).

C). **Geographic geosigmetum**. It is expressed in a geographical scale of an entire region or continent and consists of latitudinal zonation of biomes or broadly defined vegetation units arranged according to a gradient of regional climate. This category overlaps with phytogeography and is not a target for this paper.

Geosigmeta under stressing conditions. There are also other types of geosigmeta present in any territory which at the topographic scale are subjected, in varying degrees to conditions which partially or entirely mask the influence of the regional climate.

Table 2 Hydric balance in the three main sigmeta of a topographic geosigmetum

Sigmetum	Edaphoxerophile	Climatophile	Edaphohygrophile
Moisture Balance	Input from rain > Losses and gains from drainage and ETP; negative balance, net loss	Input from rain = Losses and gains from drainage and ETP; balance, no loss nor gain	Input from rain < Losses and gains from drainage and ETP; positive balance, net gain
Vegetation in relation to climate	Drier than climate	Equal to climate	Moister than climate
	Xerophytic vegetation	Climatophile vegetation	Hygrophile vegetation

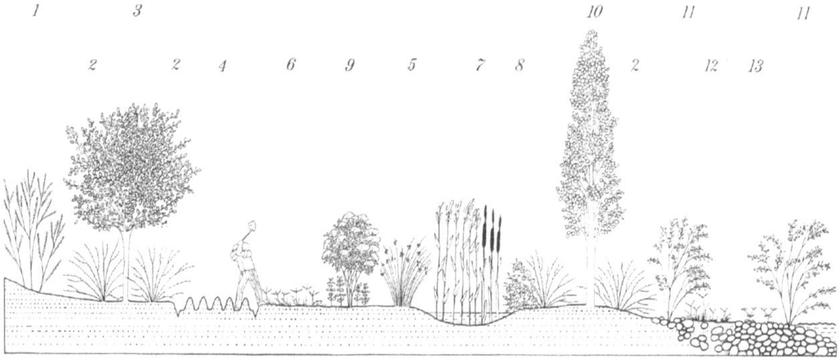

Fig. 10 A riverine geosigmetum, with all the complexity of the vegetation types associated with the soil water table and flooding gradient. 1 to 13 represent the plant communities existing in the riverine geosigmetum

Riverine Geosigmetum. This type constitutes the usual vegetation complex we have called edaphohygrophilous, which is itself not a sigmetum but a geosigmetum. It includes vegetation of the fluvial and peri-fluvial range, influenced by permanent, intermittent or sporadic streams (Fig. 10; Fig. 11, C). It does not encompass the belt of humid soils caused by a high water table unaffected by the stream (edaphohygrophilous sigmetum), usually included in the topographic geosigmetum as its most humid component. Only the fringe directly exposed to the effects of the stream—erosion, transport and sedimentation—is included in the riverine geosigmetum, i.e. the space subject to hydric stress in the soils and to longer or shorter flooding by rapidly or slowly running water. Plants bear adaptations to soil instability (periodical removal and disposal of materials) in their roots and aerial anatomy as well as in their reproductive and dispersal mechanisms. River types present high diversity all over the world; from young mountain streams to mature riverine systems or from permanent streams to those with dry river beds that only carry water when it rains, sometimes very sporadically. The fluvial regime and the flooding of exposed areas or bands determine the complexity of this geosigmetum, an essential component in any landscape [79].

Geopermasigmetum and Geominorisigmetum. Under situations of exceptional stressing conditions the mosaics of microhabitats occupied by permasigmeta and/or

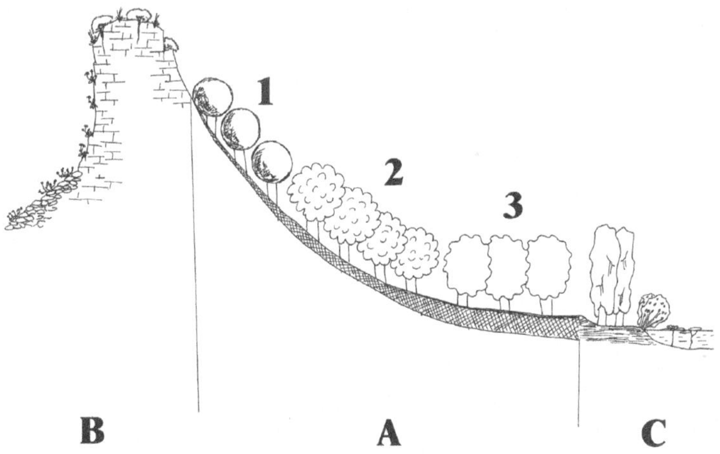

Fig. 11 Schematic representation of a complex of three geosigmeta: climatophilous geosigmetum (A), riverine geosigmetum (C) and geopermasigmetum (B). A: 1- crest or steep slope (submesophytic) sigmetum, 2- middle slope (mesophytic) sigmetum, 3- valley (tempori-hygrophile) sigmetum; B: Rupicolous permaogeosigmetum; C: Riverine geosigmetum

minorisigmeta (Fig. 11, B) are called geo-permasigmeta and geo-minorisigmeta. Complexes of this kind are usually linked to special edaphic and topographic circumstances that have a much higher influence on vegetation than climate (e.g. rock walls, ponds, mires, dunes, etc.).

Permasigmeta have received various names in the literature: **microgeosigmeta** by Rivas-Martínez [60, 62], **special geoseries** by Alcaraz [1], and the term **hipogeosigmassociation** [69, 70] is also partially synonymous. The term geopermasigmetum was finally established by Rivas-Martínez [64]. The geominorisigmetum has also been called **curtogeosigmetum** [39].

We can distinguish the following 11 types of geo-perma-geosigmeta:

- CRIOGEOPERMASIGMETA. Low temperatures are the determining factor.

1. **Polar (tundral) geopermasigmeta**
   Correspond to areas under cold climate (polar), where low temperatures are the limiting factor in the ecosystems. Sedges, small chamaephytes, bryophytes and lichens are the most common plant types.
2. **High mountain geopermasigmeta**
   In high mountains (cryo belt) cold also is the limiting factor, similarly as in the previous type. Alpine grasslands or dwarf shrublands are the most frequent types of vegetation.

  - SAXIGEOPERMASIGMETA. Outcrops of hard rock entail that soils are absent or shallow; they are particularly frequent in mountains, where glacial and periglacial erosion processes favor the existence of cliffs, wind exposed crests, cryoturbation and gelifraction phenomena favoring the accumulation of screes.

3. **Rocky cliff geopermasigmeta**
   Vertical cliffs in a rocky mountain have a number of crevices in which some very specialized stress tolerant species live. Sometimes, the tilt of the cliff is negative giving way to overhangs. Oozing rocks represent a variant as well as those cliffs with additional nitrogen nutrient inputs due to animals (bird nests) or humans.
4. **Scree geopermasigmeta**
   Screes formed by stone accumulation at the foot of cliffs, usually due to gelifraction, are an unstable substrate which only specialized plants can colonize.

- Chionophilous (snow) geopermasigmeta. Accumulated snow limits the development of the vegetation living below.

5. **Snowdrift geopermasigmeta**
   Zonation is determined by the permanence of snow, which in turn depends on its degree of accumulation. Hollows and other areas where snow accumulates are called snowfields and are found in snowy climates such as high mountains and boreal or polar climates. The catena is determined by the length of the snow coverage period. In Europe they are generally bryophytic, herbaceous or small chamaephyte communities, but in other areas they can even be woodlands.

- Hygrogeopermasigmeta. The stress is produced by the waterlogging of the soil.

6. **Freshwater inland geopermasigmeta**
   Ponds, lakes and swampy areas with still freshwater have a clear hydric gradient of anoxic stress, from the moist to the permanently flooded soil. The vegetation is composed by aquatic communities of submersed, floating and emersed plants, together with the hygrophilous terrestrial vegetation of their borders.
7. **Mires geopermasigmeta**
   Depressions that fill up with water in territories under cold climate and receive organic matter which decomposition is slowed down or stopped. This organic matter accumulates in the ponds and forms peat deposits. Such environments are hostile for many vascular plants, being dominated by bryophytes; the most frequent group of vascular plants are the *Cyperaceae*. They are very abundant in Boreal and Arctic regions but are frequent also in temperate areas. In the rest of the world, they are found only in high mountains. The influence of stagnant water soaking up the peat moss establishes the gradient which drives the plant communities zonation: the low mire rich in sedges develops at the water table level while the high mire, rich in mosses (*Sphagnum*) and with some scrubs (*Erica, Vaccinium*) rises from that level and depends more on the rainwater supply. Some reed communities at the edges of the mire and some submersed plants in the ponds (if present) complete the catena.

- Halogeopermasigmeta. High soil salinity is the main factor of stress.

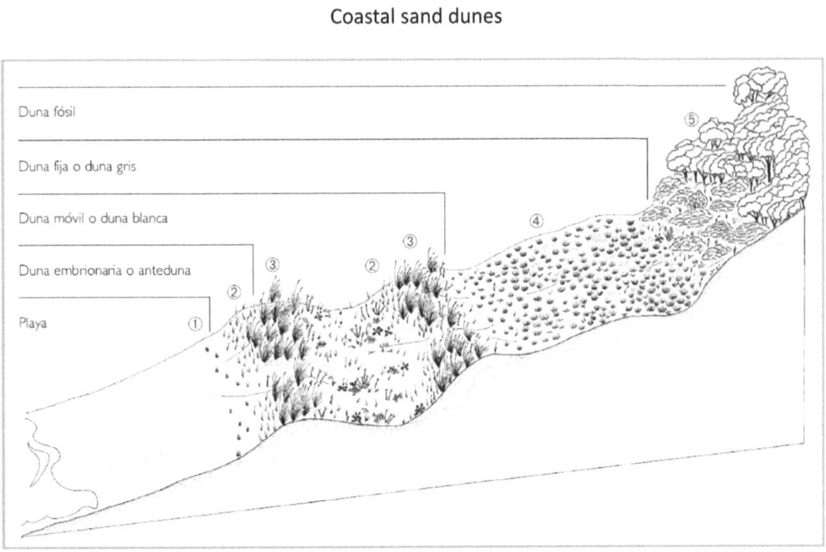

Fig. 12 Iberian Atlantic coastal sand dune system. Each of the bands represents an entire sigmetum (permasigmetum) as it occupies a permatesella. The combined gradient of salinity and wind exposure is responsible for the zonation

8. **Halophilous tidal geopermasigmeta**
   Coastal salt marshes influenced by tides and sheltered from waves are populated by vegetation adapted to the high sea water salinity. The tidal floods determine the gradient. Vascular vegetation is dominated by succulent *Chenopodioideae* in the extra-tropical countries and by mangroves in tropical countries where warm waters dominate.
9. **Halophilous inland geopermasigmeta**
   In the inland endorrheic depressions or basins, soluble salts accumulate in the soil under certain climatic and geohydrologic conditions, reproducing partially the salinity levels of the coastal salt marshes. The gradient is given by the saline concentration in the soil water.
10. **Coastal dunes geopermasigmeta**
    Between the beach and the fixed dunes, the coastal sandy deposits are modelled by the wind and salinity brought by the sea spray, creating a gradient of mobility and salinity which clearly results in the zonation of the foredunes, white or mobile dunes and grey dunes typical of this geopermasigmetum all over the world. Substrate mobility and coarseness (sand) entail dry edaphic conditions (almost immediate percolation and no water retention capacity) which are added to the more or less intense salinity brought by the sea spray (at least in the fringes nearest to the sea). The wind itself has a strong evaporation effect and a mechanical abrasive action on the plants (Fig. 12).

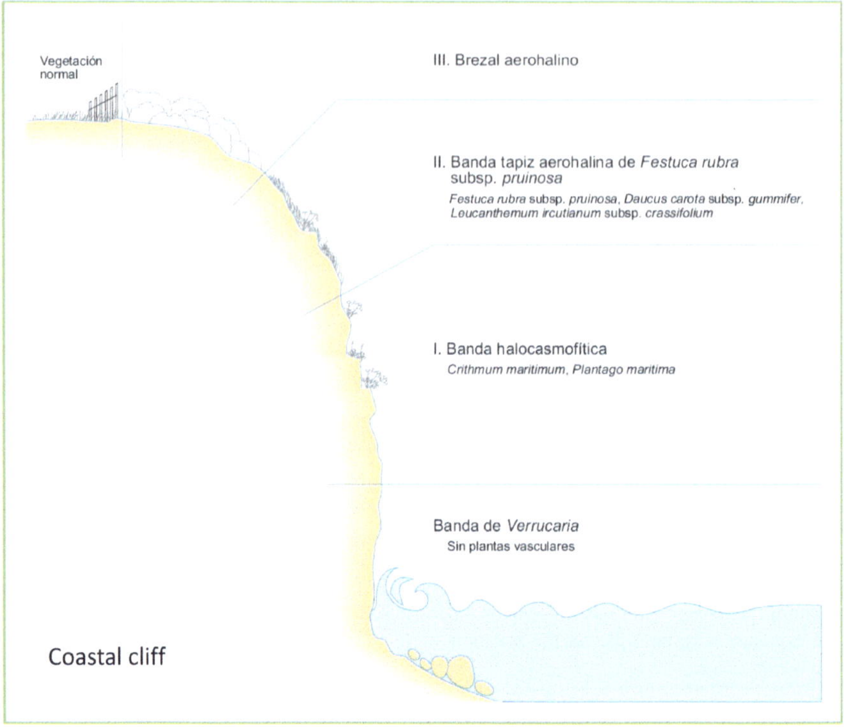

Fig. 13 Example of permageosigmetum where the permasigmeta are the vegetation belts arranged along the rockiness, slope steepness, wind exposure and salinity combined gradients

11. **Coastal halo-anemogenous rupicolous geopermasigmeta**
    Coastal areas with cliffs are influenced by sea winds which impose their mechanical and desiccating action and bring small drops of sea water spreading marine salt on the cliffs. They usually occupy a narrow strip in the upper border of the cliffs, but as the influence of the sea increases during storms, in exposed windy capes this geopermasigmetum can reach to a level of more than 100 m high and cover large areas of several kilometres inland. Topography and rockiness is the second major factor in determining the catena, as it is related to the capability of building a stable soil (Fig. 13).

Methodological Considerations in Geosynphytosociology

The geosigmetum is the most synthetic concept in dynamic-catenal phytosociology (or phytosociology *sensu lato*) and, owing to the extreme variability and complexity of the situations that occur in nature, an attempt needs to be made in order to formalize observations and sampling methodology and to facilitate a system for organizing knowledge. There is no limit to the number of sigmeta to be included in

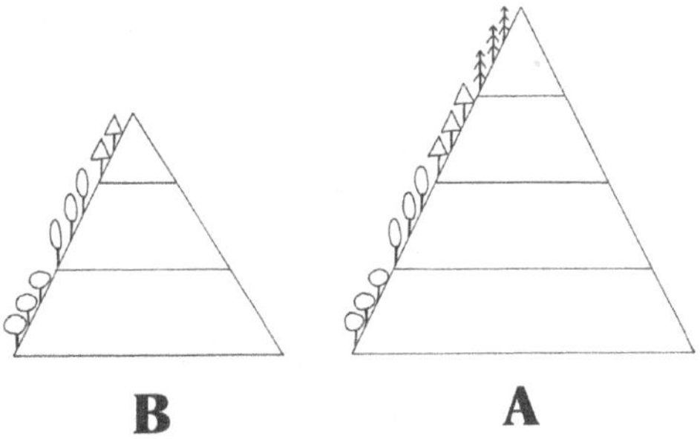

Fig. 14 Comparison of the zonation (oreogeosigmeta) between two mountains of the same biogeographical unit but different altitudes. Mountain A has a "longer" catena than B simply because it is higher. Geobotanically both are equivalent as they respond to a similar pattern of vegetation distribution and this is determined by a similar set of environmental conditions, climatic and edaphic

the sample (geosigma relevé) as long as they are defined. Only geography can impose limitations: the height of mountains, lithological and geomorphological variability, etc. So it is clear that the number of sigmeta occurring in a territory will depend on its size and biotope variability: vast but homogeneous flat countries will contain few sigmeta while mountainous areas with complicated geology and many local climatic situations will show many as an expression of their differences in biodiversity. The geosigmeta of each of the two kinds of territory will differ greatly in the number and definition of their sigmeta.

When using the concept of geosigmetum in field work, it is essential to know how to carry out the sampling procedure, which in our case means knowing where the limits of the geosigma relevé are. The comparability of the samples obtained in different territories must be ensured, as they can be used to assess the geobotanical differences or similarities between the territories. For example, consider two similar, neighboring valleys in the Alps, one of which has a higher mountain that rises up to the alpine belt, and the other with lower altitudes, with only the subalpine belt being represented at its summits (Fig. 14). The cliserial catena (cliserial geosigmetum) of both valleys will only differ in their "length", because the first one will have alpine grasslands and the second will not, but the rest of the catena is similar because the conditions, mainly climatic and geological, are the same. The comparison of both geosigma relevés can give us a misleading idea of dissimilarity but the differences are due to topographical circumstance, not to climatic, geological or biogeographical conditions. Both valleys are, using these latter criteria, geobotanically identical; one simply happens to have a higher mountain than the other.

To overcome difficulties of this kind, as Alcaraz [1] suggests that, in order to ensure that the samples are comparable, some standardizing criteria should be followed when sampling. In the case of topographic geosigmeta:

1. By placing the geosigma relevé within a single homogeneous bioclimatic area, i.e. within one thermotype and ombrotype in the typology of Rivas-Martínez [64]. For high mountains with a wide climatic gradient from the bottom to the top, segmentation in bioclimatic belts is required, sampling the topographic geosigmetum within each belt. This way each of the samples will be comparable with its bioclimatic homologue from another mountain.
2. By placing the geosigma relevé within a single type of lithological substratum, at least with regard to its trophic properties. A climatically homogeneous territory can support lithological diversity with strong influence in vegetation. The differences between base-rich and base-poor vegetation inside a homogeneous climatic region are clear.
3. By placing the geosigma relevé within a unique type of geomorphological unit in which the catena is due to a single gradient or group of associate gradients. This permits the separation of riverine geosigmeta (determined by intensity of flooding, flow of running water, etc.) from adjacent ones not directly influenced by river water, and also means that salt marsh geosigmetum, where salinity is the main factor for determining the catena, can be separated from neighboring geosigmeta.

A complete explanation of the methodology to sample coastal geopermasigmeta and geominorisigmeta is given by Demartini [20].

Geobotanical Contextualization

To the categories of dynamic-catenal phytosociology described above, can be added those of bioclimatology (e.g. those by Rivas-Martínez [59, 61, 64]). They provide the climate typological structure that orders and organizes climatic information from the perspective of vascular plants by aiming to adjust the typology to the distribution patterns of vegetation. This is done by selecting the climatic parameters with most influence on the distribution of vascular plants on earth. This bioclimatic typology provides a climatic framework within which plant communities live and develop describing the climatic scenery of any stretch of landscape, it offers a sort of description of the climatic context. For that reason, typological bioclimatology plays an important role in understanding vegetation units better. This can also be said of typological biogeography (provincialism) and dynamic-catenal phytodsociology, all of them synthetic disciplines: they summarize information on vegetation and environmental conditions.

This issue is relevant from a purely theoretical point of view. The synthetic typology provided by dynamic-catenal phytosociology at any of the integration levels, together with the typologies provided by bioclimatology, biogeography and by the earth sciences (edaphology, geomorphology, lithology, etc.), makes **geobotanical contextualization** possible, which is very important in all studies of plant communities. This is an exercise of considering all the biotic and abiotic conditions converging on a given area or site in which the plant community we are interested in lives, so that we can "put it in place" before we start with a survey. Geobotanical contextualization, in a not so formal way, has been almost routine, especially in European schools, since the early times of Geobotany [36]; today it can be done much more accurately thanks to the available typology recently developed. Geobotanical contextualization is a very convenient preliminary exercise when planning any study or research on site or area vegetation. The results of the research on any field of vegetation science should be placed in context in order to calibrate their validity and potential for extrapolation to other areas.

The aim is to build a universal typological system with explanatory power upon the distribution patterns of vegetation and valid for any portion of land on Earth. The system will enable workers to draw the outline of the geobotanical scenery of any region, using unified criteria, in such a way that results of research or observations on vegetation obtained in different territories may be correctly compared, by taking into account the similarities (or differences) of their respective geobotanical sceneries.

As an example of this comparison exercises, consider the results obtained from a survey on biodiversity of lauroid forest in central eastern China: how can we compare them to other types of forests in the world? Extrapolating to other physiognomically similar lauroid forests in the world is the initial procedure, i.e. we would compare it with the lauroid forests of Japan, eastern North America, etc.; but, what about the mountain lauroid forest in the Canary Islands (Canarian Laurisilva or Monteverde)? If we know that the first one is found in subtropical climates governed by the monsoon, which causes maximum rainfall in summer, while the second survives thank to the persistent fogs which occur on the northern slopes of the islands caused by the trade winds, in a pronounced Mediterranean context, then the comparison can be more accurately modulated. In this case of the lauroid forest, the bioclimatic study immediately shows that the annual distribution pattern of rainfall in both areas is diametrically opposed: in China, maximum rainfall occurs in summer (monsoon) whereas summer is the driest season in the Canary Islands, corresponding as it does to its Mediterranean climatic model [23]. Continentality is also much greater in China than in the extremely oceanic Canary archipelago, where frost is practically absent from the "laurisilva" level [23]. Such differences are certainly important, as they determine plant reproduction and growth cycles to a great extent. Then, when analyzing the flora and, consequently, the plant communities, from this contextual perspective, differences become quite profound. No scientific work can afford to ignore such circumstances, even if research has nothing to do with syntaxonomy or biogeography and focuses on other fields of ecology. When the results of searches for biodiversity measures are considered in this light, their interpretation, when compared to the Canarian Monteverde [63], is only relative due to the long "geobotanical distance" existing between central China and the Canary Islands. Geobotanical contextualization puts the eventual distances to similarities.

Conversely, China and Japan are closely related areas, biogeographically, bioclimatically and historically, so the comparisons are easier to interpret from the perspective of vicinity. If we compare distant territories, not only geographically but also climatically, biogeographically and historically, although they present a noticeable morphological and structural convergence in their forest types, the comparisons have to be done with much more care and taking into account that distance.

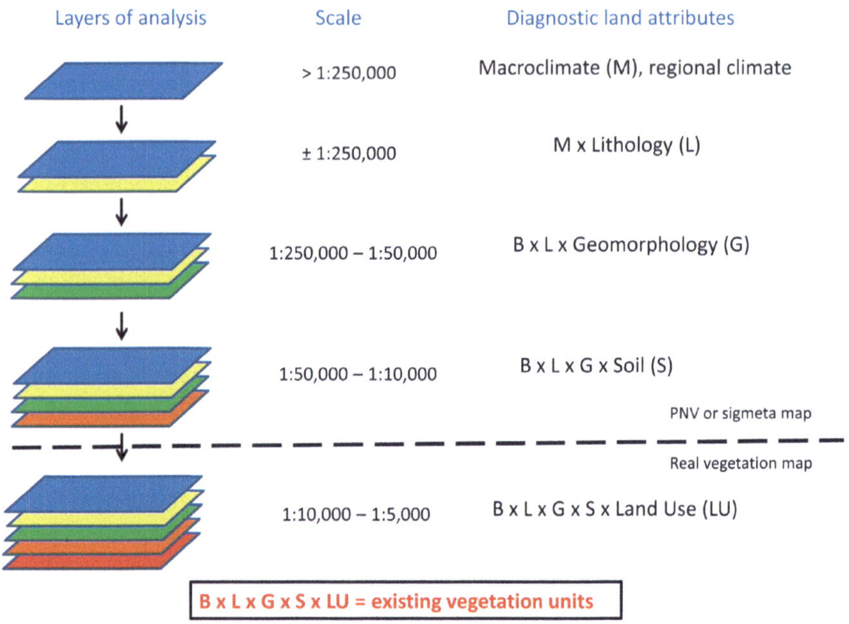

Fig. 15 After Blasi et al. 2000, cartography of sigmeta and geosigmeta over a determined region can be performed following a deductive landscape analysis in which different attributes are considered in a hierarchical order, being applied for different scales.

Thus, geobotanical contextualization should be included in the preliminary designs and planning of any experiment or observation on plant communities. This will help us to construct hypotheses and to assess the significance of the results in relation to their potential for general application. For some authors such as Alcaraz [1] the theoretical model provided by dynamic-catenal phytosociology can be considered a contribution to the formulation of a "General Theory of Vegetation Science" with sufficient capacity for general application for it to be considered a major contribution to the paradigm of Vegetation Science.

Cartography of Sigmeta and Geosigmeta

To produce a map of sigmeta and geosigmeta, in most cases an inductive process is followed, i.e. initial surveys of the flora and plant communities of the territory give way to a recognition of the units to be mapped and then the cartography is carried out. Nevertheless, there is an interesting approach done by some Italian colleagues [7, 8] in a deductive-inductive way, by which, initially, the landscape is successively analyzed after several attributes. These classifying attributes are, in hierarchical order of space and time: climate, lithology, geomorphology, human activities, soil, vegetation and fauna [22, 78]. This order of attributes moves from relatively stable factors controlling larger areas, such as climate or lithology, to more dynamic factors operating locally, so that fine-scale factors such as human land use are very variable in space and time (Fig. 15). The tessella would contain the four upper levels which ensure the ecological homogeneity, and the lowest one, with the land use, gives way to the existing plant communities.

One of the clearer, more straightforward tasks of this dynamic-catenal phytosociology is cartography. This has been done several times and using several scales in different countries of Europe (e.g. [55, 58, 47, 70, 43, 14, 7, 20]. In recent decades this type of mapping has obtained full recognition in the frame of the European Union documentation for inventories of natural resources [37]. A map representing sigmeta and geosigmeta is an excellent document summarizing the biological (vegetation) content of a territory in terms of its ecologic affinities and an expression of its potentialities. It is a map of the different vocations of the land built up by the integration of the climatic, topographic, lithologic, geomorphologic, edaphic, vegetation and human land use information. This ultra-synthetic approach enables the use of mapped units for targeting ecological restoration, if that is our intention, because they provide the referential frames for defining the Potential Natural Vegetation of any piece of territory. They also can be used to evaluate the ecologic quality of any area and, in general, as basic information for any project in land management. The interest of such cartography in the design of strategies for landscape management as well as for the protection of biodiversity and landscape quality becomes evident.

References

1. Alcaraz F (1996) Fitosociología integrada, paisaje y biogeografla. In: Loidi J (ed) Avances en Fitosociologia: 59–94. Ed. UPV/EHU
2. Asensi A (1996) Fitosociología y paisaje (Una aproximación histórica). In: Loidi J (ed) Avances en Fitosociologia. 43–58. Ed. Universidad del País Vasco
3. Béguin C, Géhu JM, Hegg O (1979) La symphytosociologie: une approche nouvelle des paysages vegetaux. Docum. Phytosociologiques NS 4:49–68
4. Béguin C, Hegg O (1975) Quelques associations d'associations (sigmassociations) sur les anticlinaux jurassiens recouvertes d'une végétation naturelle potentielle (essai d'analyse scientifique du paysage). Docum Phytosociologiques 9–14:9–18
5. Biondi E (1996) L'analisi fitosociologica nello studio integrato del paessagio. In: Loidi J (ed) Avances en Fitosociologia: 13–22. Ed. UPV/EHU
6. Biondi E (2011) Phytosociology today: methodological and conceptual evolution. Plant Biosystems 145:19–29
7. Blasi C (ed) (2010) La vegetazione d'Italia. Carta delle Serie de Vegetazione in scala 1:500.000. Palombini & Partner, Roma
8. Blasi C, Carranza ML, Frondoni R, Rosati L (2000) Ecosystem classification and mapping: a proposal for Italian landscapes. Appl Veg Sci 3:233–242
9. Blume H-P, Sukopp H (1976) Ökologische Bedeutung anthropogener Bodenveränderungen. In: Sukopp H, Trautmann W (eds) Veränderungen der Flora und Fauna in der Bundesrepublik Deutschland. Schriftenr. Vegetationskd, vol 10, pp 75–89
10. Bolòs O (1962) El paisaje vegetal barcelonés. Ayuntamiento, Barcelona, Cátedra Ciudad de Barcelona, Excmo
11. Bolòs O (1963) Botánica y geografía. Mem Real Acad Ci Artes Barcelona 34:443–491

12. Braun J, Furrer E (1913) Remarques sur l'étude des groupements de plantes. Soc Languedoc Geogr Bull 36:20–41
13. Braun-Blanquet J (1928) Pflanzensoziologie. Grundzüge der Vegetationskunde. Biologische Studienbücher. Berlin, vol 7, p 330
14. Capelo J, Costa JC, Ribeiro S, Arsénio P, Neto C, Monteiro-Henriques T, Aguiar C, Honrado J, Espirito-Santo D, Lousã M (2007) A methodological approach to potential vegetation modeling using GIS techniques and phytosociological expert-knowledge: application to mainland Portugal. Phytocoenologia 37(3–4):399–415
15. Braun-Blanquet J, Pavillard J (1928) Vocabulaire de sociologie végétale. 3 éme ed. Montpellier
16. Chiarucci A, Araújo MB, Decocq G, Beierkuhnlein C, Fernández-Palacios JM (2010) The concept of potential natural vegetation: an epitaph? J Veg Sci 21:1172–1178
17. Clements FE (1905) The development and structure ofvegetation (Studies in the vegetation on the state 111). Lincoln
18. Clements FE (1916) Plant succession: an analysis of the development of vegetation. Washington, 515 pp
19. Decoq G (2016) Moving from patterns to processes: a challenge for the phytosociology on the twenty-first century. In: Box O (ed) Vegetation structure and function at multiple spatial, temporal and conceptual scales. Geobotanic studies. Springer, pp 407–424
20. Demartini C (2016) Les végétations des côtes Manche-Atlantique françaises: essai de typologie et de cartographie dynamico-caténales. PhD. Université de Bretagne Occidentale, Brest
21. Dierschke H (1994) Pflanzensoziologie. UTB für Wissenschaft. Eugen Ulmer Verlag, Stuttgart
22. Forman R, Godron M (1986) Landscape Ecology. Wiley & Son, New York
23. Gandullo JM (ed) (1991) Estudio ecológico de la laurisilva canaria. ICONA, Madrid
24. Gaussen H (1964) Carte de la végétation de la France 1:200.000, fe u ille 77 Foix. CNRS. Toulouse
25. Géhu JM (1974) Sur l'emploi de la methode phytosociologique sigmatiste dans l'analyse, la définition et la cartographie des paysages. CR Acad Se Paris 279:1167–1170
26. Géhu JM (1976) Sur les paysages végétaux ou sigmassociations des prairies salées du Nord-Ouest de la France. Documents Phytosociologiques 15–18:57–62
27. Géhu JM (1977) Le concept de sigmassociation et son application a l'étude du paysage végétal del falaises atlantiques françaises. Vegetatio 34(2):117–125
28. Géhu JM (1978a) Les sigmassociations de la xérocére des dunes atlantiques françaises de Dunkerke a Biarritz. In: Tüxen R (ed.) Assoziationskomplexe (Sigmeten). Bericht Intern. Symposium 1977 in Rinteln: 77–82. Vaduz
29. Géhu JM (1978b) Premieres éléments pour un sigmasystéme des dunes séches holarctiques. In: Tüxen R (ed) Assoziationskomplexe (Sigmeten). Bericht Intern. Symposium 1977 in Rinteln: 267–272. Vaduz
30. Géhu JM (1979) Pour un approche nouvelle des paysages végétaux: la symphytosociologie. Bull Soco Bot Fr 126:213–223
31. Géhu JM (1986) Des complexes de groupements végétaux a la phytosociologie paysagére contemporaine. Inform Bot Ital 18:53–83
32. Géhu JM (2006) Dictionnaire de sociologie et synécologie végétales. Cramer. Berlin/ Stuttgart, p 899
33. Géhu JM, Rivas-Martínez S (1981) Notions fondamentales de phytosociologie. In: Dierschke H (ed) Syntaxonomie. Bericht Intern. Symposium 1977 in Rinteln, pp 5–33
34. Gillet F, Gallandat J-D (1996) Integrated synusial phytosociology: some notes on a new, multiscalar approach to vegetation analysis. J Veg Sci 7:13–18
35. Hardtle W (1995) On the theoretical concept of potential natural vegetation and proposals for an up-to-date modification. Folia Geobot. Phytotax. Prague 30:263–276
36. von Humboldt A, Bonpland A (1807) Essai su la géographie des plantes accompagné d'un tableau physique des régions équinoxiales. Schoell, Paris

37. Ichter J, Evans D, Richard D (eds) (2014) Terrestrial habitat mapping in Europe, an overview. Luxembourg. MNHN-EEA Technical report, p 153
38. Körner C (2003) Alpine plant life Functional plant ecology of high mountain ecosystems. Springer, p 344
39. Lazare J-J (2009) Phytosociologie dinamico-catenale et gestión de la biodiversité. Acta Bot Gallica 156(1):49–61
40. Lazare J-J, Lanniel K (2003) Une sous-association nouvelle de fourrés du Rubo ulmifolii-Tametum communis du littoral basque. J De Botanique De La Société Botanique De France 21:33–35
41. Loidi J (1991) Vegetation series: its use for small scale geobotanical mapping. Phytocoenosis (N.S.) 3:119–122
42. Loidi J, del Arco M, Pérez de Paz PL, Asensi A, Díez Garretas B, Costa M, Diáz González TE, Fernández-González F, Izco J, Penas A, Rivas-Martínez S, Sánchez-Mata D (2010) Understanding properly the "potential natural vegetation" concept. J Biogeogr 37:2209–2211
43. Loidi J, Báscones JC (1995) Memoria del mapa de series de vegetación de Navarra. Ed. Gobierno de Navarra, Depto. de Ordenación del Territorio y Medio Ambiente. Pamplona
44. Loidi J, Biurrun I, Campos JA, García-Mijangos I, Herrera M (2011) La vegetación de la comunidad autónoma del País Vasco. Leyenda del mapa de series de vegetación a escala 1.50000. University of the Basque Country (UPV/EHU). Ed. Eusko Jaularitza/Gobierno Vasco, p 197
45. Loidi J, Fernández-González F (2012) Potential natural vegetation: reburying or reboring? J Veg Sci 23:596–604
46. Matuszkiewicz W (1979) Landscape phytocomplexe and vegetation landscapes, real and typological landscape units of vegetation. Documents Phytosociologiques N.S. 4:673–688
47. Matuszkiewicz W (1984) Die Karte der potentiellen natürlichen Vegetation von Polen. Braun-Blanquetia 1:1–99. Camerino
48. Milne G (1935) Some suggested units of classification and mapping particularly for East African soils. Soil Research 4:183–189
49. Morison CGT, Hoyle AC, Hope-Simpson JF (1948) Tropical soil-vegetation catenas and mosaic. A study in the south-western part of the Anglo-Egyptian Sudan. J Ecol 36(1):1–84
50. Mucina L (2010) Floristic–phytosociological approach, potential natural vegetation, and survival of prejudice. Lazaroa 31:173–182
51. Ozenda P (1985) La végétation de la chaine alpine dans l'espace montagnard, européen. Masson, Paris
52. Ozenda P, Wagner H (1975) Les séries de végétation de la chaine alpine et leurs équivalences dans les autres systémes phytogéographiques. Doc Cart Ecol 16:49–74
53. Pignatti S (1980) Epharmonische Beziehungen zwischen struktureller und funktioneller Ordnung in der Vegetation. In: Tüxen R (ed.) Ber. Int. Sympos. Int. Ver. Vegetationskunde 1979. Vaduz, pp 385–396
54. Rivas-Martínez S (1976) Sinfitosociología, una nueva metodología para el estudio del paisaje vegetal. Anales Inst. Bot. Cavanilles 33:179–188
55. Rivas-Martínez S (1982) Mapa de series de vegetación de la provincia de Madrid. Diputación de Madrid
56. Rivas-Martínez S (1983) Series de vegetación de la región Eurosiberiana de la Penlnsúla Ibérica. Lazaroa 4:155–166
57. Rivas-Martínez S (1985) Biogeografia y vegetación. Discurso de ingreso en la Real Academia de Ci. Exactas, Físicas y Naturales. Madrid
58. Rivas-Martínez S (1987) Memoria del mapa de series de vegetación de España, 1: 200.000. ICONA. Madrid.
59. Rivas-Martínez S (1993) Bases para una clasificación bioclimática de la Tierra. Folia Botanica Matritensis 10:1–23
60. Rivas-Martínez S (1994) Dynamic-zonal phytosociology as landscape science. Phytocoenologia 24:23–25

61. Rivas-Martínez S (1996) La fitosociología en España. Discurso de investidura como Doctor honoris causa. Universidad del País Vasco: In: Loidi J (ed) Avances en fitosociología. Ed. UPV/EHU, pp 149–176
62. Rivas-Martínez S (1996) Geobotánica y bioclimatologia. Discurso de investidura como Doctor honoris causa. Universidad de Granada: 23–98. Granada
63. Rivas-Martínez S, Wildpret W, Arco M del, Rodríguez O, Pérez de Paz PL, García-Gallo A, Acebes-Ginovés R, Díaz TE, Fernández-González F (1993) Comunidades vegetales de la Isla de Tenerife (Islas Canarias). ltinera Geobotanica 7:169–374
64. Rivas-Martínez S, et al (2007) Mapa de series, geoseries y geopermaseries de vegetación de España (Memoria del mapa de vegetación potencial de España,). Parte I. Itinera Geobotanica 17:5–436
65. Rivas-Martínez et al (2011) Mapa de series, geoseries y geopermaseries de vegetación de España (Memoria del mapa de vegetación potencial de España, 2011). Parte II. Itinera Geobotanica 18(1,2):5–800
66. Russell EW (1961) Soil conditions and Plant Growth, 9th edn. Longmans, London
67. Schmithüsen J (1959) Allgemeine Vegetationsgeographie I, Ed. De Gruyter. Berlin
68. Schwabe A (1990) Stand und Perspektiven der Vegetationskomplex-Forschung. Berichte der Reinhold-Tüxen-Gesellschaft 2:45–60
69. Theurillat JP (1992a) L'analyse du paysage végétale en symphytocoenologie: ses niveaux et leur domains spatiaux. Bull Ecol 23(1–2):83–92
70. Theurillat JP (1992b) Étude et cartographie du paysage végétal (Symphytocoenologie) dans la région d'Aletsch (Valais, Switzerland). Beitr. Geobot. Landesaufn. Schweiz 68, pp 384. Geobot. Komm. Schweiz. Akad. Naturwissenschaften. Krypto, Teufen
71. Tüxen R (1956) Die heutige potentielle natürliche Vegetation als Gegenstand der Vegetationskartierung. Angewandte Pflanzensoziologie 13
72. Tüxen R (1973) Vorschlag zur Aufnahme von Gesellschaftskomplexen in potentiell natürlichen Vegetationsgebieten. Acta Bol. Acad. Sci. Hungar 19(1–4):379–384
73. Tüxen R (1977) Zur Hornogenität von Sigmassoziationen, ihrer sytaxonomischen Ordnung und ihrer Verwendung in der Vegetationskartierung. Docum. Phylosociologiques N.S. 1:321–327
74. Tüxen R (1978a) Bemerkungen zur historischen, begrifflichen und methodischen Grundlagen der Synsoziologie. In: Tüxen R (ed) Assoziationskomplexe (Sigmeten), Bericht. Intern. Symposium 1977 in Rinteln: 3–12. Vaduz
75. Tüxen R (1978b) Versuch zur Sigma-Syntaxonomie mitteleuropaischen Flusstal-Gesellschaften. In: Tüxen R (ed) Assoziationskomplexe (Sigmeten), Bericht. Intern. Symposium 1977 in Rinteln: 273–286. Vaduz
76. Vigo J (1998) Some reflections on geobotany and vegetation mapping. Acta Bot. Barc. 45:535–556
77. Zerbe S (1998) Potential natural vegetation: validity and applicability in landscape planning and nature conservation. Appl Veg Sci 1:165–172
78. Zonneveld IS (1995) Land Ecology. SPB Academic Publishing, Amsterdam
79. Schwabe A (1989) Vegetation complexes of flowing water habitats and their importance for the differentiation of landscape units. Landscape Ecol. 2(4):237–253
80. Schmithüsen J (1948) Fliesen Gefüge der Landschaft und Ökotop. Berndt. Landesk. 5:74–83

Climate Change, Wetland Management and Alpaca Pastoralism in the Bolivian High Andes Mountains

Máximo Liberman

Abstract This paper draws on an environmental management study of a community of alpaca shepherds from a high Andean region in the Apolobamba mountain range in northwestern Bolivia, at an altitude of 4300–4800 m. In this region, Aymara indigenous families face the impacts caused by persistent droughts and climate change derivatives. In the work that was carried out, a series of scientific evidences are presented that demonstrate the dramatic retreat of the glaciers based on measurements and the multitemporal interpretation of satellite images, between 1985, 2014 and 2019. The community's perceptions of climate change were assembled through interviews with alpaca shepherds and through participatory meetings with all the members of the community of Cañuma, at the Franz Tamayo province of the departmental region of La Paz. Through talking schemes and maps, all the climate-related problems they are going through were revealed, especially how decreasing water availability negatively affects sites of very high fragility, such as wetlands, known as the main source of alpaca grazing. Novel techniques for adapting to climate change are described and mapped, such as water conduction systems. These canals are built with surprising knowledge in such a way that it has allowed communities to increase wetland areas for their camelid cattle. These models of hydraulic management in the high mountains represent sustainable patterns of adaptation and effective responses to climate change that members of the community themselves have designed and implemented. These "local environmental engineering" techniques produced an increase in alpaca grazing areas and greater forage productivity in slope wetlands. This means that today, in times of climate change, the provision of food for cattle is guaranteed for this community.

Keywords Climate change · Wetlands · Pastoralism · Alpacas · High mountains · Andes · Bolivia

---

M. Liberman (✉)
Institute of Ecology, Universidad Mayor de San Andrés, Cota Cota - University Campus, La Paz, Bolivia

F. Pedrotti and E. O. Box (eds.), *Tools for Landscape-Scale Geobotany and Conservation*, Geobotany Studies, https://doi.org/10.1007/978-3-030-74950-7_5

1 Introduction

The High Andean region of Bolivia, which lies above 4000 m, is where the Aymara civilization developed. Its autonomy and originality rest on its ability to use natural resources through particular socio-cultural strategies and arrangements. Glaciers are vital for the indigenous communities that live in the high mountains. One of the skills that characterize the people from the upper part of the Apolobamba Protected Area, near the glaciers, is their use of grasslands and wet meadows, known locally as *bofedales*, for grazing the alpacas and llamas that provide them with highly valued and productive resources.

Over the past approximately three decades, climate change has begun to modify rainfall patterns and thus the availability of water in some areas. The *bofedal* wetlands regulate the flow of water, which affects access to drinking water and food for the camelid cattle. Referring to this production level, herders have made a set of environmental modifications, since they not only take advantage of natural grasslands. Peasants from the plains and valleys in the high Apolobamba mountain range, in the Bautista Saavedra and Franz Tamayo provinces of La Paz department, through different water-conduction techniques, have managed to modify the natural environment, in order to have greater areas of wet meadows or additional *bofedales* that allow them to feed their camelid herds [25]. Due to the effects of global climate change, families of camelid breeders in the high mountains are becoming more vulnerable every year [22].

Livestock populations have been disturbed by the intensifying climate phenomena (rising temperatures that decrease glacier surfaces) and microclimatic extremes (snowfalls, frosts, hailstorms and strong winds). These climatic changes have caused a retreat of the glaciers, producing melting ice and greater water runoff. New lagoons have appeared in many places where there were once glaciers. As a result there has been a gradual colonization of these sites by a range of *bofedal* species.

This study aims to assess climate-change effects on glaciers, wetlands and camelid grazing in the Apolobamba Mountain Range by evaluating a sequence of satellite images. Climate-change perceptions and experience of the local population were evaluated through a participatory workshop with an Aymara indigenous community. The goal was to propose adaptation mechanisms and strategies to minimize negative effects of climatic phenomena. Other objectives are the rescue of local knowledge, its management actions and good management practices for *bofedales* and camelid cattle.

2 Methodology

Based on multi-temporal interpretation of satellite images from 1985, 2014 and 2019, this study presents scientific evidence that demonstrates the dramatic retreat of the glaciers. Sentinel-2 satellite images were used to identify terrestrial changes on Earth. This multispectral scanner obtains information in two visible and infra-red wavelengths, allowing one to monitor changes on the earth and in the vegetation, as well as to monitor climate change itself. The scanner has a resolution of 10 m. Three scenes (optical images) from the satellite were used, from 18 July 1985, 15 August 2014, and 29 May 2019. The images cover the entire surface of the Apolobamba protected area. Each scene has 13 multispectral bands that were analyzed and combined to highlight the objective of the observation. Subsequently, a reclassification was carried out for the three different years, by separating the pixels with positive values, to identify the different glacier surfaces.

A social and biodiversity appraisal of the Cañuma community, located in the Apolobamba region, is presented. This information was collected in two field trips during June and July 2016. Several interviews were conducted with camelid herders, and observations on biodiversity management were made. In addition, the following literature was consulted: Liberman [23], SERNAP (2005), Flores Ochoa [8] and Hoffmann et al. [20].

A participatory workshop on the effects of climate change and its environmental impacts and consequences on the management of the wet meadows and the grazing of camelids in the high Andes was organized with the community. The participatory method enabled participants to rescue historical elements of peasant planning and self-assessment. In this case, so-called talking maps were used as methodological technical instruments that allow adequate communication within the community, by diagramming scenarios of what happens with natural resources on territorial maps. On these maps, which were drawn by groups of herders, they collect graphically the participants' perceptions of their territory, strengthening their peasant identity. Thus, the map reflects the most important aspects of its territory, such as glaciers, *bofedales*, grasslands, homes, water sources, and the problems that each sector has. The map making was carried out through a three-step participatory process: a) Explanation and socialization of the methodology; b) Drawing the maps, with reference to mountains, roads, paths, houses, water sources, location of wetlands, grazing areas, etc.; c) Presentation by technicians from the University of San Andrés of their results on climate and land use analyses. Once the maps were drawn, the different groups presented talking maps identifying the similarities and shortcomings of each Then the information presented by the herders was contrasted with that of the technicians, with the purpose of drawing conclusions and recommendations from the knowledge dialogue that includes the effects of climate change and the mitigation and adaptation measures developed by the farmers.

The construction of new rustic canals was highlighted, based on the people's technology to increase the surface area of *bofedales* through gravity-driven systems to optimize irrigation. Climate indicators were identified based upon the results of technical indicators of aging, such as: frost, hailstorms, rains and droughts versus mortality, morbidity, birth and fertility, as well as management of the diversity of colors in the Alpacas. Rotation of grazing areas and use of resting fences, health management of camelid cattle, fertilization of natural pastures, and establishment of Q'ochas or small lagoons, water intake and fences with acquired materials were also considered.

All this information is aimed at generating technical and scientific data belonging to the Cañuma community, for the optimization of water resources, pasture management, management of camelid cattle and adaptation of alpaca herder families to climate change. The purpose is to guarantee sovereignty with food security and conservation of life systems. Likewise, community activities are analysed in conservation, protection, restoration and integral management of waters and *bofedales*. This concerned especially the irrigation systems (main and secondary channels), by prioritizing the use of water for communities and grazing of camelid livestock.

3 Location of the Study Area

The traditional grazing system with camelids in the Apolobamba region, analyzed in this work, is located in the Ulla-Ulla high Andean plain and glacial valleys of the Apolobamba Range (Department of La Paz, Franz Tamayo province), in the Bolivian puna between 15° 00′ and 15° 10′ south latitude and 69° 05′ and 69° 20′ west longitude, above 4300 m (Map 1). The Apolobamba Mountain Range separates these plains from the eastern valleys that descend into the Yungas (montane forests). The Pleistocene glaciers of the Cordillera left formations of moraines, lakes and lagoons that today cover the Ulla-Ulla plateau and the Apolobamba mountains.

The dominant vegetation in the plains and mountains is puna grasslands of *Stipa* spp., *Festuca* spp and *Deyeuxia* spp. in different associations, and communities of *Plantago tubulosa* and *Distichia muscoides*, that dominate places saturated with water, and known locally as *bofedales* [28]. These wetlands can be natural or, instead, looks similar to the enormous extents that cover the Apolobamba plains, produced and maintained through irrigation. Water is channelled from rivers that descend from lagoons of glacial origin. The intensive study area is located in the highlands and glacial valleys of the Apolobamba mountains. Studies were concentrated in a community called Cañuma, whose members are dedicated exclusively to llama and alpaca breeding, in communal properties, since pre-Hispanic times.

Table 1 shows characteristics of the region, where the Apolobamba wetlands (*bofedales*) occur.

Figure 1 shows a schematic altitudinal profile of the study area and the location of the community of Cañuma and its relationship with the altitudinal belts and use on the eastern slope of the Apolobamba mountain range, Bolivia.

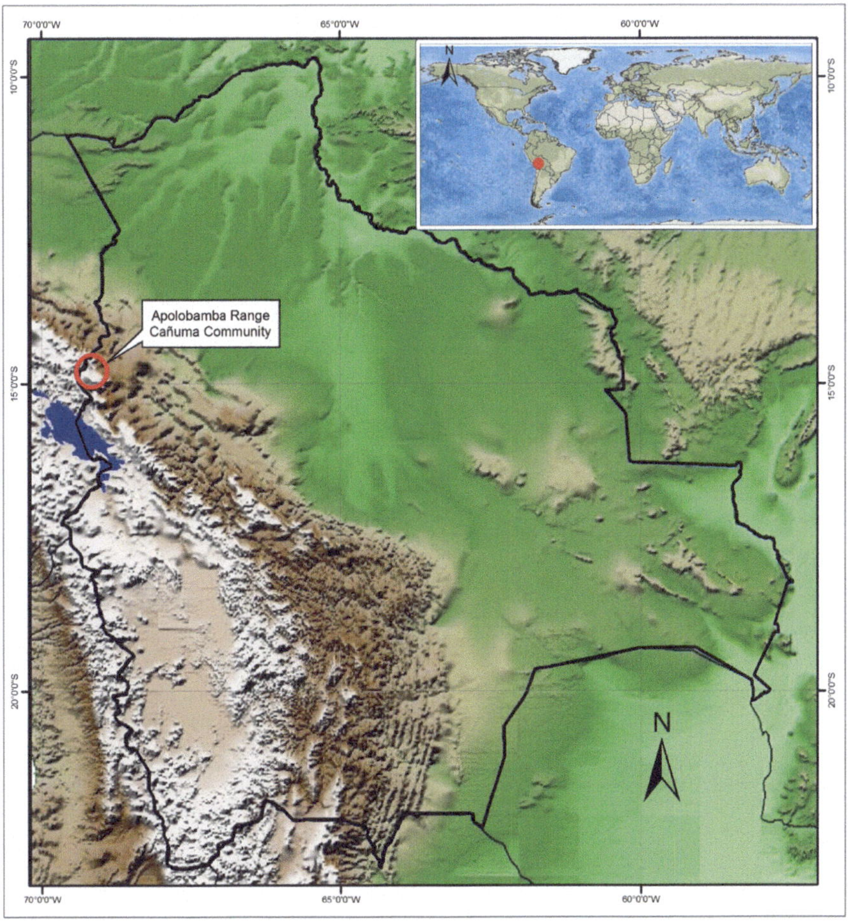

Map 1 Location of the study site Cañuma, in the Apolobamba mountains of the High Andes, northwest Bolivia

4 Results

4.1 Conditions of the Glaciers from the Andean Mountain Range of Bolivia

Tropical mountain glaciers are known for their high sensitivity to climate change, especially rising temperatures. In recent years, climate change has caused the freezing point to rise. This aspect coincides with the warming of the Andean high troposphere [11] and leads to reduced albedo and greater surface absorption of solar radiation. Unlike mid-latitude mountain ranges, as occurs in the Alps, ablation balances accumulation in the Andes, preventing the development of long-lasting

Table 1 Characteristics of the region where Apolobamba wetlands (*bofedales*) occur

Biogeographical characteristics and population of the region	High andean plain in the altiplano and mountain range of apolobamba
Geographic location: latitude and longitude coordinates	High plateau and high mountain area of the Apolobamba range: 15° 00′ to 15° 10′ S latitude; 69° 05′ to 69° 20′ W longitude
Altitude range of the territory	4300 to 5300 m
Climate characteristics: average temperature and annual average precipitation (data of the Meteorological Station of the locality of Ulla Ulla)	4.2° C; 490 mm annual average rainfall. Strong weather, more than 200 days with frost per year (4300 m above sea level). Dry and very cold winters
Wetlands (*bofedales*)	Surface: 105 km^2, located in the Ulla Ulla Altiplano and glacial valleys of the Apolobamba mountain range
Political administative entity	Dept. of La Paz, Bautista Saavedra Province, Municipalities of Curva and Pelechuco, Franz Tamayo Province
Protected area	Apolobamba. Natural Area of Integrated Management (ANMI)
Types of vegetation	Wetlands (*bofedales*), lawns, high Andean grasslands in plains and mountains
Population	Aymara indigenous communities organized in Ayllus (more than 3000 inhabitants)
Economic activity	Camelid grazing; wool and meat and fiber trade of alpacas and llamas; vicuña fiber sale; trout fishing in lagoons and high Andean rivers
Animal grazing	Domesticated camelids: alpacas (grazing in *bofedales*) and llamas (grazing in dry grasslands) Wild camelid: vicuñas (in all plant formations)
Livestock density	*Bofedal*: 2.7 alpacas/hectare; Plains: 0.3 alpacas/hectare
Major forage species	*Distichia muscoides, Oxicloe andina; Plantago tubulosa, Deyeuxia sp.*
Vegetation units with high forage quality	Wetlands (*bofedales*); lawns, grassland
Age of camelid grazing practices (Alpacas and Llamas)	Domesticated camelids between 4000 to 3500 BC (*)

Sources Molinillo [28], Liberman and Molinillo [23], Ribera and Liberman [38], Seibert [40], Meneses et al. [26, 27]; and Wheeler [47]*

seasonal snow cover [45]. Thus, there has been a retreat of the world's tropical glaciers over the last decades. In Bolivia, Soruco et al. (2009) estimated that the 367 glaciers of the Cordillera Real had suffered a setback. The average loss was 43% in volume and 48% of surface area between 1975 and 2006.

In the long term these transformations in glacier volume will cause considerable changes in the seasonal downstream flow from the glaciers. The most important changes in flow are expected to occur during the dry season, when glaciers release

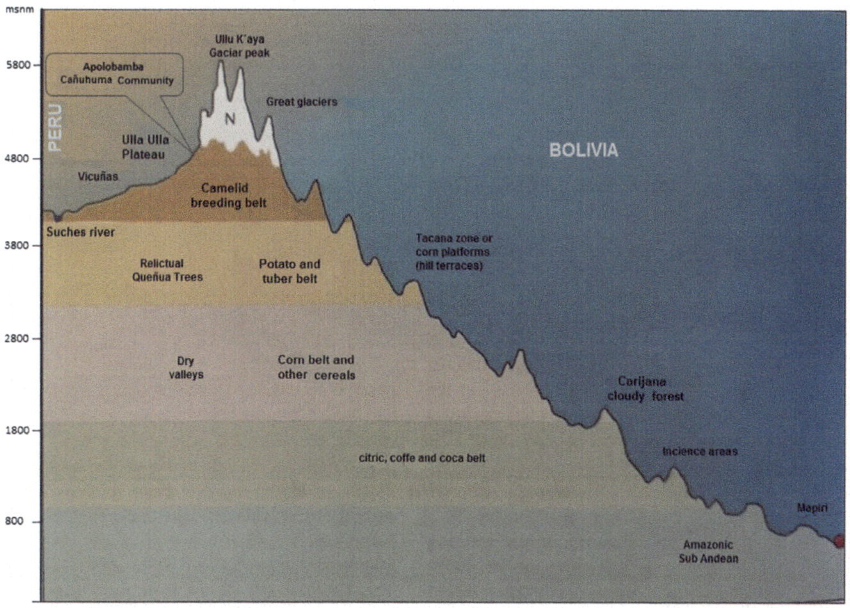

Fig. 1 Profile of vegetation belts on the eastern Andean slope, Apolobamba mountain range. The community of Cañuma is located in the high Andes, where only alpacas and llamas are bred [38]

meltwater. The snow that falls in the high Andes is stored initially as ice in the mountain glaciers before it is released after a certain time and adds to the flow during the dry season [12, 46]. Consequently, glaciers play a very important role as shock absorbers of seasonal rainfall and provide water during the dry season for numerous domestic uses. In the case of Cañuma, this flow irrigates the wetlands (*bofedales*) that feed camelid cattle (alpacas and llamas).

According to Jordan [21], 80% of the surface of the Andean glaciers has an area of less than 0.50 km^2. The melting of the glaciers began in the early 1980s, and it is presumed that many of these small glaciers have disappeared in the last 20–25 years. Some were documented (measured and characterized), but there are no records for the vast majority [19].

In the long term, glacial retreat will not only impact water for human consumption, but it will also generate stress on pastoral activities in the high mountains, thus affecting alpacas and llamas, because they will not have the necessary food in the absence of aquifers and *bofedales*. As stated by Rocha [39], this effect will occur due to the expected increase in seasonality, which will cause temporary and spatial decrease of flooded areas, as well as smaller recharge of the aquifers, where any modification in the hydrology of their basins can significantly alter their configuration [33]. Disorders in the *bofedales* are caused not only by the general drying in the region.

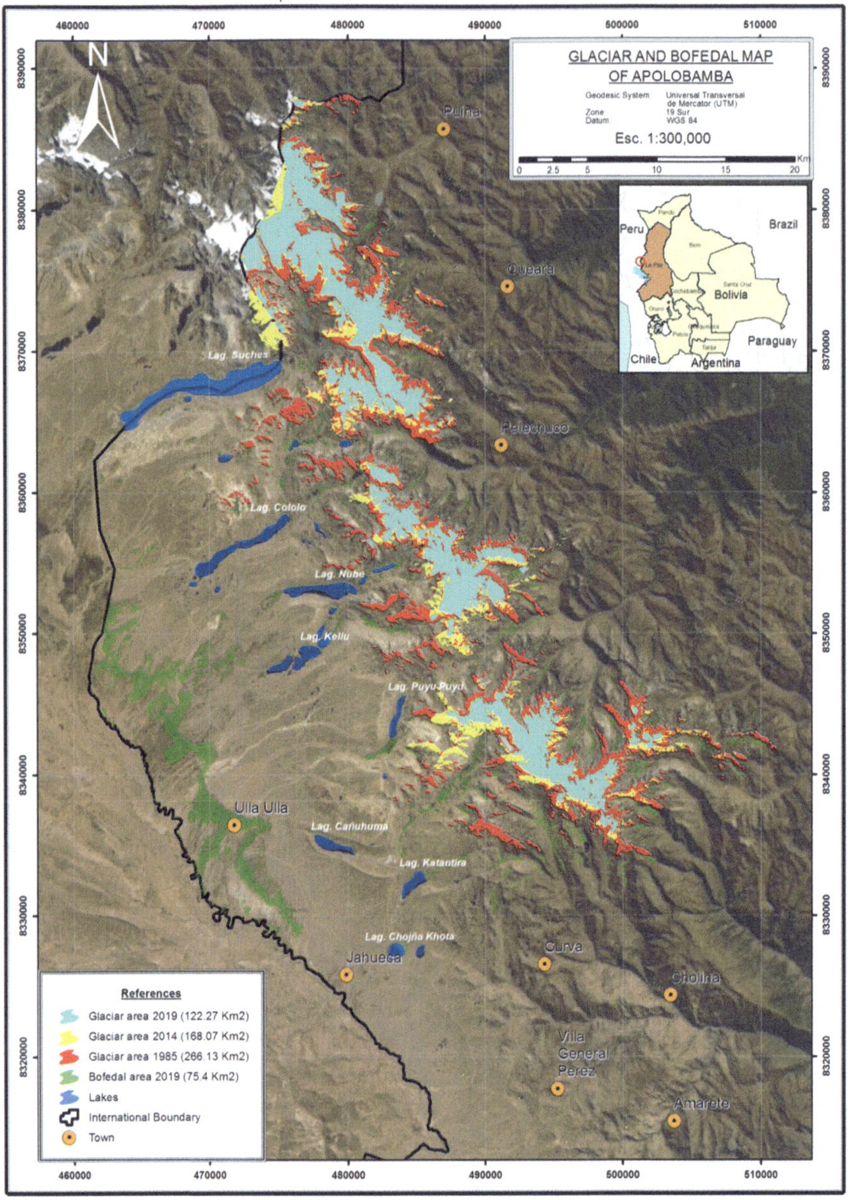

Map 2 Distribution of *bofedales*, glacier areas and their retreat from 1985 to 2019. Based on multitemporal analysis of satellite images. *Source* own elaboration

Map 2 shows the distribution and surfaces of *bofedales* and glacier areas in the Apolobamba mountain range since 1985, 2014 and 2019, based on multitemporal analyses of satellite images, in accordance with the methodology described earlier.

Photo 1 Ullu Kalli glacier in the northeastern part of the Cañuma community, Apolobamba range

The following map of glaciers shows that, from 1985 to date, their surfaces have undergone significant reductions. Indeed, over a period of 34 years, there has been a 45.94% reduction in the surface area of the Apolobamba glaciers, which is affecting the water supply for the *bofedales* Photo 1.

4.2 Puna Vegetation in the High Andean Region of Apolobamba

In the puna, the vegetation is subject to high insolation. The germination and growth of dominant plants, mainly rosettes and cushions, are hampered by extreme variations in temperature and low air humidity. This led Troll [43] to declare that the dry sectors of the Central Andes have the greatest microthermal deficit for plant growth on the planet.

In these cases plants grow very slowly, and many species function only during the morning and afternoon, between the phases of nocturnal frosts and midday overheating. The low productivity transforms the plants into radial cushions and dwarf rosettes, with small leaves and compact surfaces [6, 10]. The resulting decrease in transpiring surface area provides protection against low temperatures, wind and high evaporation [36].

The Cañuma area is located in the puna of the Bolivian Andes, where vegetation zones are distinguished above the forest limit. The following diverse belts were identified by Beck and García [16] in the puna: nival-subnival, upper and lower high Andean, and the puna belt *per se*. The eastern Andes and the Altiplano present many different climatic and topographic situations, producing several habitats in which the vegetation is rich in species and life forms. In this study, we will concentrate on the so-called wet *bofedales* (meadows) that exist in the sub-snowy and upper Andean belts from 3800 to 5000 m above sea level.

The high Andean vegetation covers mountain slopes up to 5200 m, where the nival belt and snow cover begin. According to Navarro and Maldonado [31], the altitudinal ranges for this ecological "floor" are restricted to 3900–4600 m, while the lower limit of the subnival belt occurs between 4600 and 5200 m. An important characteristic that differentiates the sub-Andean belt from the high Andean belt is that the higher presents nightly frosts throughout the year. Instead, below 4600 m, frosts decrease during the summer months (between one to three months). This area occupies the high plateaus, as well as glacial valleys, with rocky substrates. The predominant growth forms are dense bunch grasses, cushions and rosettes. Near the upper limit, isolated shrubs of *Senecio rufescens*, *Agrostis tolucensis* grasses, *Dielsiochloa floribunda*, and perennial herbs (*Senecio serratifolius*) grow on rocks that accumulate heat during the day and attenuate temperature drops at night [2, 3, 13–16].

In depressions near lagoons or other water sources, the vegetation of wet swards is characterized by the presence of *Plantago tubulosa, Azorella biloba, Gentiana sedifolia*, *Oritrophium limnophilum* and Mexican *Cotula*. Between 4300 and 4600 m, on semi-arid soils, a low sward develops. It is composed of isolated communities of grasses, evergreen dwarf shrubs and perennial herbs in rosettes or cushions. The height of the plants does not exceed 5 cm, and coverage is estimated between 5 and 30% [3]. These authors estimate that poorly drained areas promote the formation of compact herbs and hard cushions that form the *bofedales* of the puna. *Bofedales* are vegetation formations with permanent surface water throughout the year. In the area of Apolobamba they cover an altitudinal of 4000 to 5000 m and characterize the high Andean landscape and the puna. The *bofedales* of the puna have cushions and swards of *Plantago rigida*, *Werneria pigmaea* and *Lucilia tunariensis*. These *bofedales* develop on climatically humid slopes, around ponds, lagoons, streams and outcrops of water or springs, in the eastern part of the mountain range. Part of the high Andean vegetation includes *bofedales* of *Distichia filamentosa* and *Distichia muscoides* (Photos 2 and 3). These are located mainly around 5000 m but not below 4200 m [40]. These *bofedales* are different from those in the puna, which are usually found in depressions, underground water springs, on moist soil flooded by running or stagnant water, in which they form hard cushions (García et al. 1995). *Oxycloe* pads are found in contact with *Distichia*, but form separate structures, never overlapping [40].

Photo 2 Glacial valley in Cañuma, Cordillera de Apolobamba, with wetlands (foreground) and the glaciers of Ullakalli (background)

Photo 3 Wetland (*bofedal*) of *Distichia muscoides* surrounded by water courses and pools

Bofedales typically represent azonal formations. They form a mosaic of elevated and hard cushions, patches of dense or open sward, with small patches of grasses in bushes on the edge, where numerous small wells, lagoons and streams appear [10]. Their largest extent is observed at the bottom of U-shaped glacial valleys. There are also hillside *bofedales*, just where water sources appear on the slopes [23, 28]. The determining factor for the formation of *bofedales* is the surface and underground water that runs permanently, that gives life to these ecosystems (Photo 3). Steep stony places are covered by *Stipa brachyphylla*, with the presence of isolated *Opuntia flocossus*. Areas with stagnant water have communities of *Plantago tubulosa*, which belongs to the community of *Aciachne pulvinata*. In open dry areas, communities of *Pynophyllum-Selaginella* can be seen, as well as the steppe of *Deyeuxia curvula*.

Peasant herders have learned to domesticate plants by converting them into useful herbs for a number of purposes. Basic relevant, important information has been generated for the sustainable use and management of biological resources in the high Andes of Bolivia, at altitudes over 4000 m. Biological resources are crucial for the economic development of peasants and indigenous people. These people often depend on plants for their livestock food and to cover their own daily needs. Peasants suffer more than other groups in society when the environment is degraded by overgrazing their pastures, by climate change, or by other associated processes. In the High Andes, most plants and their interactions with the environment are poorly understood. In many cases the real or potential economic value has not yet been evaluated.

In the high-mountain community of Cañuma, *Distichia muscoides* became useful. It is a typical wet-meadow plant that was turned into a useful plant not only as a forage source for camelids (Photo 4), but also as a utilitarian means for the transport of water in the mountains to generate new grazing fields for alpacas and llamas.

Traditional knowledge of peasant management of species such as *Distichia muscoides*, locally called "Khuli Female", is based on a novel local experience closely connected to the natural environment where it has been developed. If preserved and disseminated, it can be widely useful in the design of sustainable management schemes in other communities. It is necessary to value these examples to promote a type of development that can take advantage of the knowledge that was surely implemented over decades of testing by families living in the highly constraining high Andean ecosystem.

4.3 Alpaca and Llama Grazing Technologies

Flores Ochoa [7, 9] describes herding camelids in the high Andes, above 4500 m, as an economic system very dependent on environmental conditions. It is limited to exploiting almost parasitically the resources of nature without making significant modifications of the natural foundation.

Photo 4 Alpaca grazing in a *bofedal* of *Distichia muscoides* in the community of Cañuma

The presence of glaciers and numerous watersheds in the Apolobamba range ensures a permanent water supply for the *bofedales*. This situation has favored and partly kept the region relieved of forage deficit, even in times of great, painful droughts [32, 37]. In 2015–2016, which included an "El Niño" event, 10% of camelid cattle would have died from extreme climatic effects (drought, frost, hail and snowfall), according to Cañuma residents.

In the Apolobamba mountains there are several communities where pastoralists have transformed grassland and *bofedal* ecosystems through innovative irrigation systems with canals and water reservoirs (Q'ochas), to expand the areas of intensive grazing, thus creating excellent grasslands that improve grazing environments for both alpacas and llamas. Grazing is not only a form of exploitation, adapting to, and making the most of resources in a high Andean ecosystem. It is also a complex, imaginative world where various management models and uses of resources are established and proposed.

For the inhabitants in the region, there is no alternative to the work of grazing the camelids. They are the support for the human populations, which benefit from their fiber, skins, fresh and dried meat, or trade them in valleys or use them as exchange goods to acquire food of agricultural origin or other products. Another local product may be fertilizers, used for grassland regeneration and fuel. Other products are fat for the manufacture of medicinal ointments, and bones, from which combs and handicraft instruments are produced. Leather is also transformed into material for roofing homes.

Physiologically, camelids take better advantage of the high Andean pastures than do the species introduced from Europe by the Spaniards. Alpacas, for example, digest the pasture grasses 22% more efficiently than do sheep [5]. Ecologically, camelids, by their way of grazing, do not degrade the pastures as much as do cattle, ungulates and horses. The camelids cut the grass, while the others uproot it. Cattle and horses can be extremely harmful in drier areas, where the vegetation cover becomes impoverished and the soil degraded. Trampling by sheep, due to the structure of their hooves, is equally damaging in the high Andean pastures; camelids, which have plant like pads, do not damage the vegetation cover.

The main characteristics of the herding society in Cañuma are as follows. As in much of the rural area of Bolivia, family and community are the pillars of social organization, which determine the daily lives of people and the modes of evaluation and management of territorial resources. Albó [1] suggests that a community is a specific group of families that share a defined territory with their own system of government. It is also an institution that controls space. A community is inseparable from the space that makes the territory, which is delimited and regulated and by the group of people which is the institution. In the case of Cañuma, the community controls the territorial resources (water, grasslands, *bofedales*) that are common property, while community members are only beneficiaries. As for family-owned land, the community guarantees the security of the rights of use. With regard to grazing activities, it operates and assigns restrictions in some areas. The community as an entity is sovereign, designates its authorities and dictates the external rules [44].

Families are responsible for grazing camelids in places where they have traditionally taken them. There are traditional sectors of intensive grazing in *bofedales* where up to six herder families can graze their alpacas. This activity is mostly done by women and their young children. Males who have not migrated to large cities or to the mines may graze cattle, but also maintain the *bofedal* irrigation canals. In dryland grasslands grazing is free.

Among the different families there is the system of exchanges, assistance or mutual help and cooperation between individuals and families for the care of camelid cattle. Cooperation, in fact, is one of the foundations of the High Andean culture, where there is an exchange of favors. These may include, for example, requesting the care and monitoring of alpacas when they graze, or communal solidarity work and internal reciprocity in the maintenance of irrigation canals, *q'ochas* (lagoons) and shearing.

There is a process of seasonal transhumance of camelid cattle within the limits of a community, where herders carry out migrations or changes of lodging so that the animals have adequate grasslands in each season of the year (dry and rainy). This means that the inhabitants of Cañuma have homes in different places, usually at least three homes (one main and two temporary located in other areas of the Cañuma). Taking advantage of the pastures corresponds to a sectorial fallow system within a limited area, without carrying out large-scale or large-distance transfers. This implies seasonal and spatial rotation.

The technique of handling camelid cattle as domestic animals is one of the great achievements of Andean cultures. The oldest camelids were apparently not adapted to great heights. Wheeler [47] affirms that domestication of alpaca and llama took place in the Central Puna of Peru. Archaeological research reveals that this occurred between 4000 and 3500 years ago and began the early development of camelid livestock in the Andean region. As a result of their knowledge of the animals, care of the herds and other techniques like grazing habits, and the rotation of grassland areas, herders obtain good yields of Alpaca wool and can maintain and irrigate the *bofedales*. Surveys conducted in the area indicate that each family may have between 100 and 150 alpacas and about 6 to 10 llamas.

Llamas are used as beasts of burden and for meat. The distribution areas of camelids show that they have specialized in living at great heights, as a consequence of the dependency initiated in the sixteenth century by the domination, pressure and marginalization exerted on Aymara populations. The association of shepherds with camelids has brought several benefits, such as pasture supply, especially in times of cold or environmental degradation; care of weak animals or neonates; and protection against natural predators. This symbiosis between society and animals has transformed the landscape for the well-being of both populations [18]. The relationship between man and animals develops in such a way that llamas and alpacas are referred as "people of the family." Herders in Cañuma give names to their animals, and both animals and people recognize themselves. There is a permanent concern in the care and breeding of animals, as if they were their own children.

Herders take care of the animals permanently, so that none get lost in the mountains or on roads. It has been observed in Cañuma that there is a special language between the herder and his alpacas, allowing the recognition of alert sounds in the face of a threat or imminent risk. On one occasion it was observed that an alpaca was lost and had moved a considerable distance away, and the herder was desperately searching for the animal, giving sharp, loud screams in amid the hills and mountain slopes. The entire cattle herd also felt nervous, until finally the alpaca heard the screams and came in a hurry to join the rest of the cattle. After this event, we observed tranquility in the group of alpacas. The herder smiled as she met her lost alpaca again.

Every day herders plan the grazing activities of their alpacas and llamas, early in the morning before the sun rises, selecting the places where they will take their animals. They know the places that have been previously grazed and the sites at different heights that are reserved for the dry or wet season. Before going out, herders wait for the sun to rise and melt the frost or snow, since wet or frozen grass produces diarrhea in the animals.

Alpacas are more territorial, so they avoid long walks. They prefer places with the highest humidity or *bofedales* that have green, hard, fresh pastures. In most cases, these are located close to homes that are located in relatively restricted, contiguous spaces. Llamas, on the other hand, are more resistant, walk long distances through the steep hills, and prefer places where dry land with yellow, hard, rough or thorny pastures. They feed themselves on their itinerant way, moving away from the place where they started their walk. Each family has developed different grazing strategies according to the season and site, making the most of what nature offers. Every day herders move to different grazing sites, rotating the feeding sites, and thus taking advantage of the diversity of forage and production seasons in an effective way. In many cases, the *bofedales* are created by the same herders over many years, building irrigation canals that transport water for the irrigation of grasslands, in such a way that they have increased the grazing area and also the areas of water reservoirs.

The animals are slaughtered in the beginning of the dry season, when the cattle have good weight and there are adequate climatic conditions (solar radiation and warmth). Camelid meat is used to produce salted dried meat (charqui), which has good profitability, occupies less space and has a longer storage life. Charqui and wool, which are surplus products of the herding work, are transported by the llamas to the valleys of Charazani, where herders exchange food products (corn, coca, potato, goose, etc.) with farmers from the lowlands. This work is done mainly by men, who guide the llama caravans. The travelers, merchants or llama herders depart from Cañuma (4500 m) to the hot valleys of Chari, Lunlaya, Charazani and other sites that are located at an average height of 3600 m. They move to the eastern slope of the Andes to exchange their products within the different ecological areas, with the sense of control of economic verticality and using different ecological spaces that are exploited interdependently. This theory explains the socio-environmental management model of the Andean space [29, 30].

Alpaca breeding programs are mainly aimed at gaining and producing fiber, which has a high commercial value. For this, alpacas are sheared between 5 and 8 times in a life time. The production of fabrics from alpaca and llama fiber is an artisanal, manual activity carried out by women while grazing their camelid cattle. Through this they produce extraordinary fabrics that express their culture that comes from pre-Hispanic periods and that show high levels of quality. They shear the animals, spin the fiber and weave beautiful clothes such as: dresses, pants, ponchos, blankets, winches for the head, *chullos* (hats), sashes, socks, scarves, blankets, etc. All the garments made on looms have abundant, geometric, symbolic decoration, both natural and with a great variety of colors made from dyes provided by native plants of the puna and from the tropics (Quisbert et al. 1984). The most common representations are alpacas, llamas, birds, flowers and daily grazing activities, which express extraordinary quality in elaborate fabrics. Teaching girls how to produce woolen fabrics starts at age 12. It is very common to see mothers and daughters with their rustic horizontal looms weaving together, while herding the alpacas.

A characteristic of this peasant unit is the minimization of risk. That is, in response to the theory of limited goods (scarcity), one could say that pastoralists act rationally according to their circumstances and levels of information. In this sense camelid cattle represent a safe source of savings and investment, both for their convertibility in times of need into meat or money, and for the increase in their value over time. This value is not only based on work; animals take advantage of resources that are not accessible to humans. They also contribute with their manure, for example, to increase the fertility of *bofedales* or as fuel (tachy) for cooking [41].

4.4 Management of Bofedales *and Climate-Change Adaptation*

Bofedales are high Andean plant formations of great importance for camelids, their daily activities and their life cycles, because they are food source for alpacas. They are azonal formations established in an organic environment characterized by permanent water saturation. They are heterogeneous depending on the physiography of the site, as well as the altitude above sea level. In general, they are dominated by the presence of the female Khuli (*Distichia muscoides*) and the male Khuli (*Oxicloe andina*) that form a set of hard, compact cushions (*Distichia*) and lax, spiny, hard formations (*Oxicloe*). They make up a wavy micro-relief that highlights the presence of small water reservoirs among the Distichia plants. Also common are *Plantago tubulosa, Eleocharis albibracteata, Carex* spp. and *Werneria pymaea* (Alzerreca et al. 2001, Castro et al., 2014, Coronel et al. 2009). Depending on the water gradient, these species may be floating, emerged or submerged, up to the edge of the zonal grassland vegetation [4].

At these sites, plants take advantage easily of the nutrients from the decomposition of organic matter at shallow depth. The *bofedales* have a great variety of aquatic plants, some growing their roots in the water and the leaves on the surface. Prieto et al. [35] report variable of yields of fodder phytomass in *bofedales* from Ulla and Huacuchani, ranging from 854.7 to 2,638.8 kg dry matter/ha. These sites are very close to Cañuma. In Cañuma, *bofedal* plants form compact cushions arranged in many layers of dead plants that can be several meters deep (Dangles et al. 2009). The source of water comes from the glacier and snows of the Ullu Kalli, as they melt.

The vision of the alpaca herder families is a deep relationship between the land (Pachamama), the pastures (Jach'u), the alpacas (Allpachus), the water (Uma), the mountain deities or spirits (Apus) and the herders (*pacocheros*) [7–9]. One cannot value one element separately from the others. They are all interrelated, forming a global total unit.

In general, during the rainy-season grazing period, herders take the alpacas to areas where they have dry green grass pastures. These are dry, hard, and of low productivity during the winter season (Palacios 1977) [34]. In winter the alpacas are taken to *bofedales* where plants are green. During this period the herders carry out activities for diverting water from the Ullu kalli glacier to irrigate the *bofedales* that have better pastures. Due to its location adjacent to the Ullu Khalli glacier, the Cañuma valley has a permanent water supply. This is an advantage for grazing, in contrast to areas where water is scarce during most of the year.

In the region, natural *bofedales* are located mainly in areas that have a flat or almost flat topography, at the bottoms of the glacial U-valleys. There are also slope *bofedales* where aquifers or springs appear. They are also called *ojos de agua* (water springs) and are permanent. Where water flows from a certain point on a slope, a *bofedal* develops and opens out downslope like a fan.

As stated earlier, many *bofedales* are artificial. In many places the wetland area was enlarged to improve pastures that provide better food for camelids. The construction of these *bofedales* is a response and an adaptation to the climate change that has developed in recent years. Herders have a set of climate-change adaptation strategies to improve grasslands and their forage yields, which ultimately determine better conditions for alpaca livestock [24].

The strategies that have been created and developed with imagination and skill by alpaca herders have managed to create ingenious "bioengineering works" and have generated creative solutions for the transfer and transport of meltwater through channels that cross steep rocky slopes. Thus, they have been able to create huge areas of *bofedales* for the benefit of their alpacas and llamas. All these works constitute a true legacy to confront climate change. It is true that the Cañuma Valley herders have historical information on the flows and availability of water throughout the Ullu Kalli glacier area, throughout the year. Their experience plays an important role because it has allowed them to manage data, based on knowledge of situations at other, similar sites. These herders manage climate risk and minimize losses to their camelid cattle.

For the construction of infrastructure works, labor is required in places where there are no access roads. When an irrigation channel for the *bofedales* has to be built, families ask their members to work within conceptions of reciprocity of Andean work, known as *Ayni.* In the Cañuma area, three types of *bofedal* irrigation canals are used to extend the borders of wetlands, to maintain their large camelid herds:

a) rustic canals of Khuli (*Distichia muscoides*);
b) rustic canals combined with *q'ochas*;
c) rustic canals in high Andean grasslands.

In these cases, shepherds have developed irrigation techniques to improve the feeding of alpacas. According to Palacios (1977), alpacas that are bred in a *bofedal* with enough soil moisture can provide 5–6 kg of wool in two years, while in dry places they give only 1.5 to 2 kg of wool during the same period. However, this type of wool is usually as dry as of that of llamas, which is why shepherds value the *bofedales*. In general a *bofedal* can support about 3 alpacas per hectare per year.

The construction and maintenance methodologies of the irrigation canals in the high mountains of Cañuma are described below:

4.4.1 Rustic Canals of Female Khuli (*Distichia Muscoides*)

In the highest part of Cañuma at about 4850 m, a set of channels was identified that conduct the melting waters of the Ullu Kalli glacier through main ditches of very slight slope along the hillsides of the glacial side moraines. From the main ditches, at certain intervals, there are breaks where the water flows to the lower slopes to irrigate the hillside *bofedales*. In this way herders increase the area of *bofedal* intensive grazing sites on the slopes of the glacial valley. Periodically, herders carry out their maintenance activities.

The interesting novelty of this irrigation system is that the walls of artificial canals are constructed using *tepes*, which are compact cushions of *Distichia muscoides* (female Khuli), i.e. the same species that forms the wetlands of the valley bottoms. This is represents extraordinary ecological knowledge, adaptation, and technological innovation to construct the hillside channels taking advantage of the cushion life form. Each cushion plant is moved at the same time to the site where a channel is formed to take water to other sites and the irrigation frontier of the grazing areas is extended. In this way the negative effects of climate change in the community are minimized. This is an example of ingenuity and work by a family of herders to increase the productivity of the vegetation and consequently the greater yield of their camelid herd. One family member affirms that his family took several years of work and intense observations of the plant, involving many difficulties for his parents and grandparents.

All the members of the family of alpaca herders from this area handle the care of their *bofedales* and their ditches skillfully, and especially the grazing trade of their alpacas. This labor and knowledge are transmitted from one generation to the next through methodical observation since childhood. This helps to perpetuate herder occupation within families. In addition, this activity is very gratifying spiritually, for herders experience special moments in the mountain and with their animals. During childhood these activities mark children's personalities, which usually bring two clear consequences. The first is that a child may reject the activity and will not want to know anything more about it. This possibility is not uncommon, because more and more children want to study and be "more than their parents." The second consequence is that the child loves herding and might never wish to separate from the camelid herds. Rustic canals were identified on the southern slope of the glacier valley where the *Laguna Verde* is located. In this area, there is a large valley-bottom *bofedal* that is irrigated with waters of the Ullu Kalli glacier, located at an elevation from 4690 to 4742 m (52 m difference; Fig. 2). The female Khuli canals were initially conceived for the provision of water for housing. Herders living at the site suffered fatigue moving water from the lake to their dwellings at the top of the hillside.

Building the channel involved several topographic, hydraulic and construction challenges. Its builders also had to study the Khuli plant's ecology, physiology and lifestyle. The channel that was designed moves water from a slope *bofedal* on the upper part of the glacial valley. From the *bofedal*, water intake is achieved through an excavation.

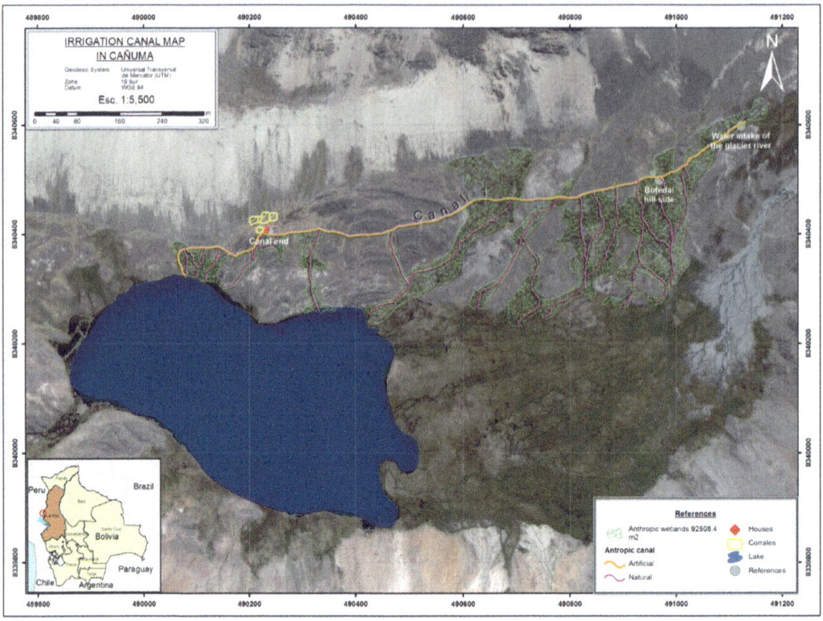

Fig. 2 Satellite image showing the Khuli canal (*Distichia muscoides*). The image reveals how the area of *bofedales* has increased on the slope of the glacial valley

The water is conducted 1231 m through a firm rustic canal, the walls of which are lined with cushions of the same *Distichia muscoides* (Khuli) plant. At the same time, this guarantees that there is no water shortage, for the survival of the plant depends on permanent flooding. If it lacks water, the plant inevitably dries up. According to the herders, building this channel took several years of work by a father and son. The channel passes along the upper part of the hillside, which in several areas has a rocky substrate where excavation work was not possible.

For construction of the canal two parallel tasks had to be carried out: construction of the canal itself and permanent provision of water to the *bofedales*, especially during the dry season. At certain intervals there are points of water distribution and regulation, where some holes were made. From these points small secondary channels are created, allowing flooding to occur on the lower slope. In this way, the surface area of the slope *bofedal* is increased. When water is not provided to the hillside, the holes are covered with stones or champas (lumps of soil with grasses). When a secondary channel is opened, the main channel is blocked, so that the entire flow waters a portion of the slope *bofedal*. When irrigation of that portion of the *bofedal* is finished, the secondary channel is blocked and water is allowed to pass through the main channel.

From the water intake, water runs through a channel with a very smooth average slope of 4.22%, following the surface contours. When the channel passes through a rocky area, it is reinforced with large cushions of Khuli, grass clusters and rock,

Photo 5 A slope canal built with Khuli *tepes* (*Distichia muscoides*) for creating artificial *bofedales* on hillsides

which support the hillside canal as real water retaining walls. The main channel decreases progressively in size towards the furthest point in the system (Photos 5, 6, 7 and 8).

To make sure that water does not drag along the plants and substrate, herders must clean and repair the channels continuously, especially after rainy-season landslides or obstructions due to excessive growth of Khuli cushions that block water flow. In general, the water intake in the main channel is maintained during August, in accordance with circumstances. Channel maintenance activities include: cleaning out plants that obstruct water flow, arrangement of stones, erosion control, and opening of parallel or secondary channels.

The result of these processes is that, for many years, the hydraulic and irrigation works on the unstable mountain slopes function properly, because they are adapted to and compatible with the environmental conditions as well as the maintenance and capacity requirements of users. Therefore, a sustainable system has been established with a continuous flow of water for the *bofedales*. The area of artificial *bofedales* in this case reaches 9.25 ha, which represents an outstanding increase in the area of the *bofedal* that provides nourishment to 27 additional alpacas for the herder.

Photo 6 Channel built with Khuli *tepes* (*Distichia muscoides*), with secondary channels for watering slope *bofedales* enabled by moving the rocks

4.4.2 Rustic Canals Combined with *Q'ochas* (Lagoons)

This type of rustic canal infrastructure, combined with *q'ocha* lagoons, is used in the middle part of the Cañuma Valley (4643–4633 m), in an area where there is no dominance of Khuli plants (*Distichia muscoides*) but where herders have built irrigation channels with a total length of 1040 m.

A major characteristic of this adaptation is that canals run along the slopes of a series of hills, basal moraines of glacial origin, which in turn are located in the enormous glacial valley of Cañuma. In this place herders designed their canal to

Photo 7 Local bioengineering work showing the path of the irrigation canal with Khuli *tepes* on a rocky hillside

begin in a small lagoon that is the overflow of the *Laguna Verde* (Green Lagoon), which is nourished with the melting waters of the Ullu Kalli glacier. This lagoon has a permanent water supply. In this case the rustic canals were excavated on the high mountainsides and have a gentle slope. The walls of the main canal are composed of stones and *champas* (grasses with lumps of soil). The canals include points of water distribution that irrigate the pasture slopes (not the same as in *bofedales*). The flow of irrigation water is so well calculated that continuous water currents on hillsides do not generate serious soil erosion. In other areas, irrigation is accomplished through overflow of the main channel. Where the canals have sharp curves, it is interesting how herders design them with a wider radius and more robust walls, to avoid collapse when there is too much water. Slope adjustments were made such that the water circulates at more slowly, avoiding useless loss and at the same time avoiding deposition of solids in the canal.

Another issue that should be highlighted is that this type of canal has improved a variety of special works that solve technical problems for better water conduction, such as waterfalls, regulation ponds and obstacle passes. All these works were undertaken with ingenuity, in an environment with rugged topography, difficult excavations, and long canals bordering hillsides. The canals are well designed in such a way that they ensure infrastructure that fulfills its functions over time. In this system we observed the presence of about three levels of secondary canals located below and parallel to the main canal. These canals receive water from the main

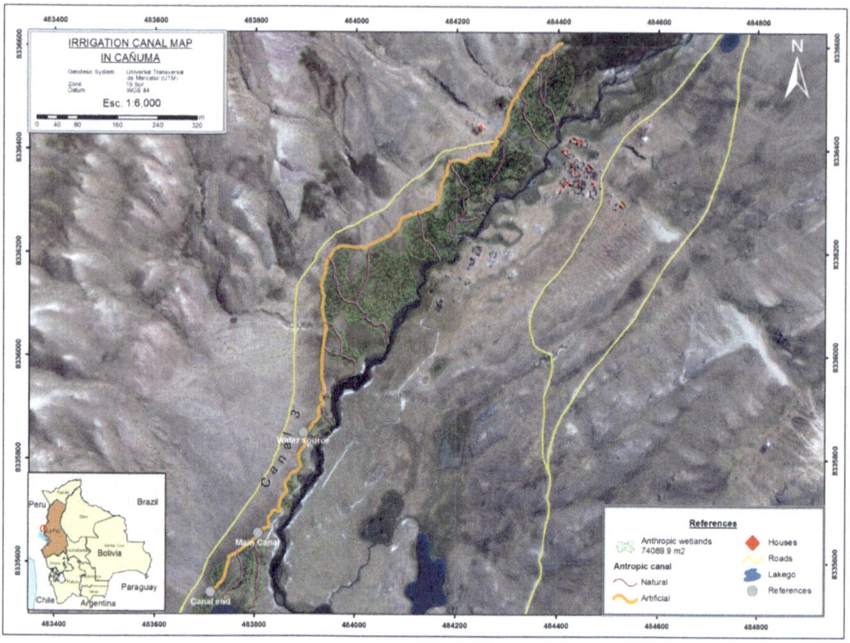

Fig. 3 Route and altitudinal variation of a rustic canal combined with *q'ochas* for grassland irrigation

canal by overflow, and retain the water in order to provide a better water supply for the hillside pastures.

In the middle part of the canal route, where hillside grasslands are irrigated, the water arrives at a depression where herders accumulate it to form lagoons. Marsh vegetation similar to parsley grows here and is craved very much by alpacas. Animals graze on this area have learnt to submerge their heads in the water to feed on the herbs that grow in the lagoon. Such lagoons are called *q'ochas* and are water reservoirs, like ponds, that regulate the flow of the irrigation canal. This irrigation system, involving canals and *q'ochas*, finishes at last part of the segment, where it reaches an area of reserve pastures or an intensive grazing site, which is fenced with wire. Here, herders gather gestating alpacas or those weak from lack of fodder, when extreme droughts occur (Fig. 3 and Photos 7, 8, 9 and 10). In this sector an additional area of about 13.8 ha was increased, providing grazing for 41 additional alpacas.

4.4.3 Rustic Canals for Irrigation of High Andean Grasslands

Rustic canals in areas of high Andean grasslands are located in the lower part of the Cañuma glacier valley, between 4493 an 4499 m, and with a length of 1602 m.

Photo 8 Start of the irrigation canal that transports water to grazing sites and small lagoons

Photo 9 Irrigation of slope *bofedales* and grasslands increases the grazing area of camelids

Photo 10 Along the canals there are several lagoons serving as water reservoirs for alpacas and llamas

(Fig. 4) The area is almost flat. Water in it is captured from the Cañuma River, which runs through the plain and has an abundant flow throughout the year. One obstacle in the area is that the Cañuma River eroded the base of the river bed, to more than 3 m. For this reason there is insufficient water within the irrigation sector. To solve this problem, herders capture water in areas where surplus water is available, almost at ground level. From there they built canals with a minimum slope of 0.35%. In this case, the herders increased the grassland by 7.4 ha, which can support an additional 22.2 alpacas.

The system consists of multi-flow irrigation canals between mono-flow sectors. High-Andean grasslands in Bolivia normally receive seasonal rainfall between November and March, so they do not require additional water (Photos 11, 12, 13 and 14). But when there is an extreme drought, like the one that occurred in 2015–2016, the need to build irrigation infrastructure to mitigate drought effects began. The infrastructure consists practically of canals excavated in clay soils to conduct water along a minimal slope towards dry grasslands.

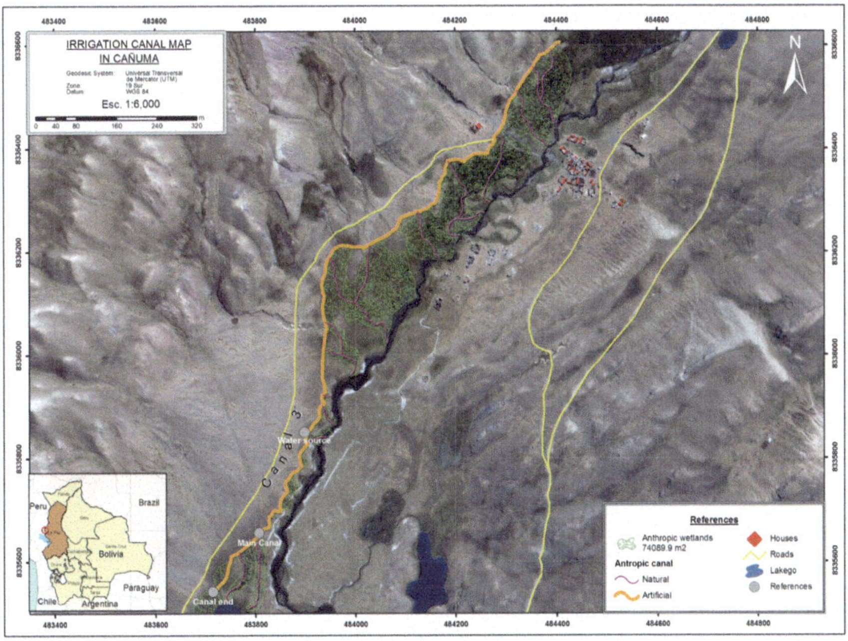

Fig. 4 Route and altitudinal variation of the rustic canals for irrigation of high Andean grasslands

Photo 11 Irrigation canals that pass through *bofedales*

Photo 12 Detail of the main channel that derives its water to a secondary channel for irrigation of *bofedales*

Photo 13 Aymara herder children in the lower part of the community

Photo 14 Cañuma herder enabling a grassland irrigation canal

When a herder decides to irrigate his grazing sector, he places stones or a *tepe* of the pastures from the area, and in this way generates flooding. Once the grassland has been watered, the obstacles are removed so that the water flows to other places that need to be irrigated.

4.5 Perceptions of Climate-Change Effects, from Alpaca and Llama Herders in Cañuma

Table 2 shows the results of the participatory workshop between technicians from the Universidad Mayor de San Andrés, La Paz (Bolivia) and members of the high Andean community of Cañuma, in the Apolobamba mountain range. In that workshop, alpaca herders presented their perceptions of climate-change effects on activities related to pastoralism and the measures that are being implemented to minimize negative effects. In that workshop, so-called talking maps were used, where farmers wrote and drew the effects of climate change.

Table 2 Perceptions of alpaca and llama peasant herders on climate change

Factors	Perceptions of peasant herders from cañuma on climate change
Rainfall	• In recent years there is little rain that affects *bofedales*. Now there are no pastures, and the alpacas are becoming thin and get sick easily • With droughts there is less water and fewer *bofedales* • The river and water springs have dried up now • In the past there were many lagoons and puddles that are now dry • Before there was more water, now there is little water • With the drought, worms that eat plants appear • Now there are changes in the rainy season • The rains before fell from September until April • Today the river has little water, downstream the river has dried and frozen
Temperature	• Now the weather is warmer and the glacier is melting • The sun is burning very strong
Hailstorms	• In recent years hailstorms have become strong • Hail comes that kills animals
Frosts	• Frosts are now more frequent • Frosts burn pastures • Water freezes • A strong frost affects alpacas and llamas • Abortions occur in our animals
Winds	• Now there are more strong winds that affect the eyes of alpacas
Snowfalls	• Heavy snowfalls occur • In 2015 there was a heavy 50 cm snowfall • The snow lasted a month and a half. It was the death of alpacas, llamas and vicuñas. We all cried for our animals
Alpacas	• Alpacas become thin due to lack of pastures. Frost affects grasses. There is much animal mortality due to lack of water • Alpacas get sick easily. They have many external and internal parasites (Chaymalako and Tenia). They also get fever • Alpacas get conjunctivitis due to strong winds • Young alpacas and llamas are attacked by foxes, pumas and condors • Worms pests emerge and feed themselves on leaves in the *bofedales* • In 1998 a hare appeared that eats *bofedales*. The food of the alpacas is eaten
People	• There is migration of peasant families to the cities and gold mines • People have no profits (in the field) due to lack of resources • Smoke from burnings affect people and children • The river freezes and dried up downstream. No water reaches the cabin • Neighbouring communities suffered from drought

5 Conclusions

- All camelid herders in the high mountains agree that climate conditions in the region have changed, and recognize that the weather has changed. There is an irregular distribution of rainfall, drought, hailstorms, snowfall, frost, more wind, high evaporation by solar irradiation, high temperatures and ultraviolet radiation.

- In order to mitigate the retreat of glaciers and the absence of rainfall, alpacas and llama herders have identified strategies to reduce vulnerabilities and to increase the resilience of the ecosystems of the alpaca herder community. The adaptation measures developed are intended to ensure the provision of water for the grasslands by increasing the surface area of hillside *bofedales* and the conduction of water through canals. In this way they ensure food security and improve responsiveness to extreme effects.
- Traditional wetland management technologies with innovative irrigation systems are considered to be forms of wetland expansion, as well as ways of organization for water management in the Cañuma area. Implementing these technological innovations for irrigation make the peasants proud, for they value their identity as alpaca herders in ecosystems of extreme conditions. These peasants also represent the legacy of herders who have created climate-change adaptation technologies. These should be rescued and transferred to other communities in the high Andean mountains.
- New rustic canals, based on innovation and their own technologies to increase of wetland areas, are constructed by herders through gravity irrigation systems to irrigate the wetlands. These must be improved with social innovations that allow the diffusion of ancestral knowledge, complemented with modern technologies.
- It is necessary to monitor the meteorological phenomena of the Cañuma station, together with the ethno-climatic knowledge through observers. This helps provide an early warning system of climate phenomena that may be occurring. It is necessary to have reports of climatological indicators for the results of technical indicators of alpaca breeding and management of *bofedales*, under conditions of: frost, hailstorms, rains and droughts; against mortality, morbidity, birth, fertility, management of color diversity in alpacas, rotation of grazing areas and use of resting fences, sanitary management of camelid cattle, fertilization of natural pastures and fences with acquired materials.
- All these climate factors must be acknowledged and investigated more deeply. This will enable the construction of adaptive processes that respond to the local deficiencies of risk management of alpaca livestock activity, as well as the acquisition of solid instruments to minimize the effects of current and future risks, within the climate-change framework.

References

1. Albo X (1994) Instituciones y alternativas organizativas para la producción. In: Albo X, Ramón G (eds) Comunidades andinas desde adentro. Dinámicas organizativas y asistencia técnica. CECI/Abya – Yala, Quito, pp 91–135
2. Anthelme F, Beck S, Ginot P, García Lino MC, Meneses RI (2015) Ecosistemas y plantas altoandinos de la Cordillera Real; únicos y frágiles. In: Meneses RI, Beck S, Anthelme F (eds) La Cordillera Real y sus Plantas. IRD & Herbario Nacional de Bolivia, La Paz, pp 17–30

3. Beck S, García E, Zenteno F (2002) Flora del Parque Nacional y Area Natural de Manejo Integrado Madidi. Proyecto de Apoyo al Parque Nacional Madidi. SERNAP-CARE-HNB-CE-WCS, La Paz, Bolivia
4. Castro L (2007) Humedales de la Puna del Norte de Chile. In: Lucic MC, Reyes LF (eds) Gestión Sostenible de Humedales. CYTED, Programa Internacional de Interculturalidad, pp 15-46
5. Comisión de medio ambiente y desarrollo de América Latina y el Caribe (1996) Amanecer en los Andes. CAF, BID PNUD, Caracas, 177 p
6. Em P, Moreira-Muñoz A, Richter M (2006) El Clima y sus efectos sobre la vegetación. In: Moraes RM, Ollgaard B, Kvist LP, Borchsenius F, Balslev H (eds) Botánica Económica de los Andes Centrales. Universidad Mayor de San Andrés, La Paz, pp 11–36
7. Flores Ochoa J (1968) Los pastores de Paratía. Una introducción a su estudio. Serie antropología social, vol 10. Instituto indigenista interamericano, México
8. Flores Ochoa J (compiler) (1977) Pastores de puna. Uywamichiq Punarunakuna. Instituto de estudios peruanos. Estudios de la sociedad rural 15, Perú
9. Flores Ochoa J (1997) Pastores de alpacas de los Andes. In: Flores Ochoa J (compiler) Pastores de la Puna. Instituto de Estudios Peruanos, pp 14–52
10. Flores Y, Villagómez JR, Blanco NM, Salcedo L, Sterner O, Almanza GR (2004) Defense mechanisms of Bolivia highlands plants. Revista Boliviana De Química 21(1):36–41
11. Francou B, Vuille M, Wagnon P, Mendoza J, Sicart JE (2003) Tropical climate chage recorded by glacier in the Andes during the last decades of the of twentieth century: Chacaltaya, Bolivia, 168S. J Geophys Res 108(D5):4154
12. Francou B (2013) El rápido retroceso de los glaciares en los Andes Tropicales: Un desafío para el estudio de la dinámica de los ecosistemas de alta montaña. Ecología En Bolivia 48 (2):69–71
13. García E., Valenzuela E, López R (1985) Vegetación y Flora. In: CECI/centro de Estudios Canadienses y Cooperación Internacional. Plan de Manejo Reserva Nacional de Fauna Andina Ulla Ulla. Informe técnico MDSP (Ministerio de Desarrollo Sostenible y Palnificación). DNCB (Dirección Nacional de Conservación de la Biodiversidad. La Paz, Bolivia
14. García E, Beck S, Zenteno F (2002) Plan de Manejo del Parque Nacional y Área Natural de Manejo Integrado Madidi. Departamento de La Paz, Bolivia. Herbario Nacional de Bolivia
15. García E, Zenteno F, Beck S, Nagashiro N (2003) Identificación y caracterización de especies de uso forestal del Area Natural de Manejo Integrado Nacional Apolobamba. Estudio de usos potenciales para un manejo racional de recursos veg et al.es. Herbario Nacional de Bolivia. La Paz, Bolivia
16. García E, Beck S (2006) Puna. In: Moraes RM, Ollgaard B, Kvist LP, Borchsenius F, Balslev H (eds) Botánica Económica de los Andes Centrales. Universidad Mayor de San Andrés, La Paz, pp 51–76
17. Gisbert T, Arce S, Cajias M (1984) Los Textiles de Charazani en su contexto histórico y cultural. In: Espacio y tiempo en el Mundo Callahuaya
18. Greslou F (1989) Visión y crianza campesina de los animales andinos. In: Crianza de llamas y Alpacas. PAL – PRATEC. Proyecto Andino, pp 11–44
19. Hoffmann D, Requena C (2012) Bolivia en un mundo 4 grados más caliente – Escenarios sociopolíticos ante el cambio climático para los años 2030 y 2060 en el Altiplano Norte. Fundación PIEB, Instituto Boliviano de la Montaña, La Paz, 160 p
20. Hoffmann D, Tarquino R, Corro Ayala J, Lavadenz L (2014) Métodos para caracterizar la dinámica de los sistemas sociológicos asociados a los bofedales altoandinos (Cordillera Real, Bolivia). Ecología en Bolivia 49(3):132–140
21. Jordan E (2015) Patrones climáticos de la Cordillera Real: Efectos sobre los glaciares y la vegetación altoandina. In: Meneses RI, Beck S, Anthelme F (eds) IRD & Herbario Nacional de Bolivia, La Paz, Bolivia, pp 42–47
22. Liberman M, Baied C (eds) (1997) Desarrollo Sostenible en ecosistemas de montaña. Manejo de áreas frágiles en los Andes. Artes Gráficas Latina, La Paz, 383 p

23. Liberman M, Molinillo M (2015) Bofedales de Apolobamba. Estudio de línea base de bofedales priorizados en el Área Natural de Manejo Integrado Apolobama, Bolivia. Soluciones Prácticas, La Paz, 160 p
24. Lino Pimentel F, Padilla Callejas V (2015) Gestión de Riesgos a Partir de Saberes Ancestrales. Instituto de Ecología – Nación Jach'a Suyu Pakajaqi, La Paz, Bolivia, 156 p
25. Maldonado M, Navarro G, Acosta F, Aguilera X, de la Barra N, Cadima M, Coronel J, Fernández E, Ferreira W, Goitia E (2012) Humedales y cambio climático en los Altos Andes de Bolivia. UMSS – RUMBOL, 40 p
26. Meneses RI, Beck S, Anthelme F (eds) (2015) La Cordillera Real y sus Plantas. IRD & Herbario Nacional de Bolivia, La Paz, Bolivia, 364 p
27. Meneses RI, García Lino MC, Beck S, Anthelme F, Ibañez D, Lliully AE, Loza Herrera S, Palabral-Aguilera AN, Pfanzelt S, Villavicencio X (2015) In: Meneses RI, Beck S, Anthelme F (eds) La Cordillera Real y sus Plantas. IRD & Herbario Nacional de Bolivia, La Paz, Bolivia, pp 91–364
28. Molinillo M (2003, Unpublished) Patrones de vegetación y pastoreo en ecosistemas alto andinos: Una comparación de casos de estudio en Páramos y Punas. Universidad de los Andes - Merida, Facultad de Ciencias, Postgrado en Ecología Tropical. Tesis, 177 p
29. Murra J (1972) El control vertical de un máximo de pisos ecológicos en la economía de las sociedades andinas. Ensayo publicado en el tomo II de la visita de la Provincia de León de Huanuco (1562). Iñigo Ortiz de Zúñiga, Visitador. Universidad Hermilia Valdizan, Huanuco, Perú
30. Murra J (2002) El mundo andino. Población, medio ambiente y economía. IEP/Pontificia Universidad Católica del Perú, Lima. Historia Andina, vol 24
31. Navarro G, Maldonado M (2002) Geografía Ecológica de Bolivia: Vegetación y Ambientes Acuáticos. Centro de Ecología Simón I. Patiño – Departamento de Difusión, Cochabamba, Bolivia
32. Nina Huanca GS, Loza Herrera S, Gonzales Pomar RK, Rebaudo F (2015) El ser humano: un actor de la dinámica de los ecosistemas altoandinos de la Cordillera Real. In: Meneses RI, Beck S, Anthelme F (eds) IRD & Herbario Nacional de Bolivia, La Paz, Bolivia, pp 64–76
33. Olmos C (2005) Résponse des glaciers Boliviens á la variabilité clamátique recente et aux évenements ENSO – Influence du taux d'englacement et de l'orientation, D.E.A. thesis. Université de Montpellier II, Montpellier, 61 p
34. Palacios F (1988) Tecnología del pastoreo. In: Ochoa F (ed) Llamichos y Paqocheros: Pastores de Llamas y Alpacas. Ed. UNSAAC, Cusco, pp 87–100
35. Prieto G, Alzerreca H, Laura J, Luna D, Laguna S (2003) Características y distribución de los bofedales en el ámbito Boliviano del Sistema T.D.P.S. In: Rocha O, Sáez C (eds) Uso pastoril en humedales altoandinos. Dirección General de Biodiversidad, La Paz, Bolivia, pp 13–38
36. Rauh W (1978) Die Wuchs-und Lebensformen der tropischen Hochgebirgsregionen und der Subantarktis – ein Vergleich. In: Troll C, Lauer W (eds) Geoökologische Beziehungen zwischen der temperierten Zone der Südhalbkugel und den Tropengebirgen. Erdwiss. Forschung 11:62–92
37. Ribera MO, Rocha O (2003) Uso pastoril en humedales de áreas protegidas altoandinas de Bolivia. In: Rocha O, Sáez C (eds) Uso pastoril en humedales altoandinos. Talleres de capacitación para el manejo integrado de los humedales altoandinos de Argentina, Bolivia, Chile y Perú. Sitio RAMSAR – Lago Titicaca, Huarina 28 de Octubre al 1 de Noviembre 2002. Convención RAMSAR, WCS/Bolivia, La Paz, Bolivia, 135–156 pp
38. Ribera MO, Liberman M (2006) El uso de la tierra y los recursos de la biodiversidad en las áreas protegidas de Bolivia. Un análisis crítico con propuestas para su conservación y manejo sostenible. SERNAP – GEF II, La Paz, 520 p
39. Rocha O (2014) Primer Diagnóstico de la Zona Montañosa del Estado Plurinacional de Bolivia. Ministerio de Relaciones Exteriores, La Paz, 140 p
40. Seibert P (1993) La vegetación de la región de los Kallawaya y del Altiplano de Ulla Ulla en los Andes Bolivianos. Ecología En Bolivia 20:1–84

41. SERNAP (2016) Plan de manejo del Área Natural de Manejo Integrado (ANMI) Apolobamba. Servicio Nacional de Áreas Protegidas de Bolivia (SERNAP)
42. Soruco A (2012) Medio Siglo de Fluctuaciones Glaciares en la Cordillera Real y sus efectos Hidrológicos en la ciudad de La Paz. Institut de Reserche pour le Dévelopment (IRD), La Paz, Bolivia, 228 p
43. Troll C (1959) Die tropischen Gebirge. Bonner Geogr. Abh. 25:1–93
44. Vassas Toral A (2016) Partir y cultivar. IRD-CIDES-UMSA/Plural Editores, La Paz, 306 p
45. Vuille M, Francou B, Wagnon P, Juen I, Kaser G, Mark BG, Bradley RS (2008) Climate change and tropical Andean glaciers – past, present, future. Earth Sci Rev 89:79–96
46. Vuille M (2013) El cambio climático y los recursos hídricos en los Andes Tropicales. Banco Interamericano de Desarrollo (BID). Nota técnica IDB – TN-517, 29 p
47. Wheeler JC (1984) La domesticación de la Alpaca (Lama pacos L.) y la Llama (Lama glama L.) y el desarrollo de la Ganadería Autóctona en los Andes Centrales. Boletín de Lima 6 (36):74–84

Essays on Geobotanic Mapping in the Andes of Bolivia, with Particular Reference to the Conservation Status of the Vegetation

Franco Pedrotti

Abstract The purposes of this contribution is to present some maps of the vegetation of Andes of Bolivia, with particular reference to the conservation status of vegetation. The study areas are the following: Capinota Valley, Cochabamba (1: 50,000), the Camacho river basin, Tarija (1: 50,000), Navado Sajama, Oruro (1: 130,000) and Isla del Sol in Lake Titicaca, Bolivian plateau (1: 16,000). For each map there is a description on the environnement and vegetation (phytocoenoses, vegetation series and vegetation geoseries), with indication of dynamic tendencies of vegetation: natural fluctuation, regeneration, secondary succession, regression. The conservation status of vegetation (understood at a phytocoenoses, series and geoseries) is very different in the four cases studied. In Capinota, Isla del Sol and Rio Camacho, the forest communities were eliminated and only a few isolated fragments remain today; in Capinota forests of *Schinopsis haenkeana* and *Aspidosperma quebracho-blanco*; on the Isla del Sol matorrales of *Polylepis incarum*; and in the Rio Camacho river forests of *Tipuana tipu*. On the Nevado Sajama volcano, despite a notable anthropogenic influence, both phytocoenoses (matorral of *Polylepis tarapacana*) and geoseries (including the vegetation series) have been maintained, although with some areas in regression due to overgrazing and fire.

Keywords Vegetation mapping · Phytocoenoses · Vegetation series · Vegetation geoseries · Andes of Bolivia · Capinota · Nevado Sajama · Isla del Sol · Lake Titicaca · Camacho river · Dynamical tendencies of vegetation · Conservation status of vegetation

F. Pedrotti (✉)
Università di Camerino, Palazzo Castelli - Via Pontoni 5, 63032 Camerino, MC, Italy
e-mail: franco.pedrotti@unicam.it

F. Pedrotti and E. O. Box (eds.), *Tools for Landscape-Scale Geobotany and Conservation*, Geobotany Studies, https://doi.org/10.1007/978-3-030-74950-7_6

1 Introduction

The purpose of this contribution is to present some maps of the vegetation of the Andes of Bolivia with particular reference to the conservation status of vegetation. The term "vegetation" is quite generic and refers to the vegetation cover in a broad sense or plant cover or plant mass, as most botanists refer to it (see, for example, [9, 22, 24, 47] etc.). More specifically, it can be said that the vegetation cover is formed by the set of plant communities present in a territory at a given time [8]. These communities (plant associations) are made up of concrete units, the phytocoenoses, each of which corresponds to an individual of association or elementary population. The higher levels of synthesis consist of vegetation series and vegetation geoseries [22, 58]. The conservation status of the vegetation is the result of the influence exerted by man on the vegetation cover and is evaluated with different methodologies depending on the authors. It is proposed here to refer to two parameters: the phytosociological typology (phytocoenoses, vegetation series, vegetation geoseries) and dynamic processes affecting vegetation.

2 Methodology

A vegetation map is a topographic map in which the specific units that form the vegetation of a specific territory are summarized graphically, that is, the phytocoenoses and their spatial relationships [48]. It represents the last stage of a research process that begins with the recognition of phytocoenoses in the field, their delimitation, their detection and the definition of a typology that may be different according to the different geobotanical schools. The Braun-Blanquet-Tüxen School of Phytosociology describes associations of plants and their classification in a hierarchical system: associations, alliances, orders and classes [22].

A map of vegetation is composed, therefore, of the following elements: topographic map, cartographic units indicated with different colors and delimited by a line called the "vegetation limit", and a list of the vegetation units represented on the map.

Mapping dynamical tendencies involves producing a plant community map and specifying two labels in each patch, i.e. one referring to the vegetation type and the other to the dominant dynamical tendency. The first example of such a map was the Map of Bosco Quarto, Gargano (Italy), printed at the scale 1:10,000 [20]. Although the dynamical tendencies distinguished have often been reported and used since then, their nomenclature has not yet been systematised.

Conservation status is evaluated relative to certain units of vegetation, which are the following: phytocoenoses, that is the concrete units of the associations (identified by phytosociology); vegetation series (according to synphytososociology); and vegetation geoseries (according to geosynphytosociology). The dynamic processes are the following: fluctuation (of natural, semi-natural and anthropogenic origin), degeneration, regeneration, primary succession, secondary succession, and regression [19].

Phytocoenoses are evaluated through intrinsic characteristics of the vegetation, such as floristic composition, vertical and horizontal structure, and dynamic processes. The individual phytocoenoses of the same association are evaluated separately, because they can be affected by different ecological processes. When the associations are not known, reference can be made to vegetation formations. The starting point for the evaluation of phytocoenoses is the phytosociological map of the actual natural vegetation. This map must be completed with the map of dynamic vegetation tendencies. Different levels of conservation [status] are associated with each dynamic tendency.

Vegetation series are evaluated with reference to the space they occupy, the homogeneous fragment of the landscape, i.e. the ecotope or "patch" (called *tesela* in Spanish). The reference parameter is the relationship forest/territory evaluated in a concrete form, i.e. within the space of the patch; we also define the current ratio of actual vegetation to potential vegetation. The series is evaluated in the space in which it develops, that is its patch (tesela), which can be completely occupied by the series association or occupied to a decreasing extent, up to that of a patch formed exclusively by secondary vegetation. With this procedure, it is possible to calculate the percentage of area covered by forest with respect to secondary vegetation. The reference for evaluation of the whole series is the integrated phytosociological map (synphytosociological map).

Vegetation geoseries occupy more or less vast territories, which have undergone human interference to varying degrees. The reference for evaluation of geoseries is the geosynphytosociological map. Also, in this case, evaluation consists of quantifying the ratio of primary vegetation to secondary vegetation, but on a broader scale than in the previous case. This leads to a global assessment of the landscape. It is thus possible to evaluate wider areas, at the level of the landscape.

For each area of study, a scale of values relative to the state of conservation of the vegetation has been proposed, without reaching a quantification. The criterion followed was to place natural vegetation (where existing) at the first place on the scale, affected by the processes of fluctuation and regeneration, and therefore the shrub associations resulting from secondary succession, the pastures obtained from the elimination of primary vegetation and finally the synanthropic units, which are all affected by the process of anthropic fluctuation, due to periodic intervention by man. This scale of values, which is based on the series of vegetation (i.e. on the different stages that compose them) and their dynamic tendencies, can be considered of general value. On the other hand, the scales proposed for the four areas to be studied (see below) detected have only a local meaning relative only to themselves. A map of the conservation status of the vegetation was also made.

Among the many authors who dealt with the conservation status of vegetation, Carboni et al. [10], Asensi et al. [1, 2], and Pinna et al. [55] refer to phytocoenoses and landscape; Slater et al. [59], Davis et al. [16] and Miles et al. [42] to a single type of vegetation (vegetation of river corridors in Wales, coastal scrub in California and dry forests of the world, respectively), Del Arco Aguilar et al. [17] refers to actual and potential vegetation; and Jewitt [25] to vegetation types.

3 The Study Areas

The study areas are the following: Capinota (Cochabamba), the Camacho River (Tarija), Nevado Sajama, and Isla del Sol, in Lake Titicaca (Bolivian plateau). Capinota and the Camacho River involve mesothermic valleys, in which the vegetation is tall forest of different types [44]. The Nevado Sajama and Isla del Sol belong to the *puna*, a term used to indicate the high-mountain environment and vegetation of the Andes between 3000 and 4500 m. The emblematic species of the puna is *Puya raimondii*, the largest known bromeliad, with a height of 9–10 m [50].

For each map there is a description of the environment and vegetation, indication of dynamic tendencies with an acronym, a summary of the general characteristics of the map, and the reproduction of the map (Map of Nevado Sajama) or of a section of the map (Maps of Capinota, Isla del Sol and the Camacho river basin). The maps reproduced have been reduced from their original scale, in order to adapt them for printing. For each map, a legend for the vegetation units is also given.

4 Vegetation Map of the Capinota Valley (Cochabamba)

The study area (Capinota Valley) is located in the lower valley of the Department of Cochabamba, between the mouth of the Tapacarí River and the town of Orcoma (2250–3250 m). Orographically, this area belongs to the Cordillera de los Andes and is part of the ramification of Tunari and Mazo Cruz.

This valley is one of the dry, mesothermic inter-Andean valleys, whose vegetation is characterized by the dominant presence of xerophytic, deciduous or evergreen, microphyllous species. It belongs to the Boliviano-Tucuman province of the Andean region [45].

The vegetation has been altered significantly by the action of man and is secondary vegetation, in which there are some elements of primary vegetation, such as the vegetation of the rivers (*Cortaderia atacamensis* and *Tessaria dodonaeifolia* with the presence of isolated saplings of *Salix humboldtiana*) and the forests of *Schinopsis haenkeana* and *Aspidosperma quebracho-blanco*. These forests correspond to the series of *Carica quercifolia* and *Schinopsis haenkeana*, described by Navarro [44] for the Cochabamba basin. In the areas along the rivers (alluvial deposits) there is intensive cultivation of crops, and in the other areas there are annual dry-land crops.

The vegetation map represents the actual natural vegetation and refers to the plant communities gathered in physiognomic categories: herbaceous communities (*Neocracca heterantha* and *Onoseris gnaphalioides*, *Cortaderia atacamensis*, *Viguiera lanceolata*), scrubs *(Tessaria dodonaeifolia, Dodonaea viscosa)*, chaparral (tall matorral) of *Acacia macracantha* and *Prosopis laevigata* var. *andicola*), forests (*Schinopsis haenkeana* and *Aspidosperma quebracho-blanco*), and synanthropic vegetation (of intensive and annual crops). In most of the Capinota Valley the potential vegetation is forest of *Schinopsis haenkeana* and *Aspidosperma quebracho-blanco* on the hillsides and riparian forest of *Salix humboldtiana* along the rivers.

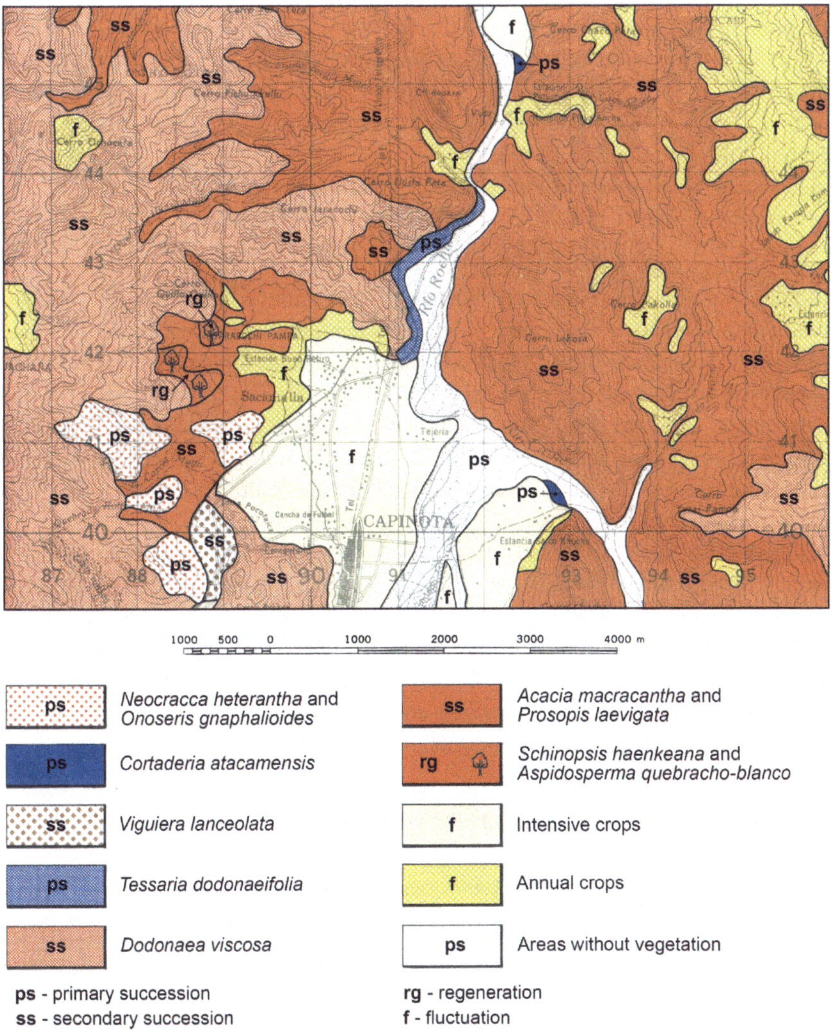

Fig. 1 Map of communities in a sector of the Capinota Valley with indication of the relative dynamic tendencies (*from* [35], *modified*)

General characteristics of the map (Fig. 1)
Size: 78 × 60 cm
Scale: 1: 50,000
Topographic map: map of the Instituto Geografico Militar (Bolivia).
Vegetation: herbaceous, shrub vegetation, chaparral (tall matorral), matorral, forest, and synanthropic communities.
Number of communities represented on the map: 11.
Cartographic representation of the vegetation: cartographic units and symbols (the forests of *Schinopsis haenkeana* and *Aspidosperma quebracho-blanco*).
Vegetation limits indicated in black.
Type of map: map of communities, actual natural vegetation.
Bibliography: Pedrotti et al. ([35, 54]),Pedrotti and Venanzoni [51].

Conservation Status of the Vegetation
The map of the actual natural vegetation of the Valley of Capinota (1: 50,000) served for the detection of dynamic vegetation tendencies. The remaining forests of *Schinopsis haenkeana* and *Aspidosperma quebracho-blanco* are affected by regeneration processes; the matorrales and chaparrales by the processes of secondary succession; the riparian vegetation (*Cortaderia atacamensis* and *Tessaria dodonaeifolia,* with isolated trees of *Salix humboldtiana)* by primary succession; and the anthropic fluctuation manifests itself in the cultivated areas and in the ruderal and nitrophilous areas (Fig. 1).

The action of man has caused the following modifications to the vegetation: 1) elimination of forest vegetation (*Schinopsis haenkeana* and *Aspidosperma quebracho-blanco*), of which only two residual nuclei remain; 2) development of secondary communities of chaparral and matorral, *Prosopis laevigata* ssp. *andicola* at lower elevations (Fig. 2) and *Dodonaea viscosa* higher; and 3) transformation of the most favorable areas into cultivated areas.

The scale of Capinota vegetation values is as follows:

I) Primary vegetation: rocky outcrops (*Neocracca heterantha* and *Onoseris gnaphalioides)* and streambeds *(Cortaderia atacamensis* and *Tessaria dodonaeifolia*)
II) Forests of *Schinopsis haenkeana* and *Aspidosperma quebracho-blanco*
III) Matorrales and chaparrales (*Prosopis laevigata* ssp. *andicola*; *Dodonaea viscosa*)
IV) Agricultural crops.

5 Map of the Vegetation of the Nevado Sajama (Oruro)

The study area is located in the northern part of the Bolivian plateau and covers approximately 400 km^2; this area can be divided into three physiographic units according to relief: a volcanic cone, glacial valleys and piedmont plains. It belongs to the plateau province [provincia altiplanica] of the Andean region [45].

The volcanic cone is that of the Nevado Sajama, reaching 6,542 m of altitude. The upper part of the Nevado is affected by recent glaciations, with large accumulations of snow and snowdrifts that have cracks in the upper part. On the slopes and around the Nevado Sajama, there are large accumulations of volcanic material (glacial cirques and glacial valleys). The piedmont plains are between 4,300 and 4,500 m of altitude.

In the Nevado Sajama area we can distinguish the following vegetation formations: high-Andean herbaceous vegetation (*Calamagrostis curvula, Festuca orthophylla, Azorella compacta, Pycnophyllum molle*); matorrales (at the upper limit of distribution) and forests of *Polylepis tarapacana*; bofedales (wetlands of *Oxychloe andina, Calamagrostis ovata, Werneria pygmaea*); grass-shrub steppes (*Parastrephia lepidophylla, Baccharis incarum, Festuca orthophylla, Poa asperiflora, Pycnophyllum tetrastichum*); grass steppes resulting from burning or natural fires (*Festuca orthophylla, Calamagrostis curvula, Calamagrostis antoniana, Poa asperifolia, Muehlenbergia peruviana*); matorrales of *Parastrephia lepidophylla*; and vegetation of saline areas (*Distichlis humilis, Puccinellia* sp., *Salicornia* sp.).

The matorrales and open forests of *Polylepis tarapacana* develop only on the slopes of the volcanic cone, from the base of the volcano (4,100 m) to 5,200 m, followed by a strip of herbaceous vegetation of high-Andean steppe. These form a concentric ring interrupted by lava flows and deposits of volcanic slag. Martín Cárdenas called them "the highest forests in the world" by altitude [11]. Today these forests have a degree of coverage between 60 and 80%, but some better-preserved nuclei reach 100%. The age pyramid of two populations of *Polylepis tarapacana* shows the low proportion of seedlings, which have great difficulties to develop in volcanic waste [29]. More frequent is the renewal of this species in forest clearings caused by cutting by man and by fires. Today these clearings are occupied by secondary grasslands of *Festuca orthophylla*. In this case, plants of *Festuca orthophylla* facilitate the vegetative development and germination and establishment of seeds of *Polylepis tarapacana* (facilitation phenomenon). These forests belong to the *Mutisio lanigerae-Polylepidetum tarapacanae* association [45, 46].

Grazing by alpacas and llamas is intense [29, 31, 32, 57].

General characteristics of the map (Fig. 3)
Size: 54 × 35 cm
Scale: 1: 130,000
Topographic map: map drawn in reduced form with these elements: hydrographic network, mountains, towns.
Vegetation: plant formations (high-Andean herbaceous vegetation, herbaceous vegetation, bofedales, matorrales, forests, steppes).
Number of formations represented on the map: 7.
Cartographic representation of the vegetation: cartographic units.
Vegetation limits are indicated in black.
Type of map: map of the formations, actual natural vegetation.
Bibliography: Liberman Cruz [26], Liberman et al. [29, 37].

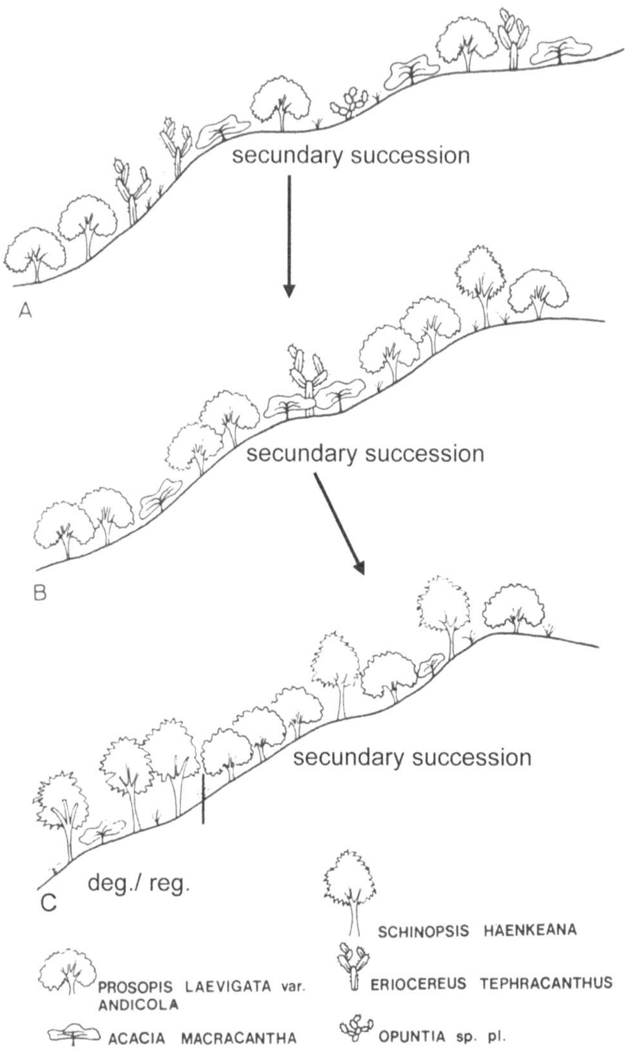

Fig. 2 Capinota, profiles of the *Acacia macracantha* and *Prosopis laevigata* var. *andicola* chaparral (tall matorral); a: stage with prevalence of *Cactaceae*, secondary succession; b: stage with prevalence of *Prosopis laevigata* var. *andicola*, secondary succession, more advanced stage; c: on the right, stage with plants of *Schinopsis haenkeana*, secondary succession, more advanced stage; on the left, residual forest of *Schinopsis haenkeana*, degeneration/regeneration. The arrows indicate the progressive evolution of the vegetation (*from* [53] *modified*)

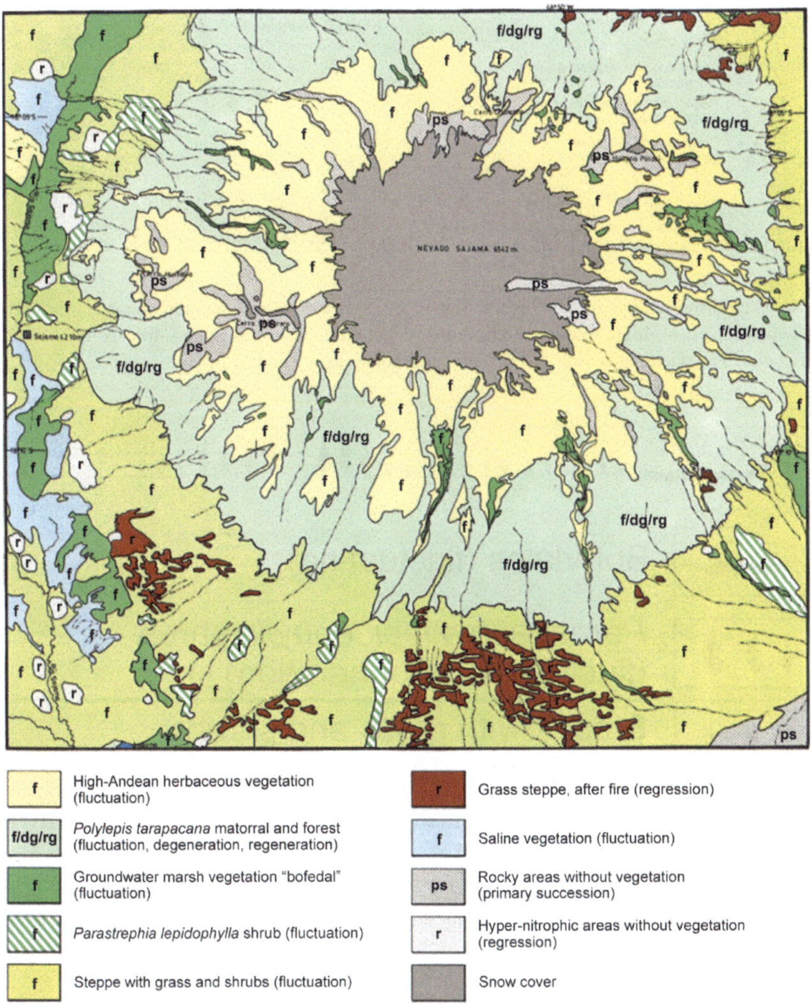

Fig. 3 Map of formations of the Nevado Sajama with indication of the relative dynamic tendencies (*from* [26], *modified*)

Conservation Status of Vegetation

In the area of the Nevado Sajama grazing is by alpacas and llamas, which has led to various changes in vegetation, as shown by the map of dynamic tendencies (Fig. 3). The high-Andean herbaceous vegetation is affected by this fluctuation; the *Polylepis* matorrales and forests have different aspects, degeneration in those parts where grazing takes place, regeneration and even fluctuation in some well-preserved nuclei. Some slope areas are in regression, due to fires. The slopes of the Nevado Sajama correspond to a geoseries, composed by 4 vegetation series (Fig. 4); on the whole, it should be noted that the belt of the matorrales and forests of *Polylepis*

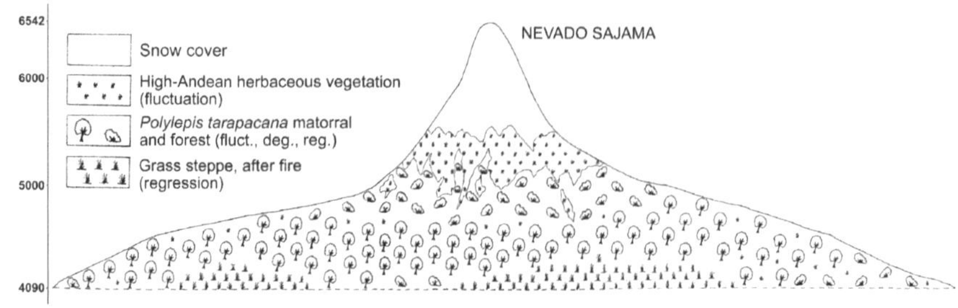

Fig. 4 The northern slope of the Nevado Sajama corresponds to a vegetation geoseries formed by 2 vegetation series: high-Andean herbaceous vegetation, matorrales and forests of *Polylepis tarapacana*. For each formation the dynamical tendencies are indicated

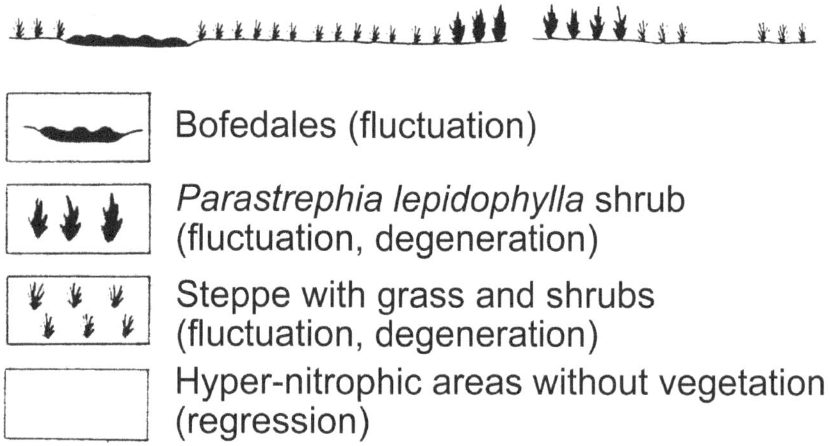

Fig. 5 Formations of the plains surrounding the Nevado Sajama, with indication of the relative dynamic tendencies

tarapacana is well preserved, although regression and degeneration are occurring. The same applies to the piedmont plain (Fig. 5).

The scale of vegetation values of the Nevado Sajama is as follows:

I) High-Andean vegetation
II) Matorrales and forests of *Polylepis tarapacana*
III) Bofedales, matorrales of *Parastrephia lepidophylla* and others
IV) Grass steppe after fire, hyper-nitrophic areas.

6 Map of the Vegetation of Isla del Sol (Lake Titicaca)

The Isla del Sol is located in Lake Titicaca (3810 m); it has a surface area of 20.9 km^2 and an irregular shape, with a NW–SE orientation. The island is formed by two mountain chains with elevations that go up to 4045 and 4076 m. In the past, the island was a sacred place for the people who lived near the lake, and it is believed that the Inca Empire originated here [60]. Today it is possible to observe remnants of the ruins of ancient Inca temples and terraces that are still used by the Aymara peasants for their crops (Fig. 6). The actual vegetation of the Isla del Sol is the result of man's influence on the environment (Figs. 7 and 8). The original forests of keñua (*Polylepis incarum*) [21, 23] were completely felled, and today there are only two small remnants left. This area belongs to the Peruvian Puna province of the Andean region [45].

Herbaceous Communities. There is a group of three communities, characterized by the presence of *Plantago sericea* ssp. *polyclada*, which develop on rocky slopes, rounded peaks and on the top of a mountain above 4,000 m. Another community is characterized by *Plantago sericea* ssp. *sericans* and forms a continuous grass cover on some hills. Flooded areas are characterized by different communities of Juncaceae (*Juncus stipulatus*, *J. ebracteatus*, *J. arcticus* var. *andicola*) and other species (*Carex boliviensis*, *Hydrocotyle ranunculoides*).

Fig. 6 Isla del Sol, with slope terraces that date back to the Inca epoch; in the foreground the remains of the Inca palace, from a drawing from the years 1863–1865 (*from* [60])

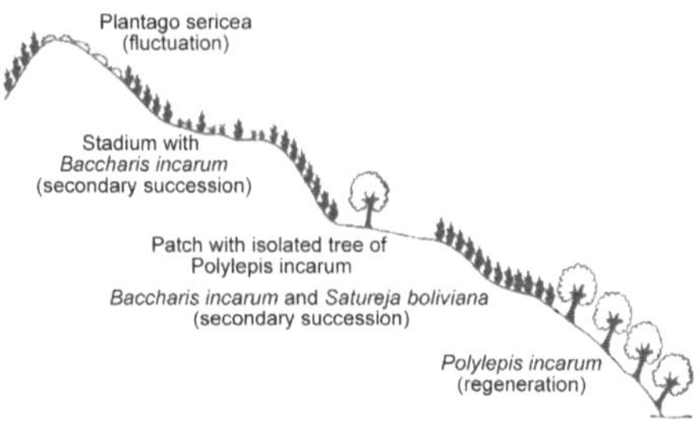

Fig. 7 Vegetation profile of the northeastern slope of Cerro Palla Khasa (Isla del Sol), with indication of the dynamical tendencies of vegetation

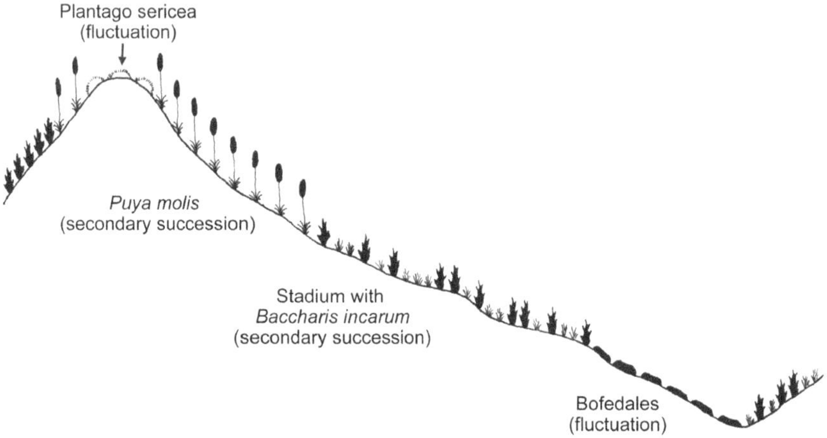

Fig. 8 Vegetation profile of the northeastern slope of Cerro Chequesani (Isla del Sol), with indication of the dynamical tendencies of vegetation

Shrub Communities. The most frequent community is the "tolares", which occupy great extents and involve *Baccharis incarum* (tola) and *Satureja boliviana*. This community also develops on the walls of the terraces and on fallow lands, forming open shrub thickets. In two areas of this community, *Polylepis incarum* was observed, in its shrubby form, with a major renovation. In rocky and humid places, a community of *Pernettya prostrata* was observed. Finally, the community of *Puya mollis* should be noted, occurring on slopes with NE exposure and a rocky sandstone substratum.

Synanthropic Communities. These are the communities associated with intensive and extensive crops. The main crops of Isla del Sol are: different varieties of *Solanum* sp. (potato), *Oxalis tuberosa* (oca), *Ullucus tuberosus* (ullucu), *Chenopodium quinoa* (quinoa), *Zea mays* (maize), *Lupinus mutabilis* (lupine) and *Hordeum vulgare* (barley). When the crop areas are abandoned or are resting for long periods of time, different shrubs develop, of which *Baccharis incarum* is usually dominant, in other cases co-dominant with the *Lupinus prostratus*. These matorrales, and their herbaceous companion species, constitute stages of development towards the matorral of *Baccharis incarum* and *Satureja boliviana*.

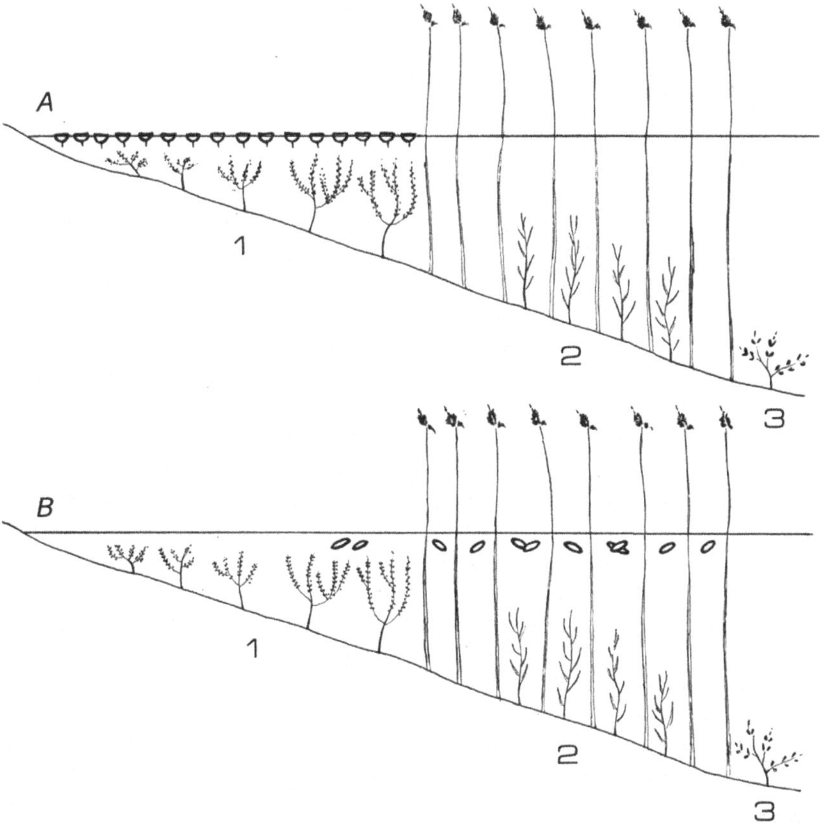

Fig. 9 Profile of the vegetation of the shores of Lake Titicaca; A: *Lemnetum minusculae-gibbae*; B: *Lemnetum valdivianae*, with 1 - *Myriophyllum elatinoides*, 2 - *Schoenoplectus californicus* ssp. *tatora*, and 3 - *Chara* sp. The vegetation is affected by fluctuation processes (*from* [35])

Lake Titicaca Communities (Fig. 9). Lake Titicaca is a large, tropical high-mountain lake with slightly mesotrophic-oligotrophic water [56]. The shallow water near the shoreline shows eutrophication almost everywhere, due to pollution from the human settlements [27]. The vegetation on the banks has the following zonation: shallow water with *Myriophyllum elatinoides* and *Elodea potamogeton*, inland water with *Schoenoplectus californicus* ssp. *tatora* (totora), and deep water with *Chara* sp. pl. [12, 13, 14]. The floating vegetation involves two associations: *Lemnetum minusculae-gibbae* in the shallow water along the shores, with *Lemna gibba, Lemna minuscula* and *Azolla filiculoides*; and *Lemnetum valdivianae* in the water far from shorelines, between the totora plants, with *Lemna valdiviana, Azolla filiculoides* and *Utricularia obtusa* [52]. The association *Lemnetum minusculae-gibbae* was also found in Lake Uru-Uru, still in the Bolivian plateau south of Lake Titicaca [36]. Along the shores of the Isla del Sol, the presence of the *Schoenoplectus californicus* ssp. *tatora*, present in only one place, has been mapped with a symbol.

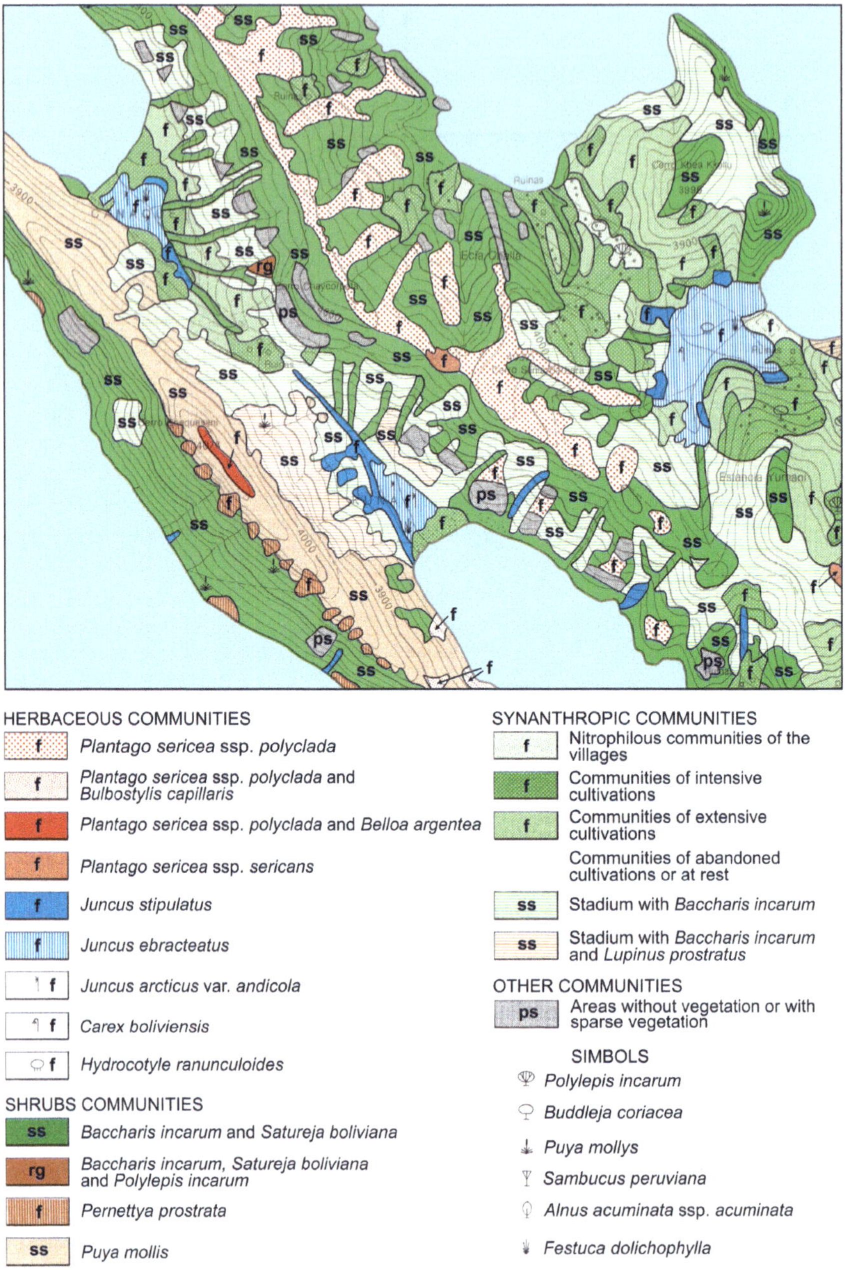

Fig. 10 Map of communities of the Isla del Sol, with indication of the relative dynamic tendencies (*from* [33] *modified*)

General characteristics of the map (Fig. 10)
Size: 82 × 68 cm
Scale: 1: 16,000
Topographic map: map of the Instituto Geografico Militar (Bolivia).
Vegetation: plant communities (herbaceous, scrubland, synanthropic)
Number of communities represented on the map: 23.

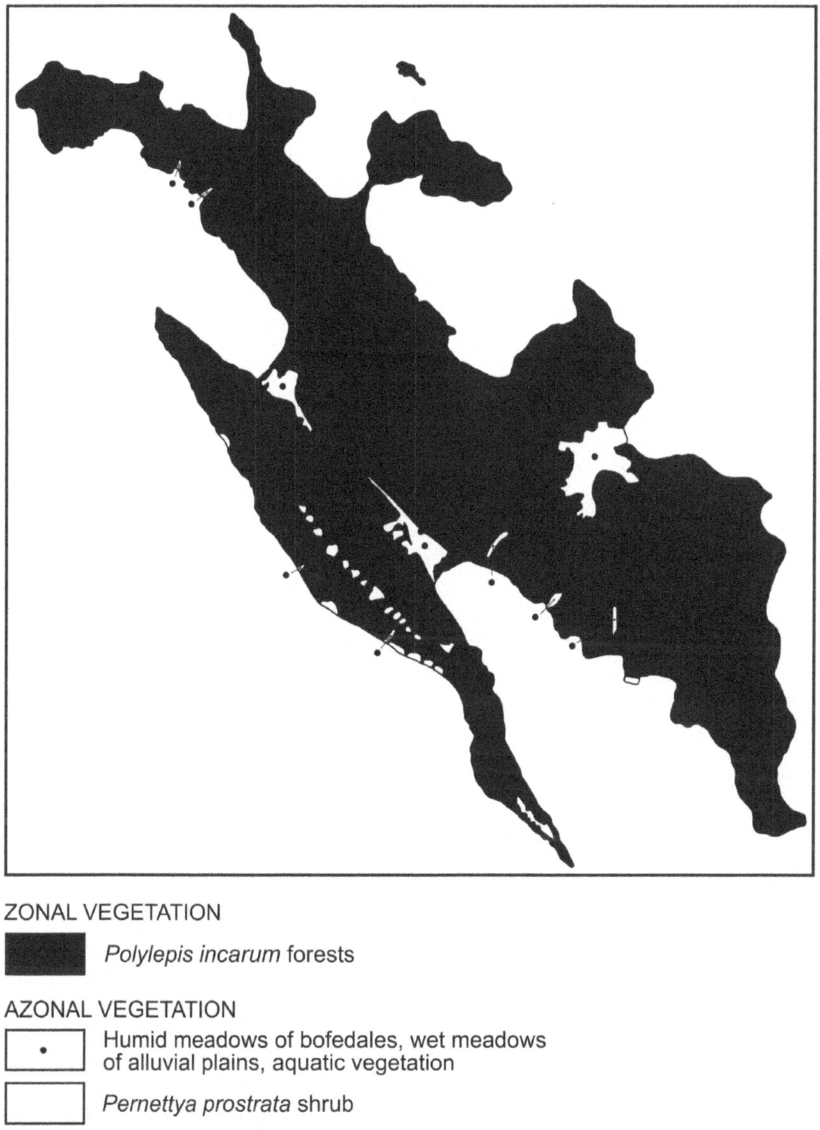

Fig. 11 Map of the potential vegetation of the Isla del Sol

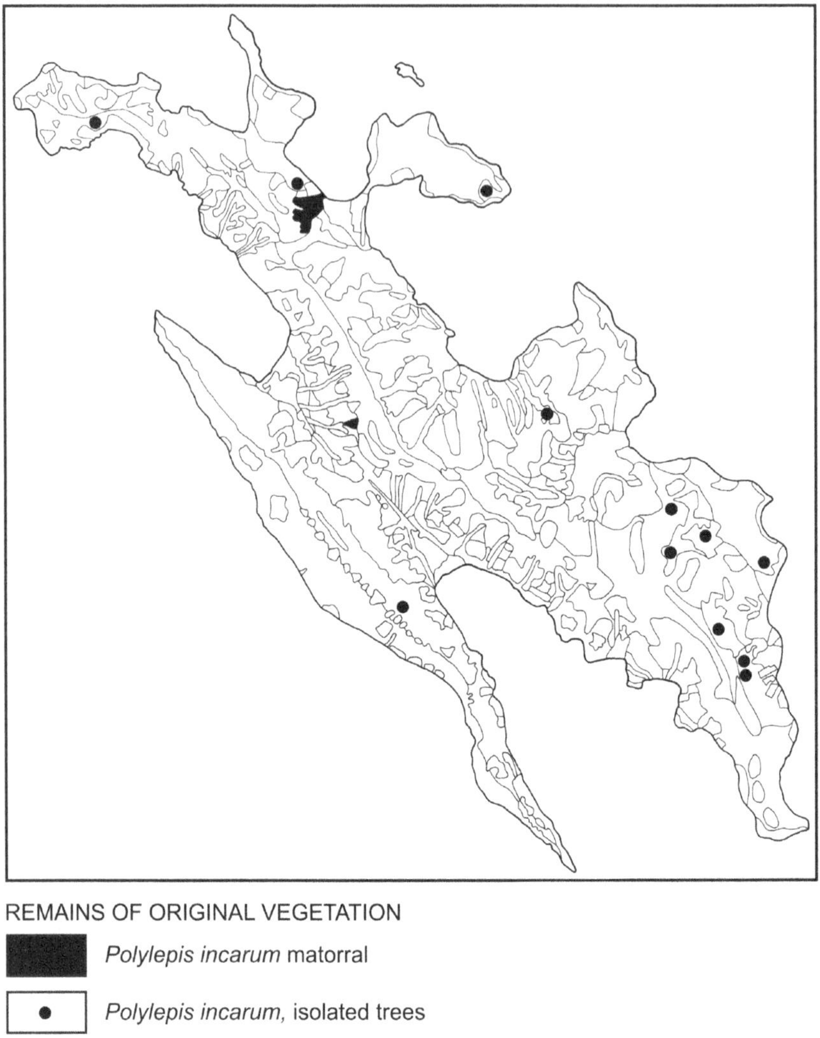

Fig. 12 Remnants of original vegetation of the Isla del Sol (from [33], integrated with [18] and [41])

Cartographic representation of vegetation: cartographic units and symbols. Symbols have been used to indicate less extensive communities (such as the communities of *Carex boliviensis*, *Hydrocotyle ranunculoides* and *Schoenoplectus californicus* spp. *tatora*) and some rare tree species (*Polylepis incarum*, *Buddleja coriacea*, *Sambucus peruana*, *Alnus acuminata* ssp. *acuminata*) and other species (*Puya mollis*, *Festuca dolichophylla*).

Vegetation limits are indicated in black.
Type of map: map of communities, actual natural vegetation.
Bibliography: Liberman Cruz et al. [28, 33, 34]; Pedrotti [49].

Conservation Status of Vegetation

The map of the actual natural vegetation has allowed us to make the map of potential vegetation, distinct in zonal and azonal vegetation (Fig. 11). In the past the whole island was covered by *Polylepis incarum*, today limited to two residual nuclei and to groups of plants in some patches (Fig. 12). Martínez and Villarte [41] report the presence in two patches of centenarian trees of *Polylepis incarum* more than 12 m high and with a diameter of 69 cm. The azonal vegetation is represented by the vegetation of humid environments (bofedales and alluvial plains, Lake Titicaca) and by *Pernettya prostrata* shrubs limited to rocky stations with northern exposure.

On the map of the actual natural vegetation are indicated the dynamic tendencies from which the phytocoenoses are affected, which are the following (Fig. 10): regeneration (*Polylepis incarum* community); secondary succession (*Baccharis incarum* and *Satureja boliviana* community, *Puya mollis* community); forced secondary succession (reforestation with *Eucalyptus* sp.); natural fluctuation (humid meadows of bofedales, wet meadows of alluvial plains, aquatic vegetation, lake vegetation); and fluctuation of anthropogenic origin (pasture communities; agricultural crops - intensive, extensive, at rest, abandoned; and ruderal and nitrophilous vegetation of the villages).

The scale of the vegetation values of the Isla del Sol is as follows:

I) *Polylepis incarum* scrub, *Pernettya prostrata* scrub, humid meadows of bofedales, wet meadows of alluvial plains, aquatic vegetation, lake vegetation
II) *Baccharis incarum* and *Satureja boliviana* scrubs
III) *Puya mollis* scrubs
IV) Pastures (*Plantago sericea* ssp. *polyclada*, *P. s.* ssp. *sericans*)
V) Agricultural crops (intensive, extensive, rest, abandoned), ruderal vegetation of the villages
VI) Reforestation of *Eucalyptus* sp.

Six categories have been reported on the map of actual natural vegetation, so as to obtain the map of vegetation conservation status values (Fig. 13). On the basis of this map it can be seen that, on the Isla del Sol, the vegetation has undergone the following changes:

1) elimination of *Polylepis incarum* forests
2) formation of different types of secondary vegetation (matorrales and pastures)
3) large fragmentation of secondary vegetation
4) maintenance of some types of vegetation (bofedales, *Pernettya prostrata*).
5) transformation of the most favorable areas into agricultural areas.

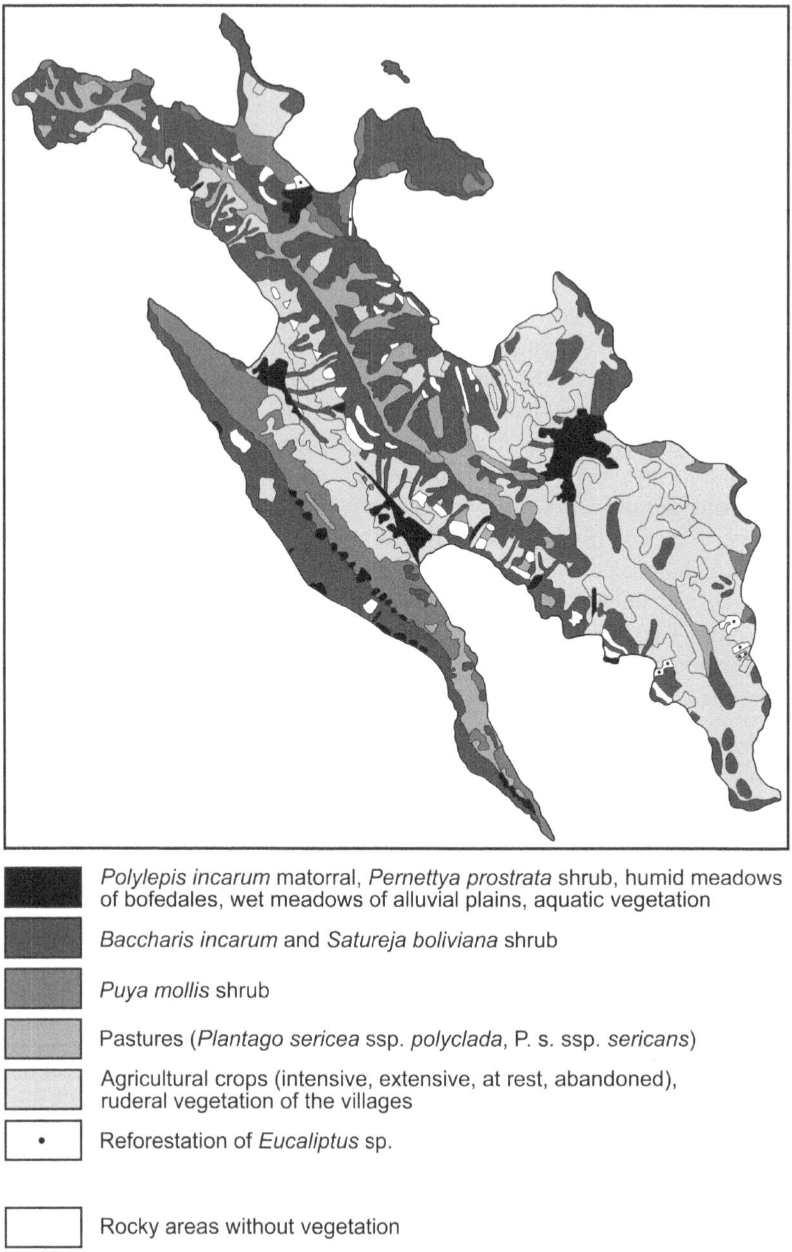

Fig. 13 Map of the values of the communities of the Isla del Sol

7 Map of Forests of the Camacho River Basin (Tarija)

The Camacho river basin is located in an inter-Andean valley in the extreme south of Bolivia, in the Department of Tarija. The approximate area is 1,050 km^2, with altitudes that go from 1600 m, at the confluence of the Camacho river with the Guadalquivir river, up to 4500 m, in and near the Taxara mountains. The basin of the Central Valley of Tarija is a mesothermic valley in which the rainfall is seasonal, concentrated between December and March. In the Camacho river basin, average annual precipitation varies from 300 mm in the area of the Angustura to 1400 mm in the cordilleran zone in the southwestern part of the basin. This area belongs to the Boliviano-Tucuman province of the Andean region [45].

The original vegetation has been strongly transformed by man in the past centuries, going back mainly to the arrival of the Spaniards [15, 43]. The few forests in acceptable condition are found in areas almost inaccessible to man, such as ravines or wet gorges, hillsides and rocky areas with steep slopes and riverbeds. The original forests were in many cases replaced by secondary formations of matorral and shrub scrub. Isolated trees appear in extreme cases, testimony to what the original vegetation was like.

White Alder Forests (*Alnus acuminata* ssp. *acuminata*). Alder forests occupy the upper part of the humid montane forest belt in the sub-Andean strip of the Andes. The dominant species is white alder, which occurs from the base of the mountains up to the subalpine belt, on slopes of southern exposure, between 2300 and 3000 m, in areas where humid winds and clouds, originating in Argentina.

Keñua Forests (*Polylepis crista-galli* and *P. hieronimi*). These correspond to the areas of semi-humid to semi-dry forests in the mountainous Andes. Forests of keñua occur on mountains with slopes from 30 to 40°. At the southern end of the basin, where there are the most mountains and valleys with northern exposure, the keñua forests reach their highest elevations, due to the Massenerhebung effect.

Pino del Cerro Forests (*Podocarpus parlatorei*). These belong to the area of humid high-montane forests, occurring in a vegetation belt below to that of the alder forests. These podocarp forests develop in areas with variable slopes, from almost flat to steep (10–15°) to even steeper (30–40°).

Tipa - Jacaranda Forests (*Tipuana tipu* – *Jacaranda mimosifolia*). This is a climax community that has been practically destroyed, because its distribution area coincides with areas where man has developed intensive agriculture, especially in areas located on the fluvio - lacustrine sediments of the Camacho river basin. Currently, there are only some groups of trees along streams or ledges where humidity accumulates. The tipa trees are the most conspicuous in the whole basin. These forests correspond to the series of *Jacaranda mimosifolia* and *Tipuana tipu*, described by Navarro [44].

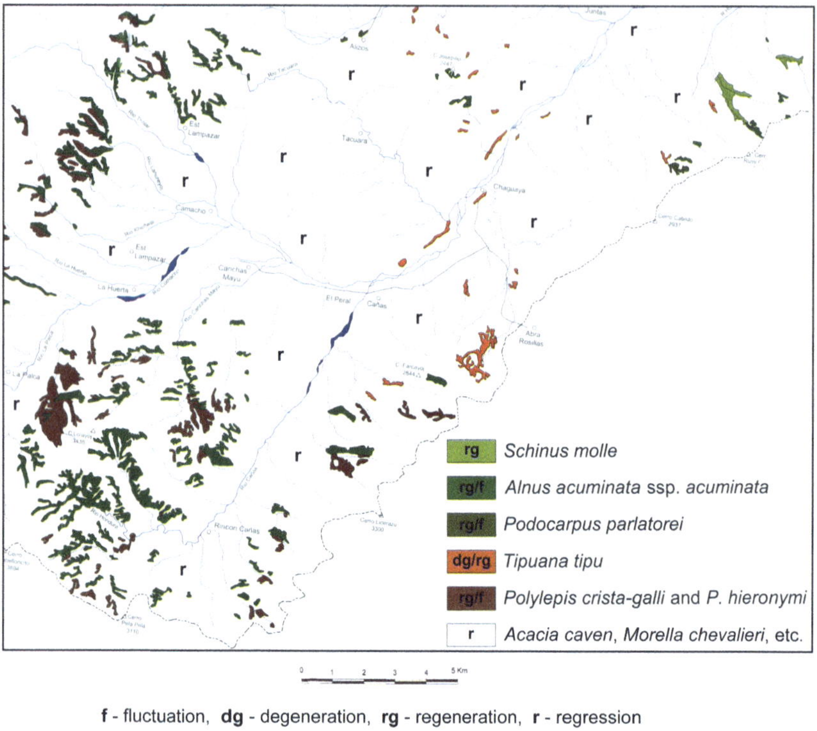

Fig. 14 Map of forests in a sector of the Río Camacho basin, with indication of the relative dynamic tendencies (*from* [30] *modified*)

Molle Forests (*Schinus molle*). These are present as very open formations with a variable coverage of 30–60%, in areas with much soil moisture, such as the alluvial plains, and the banks and terraces of rivers and streams.

Areas Without Forest Vegetation. Secondary formations of matorral and shrub thickets: red alder scrub (*Morella chevalieri*), chañar matorral (*Geoffroea decorticans*), tussock matorral (*Acacia aroma*), scrub (*Prosopis alpataco*), churqui matorral (*Acacia caven*), and others. The churqui scrub in the Camacho basin is the unit with the greatest geographic extent, colonizing various ecosystems due to its great capacity for adaptation.

General characteristics of the map (Fig. 14)
Size: 43 × 43 cm
Scale: 1: 50,000
Topographic map: map drawn in reduced form, with urban areas, towns, roads, mountain tops, streams.
Vegetation: forests, vegetation series.
Number of formations represented on the map: 5.

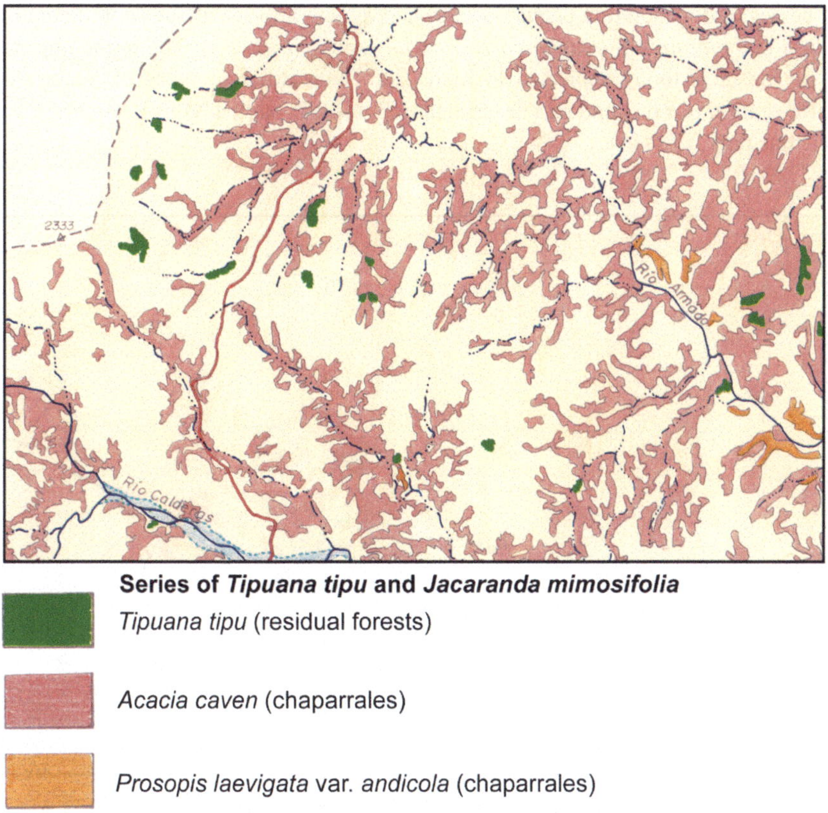

Fig. 15 Regression of the forests of *Tipuana tipu* and *Jacaranda mimosifolia* in the basin of the Río Camacho, Puesto Tunal. We observed an isolated plant of tipa (*Tipuana tipu*) on a soil column as well as several churqui (*Acacia caven*) plants: a - original level of the soil; b - intermediate level; c - level in the year of detection (1988). The arrow indicates the thickness of the eroded soil, about 3.50 m

Cartographic representation of the vegetation: cartographic units.
Vegetation limits indicated in black.
Type of map: partial map, showing only forests (Fig. 14); two maps represent vegetation series (Figs. 15 and 16).
Bibliography: Beck et al. [3–5]; Liberman Cruz et al. [30, 38].

Conservation Status of Vegetation
For the Camacho river basin, a partial map was surveyed, at a scale of 1: 50,000, on which only the tree formations listed above appear. The primary forest vegetation is now strongly fragmented and covers only 4.3% of the entire basin, corresponding to 45 km^2. Of this area, 2.9% is covered by *Tipuana tipu,* 12.4% by *Podocarpus parlatorei*, 16.9% by *Polylepis crista-galli* and *P. hieronymi*, 20% by *Schinus molle* and 45.8% by *Alnus acuminata* ssp. *acuminata*. These residual forests are in the

processes of degeneration (*Tipuana tipu*) and regeneration, with areas of fluctuation (*Alnus acuminata* ssp. *acuminata* and *Podocarpus parlatorei*). The remaining part of the basin is today subjected to regression, with different types of matorrales. Deforestation has resulted in a desertification process, originated by destruction and soil transport (Fig. 15).

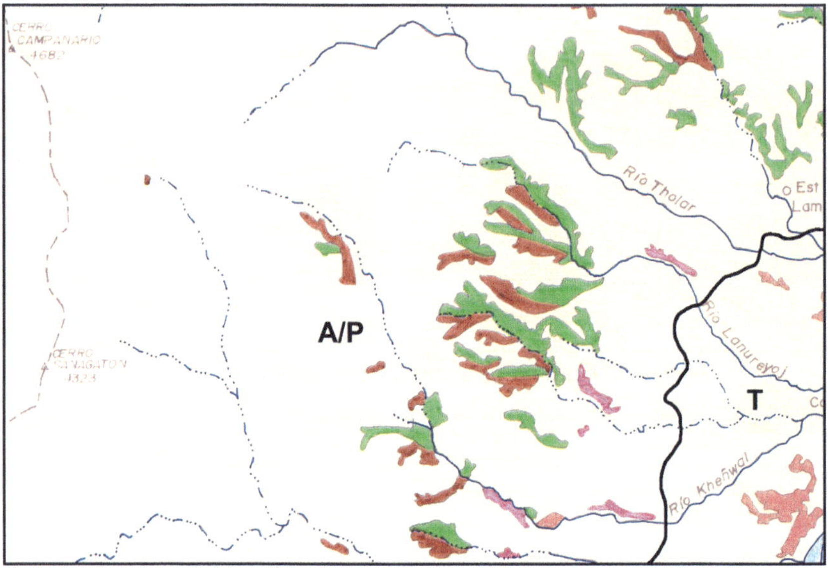

Series of *Alnus acuminata* ssp. *acuminata*

Alnus acuminata ssp. *acuminata* (residual forests)

Morella chevalieri (matorrales)

Series of *Polylepis crista-galli, P. hieronymi*

Polylepis crista-galli, P. hieronymi (residual forests)

Series of *Tipuana tipu* and *Jacaranda mimosifolia*

Acacia caven (chaparrales)

T - *Tipuana tipu* and *Jacaranda mimosifolia* series

A/P - *Alnus acuminata* series/*Polylepis* sp. series

Fig. 16 Río Camacho basin, map of the vegetation series; series of *Tipuana tipu* and *Jacaranda mimosifolia* (*survey of M. Liberman Cruz and F. Pedrotti*)

The *Tipuana tipu* forests correspond to the series of *Jacaranda mimosifolia* and *Tipuana tipu*, described by Navarro [44]. This series consists of the forest of *Tipuana tipu* and *Jacaranda mimosifolia* (head of the series) and *Acacia caven* matorral. Regression of the *Tipuana tipu* forests led to the formation of churqui (*Acacia caven*) scrub, or churquial, which tends to form irregular groups (Fig. 16) with an almost non-existent herbaceous layer.

Fig. 17 Río Camacho basin, map of the vegetation series; A/P series of *Alnus acuminata* ssp. *Acuminata* and *Polylepis crista-galli* - *P. hieronymi*; T series of *Tipuana tipu* and *Jacaranda mimosifolia* (*survey of M. Liberman Cruz and F. Pedrotti*)

The matorrales of churqui (*Acacia caven*) correspond to a successful formation in the substitution series of the original forests of tipa and jacaranda, which existed before the arrival of man in the central valley of Tarija. This is a physiognomic formation of secondary origin. Since churqui is one of the typical elements accompanying the Chaco formations, its presence and current distribution in the Camacho river basin obeys only anthropic influence. The churqui is a species of great plasticity, which has allowed it to colonize diverse environments, both arid and humid; to the above qualities one can add its high reproductive capacity and rapid growth. These shrubs are grouped and constitute the unit with the greatest geographic extent, colonizing various ecosystems due to their great adaptability. It grows from 1,600 m to 2,300 m above sea level. The vegetation that accompanies the churqui involves weeds and grasslands that, in one way or another, have been favored by agricultural activities, such as the genus *Paspalum*, a grass that always appears with important coverage.

Regression of the forests of *Alnus acuminata* ssp. *acuminata* leads to the matorrales of red alder - aliso rojo (*Morella chevalieri*). The red alder matorral corresponds to a stage in the substitution series of the forests of pino del cerro (*Podocarpus parlatorei*) and, to a greater extent, of white alder (*Alnus acuminata* ssp. *acuminata*) (Fig. 17). Red alder is found in the shrub layer of the podocarp forests on hills and white alder, with little significant coverage. When the forests are cut, devastated by uncontrolled overgrazing and fire, red alder develops secondarily, covering areas with remnants of the original forest. These are located on south-facing slopes, forming dense pure scrub or pure cases that can reach up to a height of 3 m; in other cases, they form small isolated shrub patches.

Red-alder matorrales are substitution formations of the former forests, occurring in areas where white alders previously existed and, in some cases, hill podocarps. The replacement is more evident near the lower distributional limits of the former forests. Altitudinally, these grow between 2,000 and 2,500 m. Red-alder matorrales occur in areas with south to southwest-facing slopes of 15° and 45°, with significant coverage ranging from 95 to 100%.

8 Discussion

The four vegetation mapping trials shown here have different characteristics. The maps of Capinota and Isla del Sol were made with complete topographic maps (contour lines, hydrographic network, urban centers, roads, toponymy, etc.), which undoubtedly facilitate reading and understanding. The maps of Nevado Sajama and Río Camacho have, on the other hand, a simplified topography, limited to the hydrographic network and some other elements, but nevertheless they are equally legible. In addition, the maps differ because they refer to different units of vegetation: Capinota and Isla del Sol to plant communities, Nevado Sajama to plant formations and Camacho river only to forests.

From these maps of the vegetation of the Andes of Bolivia, and from other personal experiences, it seems that a vegetation map constitutes the graphic response to a scientific problem placed *a priori* and related to the vegetation cover of a given territory (formations, associations, series, geoseries, etc.). Therefore, a map must have two fundamental requirements: appropriate scientific content, according to a determined geo-botanical conception, and adequate graphic representation, according to the prescriptions recommended by the general science of symbols, the graphic semiology [6, 7, 40], which deals with the formal aspects of the use of symbols in cartography.

The conservation status of vegetation (understood as phytocoenoses, series and geoseries) is very different in the four cases studied. In Capinota, Isla del Sol and Río Camacho, the forest communities were eliminated and only a few isolated fragments remain today: in Capinota, forests of *Schinopsis haenkeana* and *Aspidosperma quebracho-blanco*; on the Isla del Sol forests of *Polylepis incarum*; and in the Río Camacho forests of *Tipuana tipu*. In terms of the landscape (geoseries), the elimination of vegetation has triggered extensive erosion in the Río Camacho basin, while in Capinota and Isla del Sol, secondary matorral vegetation has developed.

On the Nevado Sajama volcano, despite a notable anthropogenic influence, both phytocoenoses and geoseries (including the vegetation series) have been maintained, although with some areas in regression due to overgrazing and fire.

The results obtained with the proposed methodology are qualitative: only for the Río Camacho basin was quantification relative to the residual forest present today possible, which is only 4.3% of the entire basin, corresponding to 45 km^2 out of 1,050 km^2. In addition, the percentage of the species making up these residual woods was also calculated.

Further evaluation of the conservation status of the vegetation cover could be carried out by applying to the units chosen here (phytocoenoses, series, geoseries) the indices proposed each time by the various authors mentioned above.

References

1. Asensi A, Díez-Garretas B (2010) Empleo de los hábitats naturales y seminaturales en la evaluación de las Sierras del Levante almeriense. Braun-Blanquetia 46:103–106
2. Asensi A, Díez-Garretas B, Comino O (2017) Cartografìa y evaluación de la vegetación actual en el Parque Natural Cabo de Gata-Nìjar (Almerìa, Espana). Collana Natura Aree Protette Camerino 32:62–78
3. Beck S, Liberman Cruz M, Pedrotti F, Venanzoni R (1990) Estudio de las formaciones leñosas de un valle interandino de Bolivia como base para la recuperación de áreas degradadas. Resumenes V Congreso Americano de Botanica (Palacio de las Convenciones, Ciudad de la Habana, 24–29 de junio de 1990). La Habana, p 122
4. Beck S, Liberman Cruz M, Pedrotti F, Venanzoni R (1992) Estado actual de los bosques en la cuenca del Río Camacho (Departamento de Tarija, Bolivia). Studi Geologici Camerti vol spec, pp 41–61

5. Beck S, Paniagua Zambrana N, Yevara GarÁte M, Liberman M (2001) La vegetación y uso de la tierra del altiplano de los valles en el oeste del Departamento de Tarija, Bolivia. In: Beck S, Paniagua N, Preston D (eds), Historia, ambiente y sociedad en Tarija, Bolivia. La Paz, Instituto de Ecología, UMSA – Leeds, School of Geography, University of Leeds, pp 47–93
6. Bertin J (1967) Sémiologie graphique: les diagrammes, les réseaux, les cartes. París Gauthier-Villars
7. Bertin J (1977) La graphique et le traitement graphique de l'information. París Flammarion
8. Bioret F, Estève R, Sturbois A (2009) Dictionnaire de la protection de la nature Rennes Presses Universitaires
9. Borza A, Boşcaiu N (1985) Introducere in studiul covoruli vegetal Bucarest, Acad Rep Soc Române
10. Carboni M, Carranza ML, Acosta A (2009) Assessing conservation status on coastal dunes: a multiscale approach. Landscape Urban Plan 91:17–25
11. Cárdenas M (1969) Disertaciones botánicas y amenidades biológicas. Cochabamba Los amigos del libro
12. Collot DD (1978–1980) Les Macrophytes de quelques lacs andins (Lacs Titicaca, Lac Poopó, Lac de la vallée d'Hichu Khala et d'Ovejhuyo). Paris ORSTOM
13. Collot DD (1982) Mapa de vegetación de la bahia de Puno. Ecología En Bolivia 2:49–63
14. Collot DD, Koriyama F, Garcia E (1983) Répartitions, biomasses et productions des macrophytes du Lac Titicaca. Rev Hydrobiol Trop 16(3):241–261
15. Corrado M (1884) El colegio franciscano de Tarija y sus misiones. Noticias historicas, Quaracchi (Florenze) tipografia del Colegio San Buena Ventura
16. Davis FW, Stine PA, Stoms DM (1994) Distribution and conservation status of coastal sage scrub in southwestern California. J Veg Sci 5:743–756
17. Del Arco Aguilar M-J, Gonzáles Gonzáles R, Garzón-Machado V, Pizarro-Hernández B (2010) Actual and potential natural vegetation on the Canary Islands and its conservation status. Biodivers Conserv 19:3089–3140
18. Domic AI, Palabral-Aguiklera N, Gomez MI, Hurtado R, Ortuno N, Liberman M (2017) *Polylepis incarum* (Rosaceae) una especie en peligro critico en Bolivia: propuesta de reclasificación en base al área de ocupación y estructura poblacional. Ecologia En Bolivía 52 (2):116–131
19. Faliński JB (1986) Vegetation dynamics in temperate lowland forests. Ecological studies in Bialowieza forest. Geobotany 8:1–537
20. Faliński JB, Pedrotti F (1990) The vegetation and dynamical tendencies in the vegetation of Bosco Quarto, Promontorio del Gargano, Italy (maps at scale 1:10,000 and a comment on the maps). Braun-Blanquetia 5:1–31
21. Fjeldså J, Kessler M (2004) Conservación de la biodiversidad de los bosques de Polylepis de las tierras altas de Bolivia. Santa Cruz de la Sierra Editorial FaN
22. Géhu JM (2006) Dictionnaire de sociologie et synécologie végétales. Stuttgart Cramer
23. Kessler M, Schmidt-Lebuhn AN (2006) Taxonomical and distributional notes on Polylepis (Rosaceae). Org Divers Evol 6, Electric Suppl 1:1–10
24. Ivan D (1979) Fitocenologie şi vegetaţia Republicii socialiste România. Bucarest ed didact pedagogica
25. Jewitt D (2011) Conservation target and status for vegetation types. Ecosystem ecologist Ezemvelo KZN wildlife, Cascades South Africa
26. Liberman Cruz M (1986) Microclima y distribución de Polylepis tarapacana en el Parque Nacional del Nevado Sajama, Bolivia. Doc Phytosoc X(Il):235–272
27. Cruz L (1987) Uso de la tierra en el altiplano norte de Bolivia como base por la evaluación del impacto ambiental de un proyecto de desarrollo rural. Riv Agricolt Subtrop Trop 81(1–2):207–235
28. Liberman Cruz M, Beck S, Pedrotti F, Venanzoni R (1986a) La vegetación de la Isla del Sol en el Lago Titicaca. Resumenes IV Congreso Latino-Americano de Botanica (Medellin, Colombia, 1986). Medellin, p 122

29. Liberman Cruz M, Gafta D, Pedrotti F (1997) Estructura de la población de Polylepis tarapacana en el Nevado Sajama, Bolivia. Desarrollo sostenible de ecosistemas de montaña: manejo de áreas frágiles en los Andes, pp 59–70. La Paz the United Nations University
30. Liberman Cruz M, Pedrotti F (2006) Woody formations in a mesothermic valley of Tarija Province, Bolivia. In: Gafta D, Akeroyd J (eds) Nature conservation. Concepts and practice, pp 75–86. Springer, Heidelberg
31. Liberman Cruz M, Pedrotti F, Gafta D (1995a) Los camelidos del Parque Nacional Sajama, Bolivia: distribución, pastos y problemas ambientales. In: 2nd European symposium on South American Camelids (September, 1995). Göttingen – Camerino, p 82
32. Liberman Cruz M, Pedrotti F, Molinillo M (1995b) Aspectos naturales y culturales del pastoralismo de alpacas en una region altoandina de Bolivia. In: 2nd European symposium on South American Camelids (September, 1995). Göttingen – Camerino, p 75
33. Liberman Cruz M, Pedrotti F, Venanzoni R (1986) La vegetación de la Isla del Sol en el lago Titicaca – Bolivia (Informe preliminar). Giorn Bot Ital. 120(suppl 2):159
34. Liberman Cruz M, Pedrotti F, Venanzoni R (1986c) Mapa de la vegetación de la Isla del Sol (Lago Titicaca - Bolivia). Camerino, Dipartimento Botanica Ecologia
35. Liberman Cruz M, Pedrotti F, Venanzoni R (1988) La vegetación de la Isla del Sol en el lago Titicaca – Bolivia. Ecologia En Bolivia 11:69–70
36. Liberman Cruz M, Pedrotti F, Venanzoni R (1991) La vegetación flotante (Clase Lemnetea) del Lago Uru-Uru, Depto. Oruro, Bolivia. Ecologia en Bolivia 17:47–51
37. Liberman Cruz M, Pedrotti F, Venanzoni R (1995) La Isla del Sol en el Lago Titicaca (Bolivia). Descripción de la vegetación. In: II international symposium on sustainable mountain development: managing fragile ecosystems in the Andes (Huarina - Bolivia), 2–12 April 1995, pp 71–78
38. Liberman Cruz M, Pedrotti F (2003) Deforestation and desertification in Camacho river basin (Bolivia). Abstracts 46th Symposium. I.A.V.S. (8–14 June 2003, Napoli, Italia). Camerino, Università degli Studi, p 143
39. Liberman Cruz M, Gafta D, Pedrotti F (1995) La estructura de la población de Polylepis tarapacana en el Parque Nacional del Nevado Sajama, Bolivia. In: II international symposium on sustainable mountain development: managing fragile ecosystems in the Andes (Huarina - Bolivia), 2–12 April 1995, pp 17–32
40. Martinelli M (2014) Mapas da geografia e cartografia temática. São Paulo Contexto
41. Martínez O, Villarte F (2009) Estructura dasométrica de las plantas de un parche de Polylepis incarum besseri y avifauna asociada en la Isla del Sol (Lago Titicaca, La Paz – Bolivia). Ecología En Bolivia 44(1):36–49
42. Miles L, Newton AC, DeFries RS, Ravilious C, May I, Blyth S, Kapos V, Gordon JE (2006) A global overview of the conservation status of tropical dry forests. J Biogeogr 33:491–505
43. Mingo M (1981) Historia de las Misiones Franciscanas de Tarija entre los Chiriguanos. Tarija Universidad Juan Misael Caracho
44. Navarro G (1997) Contribución a la clasificación ecológica y florística de los bosques de Bolivia. Rev Bol Ecol 2:3–37
45. Navarro G, Maldonado M (2006) Geografía ecológica de Bolivia. Cochabamba Centro Ecología Simon I Patiño
46. Navarro G, Molina J, De la Barra N (2005) Classification of the high-Andean Polylepis forests in Bolivia. Plant Ecol 176:113–130
47. Pavillard J (1935) Elements de sociologie végétale (Phytosociologie) Paris Hermann et C (new edition 2019 Paris Hermann et C by Bioret F and Pedrotti F)
48. Pedrotti F (2013) Plant and vegetation mapping. Springer, Heidelberg
49. Pedrotti F (2013) Vegetation map of the Isla del Sol. In: Plant and vegetation mapping, pp 194–198. Springer, Heidelberg
50. Pedrotti F, Venanzoni R (1986) Observaciones fitogeográficas sobre Puya raimondii en Bolivia. Doc Phytosoc X(II):183–288
51. Pedrotti F, Venanzoni R (1990) Ricerche geobotaniche in Bolivia. Inform Bot Ital 22(3):281–286

52. Pedrotti F, Liberman CM, Venanzoni R (1988) Le associazioni della classe Lemnetea del lago Titicaca (Bolivia). Riv Idrobiol 27(2–3):377–388
53. Pedrotti F, Venanzoni R, Suarez Tapia E (1988) Comunidades vegetales del Valle de Capinota (Cochabamba - Bolivia). Ecología En Bolivia 11:25–45
54. Pedrotti F, Suarez Tapia E, Venanzoni R, Crespo Callaù V (1986) Las comunidades vegetales de un valle mesotermico del interior de los Andes en el Departamento de Cochabamba - Bolivia. Giorn Bot Ital 120(suppl 2):162
55. Pinna MS, Gogoni D, Fenu G, Bacchetta G (2015) The conservation status and anthropogenic impacts assesments of mediterranean coastal dunes. Estuar Coast Shelf Sci 167:25–31
56. Quintanilla Aguirre J (1986) Sinopsis de las características fisico-químicas del lago Titicaca y Poopó (Bolivia). Riv Idrobiol 24(3):87–139
57. Ribeira MO, Liberman M (2006) Parque Nacional Sajama. In: El uso de la tierra y los recursos de la biodiversidad en la áreas protegidas de Bolivia. La Paz, Servicio Nacional Áreas Protegidas SERNaP – Proyecto Gef – ii Banco Mundial, pp 31–49
58. Rivas-Martínez S (1987) Memoria del mapa de series de vegetaciòn de España. ICONA Madrid
59. Slater FM, Curry P, Chadwell C (1987) A practical approach to the evaluation of the conservation status of vegetation in river corridors in Wales. Biol Cons 40:53–68
60. Squier EG (1974) Un viaje por tierras incaicas. Crónica de una expedición arqueológica (1863–1865). Cochabamba Los amigos del libro

Using Geobotanical Tools to Operationalize Mapping and Assessment of Ecosystem Services (MAES) in Southern Portugal

S. Mesquita, J. Capelo, I. Gama, C. Marta-Pedroso, M. Reis, and T. Domingos

Abstract As Action 5 of the EU Biodiversity Strategy for 2020 called on EU member states to map and assess ecosystems and evaluate their services (MAES) in national territories, a pilot-project was created to set up a straightforward method for survey and evaluation of ecosystem types in the south of Portugal. The project's primary aim was to produce ecosystem maps wherefrom spatial estimations of ecosystem services could be drawn, as well as biodiversity-related indicators of ecosystem condition. The resulting ecosystem typology and map were also used by others to estimate supporting, provisioning and regulating services.

M. Reis—Died 10 August 2017.

S. Mesquita (✉)
CEABN - InBio, Instituto Superior de Agronomia, Universidade de Lisboa, Tapada da Ajuda, 1349-017 Lisbon, Portugal

J. Capelo
Instituto Nacional de Investigação Agrária e Veterinária, Herbarium, Av. da República, Quinta do Marquês, 2784-505 Oeiras, Portugal
e-mail: jorge.capelo@iniav.pt

J. Capelo
LEAF Research Centre – Linking Landscape, Environment, Agriculture and Food, University of Lisbon, Tapada da Ajuda, 1349-017 Lisbon, Portugal

I. Gama
Terraprima – Serviços Ambientais, Quinta da França, Caria, 6200-710 Teixoso, Portugal
e-mail: ivo.gama@terraprima.p

C. Marta-Pedroso · T. Domingos
MARETEC – Marine, Environment and Technology Centre, Instituto Superior Técnico, Universidade de Lisboa, 1049-001 Lisbon, Portugal
e-mail: cristina.marta@tecnico.ulisboa.pt

T. Domingos
e-mail: tdomingos@tecnico.ulisboa.pt

M. Reis
CONEXA - Tecnologias e Sistemas de Informação, 2890-010 Alcochete, Portugal

F. Pedrotti and E. O. Box (eds.), *Tools for Landscape-Scale Geobotany and Conservation*, Geobotany Studies, https://doi.org/10.1007/978-3-030-74950-7_7

The proposed typology distinguishes 12 main ecosystem types based on the higher EUNIS Habitat Classification levels. EUNIS types were obtained by corresponding a *Vegetation Series Map* (Capelo et al. in Phytocoenologia 37:399–415, 2007) to land-cover/land-use datasets (COS'07—land cover map with 192 classes and MMU = 1 ha), and refined by using *elevation, geology* and *satellite imagery (NDWI) as ancillary information.* For most units, correspondence between seral stages in vegetation series and land use types allowed overall one-to-one identification of candidate groups of the corresponding *syntaxa* as EUNIS types. Auxiliary NDWI, geology and elevation allowed discriminating intrazonal special habitats such as mesic and dry grasslands. The ecological value of plant communities and phytodiversity were modelled as relevant contributions for the assessment of ecosystem condition.

Keywords Ecosystem services · Vegetation series · Mapping · Remote sensing · Portugal

1 Introduction

Action 5 of the EU Biodiversity Strategy for 2020 calls on the Member States to Map and Assess Ecosystems and their Services (MAES) in their national territories with the European Commission's assistance.

To address this call for mainland Portugal, a pilot-project was developed for mapping agricultural, agroforest and forest ecosystems and their services in the South of Portugal (Alentejo), using the best available information and resources to bring forth and test a method to accomplish the task. The project's primary aim was to produce an ecosystem map for the area wherefrom spatial estimations of ecosystem services could be drawn, as well as biodiversity-related indicators of ecosystem condition. The resulting ecosystem typology and map could also be used as a base for estimates of supporting, provisioning and regulating services.

Establishing a coherent ecosystem typology is of paramount importance to MAES. A necessary condition for MAES is that mapped ecosystem types should be unambiguous, *i.e.* clearly spatially circumscribed and expressing recurrent ecological identity as to *composition, structure* and *function*. This condition implies that ecosystem types may be recognizable as ecological entities *per se*, irrespective of changes in utilitarian approaches such as MAES. Surely, operational ecosystem attributes used for MAES should also be present or inferable from such types. In practice, a consensus seems to exist in the current literature on how to obtain ecosystem types: MAES may be primarily derived from land use [9, 31–33, 51]. According to these authors, a sufficient relationship between land-use and ecosystem types will be obtainable in most circumstances. As this assertion is debatable, Banko et al. [6] came to recommend 'thematic refinement' of land use types to approach more detailed and coherent ecosystem types. To resolve this relationship as both unique and informative, additional data must be integrated using ancillary datasets such as terrain variables, environmental regions, potential natural vegetation, satellite imagery, soil and geology [6]. Furthermore, UN et al. [51] came to rediscover the definition of 'biocoenosis' by Karl Möbius (1877) in the context of MAES. They so defined the 'land-cover/ecosystem functional unit' (LCEU) as a spatial unit, along with its ecological content, that is characterized by a unique combination of 'land-cover type, water resources, climate, altitude and soil type' (…). And they proceed: 'it should be possible consistently to differentiate an LCEU from neighbouring LCEUs based on differences in its ecosystem characteristics'. Apparently, these important calls were subsequently ignored, as the thematic refinement of land use data is seldom practised, and land-use types are straightforwardly taken as ecosystem types. Here we argue that, in most cases, land-cover and ecosystem types are not equivalent. Many of the available land-use classifications and maps are significantly too coarse to be meaningful as ecosystem types, leading to overgeneralizations and significant information loss in subsequent MAES assessments.

The second set of properties relates to ecosystem condition and expresses, in the MAES perspective, the real capacity of an ecosystem to produce services when compared to the maximum best qualitative condition possible it could be on. As an example, an ecologically mature forest can provide more and better biodiversity-related services than a degraded forest, due to disturbance and simplification of composition, structure and function in the latter case. UN et al. [51] propose that a set of key features, such as water, soil, vegetation, biodiversity, carbon, and nutrient flows, be evaluated in comparison to a reference condition. The reference condition relates either to typical, well-conserved, ecologically mature or optimal integration in landscape functions, sometimes including human activities such as traditional agricultural systems. In general, an assessment of the degree of human-induced disturbance, *i.e.* hemeroby, is taken as a surrogate of how deviant from the maximum integrity state the ecosystem is. Other features as functional stability or resilience could be considered, expressing the ecosystem's recovery capacity if human disturbance ceases, changes or reduces. Several proposals of how to find a subset of key ecological indicators to act as a surrogate of overall ecosystem condition exist. Bordt [9] recognizes that the indicators may be ecosystem-specific, and so the challenge is to find a set of key characteristics for each ecosystem type. Bordt [9] states that 'ideally, one would have a general measure of ecosystem condition that would capture the ongoing functioning of the ecosystem with respect to its capacity to generate services'. In executing MAES, few have attempted to go beyond land-use features. Most methodological proposals can be exemplified by Maes et al. [31] that propose the use of straightforward easy-to-measure indicators of ecosystem condition regarding vegetation: leaf-area index, biomass, annual net productivity (mean annual biomass increment), species richness or diversity for a given period—the accounting period—and in spatially well-delimited unit: the EAU—Ecosystem Accounting Unit [51]. Maes et al. [31] even advocate a more straightforward approach by solely assessing general physiognomic types of ecosystems: forest, lakes, grassland, crops, wetlands etc. This means not going beyond land-use features and not distinguishing between production agrosystems and natural and semi-natural ecosystems.

Conversely, as we find the latter approaches incapable of producing necessary and sufficient ecosystem types and assessing their condition, we emphasize that both are scientific problems for Ecology, namely for Vegetation Science [27, 48]. A wealth of regional or global (European and even worldwide) ecosystem typologies have been made available, namely the UICN ecosystem classification scheme [28], the EUNIS habitat classification [20, 27], the floristic vegetation classification of Europe (EUROVEG) of Mucina et al. (2016) and the crosswalk between the latter and EUNIS habitats by Schaminée et al. [48]. The assessment of ecosystem condition, namely expressing the degree of human disturbance, may follow the geobotanical concept of *Natural Potential Vegetation* (NPV) [50] and the inherent concept of 'vegetation series' [12, 43]. A vegetation series is the set of stages, including potential (NPV) and subseral, that may be found along a vegetation succession occurring on a homogeneous biotope. For details on concept operationalization in ecosystem condition assessment, please see Loidi [30] and our own proposal in 'Materials and Methods'.

It should be noted that a habitat survey is often not available or even feasible, so the thematic refinement of land-use types proposed by Banko et al. [6] will be the only least minimum standard to establish an ecosystem typology for MAES. In the latter, land use is used mostly as baseline spatial information for physiognomical types.Subsequently, many other ecological information levels are specifically associated: VNP, soil, terrain, satellite and floristic data, etc. Also, filtering and reclassification of land use may be issued by using supplementary information.

And then, what indicators should be looked at as good proxies for ecosystem services associated with biodiversity? As vegetation is ecologically dominant (more than 90% of biomass) in most terrestrial ecosystems and relatively fixed in space, the identification of habitat types resorts to vegetation features as a core proxy indicator. Plant communities, within their phytosociological framework, have been used since the early 1980s as indicators of ecosystem value in land use planning and to support the definition of nature conservation policies [3, 5, 23]. Plant communities reflect the environmental variables of the ecosystems: they integrate and summarize them in their floristic diversity, as well as the ecological relationships between organisms in the ecosystem. In the context of MAES, it should be noted that regulation & maintenance ecosystem services in the CICES classification [8] depend ultimately on biodiversity and attempting to estimate it through plant richness may be a sensible approach.

Phytodiversity comprises the plant component of an ecosystem's biodiversity. It is usually addressed using diversity indexes such as the Shannon Index, as suggested in Maes et al. [31], or the Simpson Index [21]. However, the use of such biodiversity or phytodiversity indicators is only possible when detailed information is available for the mapped ecosystem units, which is rarely the case. Moreover, these indices do not discriminate native species, exotic species, invasive species, rare species, common species, etc., which results in an overvaluation of disturbed ecosystems and an undervaluation of less diverse plant communities but where endemic or rare species of high conservation value occur. Furthermore, we propose a method of processing land-use data through several levels of floristic and vegetation information yielding an estimation of plant richness issuing from phytosociological plot databases (see Materials and Methods).

In this work, we propose the assumption of the ecosystem's plant component as a proxy for its total diversity since it reflects the entire ecosystem's complexity. Furthermore, accounting for plant communities' successional position in vegetation series reflecting their successional maturity and phytodiversity, one should obtain good proxies for ecosystem condition. We believe that the phytosociological approach to plant communities, with all its accumulated body of knowledge, is a solid methodological basis for solving ecosystem mapping and assessment of condition.

2 Materials and Methods

2.1 Study Area

The study area for the pilot-project is the NUTS II region Alentejo, one of the Territorial Units for Statistics developed by Eurostat and employed in Portugal for statistical purposes; it covers around 31 500 km^2, about one-third of mainland Portugal (Fig. 1).

2.2 Map Sources

Information available for the area encompassed maps and non-spatial information. Cartographic data relevant for the task was mainly about land cover and abiotic factors influencing vegetation development:

- Land cover map (COS'07, with 192 classes and MMU = 1 ha) [24];
- Vegetation Series map [15],
- Bioclimatic maps [34],
- Geological map [38],
- Distance to coastline, derived from the Official Portuguese Administrative Map[1] (CAOP) (Aviso n.º 2298/2018, de 7 de março de 2018);
- NDWI (Normalized Difference Water Index), derived from satellite imagery Landsat-8 R_{NIR} and R_{SWIR} [26].

2.3 Geobotanical and Phytosociological Data Sources

- Base sources for the geobotanical model for mainland Portugal:

[1]https://www.dgterritorio.pt/cartografia_e_geodesia/cartografia/carta_administrativa_oficial_de_portugal_caop_/.

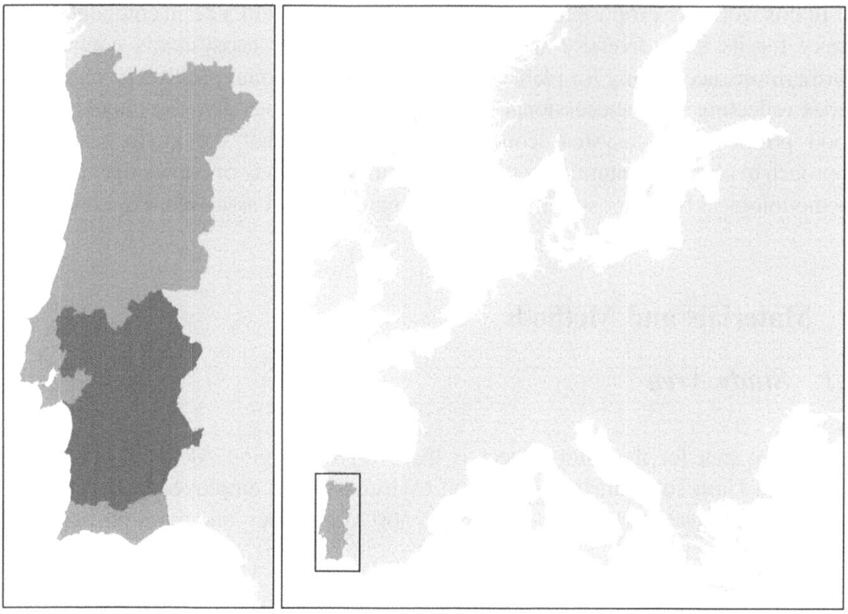

Fig. 1 Study area: NUTS II Alentejo, in Mainland Portugal

- Biogeographic model [18],
- Bioclimatic and geobotanical model of the Iberian Peninsula [41, 43, 46, 47],
- Syntaxa checklist [17],
- Vegetation Series model [15].

• Phytosociological relevés[2] on-line databases:

- SIVIM (https://www.sivim.info/sivi/);
- European Vegetation Archive (EVA; https://euroveg.org/eva-database).

• Other relevant bibliography:

- Regional geobotanical models [1, 10, 11, 13, 14, 16, 17, 36, 37, 39, 40].

2.4 *Ecosystem Mapping*

The proposed typology of ecosystems distinguishes 12 main ecosystem types based on the higher levels of the EUNIS Habitat Classification,[3] as recommended by most

[2]Lists of taxa present in a sample plot, each affected by an abundance value.

[3]https://eunis.eea.europa.eu/.

authors [48] for mapping ecosystems and their services in European countries. EUNIS is a comprehensive Pan-European typology, including natural and artificial habitats and common vegetation. It has a hierarchical structure, allowing the use of the level which is more adequate to the detail of available information within each main type. The EUNIS typology encompasses 10 habitat classes at the first hierarchical level (plus one class of mixed habitats), subdivided into as many subclasses as necessary, sometimes up to the eighth level. Moreover, it is being revised, and crosswalks have been developed between level 3 EUNIS terrestrial habitat types and vegetation *syntaxa*, which are well established for most of the Portuguese native vegetation [17].

The adopted methodology was implemented in two separate steps, generally following the approach proposed in Banko et al. [6]: first, a crosswalk between EUNIS and COS'07 land cover was performed, second, thematic refinement was carried out in a top-down approach. Since ecosystem maps were produced at a large scale, specific problems arose due to increased thematic and spatial detail, so new strategies and solutions were developed.

2.4.1 EUNIS – COS'07 Crosswalk

A first attempt to establish a formal relation between the COS'2007 land-use types and UNIS habitat types of level 2 or higher was made to detect direct 1:1 correspondence. This was successfully performed for 103 of the 142 agricultural, agroforest and forest land-use types, covering 75.3% of the study area (see Table 1, in Annexe I).

Although the main goal was to map ecosystems down to EUNIS level 2, it was considered advantageous to descend in the hierarchy as much as possible within the scope of each COS'07 class. Thanks to this conservative option in which available information is used in all its detail, it was possible to reach levels 3 or 4 in most cases, and sometimes even level 5. It is always possible to quickly revert any EUNIS class to level 2 of this hierarchy, if necessary. This detail was only possible due to the high detail of the COS'07 land use map, produced with a reference scale of 1:25,000, based on the visual interpretation of orthorectified aerial images from July to October of 2007, with the help of diverse ancillary information; it has a positional accuracy greater than or equal to 5.5 m and an overall thematic accuracy of 85.13% with a 2.00% error for a 95% confidence level.[4]

2.4.2 Thematic Refinement

The remaining 24.7% of the area corresponds to land use classes having a 1:n relation to EUNIS habitat types. These situations were solved using ancillary

---

[4]https://www.dgterritorio.pt/cartografia_e_geodesia/cartografia/cartografia_tematica.

spatially explicit datasets, which provided the information needed for addressing four distinct ecological issues.

2.4.2.1. Differentiation According to Natural Potential Vegetation
Natural potential vegetation, mapped as vegetation series [15], was used to separate different types of native heathland, scrub and woodland.

The assumed paradigm for natural potential vegetation was that of Rivas-Martinez [42] and Rivas-Martinez et al. [45], which reasons a univocal correspondence between a uniform combination of bioclimate, lithology and biogeographical context and a set of plant communities that occur successively and organize themselves in mosaics throughout the process of ecological succession.

NUTS II Alentejo vegetation series are listed in the Table 2 (Annexe II). Knowing a place's vegetation series and the vegetal cover physiognomic type that corresponds to its land-use at a particular moment, it is possible to determine the dominant plant community. Thus, for areas of undetermined native heathland, scrub and woodland, crossing current land use information with spatial explicit vegetation series data, the dominant plant community in each site can be inferred. The crosswalk between EUNIS types and COS'07 land use/Natural potential vegetation types for native heathland, scrub and woodland is presented in Table 3 (Annex II).

The EUNIS typology is currently under review by the European Vegetation Survey work group within the International Association for Vegetation Science (IAVS, see https://euroveg.org/). Since native Mediterranean scrub, mostly included in EUNIS class F, is still poorly discriminated in this typology (see [49], the most updated document at the time available from this review was used as basis for a possible crosswalk between the phytosociological typology for shrub communities and EUNIS types [27], detailed information for each type is available at https://forum.eionet.europa.eu/european-red-list-habitats/library/terrestrial-habitats/).

This task proved to be more complex than initially thought, requiring computational capacities beyond standard operator software packages. Therefore, specific programming procedures were implemented for dealing with the great amount of geographical data and topological complexity of operations for combining geographic information layers.

2.4.2.2. Differentiation According to Geology
Maritime pine and stone pine forests correspond to EUNIS level 2 class 'G3: Coniferous woodland'. However, the distinction between disclimactic pinewoods of high natural value (G3.714, G3.7312, G4.B) and productive planted pine forests (G3.F22, G4.F) is important, since these are distinct ecosystems, providing different ecosystem services.

The criteria for separating these two groups were created by Aguiar and Capelo [2]: old undisturbed pinewoods occurring in dune systems, despite their possible anthropic origin, mimic the floristic combination of the pinewoods from the Atlantic period and can therefore be taken as native forests.

For the specific purpose of mapping EUNIS types, only geological substrate was considered (as mapped in [38]), since there was no available information concerning stands' age and undercover management. Therefore, all pine forests on dunes and coastal paleodunes were considered natural forests (see Table 3 in Annexe III).

2.4.2.3. Differentiation According to Distance to the Ocean

EUNIS classification separates coastal ecosystems from all others as high as the first level. However, the available land-use maps distinguish such ecosystems only in specific conditions, such as beaches, inter-tidal zones and marshes.

Thus, an efficient methodology was conceived to separate the coastal cliffs and sparsely vegetated dunes from interior rocky outcrops and other sparsely vegetated areas. After a systematic visual inspection of the coastline, it was decided to assume as coastal all areas located less than 100 m from the coastline (see Table 5, in Annexe IV). Distance to the coast-line was obtained and computed directly from the Official Portuguese Administrative Map (CAOP), which records the state official demarcation of Portugal's administrative districts and borders.

2.4.2.4. Differentiation According to NDWI

The EUNIS classification separates natural herbaceous communities at level 2 in dry, mesic, wet and alpine grasslands, as well as woodland fringes and clearings, inland salt steppes and sparsely wooded grasslands. Except for dehesas, delimited separately in great detail in land use maps due to their economic importance in Portugal, all grasslands were mapped as permanent pasture or natural herbaceous vegetation (see Table 6, Annexe V).

Alpine grasslands and inland salt steppes are absent from the study area and woodland fringes and clearings are too small to be represented in land use maps, so grasslands included dry, mesic and wet subtypes (E1, E2 and E3), which should be separated resorting to criteria used in the definition of these classes, such as trophy degree, intra-annual variation of soil moisture and productivity (see Table 7, Annexe V).

The distinction between mesic and dry grasslands would require fieldwork, since these typologies can only be separated knowing the floristic composition of plant communities. Likewise, agriculturally improved and reseeded pastures can only be distinguished from other rainfed meadows conducting fieldwork or surveys to farmers. So, it was not possible to map E2 meadows, whose areas are included in the areas classified as E1.

E3 meadows in the study area are water-logged meadows and rush communities, which occur in alluvial areas with high soil water levels throughout the Summer. Satellite imagery obtained in the Summer was used to acquire information about the water content of vegetation during the dry season and, therefore, to map those meadows which retain high levels of water during the Summer. All areas of natural herbaceous vegetation located near water lines and with high water content were selected as E3 grasslands.

The Normalized Difference Water Index (NDWI) was used to evaluate liquid water content of vegetation, since this is a particularly effective index, less sensitive to atmospheric effects than other vegetation indexes [22]. It is calculated as $(R_{NIR} - R_{SWIR})/(R_{NIR} + R_{SWIR})$, where R_{NIR} represents the reflectance in a near-infrared band and R_{SWIR} is the reflectance in a middle-infrared band. Since Landsat-8 imagery was used, R_{NIR} and R_{SWIR} correspond, respectively, to bands 5 (0.851–0.879 μm) and 6 (1.566–1.651 μm) [26].

Cut-off values of NDWI were empirically set for separating wet meadows from dry meadows, using representative samples for each type previously known to the team. All meadow polygons with an NDWI value greater than the estimated cut value in most of its area were classified as E3 grasslands. A single cut-off value was considered unsatisfactory for the whole territory. Bioclimatic variables [34] were tested to separate the study area into two zones, corresponding to distinct areas regarding the amount of water available to plants in the Summer months. For this separation, Summer Compensated Ombrothermic Index (Ios_3) maps were used. This index is part of the "Worldwide Bioclimatic Classification System" and was created by Rivas-Martínez [44] to account for the Summer dryness in Mediterranean areas such as Alentejo, integrating information on Summer precipitation and temperature.[5]

Satellite images from 2013 were used because older images were of poor quality, so changes in land use can be a source of error, namely from grassland to scrub. Nevertheless, the results thus obtained were satisfactory.

2.5 Ecological Value of Plant Communities

Plant communities are highly visible and measurable and reflect the ecosystem's environmental variables, therefore can be assumed as reliable proxies for their diversity and condition [7, 29].

The ecological value of plant communities present in each type of ecosystem was assessed using a methodology adapted from Loidi [30]. The original method was developed in the context of detailed surveys, assuming the availability of detailed floristic information about plant communities occurring in all evaluated plots. Additionally, the method uses information about the general characteristics of the phytosociological association to which each plant community can be ascertained, as well as the region where it is located. It was later adapted to apply to less

[5] $Ios_3 = (Pp_3/Tp_3)$, where Tp_3 is the sum of the monthly average temperature of June, July and August (in tenths of degrees Celsius) and Pp_3 is the total rainfall of June, July and August (in mm).

detailed studies, particularly when field surveys are not available, as is the present case (*e.g.* [4, 35]).

'Ecological value' was achieved as the arithmetic mean of five assessed parameters, all ranging from 1 to 5: *naturalness, replaceability, threat, rarity* and *hemeroby* [4, 25, 30].

Naturalness: Expresses the intrinsic degree of human influence, reflecting, on the one hand, floristic impoverishment due to disturbance and, on the other hand, the degree to which the plant community depends on active management. The scale used for evaluating this parameter is as follows:

1. Maximum disturbance: exotic and cultivated vegetation, agricultural areas, agriculturally improved and reseeded pastures, productive forests.
2. Dehesas, natural pastureland.
3. Sparse scrub and brush (not in coastal dunes).
4. Coastal dune sparse scrub, heaths and meadows.
5. Minimal disturbance: dense native woodland, dense thickets and bushes, bare rock.

Replaceability: Capability of a vegetation type to recover itself after destruction by an external agent. It is very low in climacic communities, usually woodlands, and increases along the vegetation types' ecological succession sequence. An inverse scale of values was used to attribute a higher value to less resilient communities, as follow:

1. Exotic vegetation, agricultural land, plantations.
2. Pioneering communities: natural herbaceous vegetation, brambles.
3. Scrub, forests with exotic species, pastures.
4. Thickets, dehesas, chestnut woodland, coastal dune scrub and heaths, bare rock.
5. Climacic vegetation: native woodlands.

Threat: Reflects the community's attractiveness or its biotope for destructive human activities and is influenced by ease of access, presence of species of economic value, population density, etc.; it also reflects the threat posed by ceasing of active management in low-intensity agroforestry habitats. This criterion is highly dependable on regional specificities, particularly socio-economic context, which varies greatly throughout history. The assessment scale used was as follows:

1. Communities in biotopes of difficult to access, with no apparent economic gains and low attractiveness, therefore less threatened.
2. Brambles, thickets, pastures with permanent crops.
3. Vegetation types highly dependent on active management: dehesas, natural pastureland.
4. Vegetation types which include plants traditionally harvested for commercial exploitation.

5. Plant communities in biotopes of high economic value or high attractiveness and easy access: beaches, coastal areas[6] and areas close to urban centers.[7]

Rarity: This parameter was assessed using the typology established in Izco (1998), which integrates three distinct attributes: the size of the community's distribution area, average community size, and community occurrence frequence within its distribution area. Different combinations of the three types of rarity defined by these attributes result in seven distinct rarity types, summarized in Table 1 below.

For the pilot-project study area, geographical rarity was considered when the potential area of the community is less than 1.5% of mainland Portugal (about 130,000 hectares); small size when the average size of polygons is less than 100 ha; and low frequency of occurrence when the typology is present in the study area in less than 100 polygons.

These rarity types were translated on a scale of 1 to 5 as follows:

Without vegetation or with exotic, cultivated or cosmopolitan types of vegetation.

1. R1, R2.
2. R3.
3. R4, R5.
4. R6, R7.

Table 1 Rarity types, according to Izco (1998)

	Small geographical range	Low frequency of occurrence	Small size of plant communities
Not rare			
R1		✓	
R2			✓
R3		✓	✓
R4	✓		
R5	✓	✓	
R6	✓		✓
R7	✓	✓	✓

[6]Areas located at less than 2 km from coastline.

[7]Areas located within a circle around urban centers with a radius equal to 2/3 of their population (in meters), from the central point of its center, for all urban areas with more than 10 000 inhabitants.

Hemeroby: A measure of the distance that separates a vegetation type from its undisturbed state, namely concerning the proportion of synanthropic vegetation. The sub-parameters used for estimating this indicator were defined for each group of ecosystems in the territory according to the optimal characteristics of each, resulting in the following scale:

1. Maximum disturbance within the community, with exotic and cultivated vegetation: crops, dehesa with irrigated or permanent crops.
2. Open forests with exotic vegetation, sparse scrub, sparse herbaceous vegetation, pastures with permanent crops, bare rock.
3. Dense forests with exotic species, dense scrub, herbaceous vegetation, dehesa with rainfed crops, beaches, coastal dune sparse scrub.
4. Open forests, rush communities, chestnut woods.
5. Little or undisturbed vegetation types: dense native woodlands with no exotic species, dense scrub and thickets, dehesa with natural meadows.

The ecological value of plant communities for each mapped unit is the average of these five parameters, rounded to the nearest integer value. This semi-quantitative valuation should have the following interpretation (Table 2):

2.6 *Phytodiversity*

This indicator aims tomeasure the value of plant diversity occurring in a community using available information and avoid the disadvantages of diversity indexes commonly used. The proposed methodology assumes that EUNIS vegetation types can be further detailed using information from Vegetation Series maps; then, knowing the plant community which occurs at each location, it is possible to estimate its average or characteristic floristic composition by consulting a phytosociological table of relevés of similar communities, which reflects its floristic richness and originality, as well as the presence of endemic or threatened plant species.

The adopted approach consisted primarily of establishing a correspondence between each type of EUNIS ecosystem and the corresponding phytosociological unit, using the information provided by a Potential Natural Vegetation map [15]. For each vegetation series, correspondence between EUNIS vegetation types and the most common successional stages (woodlands, thickets, scrub, perennial

Table 2 Interpretation of ecological value

Ecological value of plant communities	Range	Interpretation
1	[1, 1.5[	Very low ecological value
2	[1.5, 2.5[	Low ecological value
3	[2.5, 3.5[	Average ecological value
4	[3.5, 4.5[	High ecological value
5	[4.5, 5]	Very high ecological value

meadows) was established using phytosociological criteria and expert knowledge, including some types of azonal vegetation, such as riparian, chasmophytic and ruderal.

Next, an evaluation of the phytodiversity for each phytosociological unit was conducted, using its floristic composition from available bibliography and information about the distribution of flora species produced in the framework of the report for 2013 under Article 17 of the Habitats Directive, on the conservation status of Natura 2000 Network in Portugal. Around 3,500 phytosociological relevés were analysed. For calculating phytodiversity in each plant community, estimates were made of mean values of: i) protected plant species; ii) other endemic species; iii) other rare plant species; iv) characteristic species. The final value of phytodiversity is a weighted average of these parameters using, respectively, values of 4, 3, 2 and 1. Estimations were carried out as follows:

1. For each phytosociological unit, a significant set of phytosociological relevés was compiled from phytosociological literature, online databases, and the authors' archives, including, whenever possible, the protologue (original table) of the association.
2. Within the set of phytosociological relevés obtained for each vegetation unit, all taxa legally protected in Portugal (including those in Annexes II and IV of Directive 92/43/EEC) were counted. This number corresponds not to protected taxa potentially occurring in the habitat type, but to those that were effectively observed in the plant communities floristically described in the relevés. The total number of taxa endemic to Portugal but not legally protected was also obtained, and the total number of rare species (not endemic or legally protected).
3. Finally, the average number of characteristic species was estimated. These correspond to a statistically reliable combination of diagnostic species or bioindicators that uniquely identify the plant community (type of vegetation) and its environmental conditions. For this purpose, companion species in the relevés were discarded, since they are interpreted as corresponding to a set of plants occasionally present, with an ecological optimum in other types of vegetation, in adjacent biotopes or mosaics (either due to catenal or successional phenomena). The average number of characteristic species for a vegetation type representats its average floristic richness and, indirectly, of its ecological integrity.

Whenever one particular stage of a vegetation series corresponds not to a single community but several territorial variants, the total set of relevés of these communities was considered. In such cases, the maximum value was considered for parameters i), ii) and iii) and its arithmetic mean in the case of parameter iv).

For major azonal vegetation types, analogous procedures were implemented. However, as these vegetation types correspond to permanent complexes along topographically explicit environmental gradients without serial dynamics, the number of characteristic species considered was the sum of characteristics of all plant associations present. Likewise, parameters i) to iii) were calculated as the maximum values of all plant associations involved.

Table 3 Interpretation of phytodiversity value

Phytodiversity value of plant communities	Range	Interpretation
0	0	No value
1	]0, 5,5]	Very low value
2	]5,5, 16]	Low value
3	]16, 26]	Average value
4	]26, 49]	High value
5	]49, 111]	Very high value

For semi-natural habitat types resulting from anthropic disturbance, pioneer vegetation types and degraded or sparse vegetation, relevés corresponding to the pioneer vegetation were used.

Phytodiversity value was calculated as the weighted arithmetic mean of these parameters, assigning different weights to each one: 1 for the total number of characteristic species, 4 for protected species, 2 for other endemic species and 3 for other rare plants. In agricultural and forestry ecosystems, where phytodiversity is unknown but generally expected low, a value corresponding to 5% of the parameter's range was assumed (5.5). Finally, the value 0 was assigned to areas without vegetation.

Phytodiversity for each mapped unit is the weighted average of these five parameters, rounded to the nearest integer value. This semi-quantitative valuation obtained should have the following interpretation (Table 3):

Since the dominant vegetation is zonal in most of the territory, and accounting for the major types of azonal vegetation, estimates of phytodiversity produced have a high degree of reliability and translate, in relative terms, the floristic and phytochoenotic importance of mapped EUNIS habitats.

3 Results

Ecosystem Map

An Ecosystems map of Alentejo was produced, with great detail, both thematic (EUNIS types mapped at levels 2 to 6) and spatially (scale 1:25,000), as presented in Fig. 2.

This map reproduces the thematic heterogeneity of land-use types in COS'07, with great detail in forest and agricultural typologies contrasting with very low detail in herbaceous and scrub vegetation types.

All 68 forest types are directly attributable to level 2 EUNIS types, but greater discrimination was developed within these types, using ancillary information. 19 EUNIS classes were obtained, from levels 2 to 6, most at levels 3 and 4.

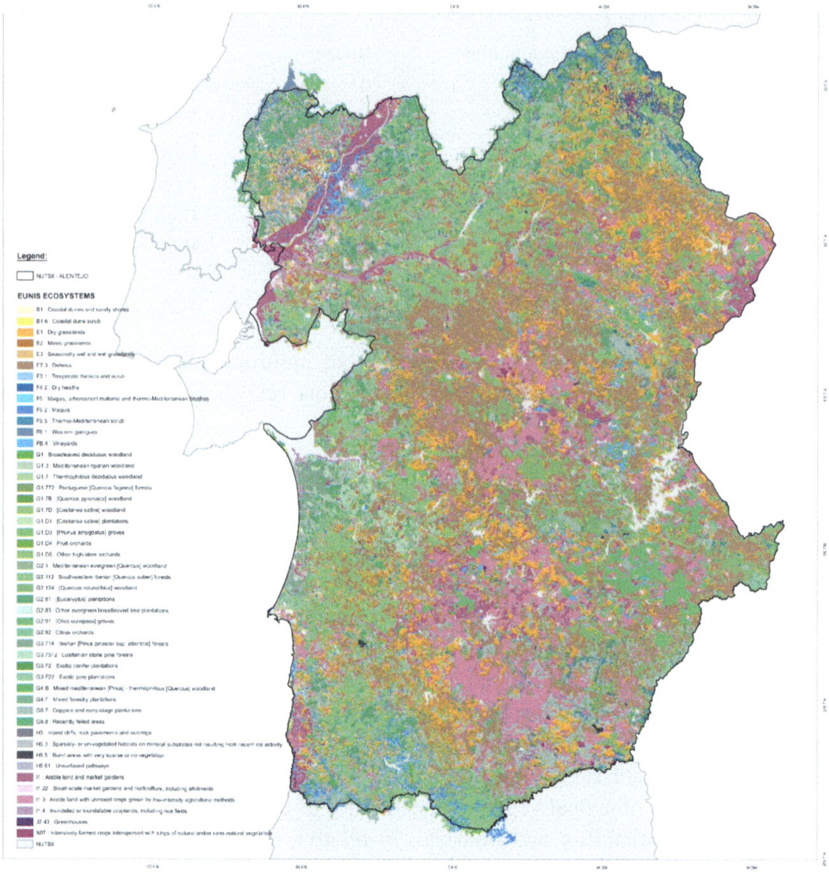

Fig. 2 Ecosystem map

Likewise, 64 agricultural land-use types were crosswalked to 14 EUNIS ecosystem types at levels 2 to 4. Correspondence between grasslands in land use map and EUNIS typology at level 2 could only be achieved using ancillary data.

Finally, a correspondence between the remaining 10 land-use types—encompassing scrub, sclerophyle and herbaceous vegetation, as well as sparsely vegetated surfaces and bare rock—and EUNIS ecosystem types was achieved using ancillary information and 13 classes were obtained, at levels 2 and 3.

Thematic detail in the produced Ecosystems map is very satisfactory, with spatial detail appropriate to the 1:25,000 scale. This map is adequate to support further analysis, since mapped units are complex entities containing a high amount of information, helpful in evaluating ecosystem services and for landscape planning and management. Thematic detail in some ecosystem types is of paramount importance for assessing ecosystem services related to biodiversity. Since EUNIS is a hierarchical classification system, working at less detailed levels is always possible and easy to perform.

The production of this map in a relatively short time was possible due to the accomplished methodological robustness and the good operational integration of ecological, GIS and computational expertise. This map was produced using an explicit, replicable methodology, allowing updates, at any time, if in the presence of updated or additional georeferenced information and the necessary processing capacity. This results in a significant ability to update information and correct it if new data is available, incorporating minor methodological rectifications if necessary.

Ecological Value of Plant Communities

A map of the ecological value of plant communities was produced for the study area, using the described methodology (Fig. 3). The chart in Fig. 4 presents the contribution of each type of ecosystem to the ecological value map, considering EUNIS classification at level 2. Some patterns stand out in this analysis:

i. High ecological value corresponds to protected and successionally evolved vegetation stands in dunes, paleodunes and coastal cliffs; and relict woods, of pristine character.

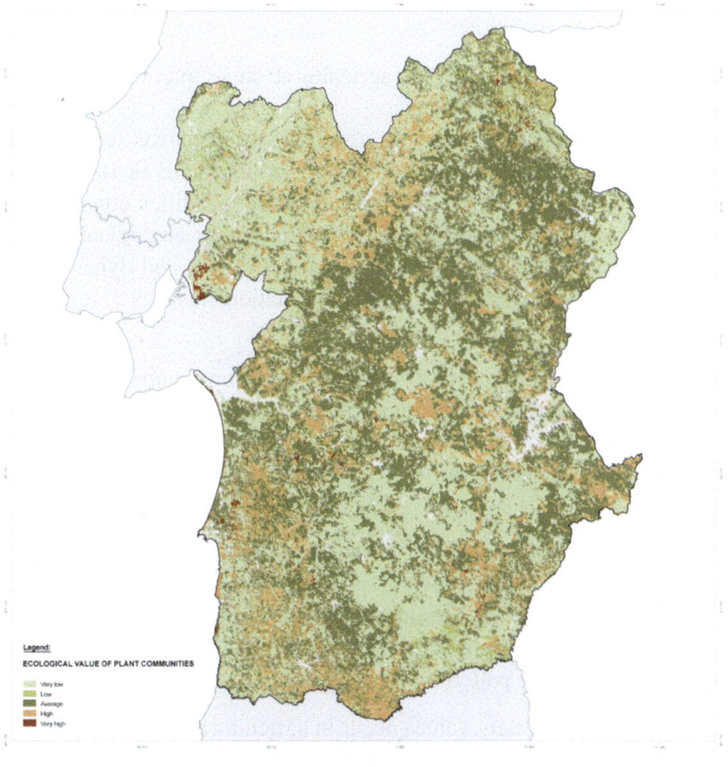

Fig. 3 Map of ecological value of plant communities

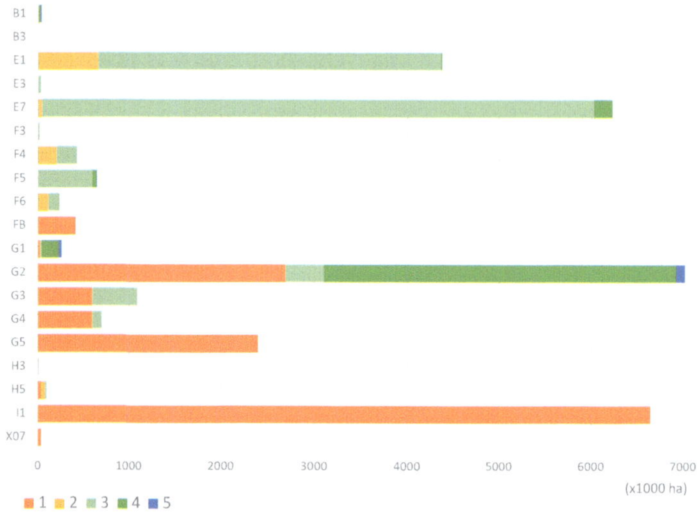

Fig. 4 Areas of ecosystem types at level 2 in the study area and their ecological value (thousands of ha)

ii. Low ecological values occur in agricultural areas, particularly in Southern Alentejo.
iii. Intermediate ecological value is associated with areas of dehesa, with low-intensity agricultural practices, usually in poor soils or steep slopes.
iv. Areas with cork oak woods as climacic vegetation differ considerably from those with holm oak climacic woods, having higher ecological value due to occupying a richer biotope—with higher precipitation and dynamical relief—facilitating the development of richer vegetation mosaics (*i.e.*, with higher beta-diversity).

'Ecological value' integrates information on plant community structural quality, phytocoenotic integrity and successional maturity. Geobotanic models reflect the essence of this variability and the variations in functional diversity with an impact on the community's capacity to provide ecosystem services.

Phytodiversity

Applying the described methodology to the study area resulted in a map of phytodiversity (Fig. 5). Total variability of phytodiversity is presented in 5 classes, defined as ranges of increasing amplitude. Figure 6 presents a chart showing, for each level 2 EUNIS ecosystem, the areas corresponding to each phytodiversity class. The most interesting patterns in this analysis are:

i. The lowest phytodiversity values occur in agricultural ecosystems on siliceous soils with annual crops, even if integrated with dehesas. The same agricultural systems on limestone soils tend to have higher phytodiversity.

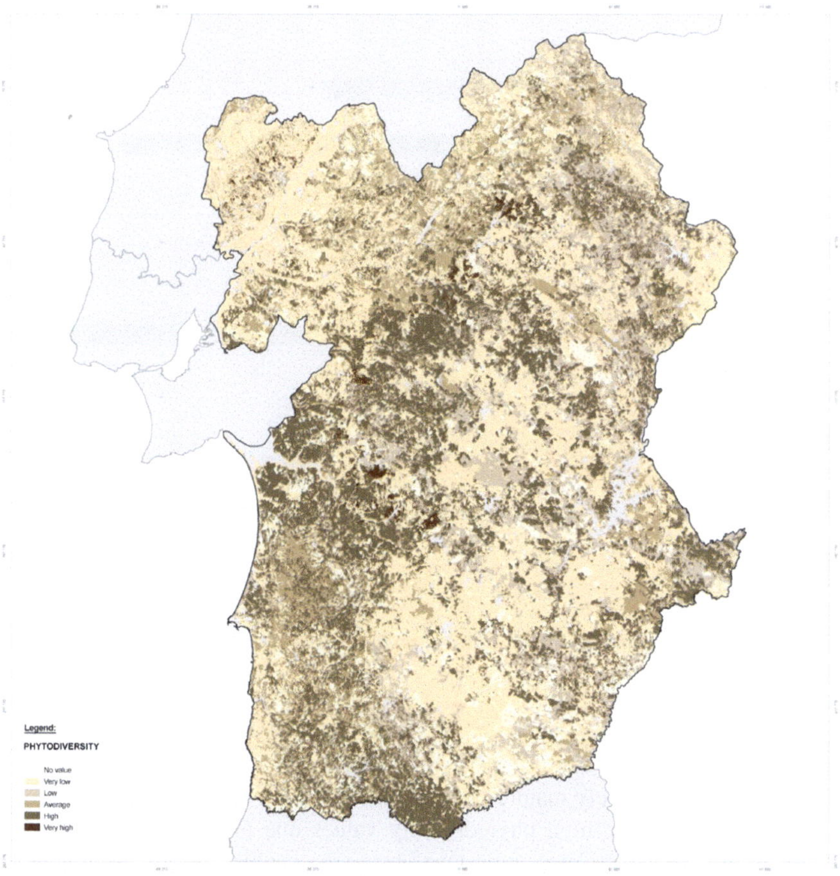

Fig. 5 Map of phytodiversity

ii. Dehesas have medium to high phytodiversity, primarily when associated with low-intensity management and steep slopes that allow for a richer vegetation mosaic and flora.
iii. Higher phytodiversity values occur in agricultural or forest ecosystems on sandy soils (arenosols and podzols on paleodunes in river Sado water basin), higher than those of the same land-use typology on siliceous soils (those derived from schist, greywacke, granite and syenite). In this particular case, biogeography is crucial, for this unit has a considerable pool of regional endemisms.
iv. High phytodiversity values do not necessarily correlate to known distribution areas of rare and protected plant species. These plants often occupy very specific biotopes with low biodiversity, to which they are narrowly specialized, such as active dunes and coastal cliffs.

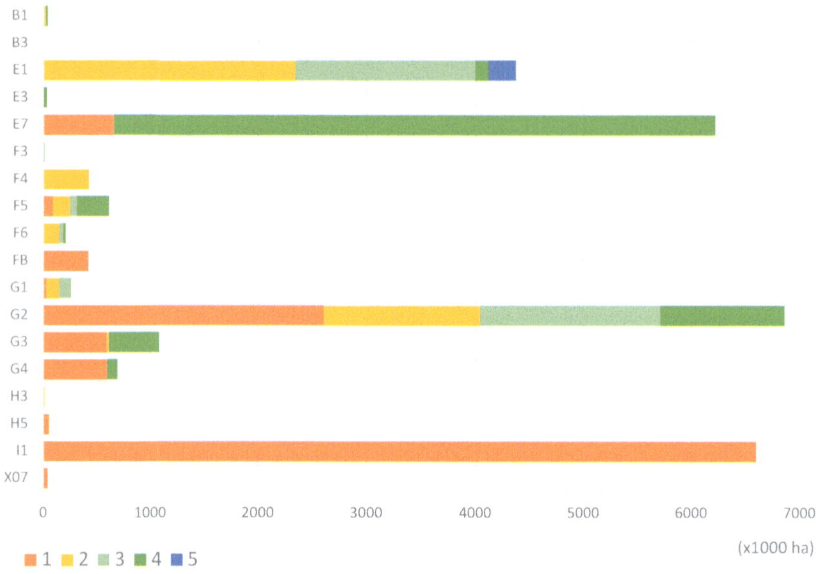

Fig. 6 Areas of ecosystem types at level 2 in the study area and their phytodiversity (thousands of ha; zero values not shown)

v. Areas of successively mature vegetation, complex mosaics or annual pastures on limestone have high phytodiversity values due to the soils' mesotrophic character. However, most limestone areas are occupied by intensive agriculture, with low phytodiversity values.

'Phytodiversity' was estimated using an entirely original methodology developed specifically for this task. It is based on the phytosociological concept of characteristic species—a set of plants that have their ecological optimum in a given community—which can be assumed as proxies for global phytodiversity. Species that, although present, have a stochastic nature are excluded since there is no point in considering them as members of the plant community.

4 Discussion and Conclusions

The EUNIS habitats typology is particularly suitable for ecosystem mapping in the context of ecosystem services assessment. It is a comprehensive hierarchical pan-European system established for harmonising habitat identification and description across Europe, covering all types of habitats from natural to artificial.

'*Habitat type*' is defined for the purposes of EUNIS classification as '*Plant and animal communities as the characterising elements of the biotic environment, together with abiotic factors operating together at a particular scale*.'[8]

The EUNIS classification is suitable for this task thanks to other characteristics worth mentioning:

i. The adopted definition of 'habitat type' is, in fact, the standard definition of ecosystem.
ii. EUNIS description of types indicates a vegetation type, the abiotic conditions in which it thrives and associated succession phenomena. This is no surprise, since plant communities are actually most of an ecosystem's biomass and, therefore, its most easily assessed and mappable component.
iii. Although covering all types of habitats, from natural to artificial, this typology is focused on defining criteria for the identification and classification of natural ecosystems. These criteria are fundamentally ecological: dominant physiognomy, biotope characteristics, biogeographic units, bioclimatic limits.

The methodology used for ecosystem mapping was based on interpreting land-use units as territorial units of ecological succession, an innovative approach in Portugal although in line with procedures developed in other EU countries [19].

The core methodological innovation of the approach is the use of phytosociological typologies as a basis for interpreting and establishing a crosswalk between vegetation types and EUNIS habitat types. A similar procedure, although using the Habitats Directive Annexe I typology, was used in France (*CarHAB*[9]) and Spain (*Hábitat (Directiva 92/43/CEE)*[10]). It is a 'rule-based' modelling procedure that integrates successional phytosociological models and land use data, incorporating ancillary information of potential natural vegetation, water indexes applied to satellite imagery and geology.

EUNIS habitat type mapping of Alentejo at levels 2 to 5 and a detailed 1:25,000 scale in a relatively short time was possible thanks to methodological robustness and optimized integration of a team of ecology, GIS and computer science experts. The methodology may efficiently be applied to all mainland Portugal and adapted to other areas, depending on the availability of spatially explicit data.

Ecological value integrates indicators of structural quality, phytocoenotic integrity and successional maturity of plant communities and reflects the pattern of territorial representativeness, or rarity, of the plant communities.

This parameter is particularly useful in managing protected areas, for instance, conveying information beyond the mere presence or absence of rare species, such as functional and structural ecological values, which are not evident when conventional biodiversity indexes and parameters are considered. Ecological value is a

[8]https://eunis.eea.europa.eu/about.

[9]https://www.cbn-alpin.fr/actions/habitats/carhab.html.

[10]https://www.miteco.gob.es/es/biodiversidad/servicios/banco-datos-naturaleza/informacion-disponible/habitat.aspx.

relevant indicator for detecting ecosystems of high ecological value but low biodiversity, such as native cork oak woodlands.

This is an efficient methodology, easy to interpret and extrapolate. Its main difficulty lies in adapting the parameter ranges defined by Loidi [30] and Izco [25] to the available information and different scales of approach. However, both publications are quite explicit in the definition of sub -parameters so, whenever a phytosociological successional model is available, the adaptation of this methodology to other study-cases is entirely possible.

Phytodiversity estimates encompass two distinct components of plant diversity. On the one hand, intrinsic species richness of ecosystems is estimated as a function of their set of characteristic species (*i.e.*, with their ecological optimum in this ecosystem), reflecting its ability to provide ecological niches for a variable number of taxa. This variability is a function of the ecosystem's structural complexity, its trophy, bioclimatic regime, etc. and of the regional pool of species.

On the other hand, phytodiversity also reflects the ecosystem's richness in regional endemisms and legally protected plants, which are extrinsic characteristics, representing a political and societal perspective on the plants' conservation value. An endemism is defined in relation to a political frontier and is not, therefore, a strictly biogeographical characteristic, although, in some cases, these two concepts may coincide.

Together, these two components may well express the habitat's ability to provide ecosystem services associated with biodiversity.

As stated before, an original methodology was developed specifically for this task, based on the phytosociological concept of characteristic species, which can be used as proxies for the global plant diversity of an ecosystem.

The main difficulty in applying the methodology is choosing which vegetation unit (*i.e.* plant association) corresponds to each type of vegetation, when dealing with low naturality ecosystems. It may be a time-consuming task, since it requires the compilation of a large number of relevés, which must necessarily be representative of all types of vegetation present.

The quality of estimated values can only be assessed by qualitatively and subjectively comparing the map produced with available information on species distribution, considering the various types of errors that may be associated with them. Globally, the results achieved were considered very satisfactory. Again, this is an objective methodology that, as such, can be easily replicated and extended to other parts of the country and to other ecosystems.

Acknowledgements The authors wish to express their gratitude to all team members of the Pilot Project "Mapping and Assessment of Ecosystem Services" at IST, for constant support, incentive, knowledge sharing and excellent companionship. Finally, we want to remember the departed Marco Reis, whose memory we all cherish, for his irreplaceable in computer science expertise and for having been essential to completing the project.

This work was carried out under the National Authority for Biodiversity and Forest Conservation (ICNF) financial support under contract AD 288/2014/ICNF/SEDE.

Annex I

Table 1 Crosswalk EUNIS – COS'07

COS'07		EUNIS
2.1.1.01.1	Rainfed crops	I1.3 Arable land with unmixed crops grown by low-intensity agricultural methods
2.1.1.02.1	Greenhouses and Nurseries	J2.43 Greenhouses
2.1.2.01.1	Irrigated crops	I1 Arable land and market gardens
2.1.3.01.1	Rice fields	I1.4 Inundated or inundatable croplands, including rice fields
2.2.1.01.1	Vineyards	FB.4 Vineyards
2.2.1.02.1	Orchards with vineyards	FB.4 Vineyards
2.2.1.03.1	Vineyards with olive groves	FB.4 Vineyards
2.2.2.01.1	Fresh fruit orchards	G1.D4 Fruit orchards
2.2.2.01.2	Almond orchards	G1.D3 Prunus amygdalus groves
2.2.2.01.3	Orchards of chestnut trees	G1.D1 Castanea sativa plantations
2.2.2.01.5	Orchards of citrus fruits	G2.92 Citrus orchards
2.2.2.01.6	Other orchards	G1.D5 Other high-stem orchards
2.2.2.02.1	Orchards of fresh fruit with vineyards	G1.D4 Fruit orchards
2.2.2.02.6	Other orchards with vineyards	G1.D5 Other high-stem orchards
2.2.2.03.1	Fresh fruit orchards with olive trees	G1.D4 Fruit orchards
2.2.2.03.3	Orchards of chestnut with olive trees	G1.D1 Castanea sativa plantations
2.2.2.03.5	Orchards of citrus with olive trees	G2.92 Citrus orchards
2.2.2.03.6	Other orchards with olive groves	G1.D5 Other high-stem orchards
2.2.3.01.1	Olive groves	G2.91 Olea europaea groves
2.2.3.02.1	Olive groves with vineyards	G2.91 Olea europaea groves
2.2.3.03.1	Olive groves with other fruit trees	G2.91 Olea europaea groves
2.4.1.01.1	Rainfed crops associated with vines	I1.3 Arable land with unmixed crops grown by low-intensity agricultural methods
2.4.1.01.2	Rainfed crops associated with orchard	I1.3 Arable land with unmixed crops grown by low-intensity agricultural methods
2.4.1.01.3	Rainfed crops associated with olive groves	I1.3 Arable land with unmixed crops grown by low-intensity agricultural methods
2.4.1.02.1	Irrigated crops associated with vines	I1.3 Arable land with unmixed crops grown by low-intensity agricultural methods

(continued)

Table 1 (continued)

COS'07		EUNIS
2.4.1.02.2	Irrigated crops associated with orchards	I1.3 Arable land with unmixed crops grown by low-intensity agricultural methods
2.4.1.02.3	Irrigated crops associated with olive groves	I1.3 Arable land with unmixed crops grown by low-intensity agricultural methods
2.4.2.01.1	Complex cultural systems	I1.22 Small-scale market gardens and horticulture, including allotments
2.4.3.01.1	Agriculture with natural and semi-natural areas	X07 Intensively-farmed crops interspersed with strips of natural and/or semi-natural vegetation
2.4.4.01.1	Agroforestry systems of cork oak with rainfed crops	E7.3 Dehesa
2.4.4.01.2	Agroforestry systems of holm oak with rainfed crops	E7.3 Dehesa
2.4.4.01.3	Agroforestry systems of other oaks with rainfed crops	E7.3 Dehesa
2.4.4.01.4	Agroforestry systems of other tree species with rainfed crops	E7.3 Dehesa
2.4.4.01.5	Agroforestry systems with cork oak, holm oak and rainfed crops	E7.3 Dehesa
2.4.4.01.6	Agroforestry systems of other mixed trees with rainfed crops	E7.3 Dehesa
2.4.4.02.1	Agroforestry systems of cork oak with irrigated crops	E7.3 Dehesa
2.4.4.02.2	Agroforestry systems of holm oak with irrigated crops	E7.3 Dehesa
2.4.4.02.4	Agroforestry systems of other tree species with irrigated crops	E7.3 Dehesa
2.4.4.02.5	Agroforestry systems with cork oak, holm oak and irrigated crops	E7.3 Dehesa
2.4.4.02.6	Agroforestry systems from other mixed trees with temporary irrigated crops	E7.3 Dehesa
2.4.4.03.1	Agroforestry systems of cork oak with pastures	E7.3 Dehesa
2.4.4.03.2	Agroforestry systems of holm oak with pastures	E7.3 Dehesa
2.4.4.03.3	Agroforestry systems of other oaks with pasture	E7.3 Dehesa
2.4.4.03.4	Agroforestry systems of other tree species with pastures	E7.3 Dehesa
2.4.4.03.5	Agroforestry systems of cork oak with holm oak and pastures	E7.3 Dehesa

(continued)

Table 1 (continued)

COS'07		EUNIS
2.4.4.03.6	Agroforestry systems of other mixed trees with pastures	E7.3 Dehesa
2.4.4.04.1	Agroforestry systems of cork oak with permanent crops	E7.3 Dehesa
2.4.4.04.2	Agroforestry systems of holm oak with permanent crops	E7.3 Dehesa
2.4.4.04.3	Agroforestry systems of other oaks with permanent crops	E7.3 Dehesa
2.4.4.04.4	Agroforestry systems of other tree species with permanent crops	E7.3 Dehesa
2.4.4.04.5	Agroforestry systems of cork oak with holm oak and permanent crops	E7.3 Dehesa
2.4.4.04.6	Agroforestry systems of other mixed trees with permanent cultures	E7.3 Dehesa
3.1.1.01.1	Cork oak forests	G2.112 Southwestern Iberian Quercus suber forests
3.1.1.01.2	Holm oak forests	G2.124 Quercus rotundifolia woodland
3.1.1.01.4	Chestnut forests	G1.7D Castanea sativa woodland
3.1.1.01.5	Eucalyptus forests	G2.81 Eucalyptus plantations
3.1.1.01.6	Invasive species forests	G2.83 Other evergreen broadleaved tree plantations
3.1.1.01.7	Forests of other hardwoods	G1 Broadleaved deciduous woodland
3.1.1.02.1	Cork oak forests with other hardwoods	G2.112 Southwestern Iberian Quercus suber forests
3.1.1.02.2	Holm oak forests with other hardwoods	G2.124 Quercus rotundifolia woodland
3.1.1.02.4	Chestnut forests with other hardwoods	G1.7D Castanea sativa woodland
3.1.1.02.5	Eucalyptus forests with other hardwoods	G2.81 Eucalyptus plantations
3.1.1.02.6	Invasive species forests with other hardwoods	G2.83 Other evergreen broadleaved tree plantations
3.1.1.02.7	Other hardwoods forests	G1 Broadleaved deciduous woodland
3.1.2.01.3	Forests of other softwoods	G3.F2 Exotic conifer plantations
3.1.2.02.1	Maritime pine forests with other softwoods	G3.F2 Exotic conifer plantations
3.1.2.02.3	Forests of other softwoods	G3.F2 Exotic conifer plantations
3.1.3.01.2	Holm oak forests with softwoods	G4.F Mixed forestry plantations
3.1.3.01.3	Other oaks forests with softwoods	G4.F Mixed forestry plantations
3.1.3.01.4	Chestnut forests with softwoods	G4.F Mixed forestry plantations
3.1.3.01.5	Eucalyptus forests with softwoods	G4.F Mixed forestry plantations

(continued)

Table 1 (continued)

COS'07		EUNIS
3.1.3.01.7	Other hardwood forests with softwoods	G4.F Mixed forestry plantations
3.1.3.01.8	Other hardwoods and softwoods mixed forests	G4.F Mixed forestry plantations
3.1.3.02.3	Other softwood forests with hardwoods	G4.F Mixed forestry plantations
3.1.3.02.4	Other softwoods and hardwoods mixed forests	G4.F Mixed forestry plantations
3.2.4.01.1	Open cork oak forests	G2.112 Southwestern Iberian Quercus suber forests
3.2.4.01.2	Open holm oak forests	G2.124 Quercus rotundifolia woodland
3.2.4.01.4	Open chestnut forests	G1.7D Castanea sativa woodland
3.2.4.01.5	Open eucalyptus forests	G2.81 Eucalyptus plantations
3.2.4.01.6	Open invasive species forests	G2.83 Other evergreen broadleaved tree plantations
3.2.4.01.7	Open forests of other hardwoods	G1 Broadleaved deciduous woodland
3.2.4.02.1	Open cork oak forests with other hardwoods	G2.112 Southwestern Iberian Quercus suber forests
3.2.4.02.2	Open holm oak forests with other hardwoods	G2.124 Quercus rotundifolia woodland
3.2.4.02.4	Open chestnut forests with other hardwoods	G1.7D Castanea sativa woodland
3.2.4.02.5	Open eucalyptus forests with other hardwoods	G2.81 Eucalyptus plantations
3.2.4.02.6	Open invasive species forests with other hardwoods	G2.83 Other evergreen broadleaved tree plantations
3.2.4.02.7	Open other hardwoods forests	G1 Broadleaved deciduous woodland
3.2.4.03.3	Open forests of other softwoods	G3.F2 Exotic conifer plantations
3.2.4.04.1	Open maritime pine forests with other softwoods	G3.F2 Exotic conifer plantations
3.2.4.05.2	Open holm oak forests with softwoods	G4.F Mixed forestry plantations
3.2.4.05.4	Open chestnut forests with softwoods	G4.F Mixed forestry plantations
3.2.4.05.5	Open eucalyptus forests with softwoods	G4.F Mixed forestry plantations
3.2.4.05.7	Open forests of other hardwood with softwoods	G4.F Mixed forestry plantations
3.2.4.05.8	Open mixed forests of other hardwoods and softwoods	G4.F Mixed forestry plantations
3.2.4.06.3	Open forests of other softwood with hardwoods	G4.F Mixed forestry plantations
3.2.4.06.4	Open mixed forests of other softwoods and hardwoods	G4.F Mixed forestry plantations

(continued)

Table 1 (continued)

COS'07		EUNIS
3.2.4.08.1	Clear cuts	G5.8 Recently felled areas
3.2.4.08.2	New plantations	G5.7 Coppice and early-stage plantations
3.2.4.09.1	Tree nurseries	G4.F Mixed forestry plantations
3.2.4.10.1	Firebreaks	H5.61 Unsurfaced pathways
3.3.1.01.1	Inland beaches and dunes	H5.3 Sparsely- or un-vegetated habitats on mineral substrates not resulting from recent ice activity
3.3.1.02.1	Coastal beaches and dunes	B1 Coastal dunes and sandy shores
3.3.4.01.1	Burnt areas	H5.5 Burnt areas with very sparse or no vegetation

Annex II

Table 2 Vegetation series present in NUTS II Alentejo

Code	Vegetation Series
1	*Arbuto unedonis-Querco pyrenaicae Sigmetum*
2	*Arisaro clusi-Querco broteroi Sigmetum*
3	*Asparago aphylli-Querco suberis Sigmetum*
4	*Daphno gnidi-Junipero navicularis Sigmetum*
10	*Lonicero implexae-Querco rotundifoliae Sigmetum*
14	*Myrto-Querco rotundifoliae Sigmetum*
15	*Oleo sylvestris-Querco suberis Sigmetum*
16	*Osyrio quadripartitae-Junipero turbinatae Sigmetum*
28	*Phlomido-Juniperetum turbinatae Sigmetum*
19	*Pyro bourgaeanae-Querco rotundifoliae Sigmetum e Pistacio terebinthi-Querco broteroi Sigmetum*
21	*Rhamno olioidis-Querco rotundifoliae Sigmetum*
23	*Sanguisorbo agrimonioidis-Querco suberis Sigmetum*
29	*Smilaco-Quercetum suberis Sigmetum e Pistacio-Juniperetum badiae Sigmetum*
24	*Teucrio baetici-Querco suberis Sigmetum*
25	*Viburno tini-Oleo sylvestris Sigmetum*
34	Riparian geosseries

Table 3 Crosswalk COS'07 + VS – EUNIS for some types of native heathland, scrub and woodland

COS'07	Vegetation Series (VS) code	EUNIS
3.1.1.01.3 Forests of other oaks 3.1.1.02.3 Forests of other oaks with other hardwoods 3.2.4.01.3 Open forests of other oaks 3.2.4.02.3 Open forests of other oaks with other hardwoods	Mosaic of 1 and 23	G1.7 Thermophilous deciduous woodland
	19	G1.772 Portuguese [Quercus faginea] forests
	2, 3, 10, 19, 23, mosaic of 2 and 3	G1.772 Portuguese [Quercus faginea] forests
	1	G1.7B [Quercus pyrenaica] woodland
	14	G2.1 Mediterranean evergreen [Quercus] woodland
3.2.2.01.1 Dense scrub 3.2.2.02.1 Sparse scrub	16, mosaic of 4 and 15	B1.6 Coastal dune scrub
	34	F3.1 Temperate thickets and scrub
	1, 3, 23, mosaic of 1 and 23	F4.2 Dry heaths
	2, 10, 14, 15, 19, 21, 24, 25, 28, 29, mosaic of 19 and 15, mosaic of 2 and 3, mosaic of 3 and 15	F6.1 Western garrigues
3.2.3.01.1 Dense sclerophyte vegetation 3.2.3.02.1 Sparse sclerophyte vegetation 3.2.4.07.1 Other woody formations	16	B1.6 Coastal dune scrub
	2, mosaic of 19 and 15, mosaic of 2 and 3, mosaic of 3 and 15	F5 Maquis, arborescent matorral and thermo-Mediterranean brushes
	1, 3, 15, 23, 24, mosaic of 24 and 15, mosaic of 4 and 15	F5.2 Maquis
	10, 14, 19, 25, 28, 29	F5.5 Thermo-Mediterranean scrub
	34	G1.3 Mediterranean riparian woodland

Annex III

Table 4 Crosswalk COS'07 + Geology – EUNIS for pine forests

COS'07	EUNIS in dune systems	EUNIS in other substrata
3.1.2.01.1 Maritime pine forests	G3.714 Iberian Pinus pinaster ssp. atlantica forests	G3.F22 Exotic pine plantations
3.2.4.03.1 Open maritime pine forests	G3.714 Iberian Pinus pinaster ssp. atlantica forests	G3.F22 Exotic pine plantations
3.1.2.01.2 Stone pine forests	G3.7312 Lusitanian stone pine forests	G3.F22 Exotic pine plantations
3.1.2.02.2 Stone pine forests with other softwoods	G3.7312 Lusitanian stone pine forests	G3.F22 Exotic pine plantations
3.2.4.03.2 Open stone pine forests	G3.7312 Lusitanian stone pine forests	G3.F22 Exotic pine plantations
3.2.4.04.2 Open stone pine forests with other softwoods	G3.7312 Lusitanian stone pine forests	G3.F22 Exotic pine plantations
3.1.3.01.1 Cork oak forests with softwoods	G4.B Mixed mediterranean Pinus - thermophilous Quercus woodland	G4.F Mixed forestry plantations
3.1.3.02.1 Maritime pine forests with hardwoods	G4.B Mixed mediterranean Pinus - thermophilous Quercus woodland	G4.F Mixed forestry plantations
3.1.3.02.2 Stone pine forests with hardwoods	G4.B Mixed mediterranean Pinus - thermophilous Quercus woodland	G4.F Mixed forestry plantations
3.2.4.05.1 Open cork oak forests with softwoods	G4.B Mixed mediterranean Pinus - thermophilous Quercus woodland	G4.F Mixed forestry plantations
3.2.4.06.1 Open maritime pine forests with hardwoods	G4.B Mixed mediterranean Pinus - thermophilous Quercus woodland	G4.F Mixed forestry plantations
3.2.4.06.2 Open stone pine forests with hardwoods	G4.B Mixed mediterranean Pinus - thermophilous Quercus woodland	G4.F Mixed forestry plantations

Annex IV

Table 5 Crosswalk COS'07 + Distance to coastline – EUNIS for sparsely- or un-vegetated areas

COS'07	EUNIS in coastal areas (<100 m from coastline)	EUNIS in inland areas (>100 m from coastline)
3.3.2.01.1 Rock outcrops	B3 Rock cliffs, ledges and shores, including the supralittoral	H3 Inland cliffs, rock pavements and outcrops
3.3.3.01.1 Sparse vegetation	B1 Coastal dunes and sandy shores	H5.3 Sparsely- or un-vegetated habitats on mineral substrates not resulting from recent ice activity

Annex V

Table 6 Grassland types included in land use map COS'07

COS'07
2.3.1.01.1 Permanent pastures
2.4.1.03.1 Pastures associated with vineyards
2.4.1.03.2 Pastures associated with orchards
2.4.1.03.3 Pastures associated with olive groves
3.2.1.01.1 Natural herbaceous vegetation

Table 7 EUNIS grasslands types present in the study area

EUNIS	Definition according to https://eunis.eea.europa.eu/habitats/	Meadows present in the study area
E1 Dry grasslands	Well-drained or dry lands dominated by grass or herbs, mostly not fertilized and with low productivity	Annual non-nitrophile meadows (*Tuberarietea guttatae*); annual nitrophile communities (*Stellarietea mediae*); perennial or biennial sub-nitrophile communities (*Artemisetea vulgaris*); biennial and perennial tall grasslands (*Stipo-Agrostietea, Festuco-Brometea, Lygeo-Stipetea*)
E2 Mesic grasslands	Lowland and montane mesotrophic and eutrophic pastures and hay meadows of the boreal, nemoral, warm-temperate humid and mediterranean zones. They are generally more fertile than dry grasslands (E1) and include sports fields and agriculturally improved and reseeded pastures	Hygrophilous meadows (*Arrhenatheretalia*); rainfed reseeded pastures
E3 Seasonally wet and wet grasslands	Unimproved or lightly improved wet meadows and tall herb communities of the boreal, nemoral, warm-temperate humid, steppic and mediterranean zones	Water-logged meadows and rush communities (*Holoschoenetalia australis*); irrigated reseeded pastures

References

1. Aguiar C (2001) Flora e vegetação da Serra de Nogueira e do Parque Natural de Montesinho. Ph.D. dissertation. Technical University of Lisbon, 659 p.
2. Aguiar C, Capelo J (2004) Anexo às fichas dos habitats de pinhal: 2180, 2270 e 9540. In ALFA (2004). Tipos de Habitats Naturais do Anexo I da Directiva 92/43/CEE (Portugal Continental): Fichas de caracterização Ecológica e de Gestão para o Plano Sectorial da Rede Natura 2000. Relatório. Lisboa
3. Arnáiz C (1981) Utilización de criterios ecológicos para la valoración del territorio. In: Géhu JM, Pelt JM (eds) L'evaluation biologique du territoire par la méthode des indices biocenotiques. Inst. Européen d' Ecologie, Metz, pp 119–126
4. Arsénio P (2011) Qualidade da paisagem e fitodiversidade. Contributo para o ordenamento e gestão de áreas costeiras de elevado valor natural. Lisboa: ISA, 426 p.
5. Asensi A (1990) Aplicación de la fitosociología a la evaluación del territorio. Monografías Flora y Vegetación Béticas 4/5: 91-100
6. Banko G, Weiss M, Moser D, Ubach R, Abdul Malak D, HALADA L, Roerink GJ, Hazeu GW, Mucher CA, Hennekens SM, Schaminee JHJ, Brodsky L (2013) Developing Conceptual Framework for Ecosystem Mapping. Kopenhagen, EEA - European Environment Agency
7. Barbour MG, Burk JH, Pitts WD (1987) Terrestrial plant ecology. Benjamin Cummings, Menlo Park
8. BISE (2016) https://biodiversity.europa.eu/maes/common-international-classification-of-ecosystem-services-cices-classification-version-4.3. Acceseed 20 Nov 2018
9. Bordt M (2015) Land and Ecosystem Condition and Capacity - DRAFT v. 1.0. Technical note. SEEA – EEA
10. Braun-Blanquet J, Pinto da Silva AR, Rozeira A (1964) Résultats de deux excursions géobotaniques à travers le Portugal septentrional & moyen III. Landes à Cistes et Ericacées (Cisto-Lavanduletea et Calluno-Ulicetea). Agronomia Lusitana 23(4):229–313
11. Capelo J (2007) Nemorum Transtaganae Descriptio. Sintaxonomia das comunidades florestais e pré-florestais do Baixo Alentejo. Ph.D. dissertation. Technical University of Lisbon, 528 p.
12. Capelo J (2018) Evidence of a unique association between single forest vegetation-types and seral sequences: praise for the concept of 'vegetation series'. In: Greller A, Fujiwara K, Pedrotti F (eds) Geographical changes in vegetation and plant functional types. Geobotany studies (basics, methods and case studies). Springer, Cham
13. Capelo J (1996) Esboço da Paisagem Vegetal da Bacia Portuguesa do Rio Guadiana. Silva Lusitana. nº esp, pp 13–64
14. Capelo J (2007) Nemorum Transtaganae Descriptio. Sintaxonomia das comunidades florestais e pré-florestais do Baixo Alentejo. Tese de doutoramento. ISA. 528 p. Instituto Superior de Agronomia
15. Capelo J, Mesquita S, Costa JC, Ribeiro S, Arsénio P, Neto C, Monteiro T, Aguiar C, Honrado J, Espírito-Santo MD, Lousã M (2007) A methodological approach to potential vegetation modeling using GIS techniques and phytosociological expert-knowledge: application to mainland Portugal. Phytocoenologia 37(3-4):399–415
16. Costa JC, Alvarez ML, Díaz T, Lousã M, Espírito-Santo MD, Vasconcelos T, Monteiro A, Amor A (1993) Vegetação da Serra de Sintra. Guia Geobotânico da Excursão das XIII Jornadas de Fitossociologia. XIII Jornadas de Fitossociologia, Lisboa, pp 1–98
17. Costa JC, Neto C, Aguiar C, Capelo J, Espírito-Santo D, Honrado J, Pinto-Gomes C, Sequeira M, Monteiro-Henriques T, Lousã M (2012) Vascular plant communities in Portugal (continental, the Azores and Madeira). Global Geobotany 2:1–180
18. Costa JC, Aguiar C, Capelo J, Lousã M, Neto C (1998) Biogeografia de Portugal Continental. Quercetea 0:1–56
19. EEA (2014) Terrestrial habitat mapping in Europe: An Overview 2014 (EEA Technicalreport nº 1/2014), MNHN-EEA. European Environment Agency, Luxembourg (ISSN1725-2237)

20. EUNIS (2007) EUNIS Habitat Classification. https://www.eea.europa.eu/themes/biodiversity/eunis/eunis-habitat-classification
21. Felipe-Lucia MR, Soliveres S, Penone C, Manning P, Van Der Plas F, Boch S, Prati D, Ammer C, Schall P, Gossner MM, Bauhus J, Buscot F, Blaser S, Blüthgen N, De Frutos A, Ehbrecht M, Frank K, Goldmann K, Hänsel F, Jung K, Kahl T, Nauss T, Oelmann Y, Pena R , Polle A, Renner S, Schloter M, Schöning I, Schrumpf M, Schulze ED, Solly E, Sorkau E, Stempfhuber B, Tschapka M, Weisser W, Wubet T, Fischer M, Allan E (2018) Multiple forest attributes underpin the supply of multiple ecosystem services. Nature Commun 9:4839
22. Gao B (1996) NDWI — a normalized difference water index for remote sensing of vegetation liquid water from space. Remote Sens Environ 58:257–266
23. Géhu JM, Géhu-Frank J (1991) Essai d'évaluation phytocoenotique de l'artificialisation des paysages. Coll Phytosoc 17:479–515. https://doi.org/10.1038/s41467-018-07082-4
24. IGP (2011) COS' 07. https://www.igeo.pt/produtos/CEGIG/Cos2007.htm
25. Izco J (1998) Types of rarity of plant communities. J Veg Sci 9:641–646
26. Jackson TJ, Chen D, Cosh M, Li F, Anderson M, Walthall C, Doriaswamy P, Hunt ER (2003) Vegetation water content mapping using Landsat data derived normalized difference water index for corn and soybeans. Remote Sens Environ 92(4):475–482
27. Janssen J, Rodwell J, Bergmeier E, Chytrý M, Schaminée JHJ (2014) Terrestrial and freshwater typology for European Red List of Ecosystem project. Draft Version, 15 Septemer 2014. Unpublished
28. Keith DA, Rodríguez JP, Rodríguez-Clark KM, Nicholson E, Aapala K, Alonso A, Asmussen M, Bachman S, Basset A, Barrow EG, Benson JS, Bishop MJ, Bonifacio R, Brooks TM, Burgman MA, Comer P, Comín FA, Essl F, Faber-Langendoen D, Fairweather PG, Holdaway RJ, Jennings M, Kingsford RT, Lester RE, Mac Nally R, Mccarthy MA, Moat J, Oliveira-Miranda MA, Pisanu P, Poulin B, Regan TJ, Riecken U, Spalding MD, Zambrano-Martínez S (2013). Scientific foundations for an IUCN red list of ecosystems. PLoS ONE 8(5):e62111. https://doi.org/10.1371/journal.pone.0062111
29. Kent M (2012) Vegetation description and analysis: a practical approach. Wiley-Blackwell, Oxford
30. Loidi J (1994) Phytosociology applied to Nature Conservation and Land Management. In: Song Y, Dierschke H, Wang X (eds) Applied vegetation ecology. East China Normal University Press. Shanghai, pp 17–30
31. Maes J, Teller A, Erhard M, Murphy P, Paracchini Ml, Barredo JI, et al (2014) Mapping and Assessment of Ecosystems and their Services: Indicators for ecosystem assessments under Action 5 of the EU Biodiversity Strategy to 2020. (European Union Technical Report; No. 2014-080). Luxembourg: Publications Office of the European Union. https://doi.org/10.2779/75203
32. Maes J, Teller A, Erhard M, Grizzetti B, Barredo JI, Paracchini Ml, Condé S, Somma F, Orgiazzi A, Jones A, Zulian A, Vallecilo S, Petersen JE, Marquardt D, Kovacevic V, Abdul Malak D, Marin Ai, Czúcz B, Mauri A, Loffler P, Bastrup Birk A, Biala K, Christiansen T, Werner B (2018) Mapping and Assessment of Ecosystems and their Services: an analytical framework for ecosystem condition. Publications office of the European Union, Luxembourg
33. Maes J, Zulian G, Thijssen M, Castell C, Baro F, Ferreira Am, Melo J, Garrett Cp, David N, Alzetta C, Geneletti D, Cortinovis C, Zwierzchowska I, Louro Alves F, Souto Cruz C, Blasi C, Alos Orti MM, Attorre F, Azzella MM, Capotorti G, Copiz R, Fusaro L, Manes F, Marando F, Marchetti M, Mollo B, Salvatori E, Zavattero L, Zingari PC, Giarratano MC, Bianchi E, Dupre E, Barton D, Stange E, Perez-Soba M, Van Eupen M, Verweij P, De Vries A, Kruse H, Polce C, Cugny-Seguin M, Erhard M, Nicolau R, Fonseca A, Fritz M, Teller A (2016) Mapping and Assessment of Ecosystems and their Services. Urban Ecosystems. Publications Office of the European Union, Luxembourg
34. Mesquita S, Sousa AJ (2009) Bioclimatic mapping using geostatistical approaches: application to mainland Portugal. Int J Climatol 29(14):2156–2170

35. Mesquita S (2013) Subsistemas da Estrutura Ecologica Nacional – Vegetacao. Magalhaes MR (coord.) Estrutura Ecologica Nacional – uma proposta de delimitacao e regulamentacao: Centro de Estudos de Arquitectura Paisagista Professor Caldeira Cabral. ISA Press. Lisboa, pp 105–120
36. NETO C (2002) A Flora e a Vegetação do superdistrito Sadense (Portugal). Guineana. 8:1–269
37. Neto C, Arsénio P, COSTA JC, (2009) Flora e vegetação do sudoeste de Portugal Continental. Quercetea 9:43–144
38. Oliveira, Pereira E, Ramalho M, Antunes MT, Monteiro JH (1992) Cartas Geológicas de Portugal Continental 1:500 000
39. Pereira EP (2013) Análise geobotânica dos bosques e galerias ripícolas da bacia hidrográfica do Tejo em Portugal. Ph.D. dissertation. IGOT, 438 p.
40. Ribeiro S (2013) Padrão de diversidade florística de comunidades herbáceas sujeitas a diferentes usos no território interior de Portugal Continental. Ph.D. dissertation, Technical University of Lisbon. 361 p.
41. Rivas-Martínez S, Díaz TE, Fernández-González F, Izco J, Lousã M, Penas A (2002) Vascular Plant Communities of Spain and Portugal. Addenda to the Syntaxonomical checklist of 2001. Itinera Geobotanica 15(1):5–432
42. Rivas-Martínez S (1976) Sinfitosociología, una nueva metodología para el estudio del paisage vegetal. Anal Instit Bot Cavanilles 33:179–188
43. Rivas-Martínez S (1987) Memoria del Mapa de Series de Vegetación de España 1:400000. ICONA. Madrid, 268 p.
44. Rivas-Martínez S (1996) Geobotánica y bioclimatologia. Discursos pronunciados en el acto de investidura de doctor "honoris causa" de excelentísima señor S. Salvador Rivas-Martínez. Universidad de Granada, 98 p.
45. Rivas-Martínez S, Sánchez-Mata D, Costa M (1999) North american boreal and western temperate forest vegetation (syntaxonomical synopsis of the potential natural plant communities of North America II). Itinera Geobotanica 12:5–316
46. Rivas-Martínez S, Lousã M, Díaz Te, Fernández-González F, Costa JC (1990)La Vegetación del Sur de Portugal (Sado, Alentejo e Algarbe). Itinera Geobotanica 3:5–126
47. Rivas-Martínez S, Penas A, Díaz T (2001) Bioclimatic Map of Europe-Thermoclimatic Belts. Cartographic Service, University of Léon, Spain
48. Schaminée JHJ, Chytrý M, Hennekens SM, Janssen Jam, Jiménezalfaro B, Knollová I, Mucina L, Rodwell JS, Tichý L (2014) Vegetation analysis and distribution maps for EUNIS habitats. Report for the European Environmental Agency (EEA/NSV/14/006), Copenhagen
49. Schaminée JHJ, Chytrý M, Hennekens SM, Mucina L, Rodwell JS, Tichý L (2012) Development of vegetation syntaxa crosswalks to EUNIS habitat classification and related data sets. Final report EEA/NSV/12/001, Alterra, Wageningen, NL, 134 p.
50. Tüxen R (1956) Die heutige potentielle natürliche Vegetation als Gegenstand der Vegetationskartierung. Angewandte Pflanzensoziologie 13:5–55
51. United Nations, European Union, Food and Agriculture Organization of the United Nations, Organisation for Economic Co-Operation and Development, World Bank Group (2014) System of Environmental-Economic Accounting 2012—Experimental Ecosystem Accounting. UN, New York, US, 177 p.

Spatial Representation of Plant Diversity at Geographical Scale: The Italian Experience

Francesco Maria Raimondo

Abstract The first attempts to draw up plant biodiversity maps in Italy are identified, some of which represent re-workings of data collected for chorological atlases of geographic areas of different extents. These were based particularly on the UTM grid, introduced and codified in Italy by Pignatti [6]. The limited examples made in Italy concern regional, provincial or district areas, some also in protected natural areas. As examples considered original, the maps developed at the Phytotaxonomy and Phytogeography Laboratory of the University of Palermo are presented. These involve the province of Palermo, the Madonie Natural Park in Palermo province, and the eastern Sicani mountains. The map of this last differs from the first two only in the type of reticule adopted for the smaller extent of the base quadrant. The three maps presented constitute complements of more complex maps that focus on the plant landscape of the mapped areas. In these, plant diversity attributed to the individual quadrants of the grid is represented by quantitative data for the specific and infra-specific taxa surveyed, expressed in different color classes. The quantitative incidence of endemic taxa is shown on each map with different symbols (red circles of different diameters), also expressed here by diametric rather than color classes. For the Madonie Natural Park, plant diversity has been mapped with the same quantitative/color criteria, also showing the phytocoenotic richness of this important plant area.

Keywords Map · Plant biodiversity · Italy · Palermo province

F. M. Raimondo (✉)
Department STEBICEF/Section of Botany and Plant Ecology, University of Palermo,
Via Archirafi 28, 90123 Palermo, Italy
e-mail: raimondo@centroplantapalermo.org

F. Pedrotti and E. O. Box (eds.), *Tools for Landscape-Scale Geobotany and Conservation*, Geobotany Studies, https://doi.org/10.1007/978-3-030-74950-7_8

1 Introduction

As can be inferred from existing literature, several authors have been engaged in representing plant diversity of a territory, especially in studies of limited extent. On the contrary, only a few concern wider areas reaching over territorial boundaries such as a small island or mountain area. This can be explained easily, considering the difficulty of acting over wider areas, even without considering the necessary economic and professional resources. The political and administrative value which knowledge of biodiversity assumes, for management of a territory and its natural resources, often leads researchers to analyse and then to represent plant diversity within territorial and administrative limits, even municipal or provincial limits, but in some cases also regional or national.

Municipal scales are easier to carry out and they do not stir the researcher much, while intermediate scales are more satisfying and give the opportunity to find funds to produce maps quickly, if the flora of that territory is already well explored taxonomically and geographically. In Italy, numerous efforts involved areas of large, medium and restricted extent. The national context has been highlighted, since 1990, by various authors ([13]) and by others under the auspices of the INTERREG IIC MEDOCC, Project "Base de données et cartographie de la biodiversité" [5]. In Sicily, the Systematics, Phytogeography and Plant Ecology Laboratory of the Department of Botanical Sciences, University of Palermo, obtained original results that can be extended to other geographical contexts (4, 8, 11, 12). For central Italy, important contributions have been provided recently by Lucchese [1, 2].

Italy is a territory rich in plant biodiversity and notoriously complex both climatically and orographically. Apart from methodological aspects, experiences and maps in Italy were produced for defined geographical areas, such as protected areas (parks or natural reserves) and administrative territories (municipalities and provinces). The regional scale is missing, except for an example limited to taxa of conservation interest in the Lazio region [1, 2].

In this paper three case studies involving our experience are presented. The first —the most important also for developing the methodology—concerns the province of Palermo in Sicily [8]. The second concerns the area of a natural park included in the same provincial territory [11]. The third concerns a representative area in the Sicani Mountains [4]. In these three cases, the approach is similar but not the same: the first is more synthetic, the last two more detailed. What changes is therefore the scale of reference and, in the case of Sicani, also the grid.

2 Italian Examples

In Italy, summary cartographic documents based on floristic richness, calculated for areas based on the IGM or UTM grid, and therefore referable to the official cartography, are rather limited. The great work of Poldini [7] for the entire Friuli-Venezia Giulia region, that could have generated a regional map of floristic biodiversity, does not achieve a summary by basic areas of accumulated quantitative data, and therefore remains only what the author had set for himself, a chorological atlas, from which one can draw a beautiful regional map of plant biodiversity. In continental Italy, Lucchese [2] produced a map of plant biodiversity for the Lazio region, but it concerns only a portion of the regional biodiversity, i.e. the taxa of greatest protectionist interest (Fig. 1). Unlike the maps of plant biodiversity of representative areas of Sicily, which will be presented below [4, 10, 11], the Lazio map was developed following a different approach and the results are not comparable. In fact, it was obtained by means of Kriging-like spatial analysis, which is nothing more than a spatial interpolation of points.

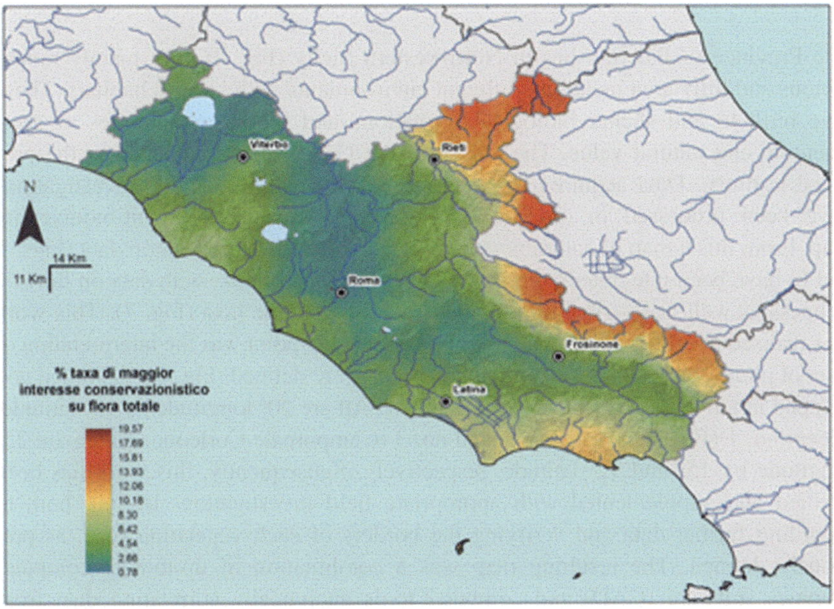

Fig. 1 Plant biodiversity map of Latium region (C Italy) (from [2])

3 Experience of the Phytotaxonomic-Phytogeographic Laboratory in the University of Palermo

The first of the three cases identified above concerned the whole province of Palermo (Sicily), about 500,000 hectares; the second case concerned only a part of it, namely the territory of the Madonie Regional Natural Park, about 45,000 hectares. The third case covered the area of the eastern Sicani Mountains (SW Sicily), rising in Agrigento province, about 36,000 hectares overall. In all three cases the territories are very different, in geomorphology and climate. The soils are mainly carbonate, but siliceous and clayey soils are also frequent. In the first and second cases, the altitude varies from sea level (approximately) to 1979 m; but in the third case from 140 to 1578 m. In the three areas, both semi-natural and anthropogenic environments prevail. Nevertheless, nearly natural and even some very restricted natural environments do occur [9].

3.1 Plant Biodiversity of Palermo Province

The Province of Palermo lies in Northwestern Sicily (Fig. 2), in a mainly mountainous and hilly area including different environments with various habitats. These give birth to and shelter biological formations and biotic communities of great scientific and natural value. The volcanic islet Ustica is also included in the province territory. Data acquired from former floristic and vegetation investigations have been processed in order to produce a landscape and plant-biodiversity map. From this a map of naturalness has been deduced. The vegetation data (Figs. 5 and 6) have been integrated—for each single unit (quadrant)—with data on floristic richness as well as distribution and incidence of endemic taxa (Fig. 7). This work was realised by a stepwise methodology. The starting point was the interpretation of a set of photographs. Then, 29 vegetation types were defined (Fig. 5), delimited and mapped in ten 1:50,000 tables (Figs. 3 and 4). All are 20' longitude by 15' latitude, except no. 1 (Balestrate-Partinico) and no. 3 (Camporeale-Corleone) which are 25' longitude by 15' and 12' latitude, respectively. Subsequently, this work has been verified and supplemented with appropriate field investigations leading both to recording further data and verifying the borders of each vegetation type, as previously defined. The resulting map was a combination of up-to-date computer graphics software (CAD) and computer tools analytically correlating data from different sources. The management of the data on the identified typologies was performed by using a GIS (Geographical Information System). Accordingly, hard questions on the available data were formulated.

The distribution of the records was digitized based on the geographical grid used in the Central European floristic cartography [6], then applied by Poldini [7] for the Friuli-Venezia Giulia region and by Lucchese and Lattanzi [3] for the Ausoni

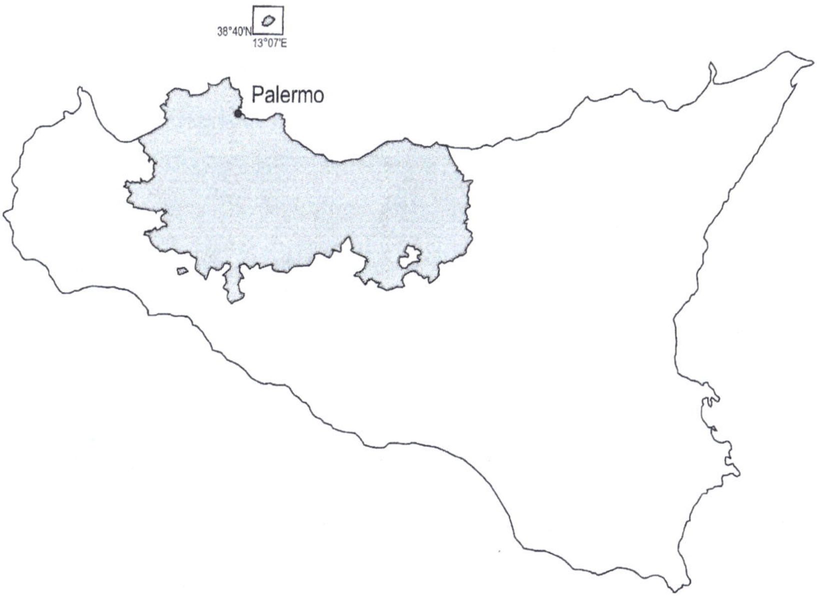

Fig. 2 Placement of the province of Palermo area (NW Sicily) (from [8])

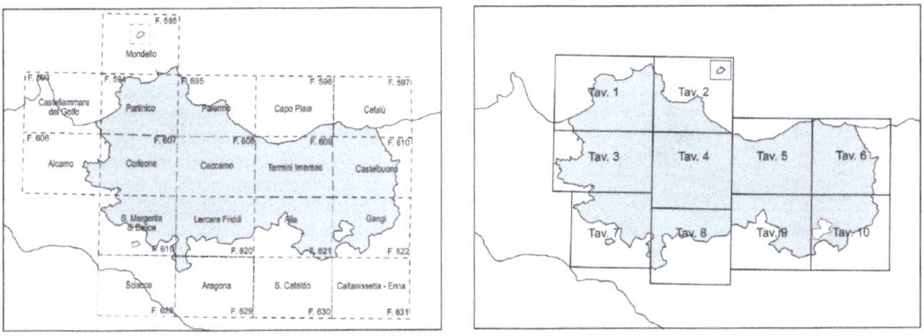

Fig. 3 On the left, geographical framework of the 19 IGM sheets (1:50000) where the territory of the province of Palermo (NW Sicily) is placed; on the right, geographical framework of the 10 tables accomplished where the territory of the province of Palermo is placed (from [8])

mountains in Lazio, and more recently by Lucchese [1, 2] for the entire Lazio region. The grid is 6' latitude by 10' longitude.

The derived unit (base area) is divided into four 3' latitude and 5' longitude quadrants with a 40.5 km^2 average extent: in total, 167 elementary quadrats were obtained. The best part of the Palermo province area is cultivated as follows: crops 39.57%; traditional extensive tree cultivation about 11% (olive-groves, to a lesser extent ash and almond-groves, etc.); and intensive shrub and tree cultivation, on the whole over 11%. The wooded areas, covering about 24,000 hectares, are mainly in

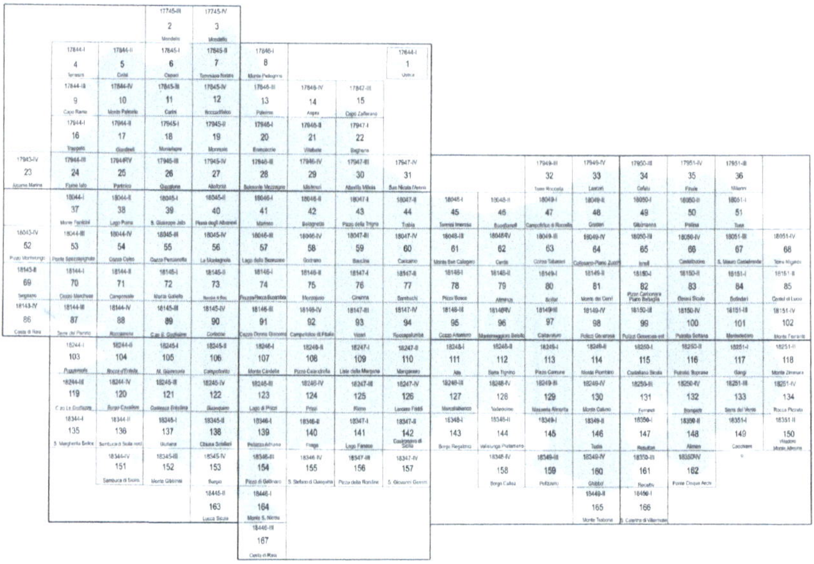

Fig. 4 Area grid of the province of Palermo (NW Sicily) divided in 167 quadrants. For each of them are shown the sequence number, the code according to Pignatti [6] and the name, corresponding to the same name of the C.T.R. (Regional Technic Map) sections in scale 1:10,000 (from [8])

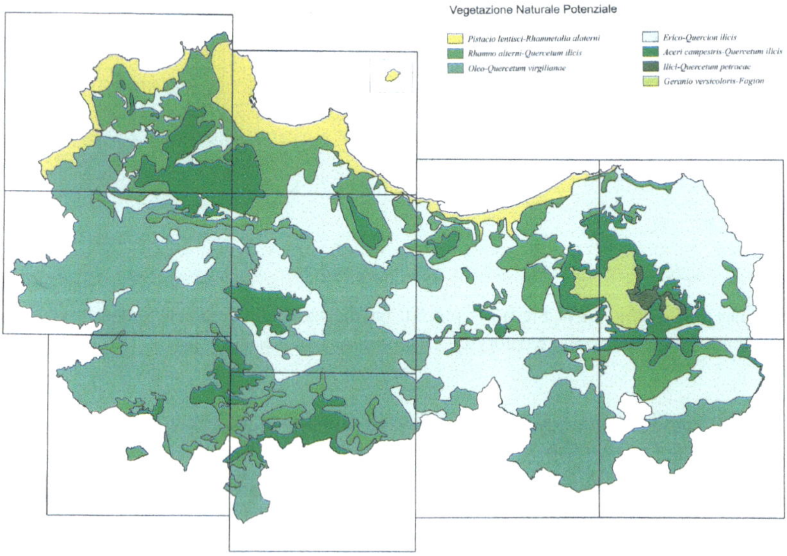

Fig. 5 A synthetic representation of the potential natural vegetation of the province of Palermo area (from [8])

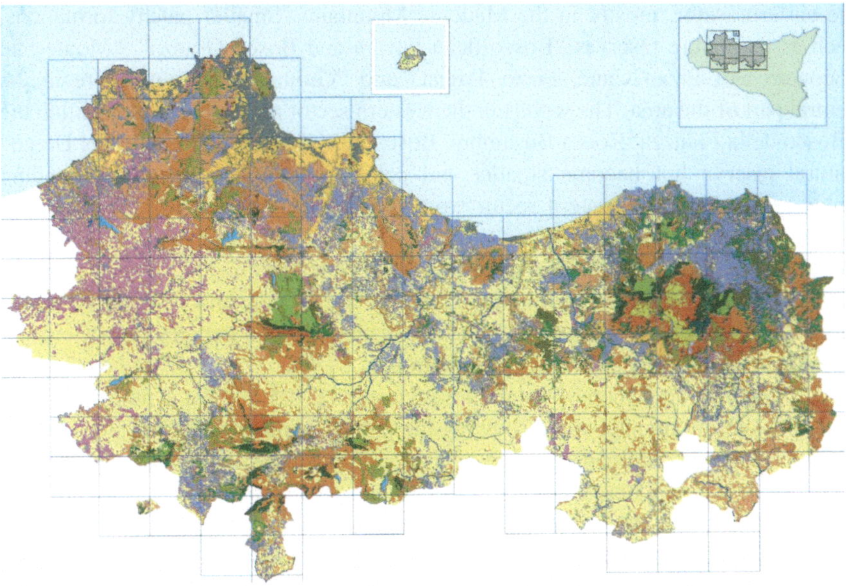

Fig. 6 A synthetic representation of the plant landscape of the province of Palermo area (NW Sicily) (from [10])

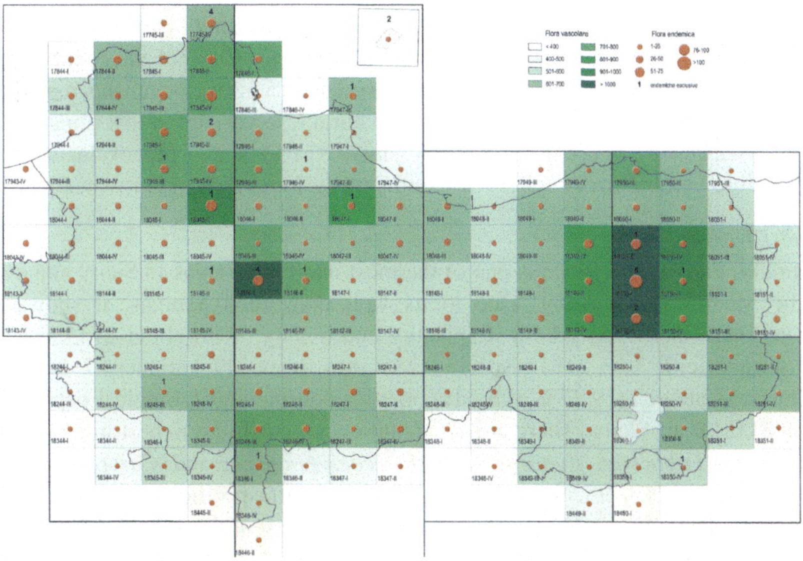

Fig. 7 Plant biodiversity map of the province of Palermo area (NW Sicily) (from [8])

the eastern sector, mostly in the Madonie Mountains. Smaller patchy formations, such as the nature reserves "Bosco della Favara and Bosco Granza", "Monte San Calogero and Pizzo Cane, Pizzo Trigna" and "Grotta Mazzamuto", are in the central part of the area. The woods in the western sector are uninterrupted within the "Bosco della Ficuzza, Rocca Busambra, Bosco del Cappelliere e Gorgo del Drago" natural reserve but become smaller and patchy on the mountains surrounding Palermo. In the southwestern sector woody formations are located on the Sicani range, mainly within the "Monte Carcaci", "Monte Genuardo and Santa Maria del Bosco" and "Monti di Palazzo Adriano and Valle del Sosio" natural reserves. Roughly 16,000 hectares of maquis, including shrublands and strongly degraded woods, occur in the whole area, with a higher incidence on the quadrants Contessa Entellina-Bisacquino and Marineo-Caccamo.

A complementary map of naturalness status, at 1:125,000—a synthesis of high biological-environmental value showing the plant communities of the area under a different degree of human pressure in time—is associated with the map of plant diversity. It is a map derived from the plant-landscape map on the basis of artificial elements occurring both in the structure and in the floristic components of the vegetation types surveyed [12]. The artificiality degrees have been digitized through a chromatic range. The map shows their incidence and distribution, allowing a comparison of the present vegetation status and the corresponding natural status. The ecological systems have been defined as: natural systems, nearly-natural systems, partly natural systems, rural systems, and highly exploited systems. Over 50% of Palermo province is rural systems mainly occurring in the hilly hinterland and lower mountain. Systems under intensive human exploitation (14%) are common in the littoral and hilly areas. The partly natural systems (29.67%) affect the whole province territory, while the nearly-natural systems (5.52%), occurring mainly in the nature park of the Madonie and in some natural reserves, are of low impact. The natural (*sensu stricto*) systems are very small (0.1%) and occur exclusively on inaccessible vertical rocky reliefs, discontinuously located in the province area.

Plant Diversity

Naturally, Palermo province is one of the most interesting areas in Sicily and even in the Mediterranean basin. Due to its well diversified habitats, it shows an extraordinary biodiversity both in quantity and quality. With about 2145 specific and infra-specific taxa (1811 species, 285 subspecies, 39 varieties, 2 forms, 8 nothotaxa), the area (19.47% of Sicily) boasts a conspicuous biological patrimony, about **81%** of all Sicily. There are 127 families, in which the richest in genera are *Asteraceae, Fabaceae, Poaceae, Brassicaceae, Apiacea*e and *Caryophyllacea*e, with 253, 225, 219, 98, 96 and 89 specific and infraspecific taxa, respectively. Within the native flora, 224 taxa are endemic, i.e. 5.6% of the vascular flora of the province and 4.5% of the Sicilian vascular flora [12]. Endemism is particularly high, both on the Madonie calcareous reliefs and on the mountains south of Palermo, such as the Rocca Busambra.

Incidence and distribution of the vascular flora in the province is shown by seven color classes (Fig. 7), each of which corresponds to a different number of taxa: <400, 401–500, 501–600, 601–700, 701–800, 801–900, 901–1000, >1000. Incidence and distribution of endemics are displayed within each quadrant through five red dots of different diameter. Each dot represents an incidence of endemic taxa as follows: 1–25, 26–50, 51–75, 77 100, >100. Each quadrant includes 588 taxa on average.

Within the 167 quadrants into which the Palermo province was divided, plant biodiversity is generally high in the less populated areas. This is especially noticeable in the quadrants of the Madonie orographic system, with a taxon number between 822 (Petralia Sottana) and 1174 (Pizzo Carbonara-Piano Battaglia), as well as 1093 in Ficuzza-Rocca Busambra, 933 in Monte Pizzuta, and 822 in Pizzo Trigna. Biodiversity is still appreciable in the quadrants that include the promontories of Monte Gallo and Monte Pellegrino, and in those of Belmonte Mezzagno, Tommaso Natale, Montelepre, Altofonte, Lago Scanzano, Mezzojuso, Palazzo Adriano and Filaga. There is a generally close qualitative correlation between endemism and plant diversity. In the quadrants of the Madonie mountains, endemic taxa oscillate from 30 (Petralia Sottana) to 118 (Pizzo Carbonara-Piano Battaglia). Endemism is also well represented in the northern and northwestern parts of the provincial territory, with highest values in the quadrants of Monte San Calogero (49), Capo Zafferano (41), Monte Pellegrino (50), Piana degli Albanesi (61); and in the southwestern mountainous reliefs of Palazzo Adriano (40). The lowest numerical values are recorded in the areas subjected to greater human impact.

The quadrant no. 82 (Pizzo Carbonara-Piano Battaglia) shows a significant peak with 1174 taxa, while the quadrant no. 158 (Borgo Callea) has the fewest, at 103 taxa. Plant biodiversity is generally higher in the area under lighter anthropogenic pressure. Qualitatively, there is a strong correlation between endemism and biological diversity. In the quadrants including the Madonie mountains (as will be shown subsequently) the incidence of endemism is at its peak with 62 endemic taxa in the quadrant no. 82 (Quacella), with the highest number of taxa.

4 Plant Biodiversity Map of the Madonie Natural Park

The landscape and plant biodiversity map of the territory of the Madonie Natural Park (scale 1:50,000) was also made by GIS, based on previous floristic and vegetation investigations, air-photo analyses, and appropriate field verifications. It was drawn up based on data obtained from the study presented earlier for the territory of the province of Palermo, which includes the area of the Madonie (Fig. 8). However, this map presents for the first time an attempt to represent not only the floristic biodiversity calculated from species richness and infraspecific taxa, but also the phytocenotic diversity calculated from richness in plant associations, as found in literature and in the field, quadrat by quadrat.

This territory has very varied geology and geo-morphology, with maximum altitude (Pizzo Carbonara) of 1979 m. The plant landscape, defined according to physiognomic and structural criteria, is expressed by 24 type units. Units related to

Fig. 8 Placement of the Madonie Natural Park in the province of Palermo (NW Sicily) (from Raimondo et al. 2004)

pasture are remarkable (about 13,000 ha), as well as to widespread agricultural crops (9,160 ha) and to forest formations (9,600 ha). Lesser extents are occupied by Mediterranean maquis cores, shrublands and very degraded woods (about 3500 ha), submontane and mountain grasslands and artificial forest plantations, as well as peculiar aspects of rock vegetation of high altitude and of lower screes, which find their maximum expression on the "Serre" of Quacella.

As for environmental integrity, it was possible to distinguish 10 natural levels, then distributed in five classes on the basis of the degree of human impact and the structure and quality of the plant systems.

Plant Diversity (Fig. 9)

Apart from the high phyto-coenotic diversity (Fig. 10), the territory of the Madonie Park also shows high biologic diversity. Within a surface of 1.55% of the entire administrative Region, about 1500 specific and infra-specific vascular taxa are found. This is a conspicuous and diversified biologic patrimony equivalent to about 57% of that occurring in the whole region. There are 170 endemic entities, which is about 46% of the endemism of the whole island. The distribution of the floristic richness of Palermo province is represented with grids having an average area of 40.5 km^2 (3' latitude by 5' longitude); this richness is divided into eight classes, the same as proposed for the map of Palermo province presented before.

Each quadrant hosts 753 entities on average, with a maximum of 1175 in Pizzo Carbonara-Piano Battaglia. Floristic biodiversity is, in general, very remarkable in the few humanized areas and in those lacking forest vegetation, where the strong relationship between endemism and floristic richness is shown.

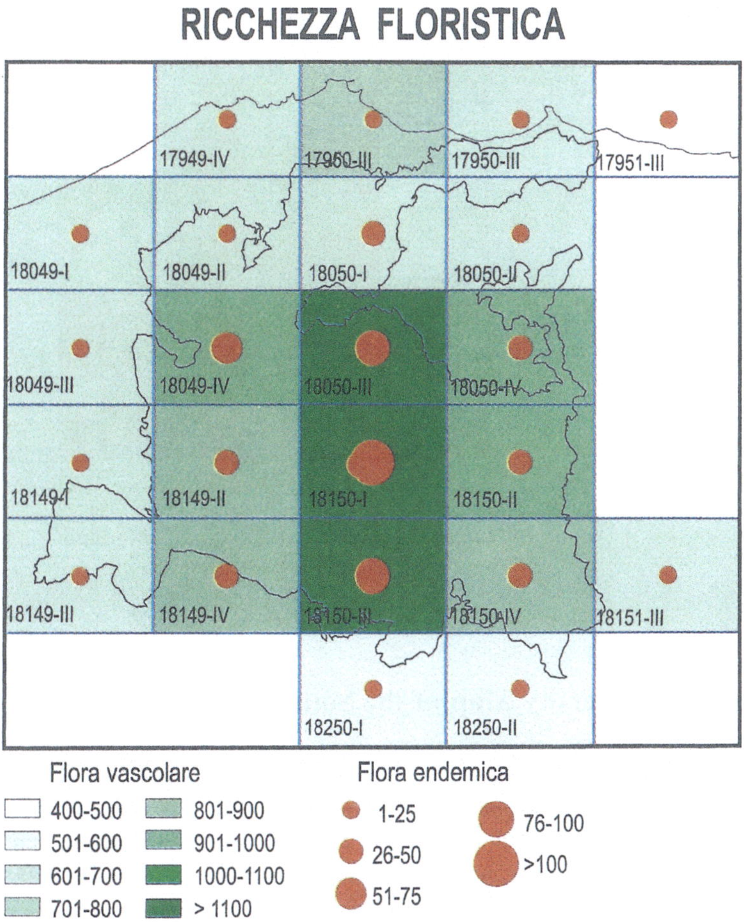

Fig. 9 Plant biodiversity map of the Madonie Natural Park (NW Sicily) (from Raimondo et al. 2004)

Plant biodiversity of the peculiar mountain area of Sicily, valued also on a phytocoenotic basis, is expressed by 127 units between syntaxonomically defined associations and floristically and ecologically well characterized groups. The richest quadrants are those of East Polizzi Generosa and Petralia Sottana (Fig. 9).

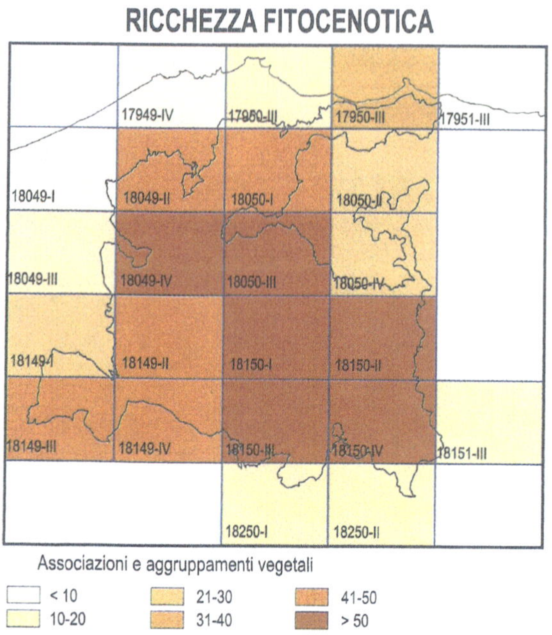

Fig. 10 Phytocoenotic biodiversity map of the Madonie Natural Park (NW Sicily) (from Raimondo et al. 2004)

5 Plant Biodiversity Map of the Southeastern Sicani Mts

The Sicani Mountains are a mostly calcareous system in southwestern Sicily, rising in the hinterland of Palermo and Agrigento provinces (Fig. 11). The map shows the eastern part of the Sicani geographic range, which is entirely in Agrigento province (N latitude 37° 32’ 26” to 37° 41’ 35”; E longitude 13° 27’ 02” to 13° 50’ 00”). Its maximum altitude is 1578 m (Monte Cammarata).

The map of the landscape and plant biodiversity of this area was made by Marino et al. [4] through a Territorial Information System (S.I.T.) in a GIS 3-D environment. Analysis of this document, which contains 32 vegetation types gathered into 10 groups, and the data reported by the authors in Table 20 of their article [4], shows that the autochtonous forest community suffered a remarkable reduction; they point out several aspects of degradation. Nevertheless, remaining *nuclei* of particular importance occur, frequent especially inside the territory of the natural reserve “Monte Cammarata” and that of SIC “M. Cammarata-Contrada Salaci”, “Pizzo della Rondine-Bosco di S. Stefano Quisquina”, “La Montagnola e Acqua Fitusa”, and ZPS “Serra del Leone e M. Stagnataro” (Fig. 12).

These are mainly holm-oak (*Quercus ilex*) woods or thermophilous deciduous oak-woods which account for about 11% of the complex. The shrublands and the brushwoods occur at the margins of wooded formations, while a few edges of Mediterranean maquis of *Oleo-Euphorbietum dendroidis* occur on semi-rupestrian calcareous sites and cover 5% as a whole. Pasture communities (14%) are of major

Fig. 11 Placement of the eastern Sicani Mts area (SW Sicily) (from [4])

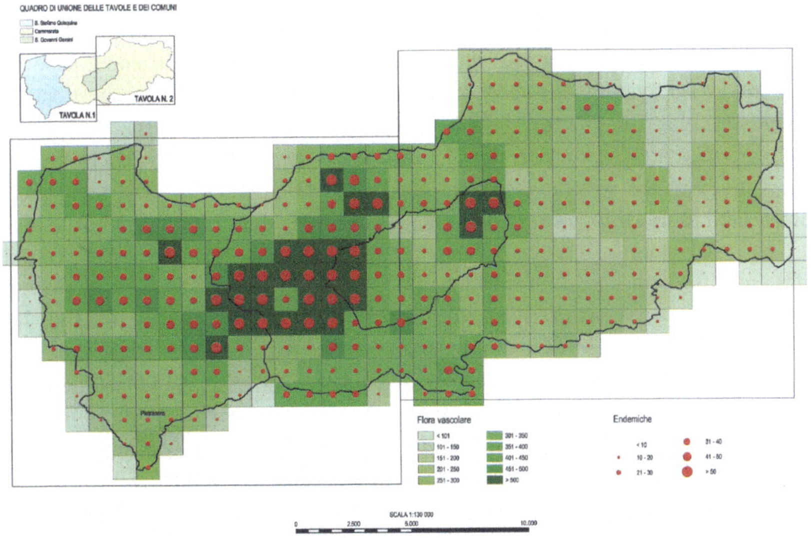

Fig. 12 Plant biodiversity map of eastern Sicani Mts area (SW Sicily) (from [4])

extent, mostly on Mt. Cammarata, and host several endemic or otherwise rare plant species. Smaller extents are evident for plant communities along water courses (3%), in marshy meadows and cane fields (3.63%), on calanques (0.9%) and sediment layers (0.55%), and on cliffs (0.33%). Remarkable is the importance of agricultural and ornamental crops (58.68%), distributed throughout almost all the territory except for mountains and characterizing the hill landscape between 140 and 900 m. Artificial forest plantations, in the hill and mountain portions of the central part of the territory, cover about 7% of the total area.

The map of the level of naturalness shows that the territory, exploited since ancient times with agro-forestry-pastoral activities, is dominated by human-rural systems (55.53%) and those having intensive use (8.46%). Semi-natural systems account for 29.07%, nearly-natural systems for 6.6%, and natural systems, *sensu strictu*, for only 0.33% of the area, only in almost inaccessible areas.

The map of potential natural vegetation suggests a trend from the present phytocoenoses toward communities of thermophilous deciduous oaks.

Plant Diversity

The floristic richness of this area includes 956 specific and intraspecific taxa, of which there are 853 species, 88 subspecies and 15 varieties, belonging to 449 genera of 95 families. The taxon-richest families are: *Asteraceae* (108), *Poaceae* (100), *Fabaceae* (90), *Apiaceae* (51), *Brassicaceae* (48), *Lamiaceae* (44), *Liliaceae* (41), *Orchidaceae* (32), *Caryophyllaceae* (31) and *Scrophulariaceae* (25).

The floristic biodiversity increases in the territory of Mt. Cammarata, Serra Quisquina and Montagnola. In these territories endemism for each base unit (quadrant) is on average between 40 and 50 specific and infra-specific taxa. The endemic vascular flora is represented overall by 90 taxa (68 species, 20 subspecies and 2 varieties), corresponding to 9.43% of the whole flora attributed to the area. Moreover, there are several rare taxa and/or taxa of particular phyto-geographic interest, including *Celtis tournefortii* ssp. *asperrima* (Lojac.) Raimondo & Schicchi and *Salvia argentea* L.

The results of this study contribute to fill gaps in information on plant biodiversity of the area. Moreover, these results could be used by the local Administration in charge of planning and management of the territory, which is very rich floristic and phytocoenotically, to prepare strategies for conservation and evaluation of the autochthonous plant patrimony. This investigation also contributes to the solution of problems connected with recovery and environmental restoration with particular regard to the extensive artificial forest plantations. Those areas, in fact, could be reconverted based on up-to-date criteria of natural silviculture, considering the potentialities of the territory and of the native forest formations, well preserved, which are optimal role models.

6 Discussion and Conclusions

The examples considered highlight how it is possible to create maps of plant biodiversity based on the floristic and phytocenotic richness of well-explored territories, using models tested widely in Europe for geographic grids compatible with official cartography available at different scales in the different countries. In the cases presented, the maps of floristic diversity do not represent the main purpose, but rather a complement of more complex maps that focus attention on the representation of the plant landscape and plant biodiversity more generally. They are therefore the result of much more complex research that has also attempted cartographic representation of the vegetation, then typified both on a physiognomic-structural and also a phytosociological basis. Putting together landscape, phytocenotic and plant diversity has not been easy. For plant diversity, associating the quantitative data with qualitative data (endemism rate) referred to elementary areas (quadrants) of different extent, based also on different grids, represents a result that is considered not only original but above all of remarkable interest for the management of the environmental and landscape heritage and therefore of biodiversity.

Ackowledgements Thanks to Franco Pedrotti for the decisive stimulus for publishing this article. Thanks to Roberta Orlando, Laura Bertini, and Edimir Murrja for their collaboration in the editing of text and figures.

References

1. Lucchese F (2017) Atlante della flora vascolare del Lazio, cartografia, ecologia, biogeografia. Parte generale e flora alloctona. Regione Lazio, Direzione Capitale naturale, Parchi e Aree protette, Roma. Digitiliab 1:352
2. Lucchese F (2018) Atlante della Flora vascolare del Lazio, cartografia, ecologia, biogeografia. La flora di maggiore interesse conservazionistico. Regione Lazio, Direzione Capitale naturale, Parchi e Aree protette, Roma. Ed. Digitiliab 2:400
3. Lucchese F, Lattanzi E (2000) Atlante della Flora dei Monti Ausoni. Reg. Lazio, Assess. Utilizzaz. e Valorizzaz. Risorse Ambientali e Univ. Del Molise, Fac. Agraria, Dip. S.A.V.A. Roma, p 461
4. Marino P, Castellano G, Bazan G, Schicchi R (2005) Carta del paesaggio e della biodiversità vegetale dei Monti Sicani sud-orientali (Sicilia centro-occidentale). Quad Bot Amb Appl **16**:3–60 [text] + map 1:50,000 (Tables 1–2)
5. Margiocco C, Mariotti M (eds.) (2001) Progetto "Basi di Dati e Cartografia della Biodiversità". Rapporto finale parte italiana. Iniziativa Comunitaria INTERREG II C/ Programma operativo MEDOCC – Università della Calabria, Centro Editoriale e Librario, p 215
6. Pignatti S (1978) Dieci anni di cartografia floristica nell'Italia di Nord-Est. Inform Bot Ital 10:212–217
7. Poldini L (1991) Atlante corologico delle piante vascolari nel Friuli-Venezia Giulia. Regione Autonoma Friuli-Venezia Giulia and Università degli Studi di Trieste, Udine, p 890
8. Raimondo FM (2000a) Carta del paesaggio e della biodiversità vegetale della provincia di Palermo. Quad Bot Amb Appl **8**(1998):3–164 [text] + map 1:50,000 (Tables 1-10)
9. Raimondo FM (2004) Landscape and floristic features of Sicily. Bocconea 17:22–64

10. Raimondo FM (2000b) Cartografie (Tables 1–10) scale 1:50000 (cover). In: Carta del paesaggio e della biodiversità vegetale della provincia di Palermo. Provincia Regionale di Palermo/Università degli Studi di Palermo
11. Raimondo FM, Schicchi R, Surano N (2004) Carta del paesaggio e della biodiversità vegetale del Parco delle Madonie (Sicilia). Naturalista Sicil, s. IV **28**(1):71–137 [text]+map 1:50,000
12. Raimondo FM, Gianguzzi L, Surano N (1994) La carta del paesaggio vegetale della provincia di Palermo (base I.G.M. scale 1:50,000): esemplificazione di elaborazione tramite G.I.S. Boll Ass It Cartografia 89:71–76
13. Taffetani F, Santolini R (1997) Un metodo per la valutazione della biodiversità su base fitosociologia e faunistica applicata allo studio di un'area collinare del Montefeltro (Provincia di Pesaro-Urbino, Italia centrale). Fitosociologia 32:245–271

Cognitive Mapping of Forest Fragments

Frédéric Roulier, Emilie Gallet-Moron, and Guillaume Decocq

Abstract The analysis of spatial cognition (i.e., people's ability to locate sites in the geographic space) can help land planning and forest management by predicting the behavior and reactions of people towards specific landscapes. To achieve this goal, it is of utmost importance to design robust, replicable methods; this is the purpose of this study. Here we develop a method which compares locations of an objective set of points (e.g. extracted from topographic maps) with locations of a homolog subjective set of points (e.g. collected by means of interviews and sketch maps) using a bi-dimensional regression analysis. After explaining the methodological flow, we demonstrate the applicability of the method to the assessment of spatial cognition of forest fragments in a 25 km^2 agricultural landscape of northern France.

Keywords Bidimensional regression analysis · Cognitive mapping · Forest fragmentation · Objective landmarks · Subjective landmarks

1 Introduction

The Sustainable Forest Management (SFM) paradigm was adopted at the Conference of the Earth in 1992 [16]. It recommends combining material values (economic and environmental) with non-material values (social, cultural, aesthetic, spiritual) [1, 13, 18]. Therefore, it now makes sense to evaluate the mental representations that people have about forests. We focused in this research on the development of a method for mapping of spatial cognition, i.e. the ability to locate places in geographic space. Analysis of this knowledge can be useful for the management of the forest fragments and predicting the behaviour and reactions of people.

F. Roulier · E. Gallet-Moron · G. Decocq (✉)
Jules Verne University of Picardie, EDYSAN, UMR 7058 CNRS-UPJV, Amiens, France
e-mail: guillaume.decocq@u-picardie.fr

F. Pedrotti and E. O. Box (eds.), *Tools for Landscape-Scale Geobotany and Conservation*, Geobotany Studies, https://doi.org/10.1007/978-3-030-74950-7_9

We propose to assess these cognitive representations of space to show differences between objective spaces (measured according to specific rules) and subjective spaces (knowledge of individuals about space). Basically, our method is based on a bidimensional regression analysis [14, 15] that compares locations of an objective set of points (collected from a topographic map) with locations of a homologous subjective set of points (collected by means of a sketch map). The deformations and statistics produced by the analysis teach us about the cognitive representations of this specific space.

However, the bidimensional regression was applied most often to analyze intra-urban cognition in Strasbourg [4–6]; Swansea [9]; Edmonton [8]; Beijing [12], and Tokyo [17]. How to exploit this powerful method best when mental representations of space are related to rural areas? To illustrate the possibilities of our method, we will map the spatial cognition of forest patches located in an area of 25 km^2 in Thiérache (northern part of the department of Aisne, France).

2 Methodological Issues

The general question is: how location of forest fragments (more generally, of landscape features in rural areas) is transformed in our minds? Three issues arise when one wants to answer this question using bidimensional regression. The first issue is related to the **designation of forest fragments**. Comparing the objective location of an object with its subjective location assumes this object is identified by the individual [10] but also that it can be referred to by a name. Unlike the urban environment where places have names and are clearly identifiable (library X, statue Y, square Z …), forest fragments, especially the smallest, have no particular designation. Even if there is no doubt that forest fragments may register in the cognitive representation of space, it is difficult to designate them with a name during data collection. The second issue relates to **the geometric nature of forest patches**. The parameters of the bidimensional regression are calculated originally for a set of homologous points. However, the subjects of research do not necessarily correspond to point features. How, in this case, assigning the information on spatial cognition of polygonal forest fragments for comparison with other data, including naturalistic data? The third issue concerns the **data resource on small space extents**. Compared to a similar surface in urban areas, the number of people likely to be questioned is necessarily limited and the number of points for bidimensional regression is much lower.

3 Methodological Solution

Our proposal seeks to resolve the three issues identified above (Fig. 1). First, we collect cognitive data using a survey. The issue of landmark scarcity in rural areas can be solved by using a pre-survey to identify the remarkable places better (villages, buildings…) or by an extension of the study area. We chose the first option because the window was predetermined in the research project. 1. We collect the image (cognitive) points using a sketch map. 2. The designation of forest fragments will be conducted indirectly during the data collection. We get the code for identified fragments by superimposing and intersecting forest areas drawn by each individual on a reference map (without forest fragments) with maps of forest fragments identified by a numerical code. 3. Then, we adjust the set of cognitive points to the set of homologous reference points using bidimensional regression that produces statistics about the cognitive models. 4. To [**affect??**] the results of adjustment to polygonal forest fragments, we will generalize the cognitive models by interpolation across the study area (using a multiquadric interpolation of a grid). This produce a space deformation. 5. The interpolation also generates statistics over the whole surface, which allows us to calculate the inaccuracy surface. 6. Then we are able to calculate statistical areas corresponding to the averages of the values contained in each forest patch. Statistics are mapped only for identified patches. We can use for mapping both interpolated and non-interpolated patches.

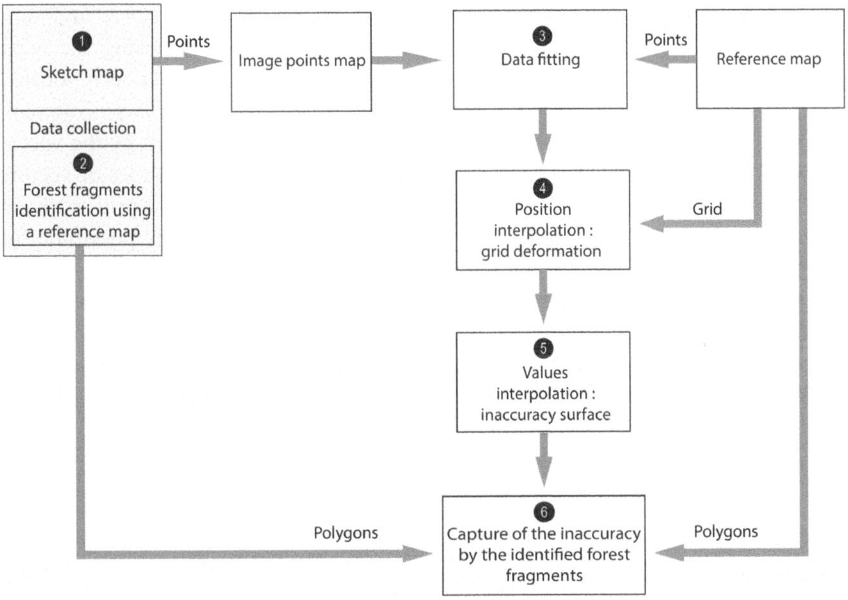

Fig. 1 General methodology

4 Application

A test survey was conducted over an area of 25 km^2 north of the Aisne[1]. The original forest there was fragmented over time by human activities. This area is crossed from north to south by the National Road 2. This context provides a framework to test our method. In order to test it in rural areas, we chose to explore a barrier effect produced by the National Road 2. The barrier effect corresponds to an overestimation of the distances between places separated by a geographical barrier [7, 11]. For example, it has been established that such distances were overestimated for places situated on either side of a river [2] or former iron curtain [3]. The data are intended to be compared and aggregated. So, only individuals who have drawn the same places have been retained for the analysis, which explains that the sample has been (drastically) reduced to a homogeneous set of 9 respondents and 11 landmarks.

For the full sample we note a relative lack of knowledge of the sector, revealed by a rather low goodness-of-fit (r^2= 0.69). The overestimation of distances suggests the reality of the desired effect. This overestimation is not uniform in intensity, nor in orientation. These variations lead to inconsistencies expressed through expansions and compressions of space (Fig. 2c). The displacement vectors (Fig. 2a), the vector field (Fig. 2b) and the deformation of the surface (Fig. 2d) show that significant distortions are mostly located west of the National Road 2 (dilatation) and to a lesser extent, east (compression). All analyses were done using morphoses software (http://morphoses.eu) [19].

We chose to map the inaccuracy of responses through the error of location. The error corresponds to the importance of the difference between objective and subjective positions, regardless of the orientation error. By superimposing the error surface over forest fragments, it is possible to identify those that are poorly located (Fig. 3b). Some fragments are very large. The average values for a large fragment could hide very different and even opposite local situations, as shown by the vector field around the fragment BT10 that combines short and long magnitudes of vectors (Fig. 3c). We have therefore chosen to transfer location errors to a grid made of square cells 40 meters on a side (Fig. 3a). This method preserves the possibility of comparing the data on knowledge with naturalistic data that can be easily associated with each cell. Finally, the low density of landmarks (fortunately well distributed here) creates areas where knowledge cannot be taken into account in the cognitive model. All these calculations were made using the plugin Morphoses for MapWindow, developed at EDYSAN research unit (http://morphoses.eu).

---

[1]*smallFOREST project*, «Biodiversity and ecosystem services of small forest fragments in European landscapes» - *EraNet-BiodivERsA 2011-40.*

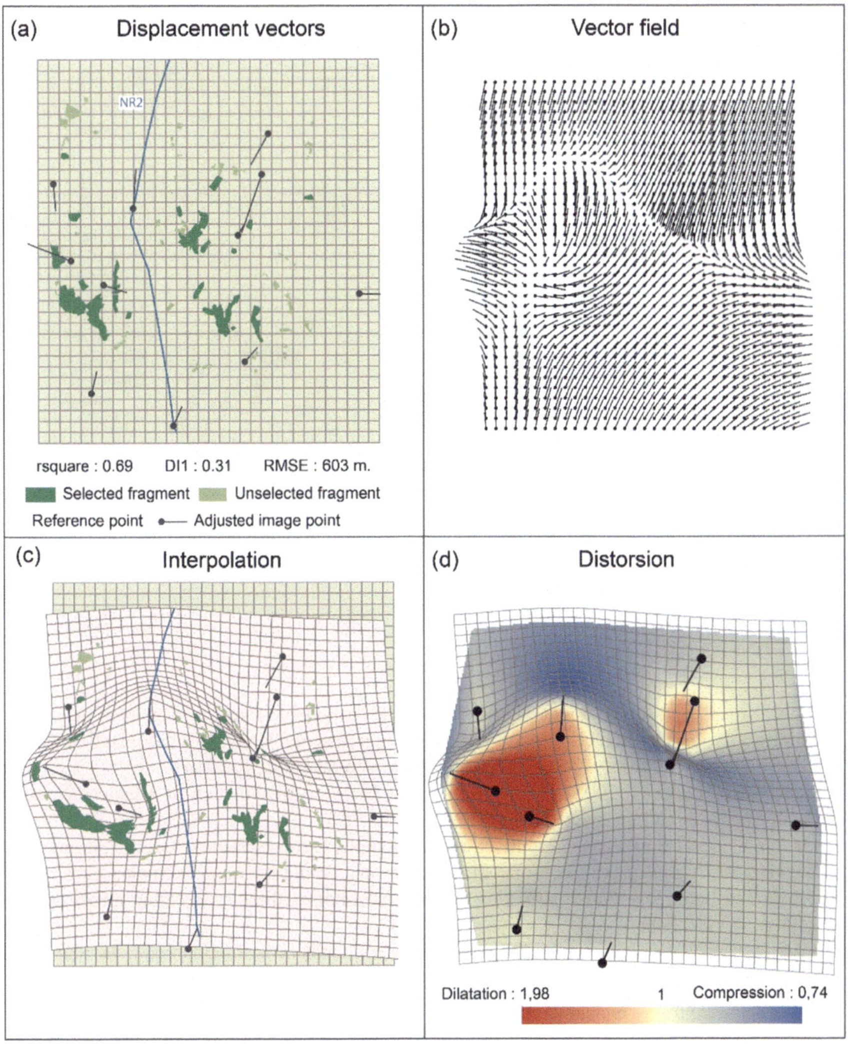

Fig. 2 Inconsistencies in the cognitive space

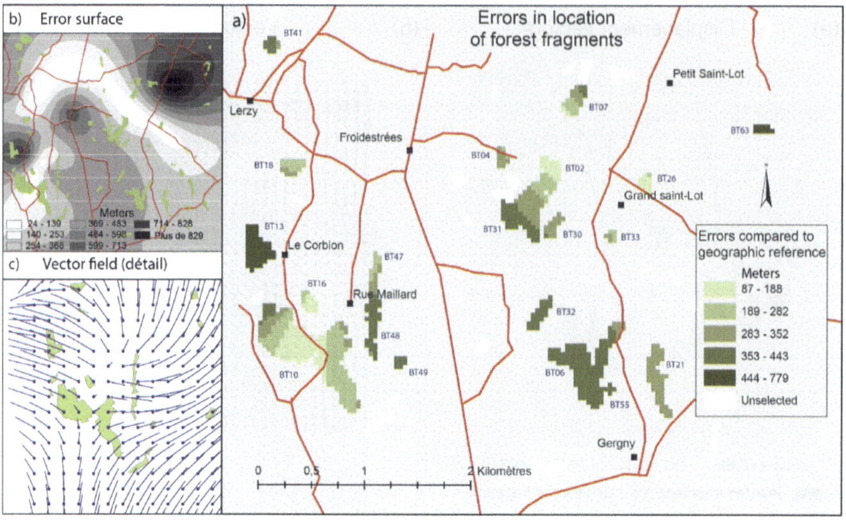

Fig. 3 Error in the knowledge of location

References

1. Angelstam P, Persson R, Schlaepfer R (2004) The sustainable forest management vision and biodiversity–barriers and bridges for implementation in actual landscapes. Ecol Bull 51:29–49
2. Canter D (1977) The psychology of place. Architectural Press, London
3. Carbon CC, Leder H (2005) The Wall inside the brain: overestimation of distances crossing the former iron curtain. Psychon Bull Rev 12:746–750
4. Cauvin C (1984a) Espaces cognitifs et transformations cartographiques. Les conditions de la comparaison des espaces cognitifs: de la carte aux configurations. Exemples de l´espace urbain strasbourgeois, Strasbourg, Université de Strasbourg, thèse de doctorat, micro-fiches à Lille n° 85 09 2134
5. Cauvin C (1984b) Une méthode générale de comparaison cartographique: la régression bidimensionnelle. Travaux et Recherches. Fascicule 4. ERA 214 (CNRS). Thèse de doctorat d'Etat, Fascicule E, Texte polycopié, Strasbourg
6. Cauvin C, Chernai Z, Daniilidis K (1998) Usagers et représentations cognitives de la ville: exemples à Strasbourg. In: Reymond H, Cauvin C, Kleinschmager R (eds) dir L'Espace géographique des villes: pour une synergie multistrates, Paris, Anthropos, coll. « Villes » , pp 301–348
7. Cohen R, Baldwin L, Sherman RC (1978) Cognitive maps of naturalistic setting. Child Dev 49:1216–1218
8. Friedman A, Kohler B (2003) Bidimensional regression: assessing the configural similarity and accuracy of cognitive maps and other two-dimensional data sets. Psychol Methods 8 (4):468–491
9. Kitchin R (1996) Methodological convergence in cognitive mapping research: investigating configurational knowledge. J Environ Psychol 16:163–185
10. Lynch K (1998) L'image de la cité. Dunod, Paris
11. Newcombe N, Liben L (1982) Barrier effects in the cognitive maps of children and adults. J Exp Child Psychol 34:46–58

12. Shen S, Xue L, Wu L, Liu Y (2007) Cognitive distortions on MDS configuration an sketch mapping: a case study in Beijing city. In: Li M, Wang J (eds) Geoinformatics 2007: cartographic theory and models, Proceeding of SPIE, vol. 6751, p 67511A
13. Tabbush P (2004) Public money for public good? public participation in forest planning. Forestry 77:145–156
14. Tobler WR (1965) Computation of the corresponding of geographical patterns. Papers Reg Sci Assoc 15:131–139
15. Tobler WR (1994) Bidimensional regression. Geogr. Anal. 26(3):187–212
16. United Nations (1992) Earth Summit 1992: The United Nations Conference on Environment and Development, Rio de Janeiro 1992, London, The Regency Press Corporation
17. Wakabayashi Y (1994) Spatial analysis of cognitive maps. Geogr Rep Tokyo Metrop Univ 29:57–102
18. Wyder J (2001) Multifunctionality in the Alps: challenges and the potential for conflict over development. Mt Res Dev 21:327–330
19. Roulier F (2018) Une adaptation de la régression bidimensionnelle aux polygones: un exemple en cognition spatiale. Cybergeo: European Journal of Geography. https://doi.org/10.4000/cybergeo.29082

Phytosociology

California Ultramafic Vegetation: A Phytosociological Update

Daniel Sánchez-Mata

Abstract The vegetation of ultramafics soils throughout the California state landscapes are summarized. Most important features related with the biogeographical distribution, structure, dynamics, floristic composition and related data of the main vegetation types are given and considered. Two appendices (taxonomic and syntaxonomic) complete this contribution; taxonomic appendix includes some nomenclatural proposals for the California flora.

Keywords California vegetation · Ultramafic vegetation · California flora · California serpentine vegetation

1 Introduction

Ultramafic rocks are a broad group that includes (besides serpentinite) peridotite, ophiolite, gabbro, dunite, pyroxenite and hornblendite, among others. Frequent minerals and elements in these rocks are olivine, chrome, nickel, antimony, cobalt, mercury, and gold [8, 10, 11]. All these ultramafic rock types cause similar stresses to plants, and all are present in the state of California (USA). Serpentinite is relatively easy to determine in the field because of its shiny or silky luster, slightly soapy feel, conchoidal fractures and greenish surface.

California ultramafic outcrops are typically associated with deep geological processes such as fault lines, which is why they occur in strips trending southeast to northwest, parallel to other neighboring fault lines (Fig. 1). Nearly all these ultramafic materials were metamorphosed while being emplaced in, or rising up through, the continental crust.

Soils developed on ultramafic areas (ultramafic soils) are usually shallow and skeletal. The steepness of the slopes and sparseness of the vegetation allow for

D. Sánchez-Mata (✉)
Department of Pharmacology, Pharmacognosy and Botany (Botany Unit), Faculty of Pharmacy, Complutense University, 28040 Madrid, Spain
e-mail: dsmata@ucm.es

F. Pedrotti and E. O. Box (eds.), *Tools for Landscape-Scale Geobotany and Conservation*, Geobotany Studies, https://doi.org/10.1007/978-3-030-74950-7_10

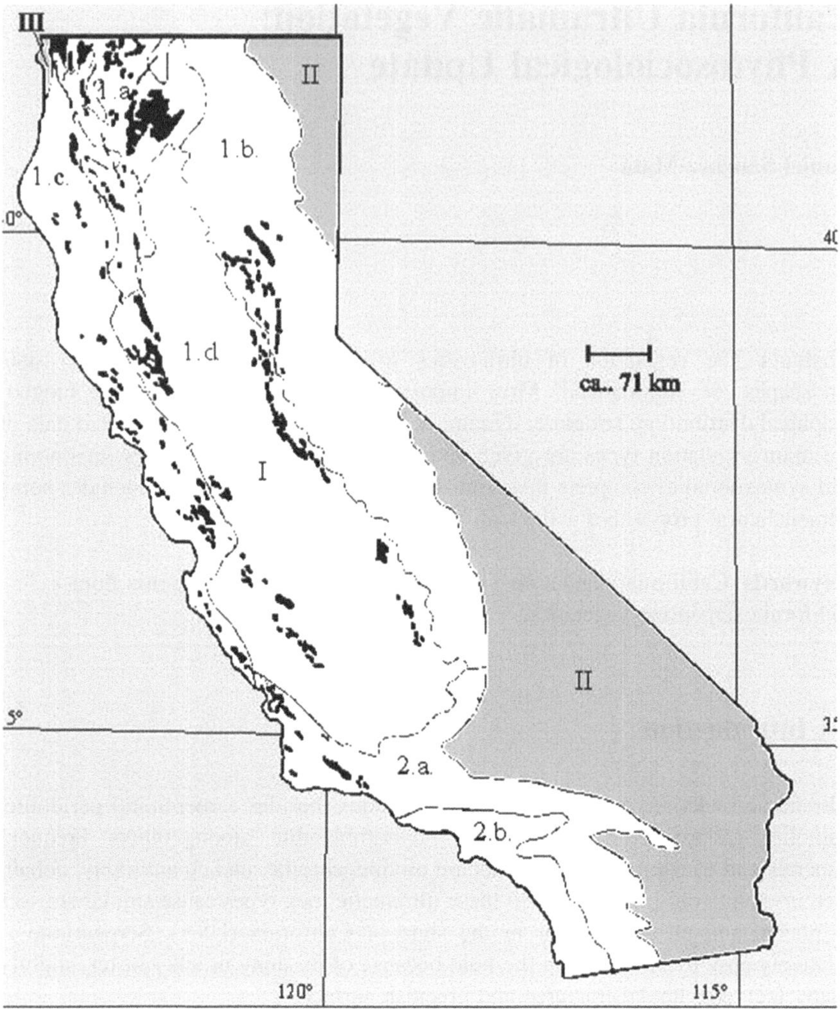

Fig. 1 Biogeographical map of the Californian Region (I) with the distribution of the main ultramafic (serpentine) outcrops (black areas), adapted from Sánchez- Mata [35]. The geological information was obtained from Kruckeberg [18] and the biogeographical limits and unit names are adapted from Rivas- Martínez et al. [25] (II. Great Basin Region; III. Rocky Mountain Region) North California Province: I.1.a. Klamath-Ciscascadian Sector; I.1.b. Sierra Nevada-Transcascadian Sector; I.1.c. North Californian Coastal Sector; I.1.d. California Great Valley Sector. South California Province: I.2.a. South Californian Ranges Sector; I.2.b. South California Coast & Channel Islands Sector

continual erosion. The most frequent soils in California's ultramafic territories are Lithic Argixerolls, Lithic Haploxerolls and Pachic Argixerolls [1], which all have low concentrations of calcium, an essential nutrient for plants and whose low availability reduces plants' ability to grow [16, 17, 20, 21, 37–39, 42]. High magnesium levels and the presence of several heavy metals add further stresses. Metals are found in large quantities in the tissues of certain plants growing on ultramafics [7]. Plants on ultramafics have heightened drought resistance [38] and tolerance to heavy metals [3].

In summary, ultramafic soils impose the following stresses on plants: calcium and magnesium imbalance, magnesium toxicity, low availability of molybdenum, toxic levels of heavy metals, occasional high alkalinity, low concentrations of such essential nutrients as nitrogen and phosphorus, and low soil water storage capacity [19].

Nomenclature: Taxonomic nomenclature follows the proposals of the second edition of *The Jepson Manual: Vascular Plants of California* [5] except those included in the taxonomic appendix at the end of this contribution. The syntaxonomical appendix includes all the syntaxa known and published from the California ultramafic (serpentine) areas, and mentions their authority and year of publication; the vegetation syntaxonomy follows the current issue of the International Code of Phytosociological Nomenclature [41].

2 Background

The state of California has > 3200 km^2 of ultramafic outcrops. These peculiar geological materials extend west-east from the North Coast Ranges to the west-facing slopes of the Sierra Nevada, and north-south from the Klamath-Siskiyou Mountains to the South Coast Ranges (Fig. 1).

A brief compilation of California's ultramafic vegetation and its diversity was published as a special chapter in the third edition of Barbour's classic book *Terrestrial Vegetation of California* [6, 32]. All the ultramafic vegetation types can be considered as xero-edaphic or hygro-edaphic climax. Sargent cypress woodlands, pine-oak woodlands, serpentine chaparrals, pine woodlands and forests, conifer forests, and riparian willow formations are the main vegetation types recognized throughout the California ultramafics [31, 34, 35].

Continentality is the most important environmental factor driving the community diversity of ultramafic vegetation in California, as revealed by our own research and field studies. In hyperoceanic areas (ultra-hyperoceanic, eu-hyperoceanic and sub-hyperoceanic) with meso-mediterranean thermotypes (Ic < 11) along the Pacific coast (from Mendocino to San Luis Obispo Counties) and in the San Francisco Bay Area the potential natural vegetation is conifer forests structured by the endemic Sargent cypress (*Hesperocyparis sargentii*). Permanent and seasonal fog, high air humidity, and the maritime oceanic realm strongly support the existence of this endemic California vegetation type, and disturbed stands of this peculiar forest vegetation allow the development of ultramafic chaparral as seral vegetation types on sunny slopes and in clearings. These climatic features are evidenced (in terms of floristic composition) by the absence of *Pinus sabiniana* in the most oceanic seral ultramafic chaparral association, *Ceanotho jepsonii-Quercetum duratae*, which is replaced by another conifer tree, *Hesperocyparis sargentii*, forming potential forest communities. However, differences in continentality are less marked toward the interior of the continent (along similar altitudinal ranges), as occurs throughout the inner Coast Ranges and the Sierra Nevada foothills (Great Valley territories, west to east sides). The differentiation of ultramafic chaparral communities can thus be explained mostly by biogeography and altitude.

Pine-oak woodlands on ultramafics at low elevations are related to formations dominated by blue oak (*Quercus douglasii*) and grey pine (*Pinus sabiniana*), which are widely distributed throughout the foothills that surround the Great Valley; all these oak-dominated vegetation types are included in the phytosociological class *Heteromelo arbutifoliae-Quercetea agrifoliae* (Fig. 2).

Fig. 2 Above, left: Sargent cypress woodland (*Hesperocyparis sargentii*) at Mount Tamalpais State Park, 750 m asl; right: mainland ultramafic chaparral (*Ceanotho albiflori-Quercetum duratae*) at Napa County, 300 m asl. Below, left: fruited branch of leather oak (*Quercus durata*), Napa County, 250 m asl; right: grey pine-blue oak woodland (*Pino sabinianae-Quercetum douglasii*) at Lake County, 350 m asl

Chaparral is a complex of many shrub-dominated vegetation types that cover 8.5% of California [6, 15]. California chaparral can be climatic climax or seral vegetation; that is, it can develop as climatic climax vegetation on sites that cannot support large forested communities mainly owing to climatic or edaphic factors.

California ultramafic (serpentine) chaparrals support the potential vegetation in oceanic ultramafic areas (semihyperoceanic, euoceanic and semicontinental territories with Ic = 11–21) following the proposals by Rivas-Martínez et al. [27] with (thermo)-meso-mediterranean thermotypes and arid to lower humid ombrotypes (Fig. 2). These chaparrals show an interesting spectrum of plant functional traits as it has been recently published [13].

Throughout the Californian ultramafic areas with a supra-mediterranean, oro-mediterranean and oro-submediterranean (oro-temperate sub-mediterranean) thermotype (mountain and high-mountain areas and Klamath-Siskiyou territories), the potential ultramafic (serpentine) chaparrals are replaced by conifer forests such as Pacific ponderosa pine forests (*Pinus ponderosa* subsp. *pacifica*) in mountain areas; Jeffrey pine forests (*Pinus jeffreyi*) at higher altitudes; northern foxtail pine woodlands (*Pinus balfouriana*) and Shasta red fir forests (*Abies shastensis*) in high-mountain landscapes in the Shasta-Klamath-Siskiyou territories; and southern foxtail pine woodlands (*Pinus austrina*) and Critchfield red fir forests (*Abies critchfieldii*) in the high mountains of southern Sierra Nevada. All these conifer vegetation units are framed within the phytosociological class *Calocedro decurrentis-Pinetea jeffreyi* (Fig. 3).

California ultramafic grasslands comprise both very diverse perennial and annual formations; the plant communities structured by annuals (therophytes) were first described and published by Rodríguez-Rojo et al. [28, 29] with the proposal of the phytosociological class *Vulpio microstachyos-Hesperolinetea micranthi*. Barrens are also devoid of woody vegetation, and occur on steep sites where the parent rock is serpentinized, as also occurs in rocky outcrops. These habitats support diverse plant communities that are rich in endemic taxa. In general, ultramafic areas, especially barrens and rocky outcrops, are refugia for endemics and native taxa as they serve as islands of native flora and are barely modified by exotic taxa that are invasive elsewhere, but intolerant of ultramafic soil [2, 12, 23, 30].

Fig. 3 Above, left: Group of mature cones, Shasta red fir (*Abies shastensis*) in Mount Eddy forests at 1800 m asl; above, right: cone of Siskiyou foxtail-pine (*Pinus balfouriana*). Below: Mount Eddy woodlands on ultramafic soils (2300 m asl) are structured by noble fir (*Abies procera*), mountain hemlock (*Tsuga mertensiana*), Shasta red fir (*Abies shastensis*), and foxtail pine (*Pinus balfouriana*)

3 A Summary of Ultramafic Vegetation Units

3.1 North Coast Ranges

Ultramafics are a frequent geological feature at low elevations in the North Coast Ranges [22]. The most extensive vegetation types are Sargent cypress woodlands (*Hesperocyparis sargentii*) toward the coast; pine-oak woodland (*Pino sabinianae-Quercetum douglasii*) and chaparrals (*Quercion duratae*) in the more continental areas, such as from Sonoma County to Mendocino and Humboldt Counties (Fig. 2); and riparian scrub (*Salicion breweri*). All are xero-edaphic climax vegetation types, except riparian scrub which is hygro-edaphic.

3.1.1 Sargent Cypress Woodland

Sargent cypress woodlands (*Hesperocyparis sargentii*) cover oceanic, maritime ultramafic sites (e.g. San Francisco Bay region and some coastal territories of Humboldt, Mendocino, Sonoma, Marin, Santa Clara, San Mateo, Monterey, San Benito, San Luis Obispo and Santa Barbara Counties, Fig. 2). These woodlands extend as a riparian formation close to Brewer willow scrub into more protected inland areas such as canyons, inland valleys, at the head of streams, and where there is a frequent occurrence of summer fog (*Calycantho occidentalis-Hesperocyparetum sargentii, Salicion breweri*); disturbed sites allow the development of seral scrub formations structured by *Frangula tomentella* and *Rhododendron occidentale* (*Frangulo tomentellae-Rhododendretum occidentale, Salicion breweri*) [35].

The floristic composition of Sargent cypress woodlands growing on ultramafics is very consistent from stand to stand. *Hesperocyparis sargentii* is usually the overwhelming or sole dominant tree in the overstory, but *Pinus sabiniana* may co-occur in northern and central California and *P. coulteri* in central and southern California. In the most mesic or oceanic situations, scattered elements from mixed evergreen forest are also present (e.g., *Arbutus menziesii, Notholithocarpus densiflorus, Pseudotsuga menziesii, Quercus agrifolia, Q. chysolepis* and *Umbellularia californica*).

Leather oak (*Quercus durata*) is the most common shrub forming seral communities in disturbed sites; *Adenostoma fasciculatum, Arctostaphylo*s spp., and *Pickeringia montana*, among others, are also frequent shrubs. Adjacent herbaceous communities include the perennial species *Calamagrostis ophitidis, Elymus multisetus, Festuca californica, Galium andrewsii, Melica californica, M. torreyana*, and *Polygala californica*, plus many ephemerals. Adjacent rocky places are dominated by *Allium falcifolium, Asclepias solanoana, Calochortus raichei* (The Cedars, Sonoma County), *Cymopterus terebinthinus* var. *californica, Galium californicum, Myriopteris gracillima, Streptanthus morrisonii* s.l., and *S. tortuosus* var. *suffrutescens*.

3.1.2 Riparian Scrub

Brewer willow formations (*Salix breweri*) grow on ultramafic streambeds, creeks and canyons in lowlands throughout the North and South Coast Ranges (*Salicion breweri*). They are vicariants of Del Norte willow groves (*Salix delnortensis*) on Klamath-Siskiyou ultramafics (*Salicion delnortensis*).

Associated shrubs include *Baccharis glutinosa*, *Calycanthus occidentalis*, *Salix breweri*, *Salix delnortensis*, and *Salix lasiolepis*. Springs and peaty communities contain such rare and threatened plants as *Carex mendocinensis*, *Cypripedium californicum*, *Delphinium uliginosum*, *Epipactis gigantea*, *Gentiana setigera*, *Helenium bigelovii*, *Parnassia palustris* and *Triteleia peduncularis*.

Another riparian scrub type is dominated by *Frangula tomentella* and *Rhododendron occidentale*, and related with the Sargent cypress woodlands as hygro-edaphic climax formations in southern areas of the North Coast Ranges and central Great Valley territories, as mentioned (*Salicion breweri*) [35].

3.2 Great Valley and Sierra Nevada

The most characteristic vegetation developed on ultramafic areas within the Great Valley is pine-oak woodlands on deep soils developed from scarcely weathered ultramafic parent rocks throughout the foothills of the inner Coast Ranges, diverse chaparral types in the more continental areas spreading into the foothills on soils derived from strongly serpentinized parent rocks, and conifer woodlands at higher elevations that also grow on strongly serpentinized soil.

3.2.1 Pine-Oak and Pine Woodlands

Pine-oak woodlands are open formations dominated by blue oak (*Quercus douglasii*) and grey pine (*Pinus sabiniana*) (*Pino sabinianae-Quercetum douglasii*). They occur from the northwest of Glenn County to the Klamath region, east into the montane zones of the Sierra Nevada, and in south to mid-elevations in Monterey, San Benito, San Luis Obispo, Fresno, and Kern Counties. These woodlands can be considered as climatic climax vegetation in lower-altitude ultramafic territories (Fig. 2).

Pacific Ponderosa pine woodlands (*Pinus ponderosa* subsp. *pacifica*) occur mainly on low-elevation ultramafics in the northwestern part of the Great Valley and along some west-facing slopes of the Sierra Nevada foothills. They grow on strongly serpentinized soil in the vicinity of Pacific ponderosa pine-California black oak woodland on deep, non-ultramafic soil (*Querco kelloggii-Pinetum pacificae*) [33]. These woodlands usually have three canopy layers. Pacific ponderosa pine is frequently associated with *Calocedrus decurrens*, *Pinus attenuata*, *P. lambertiana*, *P. ponderosa* subsp. *pacifica* x *P. jeffreyi* [4, 9], *P. sabiniana*, and *Quercus chysolepis*. The shrub layer is dominated by *Arctostaphylos viscida* and *Ceanothus cuneatus*, associated with *Arctostaphylos patula*, *A. viscida* x *A. patula*, *Ceanothus lemmonii*, *Eriodictyon californicum*, *Hesperocyparis macnabiana* (local), *Heteromeles arbutifolia*, *Pickeringia montana* and *Quercus breweri* (northwest border of Sacramento Valley). The herb layer is a mix of sparse perennials (*Allium campanulatum*, *Ceanothus prostratus*, *Elymus elymoides*, *Eriophyllum lanatum* s.l., *Monardella villosa* s.l., *Penstemon laetus* s.l., *Polygala cornuta* and *Salvia sonomensis*) and many annual grasses and forbs.

Jeffrey pine woodlands are also present on ultramafics, but growing at higher elevations, especially in the Sierra Nevada (*Querco vaccinifoliae-Pinetum jeffreyi*). At mid-elevations, they are present in Monterey, San Benito, San Luis Obispo, Fresno and Kern Counties, and also in the Sierra Nevada. Hybrid populations of *Pinus ponderosa* subsp. *pacifica*, *P. jeffreyi* and *P. coulteri* (central and south Coast Ranges) can be found in some localities [4, 9]. Associated trees include *Calocedrus decurrens*, *Pinus attenuata*, *P. coulteri*, and *P. lambertiana*.

3.2.2 Chaparral

Ultramafic chaparral types within the Great Valley are diverse, but all are dominated and structured by the leather oak (*Quercus durata*) which is also a California endemic (*Quercion duratae*). Scattered trees of grey pine (*Pinus sabiniana*) and Coulter pine (*Pinus coulteri*) can be found in central and southern areas. *Quercus durata* s.str., a Californian endemic distributed northward of Santa Barbara County, is the only American species of *Quercus* that has a restricted distribution on ultramafics (Fig. 2).

Four ultramafic chaparral types were recently recognized and published for California by Sánchez-Mata and Rodríguez-Rojo [34], covering all the xero-edaphic climatic communities developed on oceanic sites (see syntaxonomical appendix), some ultramafic chaparrals in hyperoceanic areas (Ic < 11) can also be considered as seral vegetation within the Sargent cypress forest vegetation series, as occurs throughout the California Pacific coastal territories.

These chaparrals are low open formations growing on strongly weathered (serpentinized) parent rocks that are widely distributed from Santa Barbara County northward through the Coast Ranges and Sierra Nevada foothills. The shrubs exhibit xeromorphic traits: they are dwarfed and compact, they have low productivity, and the evergreen leaves are mainly malacophyllous, hairy and with reduced size [13]. Frequent dominants include *Adenostoma fasciculatum*, *Arctostaphylos* spp., *Ceanothus cuneatus, C. jepsonii* subsp. *albiflorus*, *Frangula californica* s.l., *Fremontodendron californicum*, *F. decumbens*, *Garrya congdonii*, *Heteromeles arbutifolia*, *Quercus durata* and *Rhamnus crocea*. Scattered trees, sometimes growing as tall shrubs, include *Pinus sabiniana*, *Pseudotsuga menziesii*, *Quercus agrifolia*, *Q. breweri*, *Q. chrysolepis* and *Umbellularia californica*. There is considerable open intershrub space occupied by perennial bunch grasses, subshrubs, and annual herbs. *Arctostaphylos* spp. and *Ceanothus* spp. are the most diverse genera in California and contain many narrowly endemic taxa [14], they play an important role in differentiating community diversity.

Throughout California ultramafic areas with a supra-mediterranean, oro-mediterranean and oro-submediterranean (oro-temperate sub-mediterranean) thermotype (mountain and high mountain areas, and Klamath-Siskiyou territories), the potential ultramafic (serpentine) chaparrals are replaced by conifer woodlands and forests, as mentioned.

3.3 Klamath-Siskiyou Mountain Region

The ultramafic vegetation in this unique area is very diverse and more closely related to that of Sierra Nevada than to that of the North Coast. Typically, several main ultramafic vegetation types are recognized: Jeffrey pine woodlands (*Pinus jeffreyi*); Shasta red fir forests (*Abies shastensis*) with Brewer spruce (*Picea breweriana*), noble fir (*Abies procera*) and mountain hemlock (*Tsuga mertensiana*); subalpine woodlands; and whitebark-pine woodlands on rocky summits (*Pinus albicaulis*). Jeffrey pine and whitebark-pine woodlands can be considered xero-edaphic climax vegetation, while both fir forest formations are climatic climax vegetation. In addition to these main forest types it is worth noting some unusual vegetation types, often very rich in endemics and restricted to the ultramafic Klamath-Siskiyou high-mountain areas, frequently with a local character.

3.3.1 Jeffrey Pine Woodland

The floristic composition of higher-elevation Jeffrey pine stands depends on the type of ultramafic substrate, whether serpentinite (B1) or diorite/granodiorite (B2). Associated trees include *Abies lowiana*, *Calocedrus decurrens*, *Notholithocarpus echinoides*, *Pinus contorta* subsp. *murrayana*, *P. jeffreyi* x *P. ponderosa* subsp. *pacifica*, *P. lambertiana*, *P. monticola*, *P. sabiniana* (local), *Quercus chrysolepis* and *Q. kelloggii*, and scattered individuals of *Pseudotsuga menziesii* and *Umbellularia californica*. There are numerous shrub species: *Amelanchier alnifolia* s.l. (B1, B2), *Arctostaphylos nevadensis* (B2), *A. patula* (B2), *A. patula* x *A. nevadensis* (*A.* x*barbouri*, B2), *A. viscida* (B1), and *Ceanothus cordulatus*.

These open woodlands occur in the most xeric ultramafic sites (mainly growing on peridotite, gabbro and serpentinite-derived soils) from low to higher elevations (900–2000 m asl). *Pinus jeffreyi* and *Calocedrus decurrens* are the only common trees present in these open woodlands; *Abies lowiana*, *Hesperocyparis bakeri*, *Pinus attenuata*, *P. monticola*, *P. lambertiana*, *P. ponderosa* subsp. *pacifica* *P. ponderosa* subsp. *pacifica* x *P. jeffreyi*, and *Pseudotsuga menziesii* co-occur as scattered individuals. Open spaces are covered by shrubs and small trees such as *Arctostaphylos klamathensis*, *A. nevadensis*, *A. viscida* s.l., *A. patula*, *Ceanothus cuneatus*, *Frangula californica* s.l., *Garrya buxifolia*, *G. fremontii*, *Notholithocarpus echinoides*, *Quercus breweri*, *Q. vacciniifolia*, *Vaccinium parvifolium*, and shrub forms of *Umbellularia californica*. Perennial grasses (*Elymus elymoides* s.l., *E. glaucus*, *Festuca californica*, *F. idahoensis*, *Melica californica*, *M. geyeri* and *Stipa lemmonii*) and forbs are often present. These woodlands are framed in the association *Aspidoto densae-Pinetum jeffreyi* [24].

3.3.2 Fir forests

Fir forests with occasional Brewer spruce (*Picea breweriana*) occur in several localities in northwest California and southern Oregon on ultramafics at elevations of 1200–1800 m asl. The geographic area of Brewer spruce includes Curry, Josephine, and Jackson Counties in Oregon and Del Norte, Siskiyou, Humboldt, Trinity and Shasta Counties in California [40]. These forests are mainly structured by white fir (*Abies lowiana*) at mid-elevations and by Shasta red fir (*Abies shastensis*), noble fir (*Abies procera*) and mountain hemlock (*Tsuga mertensiana*) at higher elevations. *Pinus contorta* subsp. *murrayana*, *P. jeffreyi*, *P. lambertiana*, *P.monticola*, and *Pseudotsuga menziesii* are common associates (Fig. 3). The understory is frequently dominated by shrubs such as *Arctostaphylos canescens* s.l., *A. nevadensis*, *A. patula*, *Ceanothus pumilus*, *Quercus sadleriana*, *Q. vacciniifolia*, *Vaccinium membranaceum*, *V. parvifolium* and *Xerophyllum tenax*. Herbs include *Achlys triphylla*, *Asarum hartwegii*, *Corallorhiza maculata*, *Goodyera oblongifolia*, *Linnaea americana*, and *Prosartes hookeri* s.l., among others.

Noble fir and mountain hemlock forests grow at elevations of > 1800 m asl, sometimes with Shasta red fir (*Abies shastensis*). These highland forests should be framed in the phytosociological class *Tsugetea mertensiano-heterophyllae* [25, 26]. In a few localities, *Abies lasiocarpa* is an associate in this forest type [36] (Fig. 3).

3.3.3 Subalpine Woodland

In some high exposed rocky places that do not accumulate snow-packs there is a subalpine woodland structured by *Pinus albicaulis*. The understory contains *Cercocarpus ledifolius* s.l., *Quercus vacciniifolia* and *Purshia tridentata*, and can be considered to be a xero-edaphic climax. Somewhat more protected mesic sites support an open foxtail pine woodland (*Pinus balfouriana*); this type of vegetation is more restricted to ultramafics than whitebark pine woodland [36]. Associated trees include *Abies shastensis*, *Pinus albicaulis*, *P. contorta* subsp. *murrayana* and *P. monticola*, and common understory herbaceous perennials in openings are *Aconogonon davisiae*, *Aspidotis densa*, *Carex rossii*, *Eriogonum alpinum* and *Lewisia leana*.

The subalpine woodland that dominates the most mesic subalpine sites, e.g. at the bottom of late-melting snowpacks, is structured by *Tsuga mertensiana* in the overstory and *Phyllodoce empetriformis* in the understory. This woodland type can be considered as a permanent climatic vegetation type.

3.3.4 Port Orford Cedar Forest and Adjacent Wetlands

This unusual ultramafic forest type is dominated by Port Orford cedar (*Chamaecyparis lawsoniana*) and grows in ravines, along streams and on protected mesic slopes where there is a strong maritime influence [44]. Associated tree species in this hygro-edaphic forest include *Acer macrophyllum*, *Alnus rubra*, *Pinus monticola*, *Pseudotsuga menziesii* and *Taxus brevifolia*. Common sub-canopy trees and shrubs are *Cornus nuttallii*, *Quercus sadleriana*, *Rhododendron macrophyllum*, *R. occidentale*, *Ribes bracteosum*, *Rosa californica* and *Physocarpus capitatus*. The perennial herbs and ferns *Adiantum aleuticum*, *Darmera peltata*, *Polystichum munitum*, *Streptopus amplexifolius* var. *americanus*, and *Xerophyllum tenax* are also frequent.

Permanent water runoffs on ultramafics produce fen meadows structured by Cobra lily (*Darlingtonia californica*). *Calliscirpus criniger*, *Caltha leptosepala*, *Carex echinata*, *Castilleja miniata* subsp. *elata*, *Cypripedium californicum*, *Epipactis gigantea*, *Gentiana setigera*, *Hastingsia alba*, *Lilium bolanderi*, *L. pardalinum* subsp. *vollmeri*, *Narthecium californicum*, *Platanthera sparsiflora*, *Raillardella pringlei*, *Rudbeckia glaucescens*, *Pseudotrillium rivale*, *Triantha occidentalis*, and *Viola primulifolia* subsp. *occidentalis* are common perennials. Less acidic springs support megaforbs such as *Aconitum columbianum*, *Angelica arguta*, *Aquilegia eximia*, *Heracleum maximum*, *Lilium occidentale*, *Pteridum aquilinum* subsp. *pubescens*, *Platanthera dilatata* var. *leucostachys*, *Senecio triangularis*, *Valeriana sitchensis*, *Veratrum californicum*, *V. insolitum*, and *V. viride*.

3.4 Southern California

The ultramafic vegetation of southern California basically consists of chaparral in the more continental sites, pine woodlands at mid- and high elevations, Sargent cypress woodlands on uplands that experience some maritime influence, and pine-oak woodlands.

Ultramafic chaparral occurs at elevations of <500 m asl. It is structured by *Quercus durata*, associated with scattered *Pinus sabiniana* and *Umbellularia californica*. Shrubs include *Adenostoma fasciculatum*, *Arctostaphylos obispoensis*, *A. pechoensis*, *Artemisia californica*, *Ceanothus cuneatus* var. *fascicularis*, *C. papillosus*, *C. spinosus*, *Dendromecon rigida*, *Diplacus aurantiacus*, *Eriodictyon tomentosum*, *Hesperoyucca whipplei*, *Quercus berberidifolia*, *Q. palmeri*, *Salvia leucophylla*, *S. mellifera*, and *Trichostema lanatum*. Herbaceous perennials and annuals also have a floristic flavor of southern California.

Sargent cypress woodlands occur on slopes at <600 m asl in the Southern Coast Ranges exposed to summer fog. Frequent (but scattered) associated trees include *Arbutus menziesii*, *Lithocarpus densiflorus*, *Pinus coulteri*, *P. sabiniana*, *Quercus chrysolepis* and *Umbellularia californica*. The understory is dominated by leather oak (*Quercus durata*) and consists of the same chaparral shrubs mentioned above.

Jeffrey pine woodlands occupy xeric sites at mid- and high (>900 m asl) elevations that experience strong continentality. Associated trees include *Calocedrus decurrens*, *Pinus coulteri* and hybrid populations of Jeffrey pine and Coulter pine. These hybrids are particularly common on ultramafics in the South Coast Ranges [4, 9, 43]. Leather oak (*Quercus durata*) is a common and notable shrub in these pine woodlands.

Pine-oak woodlands can grow very locally on ultramafics throughout the South Coast Ranges. They are dominated by *Pinus sabiniana* and *Quercus douglasii*, but may be joined by *Pinus coulteri*, *Quercus john-tuckeri*, and *Juniperus californica* as the main tree and shrub species.

4 Appendices

4.1 Taxonomic Appendix

- *Abies critchfieldii* (Lanner) Rivas-Martínez & Sánchez-Mata
- *Abies lowiana* (Gordon) A. Murray bis
- *Abies shastensis* (Lemmon) Lemmon
- *Arctostaphylos* x*barbouri* nothosp. nov. [*Arctostaphylos nevadensis* A. Gray x *Arctostaphylos patula* Greene]. Erect subshrub 0.5–1 m, with intermediate characters between the parental species: twig and nascent inflorescence axis with both, short-glandular and nonglandular hairs; leaf erect, ovate to oblancelate, commonly elliptic, bright green, shiny, puberulent, glabrous in age, base wedge-shaped, tip abruptly soft pointed, margin entire, flat. Flowers whitish o weakly pink. Naturally grows between the parent species.
  Holotypus and isotypus preserved in MAF Herbaria (MAF 158,485, holotypus; MAF 158,486, isotypus): Flora of California, UC Davis Herbaria, Nevada County: Nevada City-Lake Spaulding. Tahoe National Forest. White fir forests near Casci Ranch Road (Castanopsio sempervirentis-Abietetum lowianae Rivas-Martínez & Sánchez-Mata 1997), 1220 m, July 31, 1997, D. Sánchez-Mata & P. Rodríguez-Rojo.
  Species devoted with much affection to Prof. Michael G. Barbour (Emeritus Professor at University of California, Davis), an erudite expert on California and Pacific Northwest vegetation who passed away on January 7th, 2021.

- *Ceanothus jepsonii* subsp. *albiflorus* (J.T. Howell) Sánchez-Mata & Rodríguez-Rojo
- *Linnaea americana* J. Forbes
- *Notholithocarpus echinoides* (R.Br.ter) comb. nov. Bas.: *Quercus echinoides* R.Br. ter in Ann Mag Nat Hist ser. 4, 7: 251. 1871
- *Pinus austrina* (R.J. Mastrog. & J.D. Mastrog.) Rivas-Martínez & Sánchez-Mata
- *Pinus ponderosa* subsp. *pacifica* (J.R. Haller & N.J. Vivrette) Rivas-Martínez & Sánchez-Mata
- *Pteridium aquilinum* subsp. *pubescens* (Underw.) Piper & Beattie
- *Quercus breweri* Engelm.

4.2 Syntaxonomic Appendix

Only the phytosociological units related to ultramafic vegetation types are considered.

I. HETEROMELO ARBUTIFOLIAE-QUERCETEA AGRIFOLIAE Rivas-Martínez 1997

- Quercetalia agrifolio-wislizenii Rivas-Martínez 1997
  - *Quercion douglasio-wislizenii* Rivas-Martínez 1997
    1. *Pino sabinianae-Quercetum douglasii* Rivas-Martínez 1997
- Adenostomo fasciculati-Rhamnetalia croceae Rivas- Martínez 1997
  - *Quercion duratae* Sánchez-Mata, Barbour & Rodríguez-Rojo in Rivas-Martínez 1997
    2. *Ceanotho jepsonii-Quercetum duratae* Sánchez-Mata & Rodríguez-Rojo 2016
    3. *Ceanotho albiflori-Quercetum duratae* Sánchez-Mata, Barbour & Rodríguez-Rojo [in Rivas-Martínez] 1997
    4. *Arctostaphylo glaucae-Quercetum duratae* Sánchez-Mata & Rodríguez-Rojo 2016
    5. *Arctostaphylo viscidae-Quercetum duratae* Sánchez-Mata & Rodríguez-Rojo 2016
    6. *Hesperoyucco whipplei-Quercetum duratae* Sánchez-Mata & Rodríguez-Rojo 2016

II. CALOCEDRO DECURRENTIS-PINETEA JEFFREYI Rivas-Martínez & Sánchez-Mata [in Rivas-Martínez] 1997

- Arctostaphylo patulae-Pinetalia jeffreyi Rivas-Martínez & Sánchez-Mata [in Rivas-Martínez] 1997

- *Querco vaccinifoliae-Pinion jeffreyi* Rivas-Martínez & Sánchez-Mata [in Rivas- Martínez] 1997

  1. *Querco vaccinifoliae-Pinetum jeffreyi* Rivas-Martínez & Sánchez-Mata [in Rivas- Martínez] 1997
  2. *Aspidoto densae-Pinetum jeffreyi* Rivas-Martínez & Sánchez-Mata [in Rivas- Martínez] 1997

III. SALICETEA LASIANDRO-EXIGUAE Peinado & al. 2011

- Salicetalia delnortensis-breweri Peinado & al. 2011

  - *Salicion delnortensis* Sánchez-Mata & Barbour [in Sánchez-Mata] 2018

    1. *Salicetum delnortensis* Sánchez-Mata & Barbour [in Sánchez-Mata] 2018

  - *Salicion breweri* Sánchez-Mata & Barbour 2007

    2. *Salicetum breweri* Sánchez-Mata & Barbour 2007
    3. *Solidago guiradonis-Salicetum breweri* Sánchez-Mata & Barbour [in Sánchez-Mata] 2018
    4. *Calycantho occidentalis-Hesperocyparetum sargentii* Sánchez-Mata & Barbour [in Sánchez-Mata] 2018
    5. *Frangulo tomentellae-Rhododendretum occidentalis* Sánchez-Mata & Barbour [in Sánchez-Mata] 2018

IV. VULPIO MICROSTACHYOS-HESPEROLINETEA MICRANTHI Rodríguez-Rojo, Sánchez-Mata, Rivas-Martínez & Barbour 2001

- Eriogono luteoli-Hesperolinetalia micranthi Rodríguez-Rojo, Sánchez-Mata, Rivas-Martínez & Barbour 2001

  - *Hesperevaco sparsiflorae-Hemizonion congestae* Rodríguez-Rojo, Sánchez-Mata, Rivas-Martínez & Barbour 2001

    1. *Hesperevacetum sparsiflorae* Rodríguez-Rojo & Sánchez-Mata [in Rodríguez-Rojo, Sánchez-Mata, Gavilán, Rivas-Martínez & Barbour] 2001

  - *Hesperolinion clevelendii* Rodríguez-Rojo, Sánchez-Mata, Rivas-Martínez & Barbour 2001

    2. *Hesperolinetum clevelandii* Rodríguez-Rojo & Sánchez-Mata [in Rodríguez-Rojo, Sánchez-Mata, Gavilán, Rivas-Martínez & Barbour] 2001

  - *Hesperolino micranthi-Navarretion filicaulis* Rodríguez-Rojo, Sánchez-Mata, Rivas-Martínez & Barbour 2001

3. *Hesperolino micranthi-Navarretietum filicaulis* Rodríguez-Rojo & Sánchez-Mata [in Rodríguez-Rojo, Sánchez-Mata, Gavilán, Rivas-Martínez & Barbour] 2001

– *Streptanthion polygaloidis* Rodríguez-Rojo, Sánchez-Mata, Rivas-Martínez & Barbour 2001

4. *Streptanthetum polygaloidis* Rodríguez-Rojo & Sánchez-Mata [in Rodríguez-Rojo, Sánchez-Mata, Gavilán, Rivas-Martínez & Barbour] 2001

References

1. Alexander EB, Coleman RG, Keeler-Wolf T, Harrison S (2007) Serpentine geoecology of Western North America: geology, soils, and vegetation. Oxford University Press, New York, p 512
2. Anderson RC, Fralish JS, Baskin JM (eds) (2007) Savannas, barrens, and rock outcrop plant communities of North America. Cambridge University Press, New York, p 484
3. Antonovics J, Bradshaw AD, Turner RG (1971) Heavy metal tolerance in plants. Adv Ecol Res 7:1–85
4. Armstrong WP (1977) Natural plant hybrids in San Diego County. Env South 477:14–16
5. Baldwin BG, Goldman DH, Keil DJ, Petterson R, Rosetti TJ, Wilken DH (2012) The Jepson manual. Vascular plants of California, 2nd edn. California University Press, Berkeley, p 1568
6. Barbour MG, Keeler-Wolf T, Schoenherr AA (2007) Terrestrial vegetation of California, 3rd edn. University of California Press, Berkeley, p 730
7. Bargagli R (1998) Trace elements in terrestrial plants: an ecophysiological approach to biomonitoring and biorecovery. Springer-Verlag and RG Landes Co., Berlin, p 324
8. Bates RL, Jackson JA (eds) (1984) Dictionary of geological terms (rocks, minerals and gemstones), 3rd edn. American Geological Institute. Anchor Books, Doubleday, New York, p 576
9. Callaham RZ (1956) Needle oils of three pine species and species hybrids. Forest Sci 2 (2):101–105
10. Coleman RG (1977) Ophiolites. Springer, Berlin, p 229
11. Coleman RG, Jove C (1992) geological origin of serpentinites. In: Baker AJM et al (eds) The Vegetation of Ultramafic (Serpentine) Soils. Intercept, Andover, pp 1–17, p 509
12. Harrison S, Safford HD, Grace JB, Viers JH, Davies KF (2006) Regional and local species richness in an insular environment: serpentine plants in California. Ecol Mon 76:41–56
13. Hidalgo-Triana N, Pérez Latorre AV (2018) Plant functional traits and groups in a California serpentine chaparral. Ecol Res 33:525–535
14. Keeley JE, Davis FW (2007) Chaparral. In: Barbour MF et al (eds) Terrestrial vegetation of California, 3rd edn. University of California Press, pp 339–366, p 730
15. Keeley JE, Soderstrom TJ (1986) Postfire recovery of chaparral along an elevational gradient in Southern California. Southw Nat 31:177–184
16. Kruckeberg AR (1951) Intraspecific variability in the response of certain native plant species to serpentine soil. Am J Bot 38:408–419
17. Kruckeberg AR (1954) Plant species in relation with serpentine soils. In: Whittaker RH (ed) The ecology of serpentine soils: a symposium. Ecology, vol 35, pp 267–274

18. Kruckeberg AR (1984) California serpentines: flora, vegetation, geology soils, and management problems, vol 78. University of California Publications in Botany. University of California Press, Berkeley, p 180
19. Kruckeberg AR (2002) Geology and Plant Life. The Effects of Landforms and Rock Types on Plants. University of Washington Press, p 362
20. Mason HL (1946) The edaphic factor in narrow endemism. I. The nature of environmental influences. Madroño 8:209–226
21. Mason HL (1946) The edaphic factor in narrow endemism. II. The geographic occu rrence of plants of highly restricted patterns of distribution. Madroño 8:241–257
22. McCarten N (1988) Serpentine plant communities of San Francisco Bay Region. In: Elias TS (ed) Conservation and management of rare and endangered plants. Proceedings of a California conference on the conservation and management of rare and endangered plants (1986). California Native Plant Society, Sacramento, p 630
23. Rajakaruna N (2018) Lessons on evolution from the study of edaphic specialization. Bot Rev 84:39–78
24. Rivas-Martínez S, Sánchez-Mata D (1997) Calocedro decurrentis-Pinetea jeffreyi. In: Rivas-Martínez S (ed) Syntaxonomical synopsis of the potential natural plant communities of North America, I. Itinera Geobot, vol 10, pp 54–90
25. Rivas-Martínez S, Sánchez-Mata D, Costa M (1999) North American boreal and western temperate forest vegetation (Syntaxonomical synopsis of the potential natural plant communities of North America, II). Itinera Geobot 12:5–316
26. Rivas-Martínez S, Sánchez-Mata D, Costa M (1999) North American new phytosociological classes. Itinera Geobot 13:349–352
27. Rivas-Martínez S, Penas A, del Río S, Díaz González TE, Rivas Sáenz S (2017) Bioclimatology of the Iberian penisnula and the Balearic Islands. In: Loidi L (ed) The vegetation of the Iberian Peninsula, vol 1. Plant and Vegetation 12. Springer, pp 29–80, p 676
28. Rodríguez-Rojo MP, Sánchez-Mata D, Gavilán RG, Rivas-Martínez S, Barbour MG (2001) Typology and ecology of Californian serpentine annual grasslands. J Veg Sci 12:687–698
29. Rodríguez-Rojo MP, Sánchez-Mata D, Rivas-Martínez S, Barbour MG (2001) Syntaxonomical approach for classification of the Californian serpentine annual grasslands. Lazaroa 22:83–94
30. Safford HD, Viers JH, Harrison SP (2005) Serpentine endemism in the California flora: a database of serpentine affinities. Madroño 52(4):222–257
31. Sánchez-Mata D, Rodríguez-Rojo MP, Barbour MG (2004) California ultramafic vegetation: biodiversity and phytosociological survey. In: Boyd RS et al (ed) Ultramafics rocks: their soils, vegetation and fauna (Rocas ultramáficas: sus suelos, vegetación y fauna), Proceedings of the fourth international conference on serpentine ecology, St Albans (UK), pp 177–181, p 347
32. Sánchez-Mata D (2007) Ultramafic vegetation. Chapter 3: California soils and examples of ultramafic vegetation (A. T. O'Geen, R. A. Dahlgren, and D. Sánchez-Mata). In: Barbour MG et al (eds) Terrestrial vegetation of California, 3rd edn. University of California Press, pp 93–(71)–106, p 730
33. Sánchez-Mata D (2012) Overview of the forest vegetation of the Sierra Nevada Range (California, USA). Berich RT-Gesells 24:141–153
34. Sánchez-Mata D, Rodríguez-Rojo MP (2016) Mediterranean ultramafic (serpentine) chaparrals of California (USA): a geobotanical overview. In: Box EO (ed) Vegetation structure and function at multiple spatial, temporal and conceptual scales. Geobotany Studies, Springer, pp 285–312, p 578
35. Sánchez-Mata D (2018) Studies on California ultramafic (serpentine) vegetation: riparian woody formations. Berich RT-Gesells 30:175–190
36. Sawyer JO, Thornburgh DA, Bowman WF (1970) Extension of the range of Abies lasiocarpa into California. Madroño 20:413–415
37. Walker RB (1948) Molybdenum deficiency in serpentine barren soils. Science 108:473–475

38. Walker RB (1954) Factors affecting plant growth on serpentine soils. In: Whittaker RH (ed) The ecology of serpentine soils: a symposium. Ecology, vol 35, pp 258–266
39. Walker RB, Walker HM, Ashworth PR (1955) Calcium-magnesium nutrition with special reference to serpentine soils. Plant Phys 30:214–221
40. Waring RH, Emmingham WH, Running SW (1975) Environmental limits of an endemic spruce, Picea breweriana. Canad J Bot 53:1599–1613
41. Theurillat J-P, Willner W, Fernández-González F, Bultmnan H, Carni A, Gigante D, Mucina L, Weber H (2021) International Code of Phytosociological Nomenclature, 4th edn. App Veg Sci 2021:24:e12491
42. Whittaker RH (1954) The vegetational response to serpentine soils. In: Whittaker RH (ed) The ecology of serpentine soils: a symposium. Ecology 35:275–288
43. Zobel B (1952) Jeffrey pine in the south Coast Ranges of California. Madroño 11:283–284
44. Zobel DG, Hawk GM (1980) The environment of Chamaecyparis lawsoniana. Am Mid Nat 103:280–297

Plant Associations of *Petasition officinalis* Alliance in the East Carpathians (Călimani and Gurghiu Mountains Romania)

Silvia Oroian, Mihaela Sămărghițan, and Erzsébet Domokos

Abstract This paper presents the results of phytosociological research of tall-herbs vegetation from the East Carpathians Rank Călimani and Gurghiu Mountains. A total of five plant associations were analysed with the classification in the Petasition officinalis Sillinger 1933. The study area refers to plant communities dominated by *Petasites* sp. which occur along the river banks and on the fringe of forests in the montane belt. The paper is based on field researches during 2000–2014. A data set of 70 relevés were analysed in term of flora composition and geomorphological features of the habitat. For analysis of the relationship between floristic composition and habitat variables the average non-weighted Ellenberg indicator values (for light, moisture, temperature, soil reaction, soil nutrient content) and the Shannon-Wiener indices of the relevés were used.

Keywords Phytosociology · Montane vegetation · Tall-herb plant communities

1 Introduction

The paper presents the results of the researches carried out in the mesophilous communities belonging to the Petasition officinalis alliance, which develop along the mountain valleys and the forest fringes of the Călimani-Gurghiu Mountains (East Carpathians). The paper aims to emphasize the interconditioning and mutual

S. Oroian · E. Domokos
Department of Fundamental Pharmaceutical Sciences, Discipline of Pharmaceutical Botany, University of Medicine, Pharmacy, Science and Technology of Tîrgu Mureş, 38 Gh. Marinescu Street, 540139 Tîrgu Mureş, Romania

M. Sămărghițan (✉)
Department of Natural Sciences, Mureş County Museum, 24 Horea Street, 540036 Tîrgu Mureş, Romania

E. Domokos
Department of Horticulture,Faculty of Târgu Mureş, Sapientia Hungarian University of Transylvania, 540485 Târgu-Mureş, Romania

F. Pedrotti and E. O. Box (eds.), *Tools for Landscape-Scale Geobotany and Conservation*, Geobotany Studies, https://doi.org/10.1007/978-3-030-74950-7_11

influences between the plant communities and the complex environmental factors. Features like geographical position, altitude, soil and rock types, as well as an interspecific relationship determine a high structural diversity of studied plant communities.

The study area stretches on the montane beech forest belt of the Călimani-Gurghiu Mountains, with an altitude range of 500 to 1,600 m. The mountain belt terrain is well represented in all mountains in the Romanian Carpathians. The researched areas overlap with protected areas of the Upper Mures Gorge Natural Park and ROSCI0019-Călimani-Gurghiu Natural Park Natura 2000 site. The entire level is covered by beech and spruce forests that are in a good conservation status.

Riparian communities identified in the studied territory were assigned to the Petasition officinalis alliance.

2 Materials and Methods

The field investigations carried out during the period 2000–2014 aimed to establish the qualitative structure of phytocoenoses dominated by the *Petasites* sp. identified in the area. A total of 70 surveys were analyzed which were assigned to 5 plant associations. The analysis for each plant association consisted of: the specific composition and belonging to the species of the higher coenotaxa.

For the study of vegetation the research was based on the methodology of the Zurich-Montpellier Phytosociological School (Braun-Blanquet method) adapted to the particularities of our country [2, 8]. Due to high number of relevés the obtained results are presented in form of synoptic tables.

The species complex was analyzed according to species requirements to ecological factors. The affinity of each species for different environmental conditions was characterized using the Ellenberg indicator values for vascular plants (light, temperature, humidity, soil reaction and nitrogen). The ecological indicators, as well as the name of the species were based on Sârbu et al. [31]. The national and international red lists were used [1, 3, 10, 11, 24] for the identification and classification of the species of plants at different degrees of vulnerability. The phytocoenoses were reported to Natura 2000 habitat type according to Gafta and Mountford [13] and Doniţă et al. [12].

3 Results and Discussions

The general appearance of the vegetation of the Călimani and Gurghiu Mountains on the territory of Mureş County is a reflection of a sum of factors such as past and present physico-geographic conditions, the origin of the various floristic elements and the impact of the anthropogenic factors. These anthropogenic factors induced profound changes in the vegetations' composition through extensive forestry, hydropower systems, and the change of land use by crops. In this examination we took into account some tall-herbs ("megaphorbs") with a zonal characteristic, installed after the deforestation of forests and hedges [6]. The vegetation can reach up to 50–150 cm high. Low altitude communities are often severely degraded and invaded by anthropophilic weeds, sometimes even of neophyte species.

In this analysis, we considered the megaphorbs of the Petasition officinale alliance. The studied plant associations spread over the mountain valleys or the fringe of the mesophilous forests, forming borders along the river sides. Coenoses grow in shady or semi-shady places, on more or less damp soils, that can include thick gravel, but the soil is rich in humus. Research carried out refers to five plant associations from the Petasition officinalis alliance in the habitat: 6430—Hydrophilous tall-herb fringe communities of plains in the montane to alpine levels.

A total of 70 phytosociological surveys were performed in the studied area, which were assigned to five plant associations that were reported in the Petasition officinalis alliance, Petasito-Chaerophylletalia order.

The variability of the environmental conditions as well as the floristic composition of these tall-weeds gave rise to various opinions regarding their phytosociological framing.

Morariu in 1967 described the order Petasito-Chaerophylletalia and included it in Epilobietea angustifolii Tx. et Prsg. 1950 class. This class includes the herbaceous vegetation of montane forest clearings, deforested sites. In the Petasito-Chaerophylletalia order can meet species from montane forest clearings but also species characteristic of mountane streams. Because of that, later the order was classified in montane tall-weeds of Betulo-Adenostyletea Br.-Bl. et Tx. 1943, Adenostyletea Knapp 1943 class (syn. Mulgedio-Aconitetea Hadać et Klika in Klika et Hadać 1944) [28]. Some authors include Petasition officinalis alliance into the Galio-Urticetea class based on the presence of many nitrophilic and synanthropic species [4, 7, 21, 26, 28]. Given the natural character of habitats, the Petasition association is usually ranked also in the subalpine high-herbs vegetation class Mulgedio-Aconitetea [16, 17, 19, 22].

Considering the floristic composition of the conenoses and the environmental factors that determine the studied conenoses to spread over the altitudinal gradient, we considered this last classification in our study.

Cls. **Mulgedio-Aconitetea** Hadać et Klika in Klika et Hadać 1944
Ord. **Petasito-Chaerophylletalia** Morariu 1967
All. **Petasition officinalis** Sillinger 1933
Ass. ***Telekio-Petasitetum hybridi*** (Morariu 1967) Resmeriţă et Raţiu 1974
Ass. ***Telekio-Filipenduletum*** Coldea 1996
Ass. ***Telekio speciosae-Aruncetum dioici*** Oroian 1998
Ass. ***Petasitetum kablikiani*** Szafer et al. 1926
Ass. ***Arunco-Petasitetum albi*** Br.-Bl. et Sutter 1977

1. **Telekio-Petasitetum hybridi** (Morariu 1967) Resm. et Raţiu 1974 (syn. Petasiteto - Telekietum speciosae Morariu 1967)

This endemic association of the Romanian Carpathians (Table 1, column 1, Fig. 1) is frequent along the streams, near the springs or on the edge of mesophilous forests. This association has been detailed in several mountains of the Eastern Carpathians: Maramureş [27], Baraolt [9], Rodnei [5], Calimani, Gurghiu [25], Hăşmaş [23], Gurghiu [30].

The phytocoenoses grow lushly on alluvial soil as well as on colluvial accumulations, with favorable humidity and moderate or abundant quantities of nitrates. They are very common in the study area. Due the characteristic species: *Petasites hybridus* and *Telekia speciosa*, these phytocoenoses have a regional peculiarity in the Romanian Carpathians, unlike those described in Central Europe. The coenoses have considerable coverage, due to the broad leaves of the characteristic species that shade the ground. Because of that other plant species develop only on the peripheral sides of the phytocoenosis. The coenoses form more or less narrow strips, along river banks and forest fringes. In their floral composition there are often found numerous forest elements stemming from the woods in the close vicinity. The phytocoenoses of the association were identifyed at altitudes between 660 and 1215 m. For this association, 39 phytosociological surveys were performed. These phytocoenoses are characterized by a great floral wealth, with 244 species being identified. The plant communities are layered, achieving good coverage of up to 100%. The floral composition includes the edifying species: *Telekia speciosa* and *Petasites hybridus*. The diagnostic and dominant species identified are *Telekia speciosa, Petasites hybridus, Angelica sylvestris, Cirsium oleraceum, Geranium palustre, Scirpus sylvaticus, Filipendula ulmaria, Chaerophyllum hirsutum* etc.

Regarding their belonging to different coenotic groups, there is a high share of the species of the superior syntaxa, the Petasition officinalis alliance (*Aruncus dioicus, Cirsium oleraceum, Chrysosplenium alternifolium, Cirsium erisithales*) and Petasito-Cherophylletalia aliance (*Angelica sylvestris, Chaerophyllum hirsutum, Aegopodium podagraria, Impatiens noli-tangere, Rumex obtusifolius* etc.). The endemic or subendemic species with a remarkable phytogeographical interest are: *Aconitum moldavicum, Dianthus barbatus* ssp. *compactus* and *Leucanthemum waldsteinii.* There are also Carpathian-Balkan species (*Campanula abietina, Petasites kablikianus*, and *Pulmonaria rubra*). Adenostyletalia order species like *Achillea distans, Cicerbita alpina, Deschampsia cespitosa, Gentiana asclepiadea, Senecio ovatus, Cirsium waldsteinii, Doronicum austriacum* are also present, especially in higher elevations. This is due to the proximity of these coenoses with tall-weeds of the mountain area.

2. ***Telekio-Filipenduletum*** Coldea 1996

Hygrophilic coenoses ediphied by *Filipendula ulmaria* and *Telekia speciosa* are spread in the study area in intermontane valleys on humus-rich alluvial soils at altitudes between 651 and 1063 m. For this association, 8 phytosociological surveys were performed. Phytocoenoses consist of 132 layered species, achieving a good coverage of about 80–100%. Diagnostic species are *Myosoton aquaticum* and *Scirpus sylvaticus*. Among the characteristic species, we mention *Telekia speciosa, Petasites hybridus*, and *Filipendula ulmaria* (Table 1, Column 2, and Fig. 2). The structure of the association is well represented by the species characteristic of the

Petasition officinalis alliance and the Petasitio-Chaerophylletalia order. In terms of flora, these phytocoenoses in Romania differ from those described in the Czech Republic [14] by the small number of species characteristic of the Calthion alliance, but those of the Petasition officinalis alliance are much better represented similar to those from Slovakia [18].

The phytocoenoses are distributed in more or less narrow strips along the forest fringe or water courses. Due to the proximity of the forest, a floral composition with noticeable forest elements is easily distinguishable.

Considering the coenotic groups one can notice a high weight of the species belonging to the higher syntaxa, the alliance Petasition officinalis, Petasito-Chaerophylletalia order (*Chaerophyllum hirsutum, Stellaria nemorum, Mentha longifolia, Myosotis scorpioides, Rumex alpinus, Carduus personata, Cirsium erisithales, Valeriana sambucifolia*) and the Mulgedio-Aconitetea class (*Hypericum maculatum, Rosa pendulina, Silene dioica, Veratrum album*). Additionally, Adenostyletalia species like *Deschampsia cespitosa, Senecio ovatus, Cirsium waldsteinii, Doronicum austriacum* are also present. We note that the drought and low precipitation in recent years have led to the reduction of the phytocoenoses edifyed by *Telekia speciosa* with *Filipendula ulmaria*.

Fig. 1 *Telekio-Petasitetum hybridi* (foto Mihaela Sămărghiţan)

Fig. 2 *Telekio-Filipenduletum* (foto Mihaela Sămărghiţan)

3. ***Telekio speciosae-Aruncetum dioici*** Oroian 1998

The plant groups with *Aruncus dioicus* and *Telekia speciosa* develop in semi-shady, inclined (up to 40°) slopes with northern exposure at altitudes ranging from 500 to 926 m in Sovata (Ilieşi), Bistra Mureşului on the Bistra Valley and Lunca Bradului on the Ilişoara Mare valley. The association was also described from the Mureş Gorge in the Borzia and Răstoliţa-Costeasa Valley [25]. The diagnostic species for association are *Gentiana asclepiadea, Spiraea chamaedryfolia, Aruncus dioicus, Campanula trachelium, Pulmonaria rubra.* The characteristic species of the association are *Aruncus dioicus* a circumpolar-boreal species and *Telekia speciosa*, a Carpathian-Balkan-Caucasian species, which achieves considerable coverage due to the broad leaves. In addition, *Spiraea chamaedrifolia* with constancy III is characteristic (Table 1, column 3, Figs. 3, 4). 10 surveys were performed, and 138 taxa were identified.

In the floristical composition of this association, we can observe a high participation of species belonging to the higher syntaxa which subordinates the association namely Petasition officinalis alliance (*Spiraea chamaedrifolia, Petasites hybridus, Carduus personata, Cirsium erisithales, Chaerophyllum hirsutum,* etc.), Petasito-Chaerophyletalia order (*Anthriscus sylvestris, Aegopodium podagraria, Galium aparine, Impatiens noli-tangere, Rumex obtusifolius*) and Mulgedio-Aconitetea class (*Hypericum maculatum, Rosa pendulina, Silene dioica* etc.).

Fig. 3 *Telekio speciosae-Aruncetum dioici* at Sovata (Ilieşi) (foto Silvia Oroian)

Fig. 4 *Telekio speciosae-Aruncetum dioici* at Răstoliţa (foto Silvia Oroian)

4. ***Petasitetum kablikiani*** Szafer et al. 1926

Southeast Carpathian high-weed communities of *Petasites kablikianus* form enclaves on the sides of the mountain brooks on the montane beech and spruce forest level. In Romania, from the Eastern Carpathians, this association has also been described in Rodna Mountains, Bistrița Aurie, Tazlăul Basin, Siriu Mountain, Tiştița Gorges. They are spread at altitudes between 900 and 1280 m. Climatically, phytocoenoses develop on temperatures between 5.8 and 4.5 °C with precipitation of 900–1200 mm/year. They prefer folisoils with gravel and pebbles [12].

In the study area 5 phytosociological surveys were performed, identifying 65 taxons, at altitudes ranging from 900 to 1024 m (Table 1, column 4, Figs. 5, 6). The characteristic and edifying species *Petasites kablikianus* achieve coverage of 77–95%. The diagnostic species for association are *Valeriana sambucifolia* and *Petasites kablikianus*. Other species with high constancy found in the floristic composition are *Petasites hybridus, Telekia speciosa, Geranium robertianum, Aegopodium podagraria, Eupatorium cannabinum, Glechoma hederacea, Galium aparine, Urtica dioica, Geranium phaeum, Rumex obtusifolius, Rumex alpinus.*

The floristic composition of the association is completed by numerous species belonging to higher syntaxa: Petasition officinalis alliance (*Petasites hybridus, Chaerophyllum hirsutum, Mentha longifolia, Myosotis scorpioides, Valeriana sambucifolia, Carduus personata*) and Petasito-Chaerophylletalia order (*Aegopodium podagraria, Galium aparine, Glechoma hederacea, Impatiens noli-tangere, Rumex obtusifolius* etc.) Species from Adenostyletalia order like *Achillea distans, Senecio ovatus* are also present.

Fig. 5 *Petasitetum kablikiani* at Lunca Bradului, Ilva Mare (photo Silvia Oroian)

Fig. 6 *Petasitetum kablikiani* (photo Silvia Oroian)

5. ***Arunco-Petasitetum albi*** Br.-Bl. et Sutter 1977

These coenoses meet on the montane level, from 698 to 1336 m altitude, occupying larger areas in Răstoliţa: Borzia, on Răstoliţa Valey, Costeasa Valley, Iod Valley, at Lăpuşna on Secuieu Stream, Ibăneşti on Şirodul Mare Stream. A number of 99 taxa were registered in 10 surveys. These tall-herbs are made up of species characteristic of the mountain valleys: *Petasites albus, Aruncus dioicus* (Table 1, column 5, Figs. 7, 8). Beside characteristic species of association, species belonging to Petasition officinalis alliance (*Telekia speciosa, Carduus personata, Cirsium erisithales Chrysosplenium alternifolium Epilobium montanum Leucanthemum waldsteinii, Mentha longifolia*) as well as Petasito-Chaerophylletalia order (*Aegopodium podagraria, Impatiens noli-tangere, Lamium maculatum*) record high frequencies. We also mention the presence of many transgresive meso-hygrophilic species from the Molinio-Arrhenatheretea class (incl. Molinietalia) grasslands, such as *Prunella vulgaris, Crepis biennis, Lysimachia nummularia,* or from the vegetation of the forest areas with which the phytocoenoses often bind, from Fagetalia order (*Oxalis acetosella, Acer pseudoplatanus, Aconitum moldavicum* ssp. *moldavicum, Ajuga reptans, Dentaria glandulosa, Luzula luzuloides, Salvia glutinosa, Stachys sylvatica* etc.)

Fig. 7 Coenoses with *Petasites albus* (photo Silvia Oroian)

Fig. 8 *Arunco-Petasitetum albi* (photo Silvia Oroian)

Comparative analysis shows that all five associations grow on intermontane valleys, on alluvial soils rich in humus, at altitudes between 500 and 1336 m, only association *Petasitetum kablikiani* was identified at higher altitudes, between 905 and 1025 m, near forests.

The 5 associations identified are characteristic of the habitat: 6430 - Hydrophilous tall-herb fringe communities of plains and in the montane to alpine levels [13]. It is a habitat composed of tall herbaceous grass communities, diverse in terms of species composition. The most representative communities of tall herbs (not to be confused with the weed communities that are directly related to human activities) are those of different species of butterbur, *Petasites hybridus, Petasites albus, Petasites kablikianus*, and also *Telekia speciosa, Filipendula ulmaria, Cicerbita alpina, Digitalis grandiflora, Calamagrostis arundinacea, Aruncus dioicus* etc.

Considering the characteristic species to different coenotic groups, in all 5 associations there is a high share of the species belonging to the coenotaxa that subordinate the associations Petasition officinalis alliance and Petasito-Chaerophylletalia order (*Aruncus dioicus, Cirsium oleraceum, Impatiens noli-tangere, Leucanthemum waldsteinii, Poa trivialis, Petasites hybridus, Petasites* × *reichingeri, Cirsium erisithales, Chaerophyllum hirsutum, Stellaria nemorum, Mentha longifolia, Myosotis scorpioides, Rumex alpinus* etc.), as well as Mulgedio-Aconitetea class (*Aconitum napellus* ssp. *tauricum, Hypericum maculatum, Rosa pendulina, Silene dioica)* etc.

Characteristic species of the order Adenostyletalia and class Epilobietea (*Achillea distans, Cicerbita alpina, Deschampsia cespitosa, Gentiana asclepiadea, Senecio ovatus, Cirsium waldsteinii, Doronicum austriacum*) occur in the associations probably due the proximity of mountainous tall-weeds.

In the composition of studied phytocoenoses we mention species such as *Aconitum moldavicum, Dianthus barbatus* ssp. *compactus* and *Leucanthemum waldsteinii*. They are endemic or subendemic species and have a significant phytogeographical interest. Carpathian-Balkan species are also present; out of these, we mention *Campanula abietina, Petasites kablikianus, Pulmonaria rubra.*

4 Data Analysis

Plant communities were ordered with Principal coordinate analysis (PCoA) based on Jaccard (presence/absence) and Bray-Curtis (abundance-dominance) indices. The main environmental gradients of the associations were analyzed with CCA (Canonical Correspondence Analysis). For this analysis the average non-weighted Ellenberg indicator values (for light, moisture, temperature, soil reaction, soil nutrient content) and the Shannon-Wiener indices of the relevés were used. The comparisons of the ecological indicator values and diversity indices were made with the Kruskal-Wallis test. The significance of the differences was checked with the Mann-Whitney U post hoc test. All statistics were performed with the Past Program (v. 2.17 [15]).

5 Results and Discussion

Plant communities could not be differentiated with the PCoA based on the presence/ absence of the species (Fig. 9) (PC1: 9.70%, PC2: 6.30%). The ordering of data on the basis of the Bray–Curtis index presented the grouping in two well differentiated associations: *Petasitetum kablikiani* and *Arunco-Petasitetum albi* (PC1: 32.18%, PC2: 14.27%). The other three associations sharing many plant taxa were not differentiated (Fig. 10).

The ordering of plant communities (Eigenvalues for CCA1 and CCA2: 0.346 and 0.159, respectively) on basis of ecological indicator values and the Shannon-Wiener indices presented the grouping in 5 different associations (Fig. 11). No correlations were found between the six analyzed gradients. The first CCA axis explained 51.03% variance of the species data, and the second axis 23.49%. The first axis was negatively correlated with light indicator values (correlation coefficient: −0.7633), while the second axis was positively correlated with moisture (correlation coefficient: 0.5239). Thus, the main environmental gradients in ordering of the studied plant communities were the light and the moisture. Communities that require intense light are located in the left side of the chart (*Telekio-Filipenduletum*), while those developed on shady places with northern exposure, on the right side (*Telekio speciosae-Aruncetum dioici*). On the upper part of the ordination chart are the relevés of *Petasitetum kablikiani* edified on wet alluvial soils, and some of the relevés of mesophilous, meso-hygrofilous and hygrofilous plant communities (*Arunco-Petasitetum albi*, *Telekio-Petasitetum hybridi*, *Telekio-Filipenduletum*).

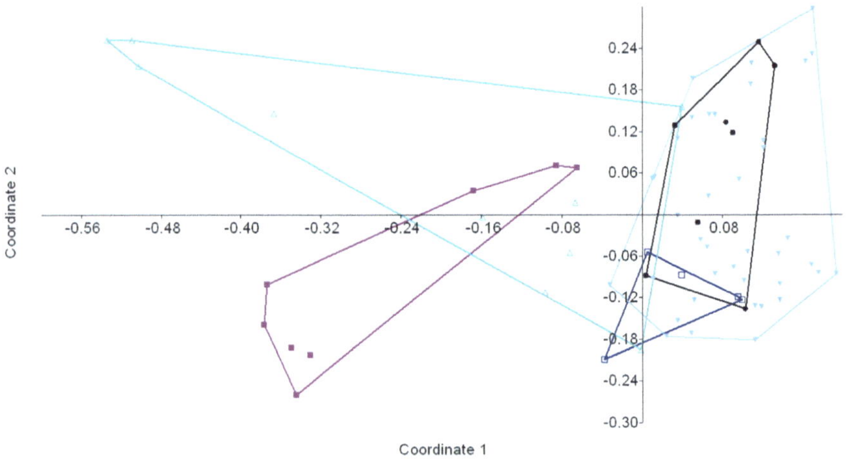

Fig. 9 PCoA of the studied plant communities based on the Jaccard index. Associations are marked as follows: □*Petasitetum kablikiani*; ■*Arunco-Petasitetum albi*; △*Telekio speciosae-Aruncetum dioici*; ●*Telekio-Filipenduletum*; ▼*Telekio-Petasitetum hybridi*

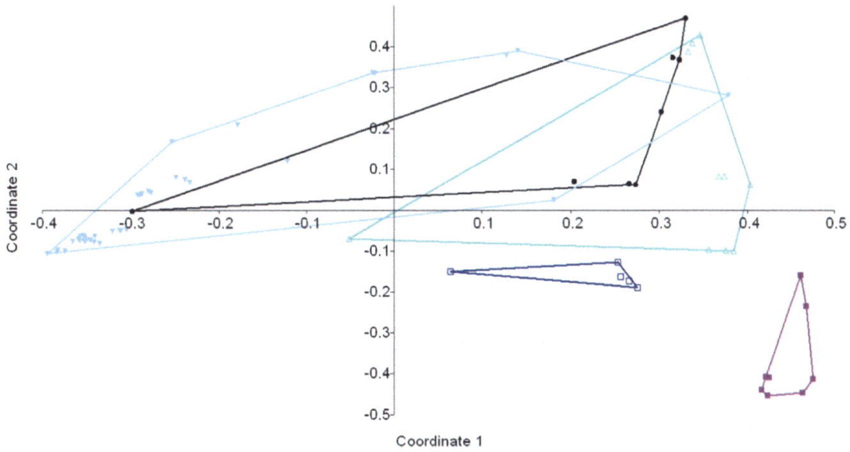

Fig. 10 PCoA of the studied plant communities based on the Bray–Curtis index. Associations are marked as follows: □*Petasitetum kablikiani*; ■*Arunco-Petasitetum albi*; △*Telekio speciosae-Aruncetum dioici*; ●*Telekio-Filipenduletum*; ▼*Telekio-Petasitetum hybridi*

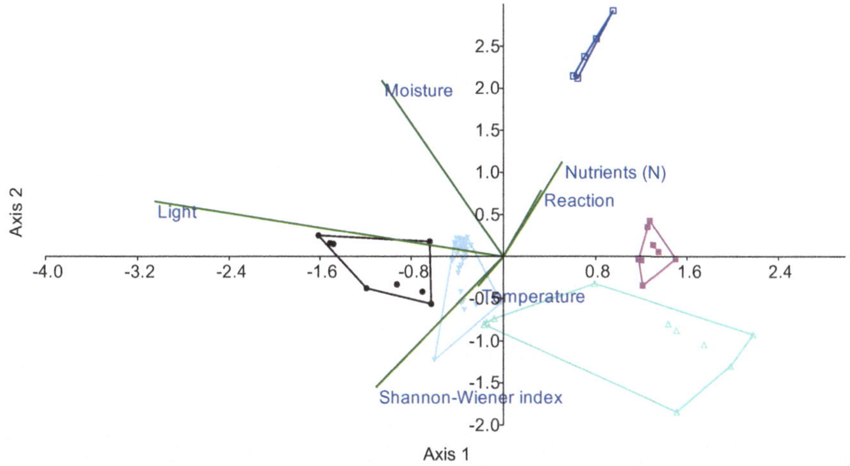

Fig. 11 Canonical correspondence analysis (CCA) of the studied plant communities. The average non-weighted Ellenberg indicator values and the Shannon-Wiener indexes of the relevés were plotted on the CCA ordination chart. Associations are marked as follows: □*Petasitetum kablikiani*; ■*Arunco-Petasitetum albi*; △*Telekio speciosae-Aruncetum dioici*; ●*Telekio-Filipenduletum*; ▼*Telekio-Petasitetum hybridi*

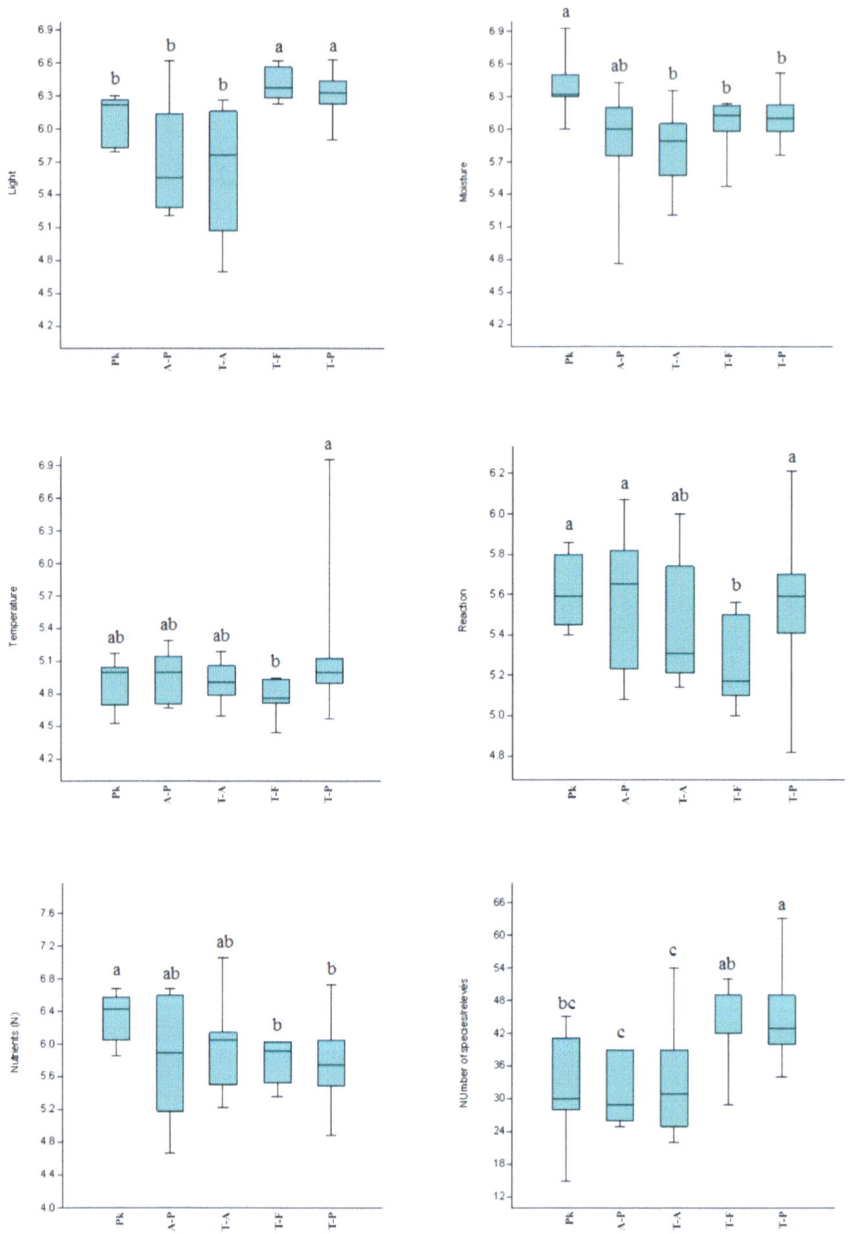

Fig. 12 Ellenberg indicator values and species richness in the studied phytocoenoses. Different letters signify $p < 0.05$ (Pk *Petasitetum kablikiani*; A-P *Arunco-Petasitetum albi*; T-A *Telekio speciosae-Aruncetum dioici*; T-F *Telekio-Filipenduletum*; T-P *Telekio-Petasitetum hybridi*)

Fig. 13 *Dianthus barbatus* ssp. *compactus* (photo Silvia Oroian)

Fig. 14 *Angelica archangelica* (photo Mihaela Sămărghiţan)

There were significant differences between the studied plant communities regarding the Ellenberg indicator values and the species number of the relevés (100 m^2) (Fig. 12). The highest values (significantly different from the other studied communities) for light intensity were obtained in case of the association *Telekio-Filipenduletum* and *Telekio-Petasitetum hybridi*, both encountered at higher altitudes. In case of moisture, high values were found in the association *Petasitetum kablikiani* (significantly higher than the associations with *Telekia speciosa*) encountered in mountain valleys, at the level of beech forests. The association *Arunco-Petasitetum albi*, specific for mountain valleys situated at lower altitudes, presented also high values for moisture. The association *Telekio-Filipenduletum* had the lowest indicator values for temperature and also for soil reaction. The demands for nutrients (N) were significantly lower for the association *Telekio-Filipenduletum* and *Telekio-Petasitetum hybridi* compared to the association *Petasitetum kablikiani*. The association *Telekio-Petasitetum hybridi* had significantly higher species number/relevés than the other communities, but outside the association *Telekio-Filipenduletum*.

Analyzing the conservation status of habitat, we conclude that almost all the described associations are in very good conservation status and presenting several endemic and protected species (Figs. 13 and 14).

Although the conservative value of the 5 plant associations, which at their origin have coenoses that spread on the edge of forests and along watercourses is low, they have an ecotonal character that provides shelter for many animal species, shelter a very wide range of invertebrates but also is a habitat representing a feeding place for many small and large mammal species, therefore careful protection is a necessity. With their rich biomass they often complement the habitat with alluvial forests (91E0*) and its role as an ecological corridor.

In the phytocoenoses of 4 plant associations: *Telekio-Petasitetum hybridi, Telekio-Filipenduletum, Telekio speciosae-Aruncetum dioici* and *Arunco-Petasitetum albi*, were noted some invasive species such as: *Erigeron annuus, Galinsoga parviflora* and *Rudbeckia laciniata* but they don't have a significant impact on phytocoenoses' short term evolution.

Table 1 Plant associations of *Petasition officinalis*

		T-P	T-F	T-A	PK	A-P
	Association number	**1**	**2**	**3**	**4**	**5**
	Relevés number	38	8	10	5	8
	Altitude (m.s.m)	660-1215	650-1063	500-926	905-1025	525-1336
	Number of species	244	132	138	65	99
	Coverage (%)	87-100%	80-100	90-100	82-94	82-100
	Dominant species (Characteristic species)					
1.	*Telekia speciosa*	V	V	V	III	V
2.	*Filipendula ulmaria*	V	V	I	IV	-
3.	*Cirsium oleraceum*	III	II	II	II	-
4.	*Carduus personata*	III	II	I	IV	I

Table 1 (continued)

	Diagnostic species					
5.	*Myosoton aquaticum*	II	III	-	-	-
6.	*Scirpus sylvaticus*	II	IV	-	II	-
7.	*Gentiana asclepiadea*	I	-	III	I	-
8.	*Spiraea chamaedryfolia*	I	-	III	-	-
9.	*Aruncus dioicus*	I	II	V	-	V
10.	*Campanula trachelium*	-	-	II	-	-
11.	*Pulmonaria rubra*	I	-	III	-	II
12.	*Valeriana sambucifolia*	I	II	I	IV	-
13.	*Petasites kablikianus*	I	-	-	V	-
14.	*Scopolia carniolica*	-	-	-	-	III
15.	*Petasites albus*	I	I	I	-	V
16.	*Cardamine pratensis*	I	-	-	-	-
	Petasition officinalis et Petasito-Chaerophylletalia	-	-	-	-	-
17.	*Aegopodium podagraria*	III	IV	I	II	III
18.	*Angelica sylvestris*	I	II	II	-	-
19.	*Anthriscus sylvestris*	I	III	-	-	-
20.	*Chaerophyllum aromaticum*	I	-	II	-	-
21.	*Chaerophyllum hirsutum*	IV	II	I	IV	IV
22.	*Chrysosplenium alternifolium*	I	-	-	-	IV
23.	*Cirsium erisithales*	I	II	I	-	II
24.	*Galium aparine*	II	IV	III	II	-
25.	*Geranium phaeum*	I	-	-	II	-
26.	*Geum rivale*	-	-	-	-	I
27.	*Glechoma hederacea*	II	-	III	II	-
28.	*Impatiens noli-tangere*	IV	II	III	IV	IV
29.	*Lamium maculatum*	I	-	-	-	II
30.	*Leucanthemum waldsteinii*	II	II	I	II	II
31.	*Mentha longifolia*	IV	IV	II	IV	I
32.	*Myosotis scorpioides*	IV	IV	I	II	II
33.	*Petasites hybridus*	V	III	II	IV	II
34.	*Poa trivialis*	I	II	I	-	I
35.	*Rumex alpinus*	I	IV	I	III	I
36.	*Rumex obtusifolius*	III	II	-	II	-
37.	*Stellaria nemorum*	III	II	I	IV	IV
	Adenostyletalia					
38.	*Achillea distans*	II	II	II	II	I
39.	*Angelica archangelica*	I	-	-	-	-
40.	*Cicerbita alpina*	I	II	-	-	-
41.	*Cirsium waldsteinii*	IV	III	II	-	I

Table 1 (continued)

42.	*Deschampsia cespitosa*	III	IV	I	II	III
43.	*Doronicum austriacum*	I	II	-	-	II
44.	*Senecio ovatus*	II	II	I	II	-
45.	*Silene vulgaris*	I	-	-	-	-
	Mulgedio-Aconitetea					
46.	*Aconitum napellus ssp. tauricum*	-	-	-	II	-
47.	*Hypericum maculatum*	III	II	I	II	III
48.	*Rosa pendulina*	-	I	I	-	-
49.	*Silene dioica*	I	II	I	-	I
50.	*Veratrum album* ssp. *lobelianum*	I	I	-	-	-
	Geo urbani-Alliarion officinalis (syn. Galio-Alliarion)					
51.	*Alliaria petiolata*	I	-	-	-	II
52.	*Campanula rapunculoides*	I	-	I	-	-
53.	*Carduus crispus*	-	-	I	-	-
54.	*Chelidonium majus*	I	-	-	-	-
55.	*Clematis vitalba*	-	-	I	-	-
56.	*Geranium robertianum*	III	II	III	IV	IV
57.	*Mycelis muralis*	II	II	II	-	II
58.	*Rubus caesius*	II	I	II	I	-
	Senecionion fluviatilis	-	-	-	-	-
59.	*Heracleum spondylium*	II	I	II	-	-
60.	*Lamium album*	I	-	I	-	II
	Epilobietea angustifolii (incl.Epilobion angustifolii)	-	-	-	-	-
61.	*Athyrium filix-femina*	IV	IV	IV	III	IV
62.	*Calamagrostis arundinacea*	II	III	III	-	-
63.	*Carex hirta*	I	-	-	-	-
64.	*Chamerion angustifolium*	II	II	I	II	I
65.	*Digitalis grandiflora*	I	-	II	-	I
66.	*Dryopteris filix-mas*	I	-	I	-	III
67.	*Eupatorium cannabinum*	III	II	I	I	II
68.	*Fragaria vesca*	II	II	I	-	I
69.	*Galeopsis speciosa*	II	II	-	-	-
70.	*Galeopsis tetrahit*	III	II	II	II	II
71.	*Galium odoratum*	-	-	-	-	III
72.	*Juncus tenuis*	I	-	-	-	-
73.	*Populus tremula*	I	-	-	-	-
74.	*Rubus idaeus*	III	IV	IV	IV	III
75.	*Scrophularia nodosa*	I	-	I	I	II
76.	*Senecio nemorensis*	III	II	-	III	I
77.	*Solanum dulcamara*	I	-	I	-	-

Table 1 (continued)

	Sambuco-Salicion	-	-	-	-	-
78.	*Salix caprea*	III	III	III	I	II
79.	*Sambucus nigra*	I	-	I	-	-
80.	*Sambucus racemosa*	-	I	-	II	II
81.	*Urtica dioica*	V	III	IV	V	III
	Arction lappae	-	-	-	-	-
82.	*Arctium lappa*	I	-	-	-	II
83.	*Tanacetum vulgare*	I	-	-	-	-
	Molinietalia (incl.Molinion)	-	-	-	-	-
84.	*Agrostis stolonifera*	I	II	-	II	-
85.	*Carex distans*	I	-	-	-	-
86.	*Cirsium palustre*	I	I	-	II	-
87.	*Equisetum palustre*	II	II	II	-	-
88.	*Euphrasia rostkoviana*	I	-	-	-	-
89.	*Galium uliginosum*	II	II	I	-	I
90.	*Juncus effusus*	III	IV	-	-	-
91.	*Lathyrus pratensis*	I	-	I	-	-
92.	*Lychnis flos-cuculi*	I	II	-	-	I
93.	*Lycopus europaeus*	I	-	-	II	-
94.	*Lythrum salicaria*	III	II	I	I	I
95.	*Mentha × verticillata*	I	-	-	-	-
96.	*Potentilla erecta*	II	I	-	-	-
97.	*Rumex crispus*	I	-	-	-	-
98.	*Stachys officinalis*	I	-	-	-	-
99.	*Stellaria graminea*	III	-	I	-	-
100.	*Symphytum officinale*	I	-	I	-	I
101.	*Trifolium hybridum*	III	-	I	II	-
	Filipendulion	-	-	-	-	-
102.	*Valeriana officinalis*	III	III	II	-	-
103.	*Veronica chamaedrys*	I	I	-	-	-
104.	*Vicia sepium*	I	-	I	-	-
	Molinio-Arrhenatheretea	-	-	-	-	-
105.	*Achillea millefolium*	II	II	-	-	-
106.	*Agrostis capillaris*	II	II	I	-	-
107.	*Alchemilla xanthochlora*	I	II	I	-	-
108.	*Campanula glomerata*	I	I	I	-	-
109.	*Campanula abietina*	I	I	I	II	II
110.	*Campanula patula* ssp. *patula*	II	I	I	-	-
111.	*Centaurea phrygia.* ssp. *phrygia*	I	I	-	-	-
112.	*Cerastium holosteoides*	II	IV	I	II	-
113.	*Crepis biennis*	I	II	-	-	I

Table 1 (continued)

114.	*Cynosurus cristatus*	I	-	I	-	-
115.	*Dactylis glomerata*	IV	III	I	IV	I
116.	*Elymus repens*	I	-	-	-	-
117.	*Equisetum arvense*	I	-	II	-	III
118.	*Holcus lanatus*	I	II	-	-	-
119.	*Lapsana communis*	I	-	I	-	-
120.	*Leucanthemum vulgare*	II	-	III	I	I
121.	*Lotus corniculatus*	II	II	I	-	-
122.	*Lysimachia nummularia*	IV	IV	III	III	II
123.	*Lysimachia vulgaris*	II	II	I	II	-
124.	*Medicago lupulina*	I	II	-	-	I
125.	*Ranunculus repens*	IV	IV	I	III	IV
126.	*Plantago lanceolata*	I	II	I	-	-
127.	*Plantago major*	III	II	I	-	-
128.	*Poa palustris*	I	-	-	-	-
129.	*Poa pratensis*	II	II	I	II	-
130.	*Potentilla reptans*	I	-	-	-	-
131.	*Prunella vulgaris*	IV	III	II	III	IV
132.	*Taraxacum officinale*	-	-	-	-	II
133.	*Trifolium repens*	III	III	II	-	II
134.	*Vicia cracca*	I	-	-	-	-
	Alno-Ulmion	-	-	-	-	-
135.	*Brachypodium sylvaticum*	II	-	I	-	-
136.	*Circaea lutetiana*	I	II	-	III	-
137.	*Elymus caninus*	II	II	II	I	II
138.	*Epilobium montanum*	III	II	II	IV	II
139.	*Equisetum hyemale*	I	-	-	-	-
140.	*Equisetum pratense*	-	-	II	-	-
141.	*Equisetum sylvaticum*	II	II	I	-	-
142.	*Equisetum telmateia*	I	-	-	-	-
143.	*Festuca gigantea*	I	II	II	-	-
144.	*Listera ovata*	-	-	I	-	-
145.	*Matteuccia struthiopteris*	II	-	-	I	III
146.	*Primula elatior*	-	-	I	-	-
147.	*Rumex sanguineus*	I	I	-	-	-
148.	*Salix alba*	I	-	-	-	-
149.	*Thalictrum aquilegiifolium*	I	I	-	-	-
	Fagetalia	-	-	-	-	-
150.	*Acer pseudoplatanus*	I	II	-	-	-
151.	*Aconitum moldavicum*	I	-	-	-	-
152.	*Ajuga reptans*	I	II	II	-	-

Table 1 (continued)

153.	*Anemone nemorosa*	-	-	-	-	I
154.	*Anemone ranunculoides*	-	-	-	-	I
155.	*Asarum europaeum*	-	-	II	-	II
156.	*Carex sylvatica*	I	-	-	-	-
157.	*Daphne mezereum*	I	-	-	-	-
158.	*Dentaria glandulosa*	-	-	-	-	II
159.	*Euphorbia amygdaloides*	I	-	-	-	II
160.	*Fagus sylvatica*	I	-	-	-	-
161.	*Fraxinus excelsior*	I	-	-	-	-
162.	*Galium schultesii*	-	-	II	-	-
163.	*Luzula luzuloides*	I	I	II	I	II
164.	*Luzula sylvatica*	II	I	-	-	-
165.	*Mercurialis perennis*	-	-	II	-	-
166.	*Oxalis acetosella*	I	-	II	-	III
167.	*Pulmonaria officinalis*	-	-	I	-	-
168.	*Salvia glutinosa*	II	IV	III	III	-
169.	*Stachys sylvatica*	IV	IV	II	IV	-
170.	*Veronica urticifolia*	-	-	II	-	II
171.	*Viola riviniana*	-	-	I	-	-
	Variae Syntaxa	-	-	-	-	-
172.	*Alnus incana*	I	-	-	-	-
173.	*Alopecurus pratensis*	I	-	-	-	-
174.	*Anthriscus cerefolium*	I	-	I	-	-
175.	*Arrhenatherum elatius*	I	-	-	-	-
176.	*Artemisia vulgaris*	I	-	-	I	-
177.	*Asplenium trichomanes*	-	-	I	-	-
178.	*Astragalus glycyphyllos*	I	-	I	-	I
179.	*Bidens tripartita*	I	-	-	I	-
180.	*Bromus commutatus*	I	-	-	-	-
181.	*Calamagrostis pseudophragmites*	I	-	-	III	I
182.	*Caltha palustris ssp. laeta*	I	I	-	-	II
183.	*Calystegia sepium*	II	II	I	II	-
184.	*Campanula bononiensis*	I	-	-	-	-
185.	*Cardamine impatiens*	I	-	-	-	II
186.	*Carex leporina*	I	II	-	-	-
187.	*Carex pallescens*	I	-	-	-	-
188.	*Carex remota*	I	-	-	-	II
189.	*Carex vulpina*	I	II	-	-	-
190.	*Catabrosa aquatica*	I	-	-	-	-
191.	*Centaurea biebersteinii subsp. biebersteinii*	-	I	-	-	I

Table 1 (continued)

192.	*Centaurea nigrescens*	I	-	-	-	-
193.	*Centaurea pseudophrygia*	I	II	-	-	-
194.	*Chenopodium album*	I	-	-	-	-
195.	*Cichorium intybus*	I	-	-	-	-
196.	*Cirsium arvense*	II	II	I	-	-
197.	*Cirsium canum*	I	II	-	-	-
198.	*Cirsium vulgare*	I	-	-	-	-
199.	*Clinopodium vulgare*	II	-	II	-	-
200.	*Convolvulus arvensis*	I	-	-	-	-
201.	*Cornus sanguinea*	-	-	I	-	-
202.	*Coronilla varia*	I	-	-	-	-
203.	*Corylus avellana*	I	-	-	-	II
204.	*Crataegus monogyna*	I	-	-	-	-
205.	*Crepis paludosa*	I	-	-	-	-
206.	*Cruciata glabra*	I	I	-	-	-
207.	*Cruciata laevipes*	I	-	-	-	-
208.	*Cystopteris fragilis*	-	-	I	-	-
209.	*Dianthus barbatus* ssp. *barbatus*	I	-	-	-	II
210.	*Dianthus barbatus* ssp. *compactus*	I	-	II	-	-
211.	*Dianthus carthusianorum*	-	II	-	-	-
212.	*Dipsacus pilosus*	I	-	-	-	-
213.	*Epilobium hirsutum*	I	-	-	-	-
214.	*Epilobium palustre*	I	-	I	-	-
215.	*Epilobium parviflorum*	I	-	I	-	-
216.	*Erigeron annuus*	II	IV	I	-	I
217.	*Euphorbia carniolica*	-	I	I	-	-
218.	*Euphorbia epithymoides*	II	-	-	-	I
219.	*Festuca pratensis*	-	I	-	-	-
220.	*Galinsoga parviflora*	I	-	-	-	-
221.	*Galium mollugo*	I	I	-	-	-
222.	*Galium verum*	I	-	I	-	-
223.	*Geranium palustre*	I	-	-	-	-
224.	*Geum urbanum*	I	-	I	-	-
225.	*Glechoma hirsuta*	I	-	II	-	I
226.	*Glyceria fluitans*	I	I	-	-	-
227.	*Glyceria maxima*	I	-	-	-	-
228.	*Glyceria notata*	I	-	-	II	-
229.	*Gnaphalium sylvaticum*	I	II	-	-	-
230.	*Gymnocarpium robertianum*	I	-	I	-	-
231.	*Hieracium aurantiacum*	-	I	-	-	-
232.	*Hieracium umbellatum*	I	-	-	-	I

Table 1 (continued)

233.	*Homogyne alpina*	-	-	-	-	I
234.	*Hypericum montanum*	-	-	I	-	-
235.	*Hypericum perforatum*	I	I	I	-	II
236.	*Hypericum tetrapterum*	I	-	-	-	-
237.	*Lathyrus tuberosus*	I	-	-	-	-
238.	*Lonicera xylosteum*	-	-	I	-	-
239.	*Lychnis viscaria*	I	-	-	-	-
240.	*Lysimachia punctata*	I	-	-	-	-
241.	*Matricaria perforata*	I	-	-	-	-
242.	*Medicago minima*	I	I	-	-	-
243.	*Melilotus albus*	I	-	-	-	-
244.	*Mentha aquatica*	I	-	-	-	-
245.	*Mentha arvensis*	I	I	I	-	II
246.	*Origanum vulgare*	I	-	-	-	I
247.	*Petasites × reichingeri*	-	-	-	I	-
248.	*Phalaris arundinacea*	I	-	-	-	-
249.	*Phegopteris connectilis*	-	-	I	-	-
250.	*Phleum montanum*	I	-	-	-	-
251.	*Phleum phleoides*	I	-	-	-	-
252.	*Picea abies*	I	II	I	-	-
253.	*Plantago media*	I	II	I	-	III
254.	*Poa nemoralis*	I	-	I	-	I
255.	*Polygonum hydropiper*	I	-	-	-	-
256.	*Polygonum lapathifolium*	I	-	-	-	-
257.	*Polygonum mite*	I	I	-	-	-
258.	*Polypodium vulgare*	-	-	I	-	-
259.	*Rhinanthus rumelicus*	I	-	-	-	-
260.	*Rorippa pyrenaica*	I	-	-	-	-
261.	*Rorippa sylvestris*	I	-	I	-	-
262.	*Rubus hirtus*	I	II	-	-	-
263.	*Rudbeckia laciniata*	I	-	-	-	-
264.	*Salix cinerea*	I	-	-	-	-
265.	*Salix fragilis*	I	-	-	-	-
266.	*Salix purpurea*	I	-	-	-	-
267.	*Saponaria officinalis*	I	I	I	-	-
268.	*Scutellaria galericulata*	I	-	I	-	I
269.	*Sedum maximum*	-	-	I	-	-
270.	*Senecio erucifolius*	-	-	-	-	I
271.	*Senecio vulgaris*	I	-	-	-	-
272.	*Seseli libanotis*	-	-	-	-	I
273.	*Silene armeria*	-	-	-	-	I

Table 1 (continued)

274.	*Silene latifolia* ssp. *alba*	I	-	-	-	II
275.	*Silene nutans ssp. dubia*	-	-	I	-	I
276.	*Stellaria media*	I	-	-	-	-
277.	*Trifolium alpestre*	-	II	-	-	-
278.	*Trifolium campestre*	I	-	I	-	-
279.	*Trifolium medium*	I	-	-	-	-
280.	*Trifolium montanum*	-	I	-	-	-
281.	*Trifolium pratense*	I	IV	-	-	-
282.	*Tussilago farfara*	IV	IV	II	-	IV
283.	*Valeriana tripteris*	-	-	I	-	-
284.	*Veronica beccabunga*	I	-	-	-	-
285.	*Veronica officinalis*	II	II	I	-	II
286.	*Veronica teucrium*	I	II	-	-	-
287.	*Vicia sylvatica*	I	I	-	-	-
288.	*Viola arvensis*	I	I	-	-	-
289.	*Viola tricolor*	-	II	-	-	-

Date and location of relevés:

1. *Telekio-Petasitetum hybridi* (Morariu 1967) Resmeriţă *et* Raţiu 1974: R1 - Ciobotani, Mermezeu (15.07.2014); R2 - Câmpul Cetăţii, valea Nirajul Mare (01.07.2014); R3 - Lăpuşna (05.07.2014); R4 - Lăpuşna, pârâul Secuieu (07.07.2014); R5 - Lăpuşna, pârâul Negru (07.07.2014); R6 - Ibăneşti, Şirodul Mare (10.07.2014); R7 - Câmpul Cetăţii, valea Nirajul Mare (10.07.2014); R8 - Câmpul Cetăţii, valea Nirajul Mare (15.07.2014); R9 - Sovata, Ilieşi (18.07.2014); R10 - Răstoliţa, Valea Răstoliţei (19.07.2014); R11 - Bistra Mureşului, Valea Bistrei (23.07. 2014); R12 - Bistra Mureşului, Valea Bistrei (23.07. 2014); R13 - Răstoliţa, Gălăoaia (25.07. 2014); R14 - Răstoliţa, Gălăoaia (25.07. 2014); R15 - Răstoliţa, Gălăoaia (25.07. 2014); R16 - Răstoliţa, Valea Vişa (28.07. 2014); R17 - Răstoliţa, Valea Iodului (30.07. 2014); R18 - Răstoliţa, Valea Iodului (30.07. 2014); R19 - Sovata, valea Sebeş (04.08. 2014); R20 - Sovata, valea Sovata (06.08. 2014); R21 - Sovata, valea Sovata (06.08. 2014); R22 - Sovata, Târnava Mică (06.08.2014); R23 - Lunca Bradului, Ilişoara Mare (11.08.2014); R24 - Lunca Bradului, Ilişoara Mică (11.08. 2014); R25 - Lunca Bradului, Ilva Mare (11.08.2014); R26 - Stânceni, Gudea Mare (12.08.2014); R27 - Stânceni, Gudea Mică (12.08.2014); R28 - Răstoliţa, Valea Rusu (13.08.2014); R29 - Sălard, Hidegag (16.08.2014); R30 - Sălard, Şolea (16.08.2014); R31 - Sălard, valea Sălard (16.08.2014); R32 - Răstoliţa, Borzia (17.08.2014); R33 - Sălard, Ţâba Mică (18.08.2014); R34 - Sălard, Valea Belciu (18.08.2014); R35 - Sălard, Jirca (18.08.2014); R36 - Neagra, pârâul Neagra (23.08.2014); R37 - Ibăneşti, Fâncel (26.08.2014); R38 - Sălard, Pârâul Wagner (26.08.2014); R.39 - Ciobotani, Zebrac (16.09.2014).

2. *Telekio-Filipenduletum* Coldea 1996: R1 - Câmpul Cetății, valea Nirajul Mare (01.07.2014); R2 - Câmpul Cetății, valea Nirajul Mare (12.07.2014); R3 - Răstolița, Valea Răstoliței (19.07.2014); R4,5 - Bistra Mureșului, valea Bistrei (23.07.2014); R6- Sălard, valea Hidegag (16.08.2014); R7 - Sălard, valea Șolea (16.08.2014); R8 - Stânceni, Ciobotani (16.09.2014).
3. *Telekio speciosae-Aruncetum dioici* Oroian 1998: 1-Răstolița, Peșcoasa Mare, 06.07.2014; R:2,3 - Sovata-pe valea Târnava Mică, Ilieși, 18.07.2014; R4: Bistra Mureșului, valea Bistrei (23.07.2014); R5 - Lunca Bradului Ilișoara Mare (11.08.2014); R6,7 - Răstolița,Borzia (07.2000); R 8,9,10 –Răstolița, Valea Costeasa (07.2000).
4. *Petasitetum kablikiani* Szafer et al. 1926: R1 - Lunca Bradului, Ilișoara Mare (11.08.2014); R2,3 - Lunca Bradului, Ilva Mare (11.08.2014); R4,5 - Stânceni, Gudea Mare (12.08.2014).
5. *Arunco-Petasitetum albi* Br.-Bl. et Sutter 1977: R1 - Răstolița, Valea Răstoliței (19.07.2014); R2 - Lăpușna, Pârâul Secuieu (07.07.2014); R3 –Ibănești, Șirodul Mare (10.07.2014); 4-Răstolița "Costeasa" (2000.06.15); 5-Răstolița, Iod (2000.06.17); 6-Răstolița "Costeasa" (2001.07.11); 7,8 -lunca Mureșului între Răstolița și Borzia (2001.07.28).

6 Conclusions

A data set of 70 relevés were analysed. They were recorded in five associations belonging to Petasition officinalis alliance.

If we compare the 5 plant associations, from the floral point of view, the coenoses resemble but are distinguished by the abundance of characteristic species.

It is worth mentioning the significant percentage of Alpine-Carpathian, Carpathian and Carpathian-Balkan elements, which give a particular aspect to this area, conferring phytocenoses a distinct regional peculiarity of that of the vicarians in Central Europe, demonstrating the intensity of the process of differentiation and preservation from the flora of tertiary and pleistocene origin and highlights the close florogenetic connections between the flora of this Carpathian area, strictly geomorphologically delimited and that of the Balkan mountains. Among these we mention: *Petasites kablikianus, Leucanthemum waldsteinii, Achillea distans, Aconitum napellus* ssp. *tauricum, Telekia speciosa, Scopolia carniolica, Cirsium waldsteinii, Pulmonaria rubra, Silene nutans* ssp. *dubia.*

We report the presence in the composition of plant associations of some endangered species found in the Red Lists*: Campanula abietina* (in all associations) and *Angelica archangelica* (association 1). A remarkable phytogeographical interest has the endemic or subendemic species: *Aconitum moldavicum* (association1), *Dianthus barbatus* ssp. *compactus* (associations 1,3), *Cirsium waldsteinii* (associations 1, 2, 3, 5) *Leucanthemum waldsteinii* (in all 5 associations) *and Silene nutans* ssp. *dubia* (associations 3, 5). There are also Carpathian-Balkan species (*Campanula abietina, Petasites kablikianus, Pulmonaria rubra).*

References

1. Bilz M, Kell SP, Maxted N, Lansdown RV (2011) European red list of vascular plants. Publications Office of the European Union, Luxembourg
2. Borza A, Boşcaiu N (1965) Introducere în studiul covorului vegetal, Ed. Acad. R.P.R., Bucureşti
3. Boşcaiu N, Coldea G, Horeanu C (1994) Lista roşie a plantelor vasculare dispărute, periclitate, vulnerabile şi rare din flora României. Ocrot. Nat. Med. Înconj. 38(1):45–56
4. Chytrý M (ed) Vegetace České republiky. 2. Travinná a keříčková vegetace Academia, Praha
5. Coldea G (1990) Munţii Rodnei. Studiu geobotanic. Ed. Acad. Bucureşti, p 183
6. Coldea G (1991) Prodrome des associations vegetales des Carpates du sud-est (Carpates Roumaines). Doc Phytosociol Camerino 13:317–359
7. Coldea G (2012) Les association végétales de Roumanie, T. 2 (Les associations anthropogènes), Edit. Presa Universitară Clujeană, Cluj Napoca
8. Cristea V, Gafta D, Pedrotti F (2004) Fitosociologie. Edit. Presa Universitară Clujeană, Cluj-Napoca
9. Danciu M (1973) Contribuţii la studiul asociaţiilor alianţei Filipendulo-Petasition din sudul munţilor Baraolt. Bul. Univ. Braşov. Seria C. Matematică-Fizică-Chimie-Şti. Nat. Braşov 15:123–131
10. Dihoru G, Negrean G (2009) Cartea roşie a plantelor vasculare din România, Edit. Acad. Române, Bucureşti, p 630
11. Dihoru GH, Dihoru A (1993–1994) Plante rare, periclitate şi endemice în flora României – Lista Roşie, Acta Botanica Horti Bucurestiensis, Bucureşti, pp 173–179
12. Doniţă N, Popescu A, Paucă-Comănescu M, Mihăilescu S, Biriş I-A (2005) Habitate din România, Edit. Tehnică Silvică Bucureşti
13. Gafta D, Mountford O (coord.) (2008) Manual de interpretare a habitatelor Natura 2000 din România, Edit. Risoprint, Cluj-Napoca
14. Hájková P, Hájek M (2007) Calthion palustris. In: Chytrý M (ed) Vegetace České republiky. 1. Travinná a keříčková vegetace Academia, Praha, pp 238–280
15. Hammer Ø, Harper DAT, Ryan PD (2001) PAST: paleontological statistics software package for education and data analysis. Palaeontologia Electronica 4(1):9
16. Jarolimek I, Kliment J, Valachovič M (2002) The syntaxonomicalrevision of the riparian plant communities dominated byPetasites hybridus in Slovakia. Biologia 57:471–492
17. Kliment J, Jarolimek I (2002) Syntaxonomical revision of the Petasites kablikianus communities (Petasition officinalis) in the West Carpathians. Biologia 57:101–118
18. Kliment J, Valachovic M (eds) (2007) Rastlinné spolocenstvá Slovenska. 4. Vysokohorská vegetátia. Veda, Bratislava
19. Kliment J, Šibík J, Šibíková I, Jarolímek I, Dúbravcová Z, Uhlířová J (2010) High-altitude vegetation of the Western Carpathians – a syntaxonomical review. Biologia 65/6:965–989, Section Botany. https://doi.org/10.2478/s11756-010-0109-4
20. Morariu I (1967) Clasificarea vegetatţiei nitrofile din România.Contr. Bot. 1967: 233–246
21. Mucina L (1993) Galio-urticetea. In: Mucina L, Grabherr G, Ellmauer T (eds) Die Pflanzengesellschaften Österreichs. Teil I, Gustav Fischer Verlag, Jena, DE, pp 203–251
22. Mucina L, Bültmann H, Dierßen K, Theurillat J-P, Raus T, Čarni A, Šumberová K, Willner W, Dengler J, Tichý L (2016) Vegetation of Europe: hierarchical floristic classification system of vascular plant, bryophyte, lichen and algal communities. Appl Veg Sci 19(Suppl. 1):3–264
23. Nechita N (2003) Flora şi vegetaţia cormofitelor din Masivul Măşmaş, Cheile Bicazului şi Lacul Roşu, Ed. "Constantin Matasă" Piatra Neamţ
24. Oltean M, Negrean G, Popescu A, Roman N, Dihoru G, Sanda V, Mihăilescu S (1994) Listă roşie a plantelor superioare din România, Studii, sinteze, documentaţii de ecologie, 1, Acad. Română, Instit. de Biologie, Bucureşti

25. Oroian S (1998) Flora şi Vegetaţia Defileului Mureşului între Topliţa şi Deda, Casa de Editură Mureş
26. Pott R (1995) Die Pfl anzengesellschaften Deutschlands. 2nd edn. Ulmer, Stuttgart, p 622
27. Resmeriţă I, Raţiu O (1974) Vegetaţia higro-şi hidrofilă din Maramureş. Contrib. Bot., Cluj-Napoca, pp 115–129
28. Sanda V, Popescu A, Barabaş N (1997) Cenotaxonomia şi caracterizarea grupărilor vegetale din România. St. Şi Com. Muz. Şt. Nat. Bacău, Biol. Veget., 14: 365 p.
29. Sanda V, Popescu A, Stancu DI (2001) Structura cenotică şi caracterizarea ecologică a fitocenozelor din România, Ed. Conphis, Piteşti
30. Sămărghiţan M (2005) Flora şi vegetaţia Văii Gurghiului, Ed. University Press, Târgu-Mureş
31. Sârbu I, Ştefan N, Oprea A (2013) Plante vasculare din România, determinator ilustrat de teren. Edit. VictorBVictor, Bucureşti

Arthrocnemum macrostachyum plant communities in the Iberian Peninsula, Balearic and Canary Islands (Spain and Portugal)

Esteban Ramírez, Lourdes Rufo, Daniel Sánchez-Mata, Irene Sánchez-Gavilán, and Vicenta de la Fuente

Abstract We present a synthesis of the halophilous fruticose vegetation structured by *Arthrocnemum macrostachyum* on the Iberian Peninsula, Balearic Islands and Canary Islands (*Arthrocnemo macrostachyi-Suaedetalia braun-blanquetii*, *Salicornietea fruticosae*). The analysis and study of over 200 selected phytosociological relevés supports the conclusions reflected in the syntaxonomical checklist of the units recognised in the study area. The floristic appendix also contains taxonomic and floristic-chorological novelties.

Keywords Arthrocnemion macrostachyi · Salicornietea fruticosae · Halophytic vegetation · Iberian Peninsula vegetation · Saline soils habitats

1 Introduction

Plants adapted to live in saline environments, a group known as halophytic plants, are found along coastlines and inland salt marshes around the world [43]. The main ecological factors that determine the habitat type and limit the development of certain species in these environments are the high concentration of salts present in the soils, and the daily and seasonal variation in the degree of soil waterlogging due to tidal action along the coast and the rainfall regime in inland environments [4]. To respond to these peculiar, extreme environmental conditions, species in the family

E. Ramírez · I. Sánchez-Gavilán · V. de la Fuente
Department of Biology (Botany), Faculty of Sciences, Autonomous University of Madrid, Cantoblanco, 28049 Madrid, Spain

L. Rufo
Department of Pharmacy, Faculty of Experimental Sciences, Francisco de Vitoria University, Pozuelo de Alarcón, 28223 Madrid, Spain

D. Sánchez-Mata (✉)
Department of Pharmacology, Pharmacognosy and Botany (Botany Unit), Faculty of Pharmacy, Complutense University, 28040 Madrid, Spain
e-mail: dsmata@ucm.es

F. Pedrotti and E. O. Box (eds.), *Tools for Landscape-Scale Geobotany and Conservation*, Geobotany Studies, https://doi.org/10.1007/978-3-030-74950-7_12

Chenopodiaceae (subfamily Salicornioideae) are among the lineages with the largest proportion of genera that mainly form plant communities in saline environments. Of particular note are the communities formed by succulent therophytes belonging to the phytosociological class *Therosalicornietea* (mainly species from the genera *Halopeplis*, *Microcnemum* and *Salicornia*), and/or by succulent chamaephytes [mainly species from the genera *Sarcocornia* (hereafter *S.*), *Halocnemum* and *Arthrocnemum* (hereafter *A.*)] in the class *Salicornietea fruticosae* [5, 13, 19, 23–26, 36, 40–42, 44, 49–52]. This last class is characterised by permanent halophilous fruticose communities and includes vegetation units structured by *Arthrocnemum macrostachyum* (hereafter *A. macrostachyum*) in territories of Spain and Portugal reviewed by [49, 50, 52].

The hierarchical classification and prior knowledge of these communities are essential prerequisites for determining, managing and protecting their diversity. These are also fundamental instruments for interpreting the landscape in reforestation programs and for recovery of vegetation cover [14, 33]. Halophilous fruticose communities (*Salicornietea fruticosae*) and their characteristic biotopes are included in Annex I of the Habitats Directive 92/43/EEC [17].

From knowledge of the taxonomy, ecology and habitats of the subfamily *Salicornioideae* in the Mediterranean [23–26, 42], we undertake and assess the diversity of the halophilous communities structured by *A. macrostachyum* and contribute a proposal for their syntaxonomical classification, based on analysis of phytosociological data by means of K-means clustering.

2 Nomenclature and References

The taxonomic nomenclature follows the Euro+Med database [16], exceptions are included in the floristic appendix with their recognized authority. The main references were the biogeographical units of Western Mediterranean Europe synthesized by Rivas-Martínez et al. [53] and the syntaxonomy of the *Salicornietea fruticosae* class proposed by Rufo et al. [26].

3 Material and Methods

3.1 Data Sources

The diversity of halophytic vegetation structured by *A. macrostachyum* on the Iberian Peninsula and Canary and Balearic Islands was analysed using phytosociological relevés collected between 1958 and 2019 from the Iberian and Macaronesian Vegetation Information System [21, 22], from other bibliographical sources, and from our own unpublished fieldnotes. Relevés compiled according to the phytosociological methodology of the Zürich-Montpellier school [7] were recovered from the following vegetation studies: [1–3, 8–12, 23, 28, 30, 32, 34, 35, 37–39, 44, 47, 48, 54, 56–58].

The total number of relevés was 293 initially. We screened the data manually to exclude duplicate and heterogeneous relevés (those that included species indicating a mixture of ecologically different plant communities) and those where the identity and presence of some species were unclear. This reduced the number of relevés to 225. Taxa determined to supraspecific, sensu lato, and uncertain taxa were eliminated. Synonyms were merged and taxa present in only one or two relevés were eliminated. After this screening, the number of taxa to analyse was 89.

3.2 Data Analysis

The following protocol was used to assess the halophytic vegetation in the territories studied. The data matrix was stratified by biogeographic sectors [53], then subjected to a heterogeneity-constrained random (HCR) re-sampling using the Bray–Curtis dissimilarity index [31]. The number of relevés selected was proportional to beta diversity within strata, based on the Bray-Curtis index [61], and the number of relevés to be re-sampled was between 10 and 40 per stratum. The matrix obtained includes 187 relevés and 89 species, and these data were analysed by the K-means method using the Hellinger Distance as a dissimilarity measure, Ward's hierarchical clustering method and a minimum of two and a maximum of seven groups. Finally, a synthetic table was created with the groups generated and rearranged based on the percentage of present and faithful species contributed by the K-means analysis. A synthesis of the initial and screened relevés, and those remaining for analysis, is given in Table 1.

Data editing and K-means analysis were carried out with the Quercus edition module in the management package Vegana for ecological data [20] and re-sampled with the JUICE 7.0 program [59]. The methodology was based on our previous study focused on plant communities of *Sarcocornia* spp. [26], with some variations adapted to the new data matrix.

Table 1 References used in the analysis, with the number of tables consulted, number of initial relevés, number of relevés remaining after filtering, number of relevés remaining for analysis and the global sum of each column

References	Table	Initial	Filtered	Analysed
[1]	12	5	5	5
[2]	2.5	5	5	4
[3]	34	4	4	4
[6]	–	12	11	11
[8]	11	8	4	4
[9]	20	16	–	–
[10]	9	10	7	5
[11]	10	17	16	11
[12]	7	19	17	15
[18]	1	15	13	10
[23]	3	5	5	4
[28]	11	5	4	4
[30]	01	3	3	3
[32]	4	4	4	4
[34]	9	14	10	10
[35]	2	8	4	4
[37]	20	6	5	5
[38]	62	10	10	9
[39]	129	4	4	4
[44]	1	12	12	6
[47]	20	5	5	5
[48]	7	12	2	2
[54]	2	22	8	5
[56]	7	4	4	4
[57]	7	4	4	4
[58]	2	8	3	3
Unpublished own data	–	56	56	42
Σ	–	**293**	**225**	**187**

4 Results and Discussion

The analysis corresponding to the separation in six groups has the highest silhouette value (K = 6; average silhouette = 0.157) of all the different K-means analyses carried out, and is the grouping that has been considered. Table 2 shows the synthetic relevés corresponding to this grouping.

Group 1: Represents the shrubland community in which *A. macrostachyum* coexists with *S. hispanica*; it contains relevés from the Spanish provinces of Albacete, Alicante, Murcia, Almería, and Granada.

Table 2 Synthetic table with the six groups generated by K-means analysis. Numbers represent the presence percentage in all relevés in each group. Grey-shaded values indicate faithful taxa

Taxon name/K-means groups	G1	G2	G3	G4	G5	G6
N° of relevés	26	50	25	15	17	53
N° of species	33	45	25	18	16	55
Arthrocnemum macrostachyum	100	100	100	100	100	100
Faithful species G1						
Sarcocornia hispanica	96	3	8	.	.	.
Limonium supinum	30	.	.	.	.	.
Lygeum spartum	46	.	8	6	5	5
Suaeda vera	88	33	4	.	5	58
Puccinellia fasciculata	11	.	.	.	.	.
Puccinellia caespitosa	11	.	.	.	.	.
Limonium majus	11	.	.	.	.	.
Phragmites australis	23	3	.	6	.	1
Limonium cossonianum	42	1	8	.	70	3
Faithful species G2						
Halimione portulacoides	7	100	40	.	5	3
Limonium algarvense	.	19	.	.	.	.
Sarcocornia pruinosa	.	23	.	.	.	3
Sarcocornia alpini	.	27	.	.	5	9
Myriolimon ferulaceum	3	21	.	.	.	3
Limoniastrum monopetalum	.	27	.	.	11	9
Faithful species G3						
Sarcocornia lagascae	.	.	100	.	.	.
Limonium bellidifolium	.	.	16	.	.	1
Faithful species G4						
Suaeda braun-blanquetii	.	.	.	93	.	.
Spergularia diandra	.	.	.	60	.	.
Frankenia pulverulenta	.	.	12	80	.	1
Hornungia procumbens	.	1	8	66	.	1
Sphenopus divaricatus	3	9	.	73	.	.
Parapholis incurva	7	.	.	53	.	3
Limonium echioides	.	1	.	33	.	.
Rostraria cristata	.	.	.	13	.	.
Aizoon hispanicum	.	.	.	13	.	.
Faithful species G5						
Frankenia corymbosa	.	7	4	.	76	9
Phragmites communis	.	.	.	.	29	.
Limbarda crithmoides	19	27	16	.	82	.
Pancratium maritimum	.	.	.	.	17	.
Faithful species G6						
Tetraena fontanesii	.	.	.	.	.	20
Other species						
Parapholis filiformis	.	5	24	.	.	5
Juncus maritimus	15	13	.	.	.	3
Frankenia thymifolia	15	.	.	13	.	.
Plantago coronopus	.	.	12	6	.	9
Atriplex glauca	15	.	4	.	.	7

Limonium delicatulum	19	3	.	.	.	3
Artemisia gallica	3	9	.	6	.	1
Limonium angustebracteatum	3	5	.	.	11	.
Polypogon maritimus	3	1	8	6	.	.
Cistanche phelypaea	.	15	.	.	.	3
Spartina densiflora	.	13	.	.	.	5
Juncus subulatus	3	3	4	.	.	7
Limonium latebracteatum	11	.	4	.	.	1
Suaeda maritima	.	13	.	.	.	3
Salsola soda	.	.	.	13	.	3
Limonium caesium	3	5	.	.	.	7
Limonium gibertii	.	.	8	.	.	7
Cynomorium coccineum	.	.	12	.	.	1
Limonium retusum	.	.	12	.	.	1
Suaeda spicata	3	.	8	.	.	1
Frankenia laevis	.	7	.	.	.	5
Mesembryanthemum nodiflorum	.	5	.	.	.	7
Plantago crassifolia	7	.	.	.	5	.
Frankenia capitata	.	.	.	.	.	11
Limonium lanceolatum	.	11	.	.	.	.
Myriolimon diffusum	.	11	.	.	.	.

In addition to *A. macrostachyum*, the faithful species in this group are *Limonium cossonianum, L. majus, L. supinum, Lygeum spartum, Phragmites australis, Puccinellia caespitosa, P. fasciculata* and *S. hispanica*. Another species often present in this group is the Murcian-Almerian endemic *Limonium delicatulum*.

The floristic composition and biogeographical distribution of the relevés included in this group lead us to consider its inclusion in the association *Arthrocnemo macrostachyi-Sarcocornietum hispanicae* [24] (Albacete, Alicante, Murcia and Almería) and in the endemic association of the Guadician-Bacensean sector *Limonio majoris-Sarcocornietum hispanicae* (Granada) [24, 26, 30]. The biogeographical distribution of these communities covers inland saline areas throughout the Central Iberian Mediterranean biogeographical province (Castilian subprovince, southwest territories of the La Mancha sector), Murcia and Almería biogeographical province (Murcia and Almería subprovince, Alicante and Murcia and Almería sectors), and the Bética province (Bética subprovince, Hoyas de Guadix and Baza sector).

These communities can withstand temporary flooding (winter, spring), summer drought and a high concentration of salts in the soil due mainly to sulphates, carbonates and bicarbonates.

The relevés dominated by *A. macrostachyum*, with the presence of *S. hispanica* [*S. fruticosa* sensu auct. pl.] in these territories in southeast Iberia have traditionally been ascribed to the association *Frankenio corymbosae-Arthrocnemetum macrostachyi* [1, 3, 35, 58]. Salazar et al. [56] include the relevés controlled by *A. macrostachyum* in the inland salt pans of Granada (Guadix, Baza) in an association structured by what they interpret as Sarcocornia *fruticosa*. The relevés we use to describe the phytosociological association *Arthrocnemo macrostachyi-Sarcocornietum hispanicae* for the eastern Iberian Peninsula also reflect this last syntaxonomical interpretation [24, 25].

Species with low presence in G1: *Puccinellia stenophylla* 3, *Elytrigia curvifolia* 3, *Helianthemum polygonoides* 7, *Lycium intricatum* 3, *Aeluropus littoralis* 7, *Sonchus tenerrimus* 3; in G2: *Limonium virgatum* 1, *Limonium daveaui* 1, *Elytrigia elongata* 3, *Tripolium pannonicum* 3, *Puccinellia stenophylla* 1, *Atriplex halimus* 3, *Polygonum equisetiforme* 5, *Triglochin barrelieri* 7, *Asparagus horridus* 3, *Spergularia media* 9, *Sporolobus pungens* 3, *Sonchus tenerrimus* 7, *Hordeum marinum* 1; in G3: *Lycium intricatum* 4, *Salicornia patula* 8, *Limonium delicatulum* subsp. *formenterae* 8; in G4: *Hordeum marinum* 6; in G5: *Halocnemum cruciatum* 5, *Asparagus horridus* 5, *Sporolobus pungens* 5; and in G6: *Limonium virgatum* 1, *Limonium daveaui* 1, *Trisetaria panicea* 3, *Anisantha madritensis* 3, *Lactuca tenerrima* 3, *Spergularia heldreichii* 3, *Elytrigia juncea* 3, *Traganum moquinii* 3, *Elytrigia curvifolia* 1, *Spergularia marina* 5, *Juncus bufonius* 5, *Suaeda vermiculata* 5, *Atriplex halimus* 3, *Polygonum equisetiforme* 2, *Limonium tuberculatum* 7, *Aeluropus litoralis* 1, *Sporolobus pungens* 1, *Hordeum marinum* 3.

Group 2: this group includes relevés from the west coast of the Iberian Peninsula (Portugal, Spain) in the biogeographical territories of the Coastal Lusitania-West Andalusia province (Divisorio Portuguese, and Cádiz and Sado subprovinces; Divisorio Portuguese, Ribatejo and Sado, Algarve and Monchique, Cádiz and Huelva coastline, and Algeciras and Aljibe sectors). It contains relevés with 46 taxa; the faithful species are *A. macrostachyum*, *Halimione portulacoides*, *Limoniastrum monopetalum*, *Limonium algarvense*, *Myriolimon ferulaceum*, *S. alpini* and *S. pruinosa*. Other species with a relevant presence are *Cistanche phelypaea*, *Limbarda crithmoides*, *Limonium lanceolatum*, *Myriolimon diffusum*, *Spartina densiflora* and *Suaeda maritima*.

This set of relevés corresponds closely to the association *Inulo crithmoidis-Arthrocnemetum macrostachyi*. The territory where these halophilous communities develop has an oceanic Mediterranean pluvi-seasonal bioclimate characterised by high precipitation, absence of frost in coastal areas, and high oceanicity. These thrive in the higher parts of Atlantic coastal marshes and in the brinier and more emerged zones of the wetlands, occupying the driest parts of salt marshes [10, 11, 34, 37, 57].

A few relevés from the east of the peninsula have joined the western block. These share the presence of coastal species with wide distributions (*Limoniastrum monopetalum* and *Halimione portulacoides*), which, due to the low biomass of the other species, may have influenced the result. It should also be noted that *Halimione portulacoides* may predominate exceptionally over *A. macrostachyum* in these communities in the southwestern Iberian Peninsula [46]. Similarly, a relevé from Lanzarote (Canary Islands) with a high cover of *Halimione portulacoides* appears in this group.

Group 3: comprises 25 relevés from the eastern Iberian Peninsula and the Balearic Islands that can be ascribed clearly to the association *Sphenopo divaricati-Arthrocnemetum glauci*. The faithful species are *Limonium bellidifolium* and *S. lagascae*. This last species has the highest presence percentage and clearly defines the community dominated by the succulent halophyte *A. macrostachyum*. Other species with significant presence are *Cynomorium coccineum*, *Halimione portulacoides*, *Limonium retusum* and *Parapholis filiformis*. The community is characteristic of eastern coastal areas of the Iberian Peninsula and the Balearic archipelago, and is included in the Valencia-Provençe and Balearic biogeographical province. The association occupies low-lying zones with high salt concentrations [8, 12, 32, 39, 47, 48].

Group 4: is formed by relevés from the Spanish provinces of Toledo, Madrid and Zaragoza, and groups in a single block the two associations described for the central Iberian Peninsula and the Ebro Valley. The most important faithful species are *Aizoon hispanicum*, *Frankenia pulverulenta*, *F. thymifolia*, *Hornungia procumbens*, *Limonium echioides*, *Parapholis incurva*, *Rostraria cristata*, *Salsola soda*, *Spergularia diandra*, *Sphenopus divaricatus* and *Suaeda braun-blanquetii*. The result of the numerical analysis supports the observations made in the field on these ecosystems: *A. macrostachyum* does not coexist with *S. carinata* in the salt flats of Toledo and Madrid, and there are no supported references from the province of Zaragoza; although *Puccinellia fasciculata* is not present in the community dominated by *A. macrostachyum*, *Suaeda braun-blanquetii* contributes biomass and singularity to the association when considered as a whole.

When Castroviejo and Cirujano [9] describe the association that they propose for the center of the peninsula (*Puccinellio fasciculatae-Arthrocnemetum macrostachyi*), they explain that it is very close to the association described for the Ebro Valley. They argue, however, that the differential species *S. carinata* (sub *Sarcocornia perennis* ssp. *alpini*) and *Puccinellia fasciculata* allow one syntaxon to be separated from the other. The table complied by the authors includes 16 relevés from the same locality (Aranjuez), in seven of which they indicate the joint presence of *A. macrostachyum* and *S. carinata*. In our field campaigns in these and neighbouring salt marshes we never observed these species growing together. For this reason it was decided to exclude these relevés from the numerical analysis. Taking into account the floristic similarities between these two communities, we propose their assimilation in the alliance *Suaedion braun-blanquetii* under the name *Suaedo braun-blanquetii-Arthrocnemetum macrostachyi*, which takes precedence due to its age [60].

The halophilous community as a whole belongs to the Central Iberian Mediterranean biogeographical province distributed throughout the territories of Lower Aragón and Upper Ebro, and the Castilian subprovinces. It thrives in endorrheic lakes with high salt concentrations in which the soil is waterlogged in winter and spring, and becomes dry and intensely saline in summer [6, 9].

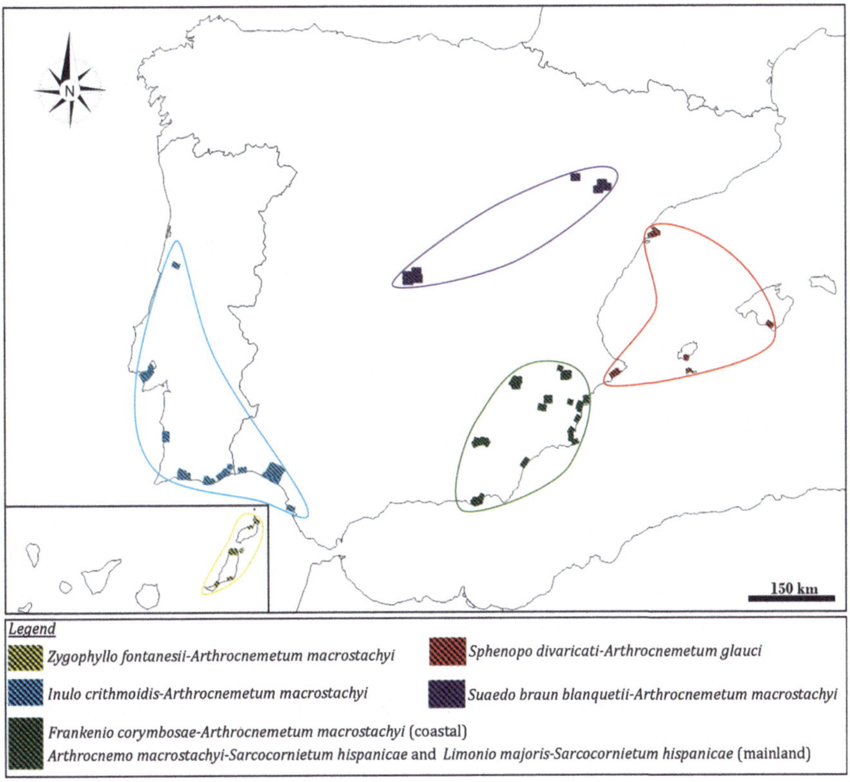

Fig. 1 Geographical location of the relevés grouped by the different plant communities recognized

Group 5: is defined by the majority presence of *Frankenia corymbosa*, *Limbarda crithmoides*, *Limonium cossonianum*, *Pancratium maritimum* and *Phragmites communis*, with relevés from the Spanish provinces of Alicante, Almería and Murcia (Murcia and Almería biogeographical province). This block groups the most clearly defined relevés in the association *Frankenio corymbosae-Arthrocnemetum macrostachyi*. This is a strictly coastal community characteristic of the Murcian-Almerian territory and grows on soils with a compact texture, occasionally flooded by saltwater and with frequent saline efflorescences [1, 2, 38, 44].

Group 6: combines a total of 53 relevés from different points throughout Spanish and Portuguese territories which have been grouped without a completely clear pattern of blocks. The presence percentage of the different taxa in this group is lower than the other groups. Most of the relevés screened from the Canary Islands, however, have been included in this block, in which *Tetraena fontanesii* is the faithful species. The Canary association *Zygophyllo fontanesii-Arthrocnemetum macrostachyi* is differentiated clearly, floristically and biogeographically, from the communities on the Iberian Peninsula and Balearic Islands, and must therefore be maintained in its entirety. Species such as *Frankenia capitata*, *Limonium tuberculatum* and *Traganum moquinii* are only found in these islands within the framework studied.

A common feature of the relevés in the group is a high presence of *Suaeda vera*, a halo-nitrophilous taxon that occurs in wetlands and salt marshes over almost all of the Iberian Peninsula and islands. Some of these relevés may represent alterations in community structure. On Lanzarote and Fuerteventura the abundance of this species is interpreted as the result of a dynamic caused by contamination and/or erosion [18, 54].

The association *Zygophyllo fontanesii-Arthrocnemetum macrostachyi* has a coastal distribution throughout the Canary biogeographical province (East Canary subprovince, Lanzarote and Fuerteventura islands). The type community occupies upper levels of salt marshes and is subjected to long periods of drought when the tides fluctuate only slightly. Saline efflorescences emerge in winter in the dried substrate and lead to the development of communities of *Frankenia capitata* and *Mesembryanthemum nodiflorum*.

The geographical distributions of the associations treated in this study are shown on Fig. 1. On the other hand, Fig. 2 shows details of *A. macrostachyum* and the different communities recognized in this study.

5 Syntaxonomy

SALICORNIETEA FRUTICOSAE Br.-Bl. & Tüxen ex A. & O. Bolòs 1950.

* *Arthrocnemo macrostachyi-Suaedetalia braun-blanquetii* Rufo, Fuente & Sánchez-Mata 2016 [in Phytocoenol 46(4):391]

- *Arthrocnemion glauci* (Rivas-Martínez in Rivas-Martínez & al. 1980) Rivas-Martínez & Costa 1984 [in Doc. Phytosociol 8:18]

1. *Inulo crithmoidis-Arthrocnemetum macrostachyi* Fontes ex Géhu & Géhu-Franck 1977 [in Acta Bot. Malacitana 3:149]
2. *Frankenio corymbosae-Arthrocnemetum macrostachyi* Rivas-Martínez, Alcaraz, Belmonte, Cantó & Sánchez-Mata 1984 [in Doc. Phytosociol 8:336]
3. *Sphenopo divaricati-Arthrocnemetum glauci* Br.-Bl. 1933 [Prodr Group Veg 1:21]
4. *Arthrocnemo macrostachyi-Sarcocornietum hispanicae* Fuente, Rufo, Teijeiro & Sánchez-Mata 2013 [in Lazaroa 34:270]
5. *Limonio majoris-Sarcocornietum hispanicae* Lendínez, Marchal & Salazar 2012 corr. Rufo, Fuente & Sánchez-Mata 2016 [in Phytocoenol 46(4):392]
6. *Zygophyllo fontanesii-Arthrocnemetum macrostachyi* Fernández Galván & Santos 1984 [in Lazaroa 5:146]

Fig. 2 **A.** Detail of *Arthrocnemum macrostachyum* in flowering state. **B.** Succulent stems in the fruiting stage. **C.** Wetlands dominated by *A. macrostachyum* with a high presence of *Halimione portulacoides* (Portugal, Tavira). **D.** Appearance of the association *Inulo crithmoidis-Arthrocnemetum macrostachyi* (Spain, Huelva, La Rábida). **E.** *A. macrostachyum* in an inland salt marsh (*Suaedo braun-blanquetii-Arthrocnemetum macrostachyi*); in the background *Suaeda braun-blanquetii* (Spain, Madrid, Colmenar de Oreja). **F.** Appearance of the association *Arthrocnemo macrostachyi-Sarcocornietum hispanicae*; in the foreground with more modest height is *A. macrostachyum*; behind and with higher branches is *S. hispanica* (Spain, Albacete, Cordovilla)

- *Suaedion braun-blanquetii* Br.- Bl. & O. Bolòs 1958 corr. Rivas-Martínez, Báscones, T.E. Díaz, Fernández-González & Loidi 1991 [in Itinera Geobot 5:407]

1. *Suaedo braun-blanquetii-Arthrocnemetum macrostachyi* (Br.-Bl. & O. Bolòs 1958) Rivas-Martínez, Báscones, T.E. Díaz, Férnandez-Gónzalez & Loidi 1991 [in Itinera Geobot 5: 407, incl. *Puccinellio fasciculatae- Arthrocnemetum macrostachyi* Castroviejo & Cirujano 1980 in Anales Jard Bot Madrid 37 (1):144]

6 Floristic Appendix and New Herbarium Records

- Sarcocornia alpini (Lag.) Rivas Mart.
- *Sarcocornia carinata* (Fuente, Rufo & Sánchez-Mata) Fuente, Rufo & Sánchez-Mata
- *Sarcocornia lagascae* Fuente, Rufo & Sánchez-Mata
- *Sarcocornia pruinosa* Fuente, Rufo & Sánchez-Mata
- A new nomenclatural combination is obliged if *Sarcocornia* is taxonomically treated as synonym of *Salicornia* sensu Piirainen et al. [41]: *Salicornia carinata* (Fuente, Rufo & Sánchez-Mata) comb. nov. Bas.: *Sarcocornia alpini* subsp. *carinata* Fuente, Rufo & Sánchez Mata in Pl. Biosyst 147(1): 170. 2013.
- *Suaeda braun-blanquetii* (Pedrol & Castrov.) Rivas Mart., Cantó & Sánchez Mata
- New herbarium records for *Sarcocornia hispanica* (SPAIN):
  ESP. **Albacete**: Hellín, Cordovilla, Saladares de Cordovilla, entre Cordovilla y Fuente García, 06.02.2019, comunidades fruticosas halófilas (*Arthrocnemo-Sarcocornietum hispanicae*). D. Sánchez-Mata, V. de la Fuente & al. (MAF 177418).
  ESP. **Albacete**: Hellín, Agramón, Saladares de Agramón, 400 m, 06.02.2019, comunidades fruticosas halófilas (*Arthrocnemo-Sarcocornietum hispanicae*). D. Sánchez-Mata, V. de la Fuente & al. (MAF 177419, 177420).
  ESP. **Almería**: Roquetas de Mar. Paraje Natural Punta Entinas-Sabinar, cerca del Charcón del Flamenco, 10.08.2017, comunidades halófilas fruticosas (*Arthrocnemo-Sarcocornietum hispanicae*). D. Sánchez-Mata & I. Sánchez-Gavilán (MAF 177277, 177278).
  ESP. **Almería**: Vera, Saladar de los Canos, entre Garrucha y Villaricos, 24–08.2016, comunidades halófilas fruticosas (*Arthrocnemo-Sarcocornietum hispanicae*). D. Sánchez-Mata & I. Sánchez-Gavilán (MAF 177279).

References

1. Alcaraz FJ (1984) Flora y vegetación del N.E. de Murcia. Secretariado de Publicaciones de la Universidad de Murcia, Murcia, p 406
2. Alcaraz FJ, Díaz TE, Rivas-Martínez S, Sánchez-Gómez P (1989) Datos sobre la vegetación del sureste de España: provincia biogeográfica Murciano-Almeriense. Itinera Geobot 2:5–133
3. Alonso MA (1996) Flora y vegetación del valle de Villena (Alicante). Publicaciones del Instituto Alicantino de Cultura Juan Gil-Albert, Alicante, p 263
4. Álvarez J (1997) Relaciones suelo-vegetación en saladares del Sureste de España. Ph Dissertation. Universidad de Murcia, Murcia, p 456
5. Biondi E, Casavecchia S, Estrelles E, Soriano P (2013) *Halocnemum* M. Bieb. vegetation in the Mediterranean Basin. Pl Biosyst 147:536–547
6. Braun-Blanquet J, de Bolòs O (1958) Les groupements végétaux du bassin moyen de l'Ebre et leur dynamisme. Anales Estac Exp. Aula Dei 5:1–266
7. Braun-Blanquet J (1979) Fitosociología: Bases para el estudio de las comunidades vegetales. Blume Ediciones, Madrid, p 820
8. Cantó P, Laorga S, Belmonte D (1986) Vegetación y catálogo florístico del Peñón de Ifach (Penyal d'Ifac) (Alicante, España). Opusc Bot Pharm Complutensis 3:3–86
9. Castroviejo S, Cirujano S (1980) *Sarcocornietea* en la Mancha (España). Anales Jard Bot Madrid 37(1):143–154
10. Costa JC, Lousã M (1989) Communautes psammophiles et halophiles du "Ria du Alvor". Coll Phytosociol XVIII:120–135
11. Costa JC, Lousã M, Espírito-Santo MD (1996) A vegetaçao do parque natural da Ria Formosa (Algarve, Portugal). Stud Bot Univ Salamanca 15:69–157
12. Curcó A (1996) La vegetación del Delta de l'Ebro (II): les comunitats halòfiles i halo-nitròfiles (Classes Puccinellio-Salicornietea i Cakiletea maritimae). Fol Bot Misc 10:113–139
13. Del Arco Aguilar MJ, Izco J, Rodríguez Delgado O (2017) The name *Salicornietum perennis* (sensu lato) and global nomenclatural issues. Lazaroa 38(2):203–206
14. Díaz-González TE (2004) Pasado, presente y futuro de la Fitosociología española. Lazaroa 25:3–13
15. DOCE (1992) Directiva 92/43/CEE del Consejo, de Conservación de Hábitats Naturales y de la Fauna y Flora Silvestres
16. Euro+Med (2006) Euro+Med PlantBase. The information resource for Euro-Mediterranean plant diversity. https://ww2.bgbm.org/EuroPlusMed/. Accessed Apr 2021
17. European Commission (2013) Interpretation Manual of European Union Habitats - EUR 28. DG Environment, Nature ENV B.3, Strasbourg
18. Fernández M, Santos A (1984) La vegetación del litoral de Canarias. I. Arthrocnemetea. Lazaroa 5:143–155
19. Flowers TJ, Galal HK, Bromham L (2010) Evolution of halophytes: multiple origins of salt tolerance in land plants. Funct Pl Biol 37:604–612
20. Font X (2005) La gestión de la biodiversidad mediante bases de datos en línea y el programario b-vegana. Rec Rurais, Serie Cursos 2:65–72
21. Font X, Rodríguez-Rojo MP, Acedo C, Biurrun I, Fernández-González F, Lence C, Loidi J, Ninot JM (2010) SIVIM: an online database of Iberian and Macaronesian vegetation. Waldök Landschafts Natur 9:15–22
22. Font X, Pérez-García N, Biurrun I, Fernández-González F, Lence C (2012) The Iberian and Macaronesian vegetation information system (SIVIM, www.sivim.info), five years of online vegetation's data publishing. Plant Sociol 49:89–95
23. de la Fuente V, Rufo L, Sánchez-Mata D (2011) Sarcocornia hispanica (Chenopodiaceae), a new species from the Iberian Peninsula. Lazaroa 32:9–13

24. de la Fuente V, Rufo L, Teijeiro A, Sánchez-Mata D (2013) Arthrocnemo macrostachyi-Sarcocornietum hispanicae, a new halophytic plant community from eastern Iberian Península. Lazaroa 34:267–272
25. de la Fuente V, Oggerin M, Rufo L, Ortúñez E, Sánchez-Mata D, Amils R (2013) A micromorphological and phylogenetic study of *Sarcocornia* A.J. Scott (Chenopodiaceae) on the Iberian Peninsula. Plant Biosyst 147:158–173
26. de la Fuente V, Rufo L, Rodríguez N, Sánchez-Mata D, Amils R (2016) A study of *Sarcocornia* A.J. Scott (Chenopodiaceae) from Western Mediterranean Europe. Plant Biosyst 150(2):343–356
27. Géhu J-M, Géhu-Franck J (1977) Quelques données dur les Arthrocnemetea fruticosi ibériques sud-occidentaux. Acta Bot Malacitana 3:145–157
28. Giménez E, Navarro J, Oña JA, Gómez F (2003) *Paraje Natural Punta Entinas-Sabinar (Almería): Flora, vegetación y ornitofauna.* Servicio de Publicaciones, Universidad de Almería, Almería, p 171
29. Kadereit G, Mucina L, Freitag H (2006) Phylogeny of Salicornioideae (Chenopodiaceae): diversification, biogeography and evolutionary trends in leaf and flower morphology. Taxon 55:617–642
30. Lendínez ML, Marchal FM, Salazar C (2012) Una nueva asociación de vegetación halófila en el sureste de la Península Ibérica (España): *Limonio majoris-Sarcocornietum fruticosae.* Lagascalia 32:229–236
31. Lengyel A, Chytrý M, Tichý L (2011) Heterogeneity constrained random resampling of phytosociological databases. J Veg Sci 22:175–183
32. Llorens L (1986) La vegetación de los saladares de la isla de Formentera (Baleares). Anales Jard Bot Madrid 42(2):469–479
33. Loidi J (2008) La fitosociología como proveedora de herramientas de gestión. Lazaroa 29:7–17
34. Luis-Dias TM (2009) Flora e vegetação dos salgados de Coina, Corrios e Alcochete. Ms Dissertation. Universidade Técnica de Lisboa, p 79
35. Molina JA, Pertíñez C, Cruz MT (2001) Dos sobre la relación suelo-vegetación en los saladares de Cordovilla (Albacete, España). Sabuco 1:217–232
36. Mucina L., Bültmann H, Dierssen K, Theurillat J-P, Raus T, Čarni A, Šumberová K, Willner W, Dengler J, Gavilán García R, Chytrý M, Hájek M, Di Pietro R, Iakushenko D, Pallas J, Daniëls FJA, Bergmeier E, Santos Guerra A, Ermakov N, Valachovič M, Schaminée JHJ, Lysenko T, Didukh YP, Pignatii S, Rodwell JS, Capelo J, Weber HE, Solomeshch A, Dimopoulos P, Aguiar C, Hennekens SM, Tichý L (2016) Vegetation of Europe: hierarchical floristic classification system of vascular plant, bryophyte, lichen and algal communities. App Veg Sci (supp 1): 3–264
37. Neto C, Arsénio P, Costa JC (2009) Flora e Vegetaçã do sudoeste de Portugal continental. Quercetea 9:43–144
38. Peinado M, Alcaraz F, Martínez-Parras JM (1992) Vegetation of Southeastern Spain. Flora et Vegetatio Mundi, Stuttgart, p 487
39. Pérez Badia MR (1997) Flora vascular y vegetación de la comarca de la Marina Alta (Alicante). Colección Técnica 16, Instituto de Cultura Juan Gil Albert, Alicante, p 566
40. Pigott CD, Ratcliffe DA, Malloch AJC, Birks HJB, Proctor MCF, Shimwell DW, Huntley JP, Radford E, Wigginton MJ, Wilkins P (2000) Annual Salicornia salt-marsh community: Salicornietum europaeae Warming 1906. In: JS Rodwell (ed) british plant communities, vol 5, pp 45–46. Cambridge, p 526
41. Piirainen M, Liebisch O, Kadereit G (2017) Phylogeny, biogeography, systematics and taxonomy of Salicornioideae (Amaranthaceae/Chenopodiaceae) – a cosmopolitan, highly specialized hygrohalophyte lineage dating back to the Oligocene. Taxon 66(1):109–132
42. Ramírez E, Rufo L, Sánchez-Mata D, de la Fuente V (2019) Arthrocaulon meridionalis, a new species of mediterranean flora. Med Bot 40(1):33–41

43. Redondo-Gómez S, Mateos-Naranjo E, Figueroa ME, Davy AJ (2010) Salt stimulation of growth and photosynthesis in an extreme halophyte, Arthrocnemum macrostachyum. Plant Biol 12:79–87
44. Rivas-Martínez S (1984) Vegetatio hispaniae Notula VI. Stud Bot Univ Salamanca 3:7–16
45. Rivas-Martínez S, Alcaraz F, Belmonte D, Cantó P, Sánchez-Mata D (1984) Contribución al conocimiento de la vegetación de los saladares del SE de la Península Ibérica (Arthrocnemion glauci). Doc Phytosociol 8:335–342
46. Rivas-Martínez S, Lousã M, Díaz TE, Fernández-González F, Costa JC (1990) La vegetación del sur de Portugal (Sado, Alentejo y Algarve). Itinera Geobot 3:5–126
47. Rivas-Martínez S, Costa M, Loidi J (1992) La vegetación de las islas de Ibiza y Formentera (Islas Baleares, España). Itinera Geobot 6:99–236
48. Rivas-Martínez S, Costa M, Soriano P, Pérez Badia MR, Llorens L, Roselló JA (1992) Datos sobre el paisaje vegetal de Mallorca e Ibiza (Islas Baleares, España). Itinera Geobot 6:5–98
49. Rivas-Martínez S, Fernández-González F, Loidi J, Fernandes M, Penas A (2001) Sintaxonomical checklist of vascular plant communities of Spain and Portugal to association level. Itinera Geobot 14:5–41
50. Rivas-Martínez S, Díaz-González TE, Fernández-González F, Izco LJ, Fernandes M, Penas A (2002) Vascular plant communities of Spain and Portugal: addenda to the syntaxonomical checklist of 2001. Itinera Geobot 15(1):5–432
51. Rivas-Martínez S et al (2007) Mapa de series, geoseries y geopermaseries de vegetación de España. Parte I. Itinera Geobot 17:5–436
52. Rivas-Martínez et al (2011) Mapa de series, geoseries y geopermaseries de vegetación de España: [Memoria del mapa de vegetación potencial de España]. Itinera Geobot 18(2):425–800
53. Rivas-Martínez S, Penas A, Díaz-González TE, Cantó P, Del Río S, Costa JC, Herrero L, Molero J (2017) Biogeographic units of the Iberian Peninsula and Balearic Islands to district level: a concise sinopsis. In: Loidi J (ed). The vegetation of Iberian Peninsula, plant and vegetation, vol 12, pp 131–188. Springer, Cham
54. Rodríguez O, García A, Reyes AJ (2000) Estudio fitosociológico de la vegetación actual de Fuerteventura (Islas Canarias). Vieraea 28:61–98
55. Rufo L, de la Fuente V, Sánchez-Mata D (2016) Sarcocornia plant communities of the Iberian Peninsula and the Balearic Islands. Phytocoenol 46(4):388–396
56. Salazar C, Torres JA, Marchal MF, Cano E (2002) La vegetación edafohigrófila del distrito Guadiciano-Bastetano (Granada-Jaén, España). Lazaroa 23:45–64
57. Tamajón R, Muñoz J (2001) La vegetación de las marismas y lagunas de la hoja cartográfica de Lebrija (Suroeste de España). Studia Bot Univ Salamanca 20:93–114
58. Valdés A, González JL, Molina R (1993) Flora y vegetación de los saladares de Cordovilla y Agramon (SE de Albacete). Publicaciones del Instituto de Estudios Albacetenses, Albacete, p 158
59. Tichý L (2002) JUICE, software for vegetation classification. J Veg Sci 13:451–453
60. Weber HE, Moravec J, Theurillat J-P (2000) International code of phytosociological nomenclature. J Veg Sci 11:739–768
61. Wiser SK, De Cáceres M (2013) Updating vegetation classifications: an example with New Zealand's woody vegetation. J Veg Sci 24:80–93

Geosynphytosociological Typology of French Atlantic Coastal Rocky-Cliff Vegetation

Charlotte Demartini and Frédéric Bioret

Abstract Coastal cliff vegetation is spatially organized into parallel belts according to ecological gradients caused by salt spray, wind exposure, relief energy, substrate nature and depth. Along the French Atlantic coasts, a recent research programme set up in 2011 is dedicated to a geosynphytosociological study of coastal vegetation. The present work focuses on permanent vegetation of the French Atlantic coastal rocky cliffs, corresponding to permaseries. Permaseries occuring within homogeneous geomorphological units are listed within geosynrelevés. The analysis of a set of geosynrelevés from northern France up to the Pyrenees-Atlantic allows us to characterise phytocoenotical diversity and to propose a preliminary typology of coastal cliff vegetation. The geopermaseries assemblies follow Dupias's and Rey's phytoecological regions. These biogeographical units are more precisely characterized by substrate and climate. This analysis completes Géhu's works about sigma-association description of French Atlantic cliffs. Eleven geopermaseries of rocky coast are defined: Sandstone or marly-cliff geopermaseries of Boulonnais; Marly and chalky cliff geopermaseries of Boulonnais; Chalky-cliff geopermaseries of Pays de Caux; Crystalline-cliff geopermaseries of the Gulf Normand-Breton; Crystalline-cliff geopermaseries of the western Massif Armoricain, under a hyper Atlantic climate; Crystalline-cliff geopermaseries of the southern Massif Armoricain, under a thermo-Atlantic climate; Crystalline-cliff geopermaseries of islands of the southern Massif Armoricain, under a thermo-Atlantic climate; Cliff-top fossil dune geopermaseries on sandy decalcified substrates; Calcareous cliff geopermaseries of the right-bank of the Gironde estuary, under a thermo-Atlantic climate; Chalky and sandstone cliff geopermaseries of the south-Aquitan Basin; and marly-chalky or sandstone cliff geopermaseries of the Corniche Basque.

Keywords Rocky cliff vegetation · French Atlantic coast · Geopermaseries · Geosynphytosociology

C. Demartini · F. Bioret (✉)
EA 7462 Géoarchitecture, University of Western Brittany, Brest, France
e-mail: frederic.bioret@univ-brest.fr

F. Pedrotti and E. O. Box (eds.), *Tools for Landscape-Scale Geobotany and Conservation*, Geobotany Studies, https://doi.org/10.1007/978-3-030-74950-7_13

1 Introduction

Financed by the French Ecology, Energy and Sustainable Development Ministry, the national habitats and vegetation mapping program (CarHAB) was launched in 2011. Based on the dynamico-catenal phytosociological approach, its purpose is to produce a map for French vegetation series and geoseries up to 2025, to answer the needs of the European Flora-Fauna Habitat Directive of 1992 [42]. Analysis of vegetation along the Channel-Atlantic coast requires a specific approach, due mainly to vegetation zonation induced by specific ecological constraints. This work started in 2012 and aims to create a typology and mapping methodology focused on geopermaseries and geominoriseries of Channel-Atlantic coastal vegetation [14]. This paper focuses exclusively on permanent vegetation developing along rocky coasts, on the supra-coastal zone. Several geopermaseries are identified by analysing géosynrelevés, each of them distinguished by regional climatic and geomorphological conditions.

2 Study Area

Our study focuses on the phytocoenotic diversity of the cliffs of the French Channel-Atlantic coastline, from the Belgian to the Spanish borders. Most of the sites inventoried are located between Blanc-Nez cape and the Corniche Basque (Fig. 1).

3 Concepts and Method

3.1 Concepts: Series and Geoseries

Analysis of coastal vegetation landscapes is based on the synphytosociological and geosynphytosociological methods. Synphytosociology, also known as serial phytosociology, has vegetation series as its fundamental unit. It studies spatially sets of plant communities that are linked dynamically, including substitute communities up to potential vegetation [24, 37, 46], on ecologically homogeneous areas called tesselas.

Geosynphytosociology (or catenal phytosociology) has geoseries as its fundamental unit. It studies homogeneous geomorphological units (e.g. crest, coastal dune, rocky coast…) composed of vegetation series belonging to sets of adjacent tesselas called a catena (Fig. 2).

In the coastal zone, ecological constraints (salt and wind exposure, dryness, oligotrophic and superficial substrates…) block vegetation succession so that plant

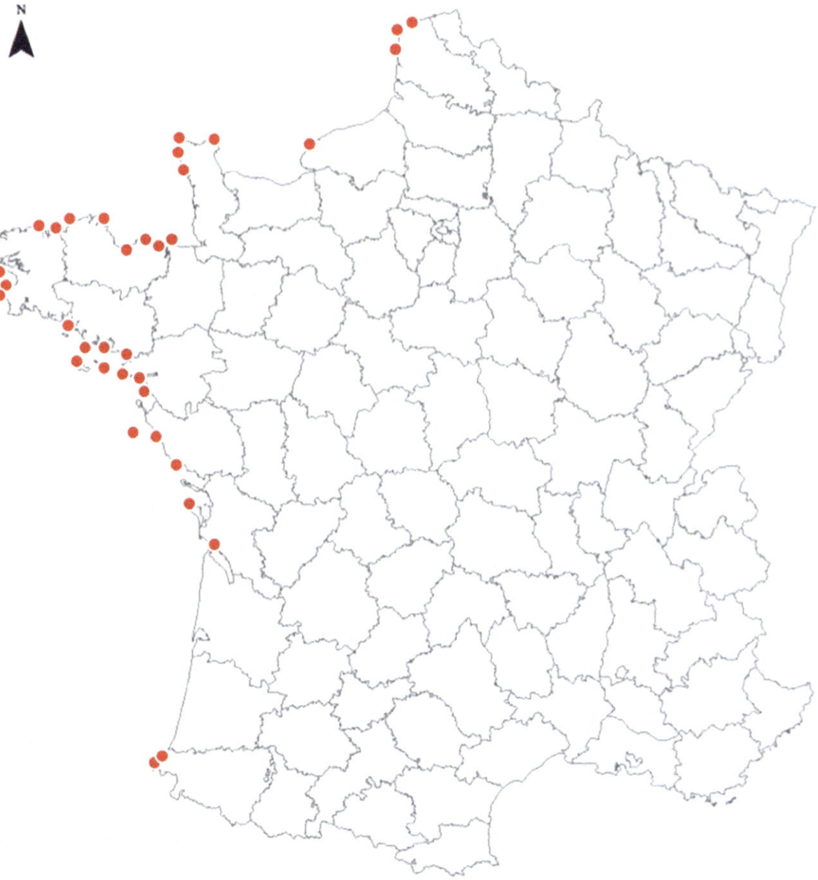

Fig. 1 Location of the study sites

communities form permanent vegetation types, called permaseries, that do not correspond to the climatic potential.

For Rivas-Martínez: "permaseries or permasigmeta are perennial, stable communities that populate microtesselas or microtessela complexes that are very similar to each other, in particular areas such as: polar regions, crests of high mountains, coastal zones, mobile dunes, cliffs, coastal reefs washed by seawater, etc. The steady mature stage, or climax, corresponds to a perennial vascular community that is generally poorly stratified (…)" [47, 48].

The new concept of mesoseries [44] has been replaced by the term 'curtaseries' or truncated series [43]. It concerns cases intermediate between permaseries and series, where the vegetation is under less extreme ecological constraints, which permit the first steps of succession but are strong enough to prevent reaching the stage of natural potential vegetation, which is forest.

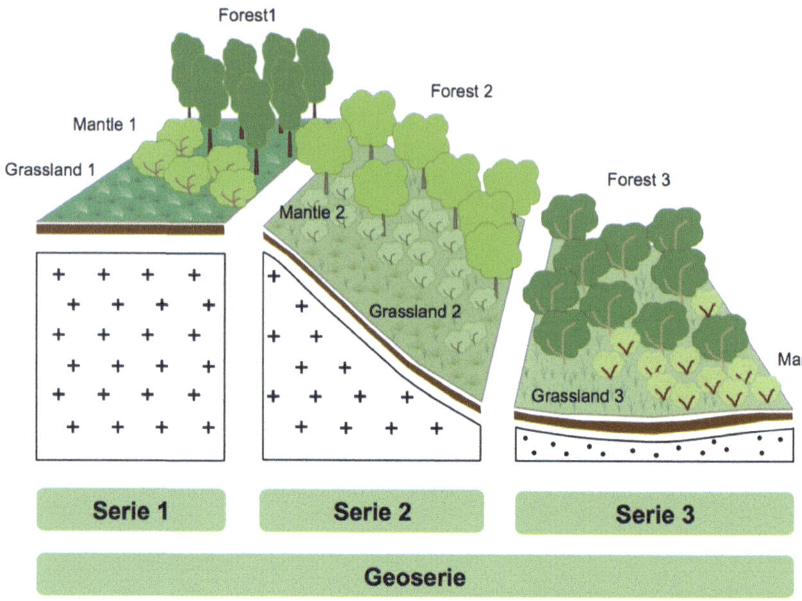

Fig. 2 Theoretical representation of a vegetation geoseries inside a catena composed of three homogeneous tesselar units (series 1, 2 and 3) (after [4])

For grammatical reasons, Rivas-Martínez [49] proposes to name this concept "minoriseries" instead of "curtaseries". For this author, these terms are synonymous.

Coastal tesselas grouping minoriseries are usually inserted, continuously or in a dotted line [45], between the microtesselas of permaseries and the tesselas of adjacent series (Fig. 3).

3.2 *Method: Characterization of Coastal Geopermaseries*

Associations of rocky coasts are identified in various phytosociological works [3, 5–8, 16, 17, 19, 25–30, 32–36, 38–41]. The synsystem was developed from different works: Géhu [27], Bardat et al. [1], Bioret and Géhu [6]; De Foucault and Bioret [12] and De Foucault and Catteau [13].

In the field, the method consists on making systematic geopermasynrelevés. The aim is to establish the list of permaseries (permanent plant communities) present in a homogeneous geomorphological and biogeographical area. (In our study this corresponds to a structurally homogeneous rocky coast) [24]. Coefficients of spatial shape and abundance-dominance are added to each permaseries. The abundance-dominance scale follows Braun-Blanquet [9–11]. The spatial shapes of syntaxa in the landscape follow Béguin et al. [2], Géhu [18], Géhu &

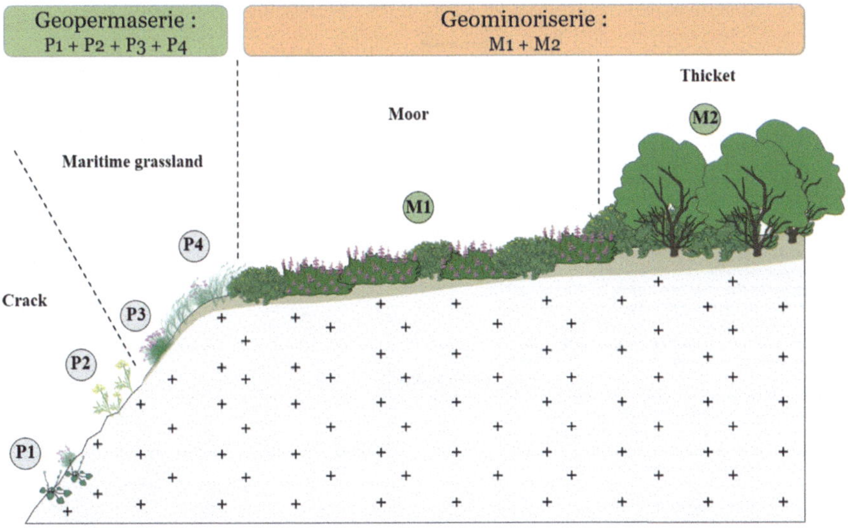

P : permaserie; **M** : minoriserie

Geopermaserie : *Armerio maritimae-Cochleario officinalis geopermasigmetum* (P1 : *Armerio maritimae-Cochlearietum officinalis* Géhu & Géhu-Franck 1984 ; P2 : *Spergulario rupicolae-Crithmetum maritimi* (Roux & Lahondère 1960) Géhu 1962 *nom. inv.* ; P3 : *Spergulario rupicolae-Armerietum maritimae* Géhu 2008 ; P4 : *Armerio maritimae-Festucetum pruinosae* Géhu 2008)

Geominoriserie : *Ulici maritimi-Pruno spinosae geominorisigmetum* (M1 : *Ulici maritimi-Erico cinereae minorisigmetum* ; M2 : *Ulici maritimi-Pruno spinosae minorisigmetum*)

Fig. 3 Theoretical zonation for permaseries and minoriseries of Finistere rocky coasts

Rivas-Martínez [37], and Tüxen [50]: O spatial; o limited spatial;/linear; 0 spatial linear, wide fringe; punctual; … mosaic.

Geopermasynrelevés have been made throughout a large part of the French coastal territory, from the cliffs of the Gris-Nez cape in the Pas-de-Calais to the Corniche Basque in the Atlantic Pyrenées. As an example, geopermasynrelevés made on a part of the Finistère rocky coast are presented in Table 1.

4 Results

Analysis of geopermasynrelevés allows one to propose a zonation of French rocky-coast vegetation. The geopermaseries assemblies follow the phyto-ecological regions of Dupias and Rey [15]: chalky or marly cliffs of the Blanc-Nez and Gris-Nez capes (Nord Pas-de-Calais) and of Etretat (Haute-Normandie) belong to the sedimentary Parisian area,western Channel crystalline rocks of the southern part of Vendée are linked to the Armorican massif; limestone or sand topped cliffs (Jard-sur-Mer, Saint-Palais-sur-Mer), chalky and sandstone cliffs of Biarritz and marly-chalky or sandstone cliffs of the Corniche Basque are typical of the Aquitaine area.

Table 1 Geopermasynrelevés realized on a part of the Finistère rocky coast (Brittany)

Géosynrelevé number	1	2	3	4	5	6	7	8	9	10	11	12
Department	29	29	29	29	29	29	29	29	29	29	29	29
Number of syntaxa	13	13	9	16	16	12	17	15	14	12	16	14
Chasmo-halophytic and halo-chomophytic vegetation												
Spergulario rupicolae-Crithmetum maritimi (Roux & lahondère 1960) Géhu 1962 *nom. inv.*	/+	/+	/1	.	/+	.	/+	/+	/+	/+	/+	/+
Armerio maritimae-Asplenietum marini Géhu & Géhu-Franck 1984	.r	/+	.	/+	/+	/+	/+	/+	/+	/+	/+	/r
Armerio maritimae-Inuletum crithmoidis Géhu 2006	.	.	/r	/r	/+	.	/+	/O+	.r	.	.	/+
Armerio maritimae-Cochlearietum officinalis Géhu & Géhu-Franck 1984	/+	.	/+	/r	/+	.+	/+	.r	/+	/+	.	.
Umbilico rupestris-Asplenietum obovati Géhu & Biondi 1994	.	.	.	.	.	/+	.	.	.r	.+	.	.
Crithmo maritimi-Juncetum maritimi Géhu & Duquef 2009	.+	.	.	/r	.	.	.	.	.	.	.	.
Spergulario rupicolae-Frankenietum laevis Bioret & Lahondère 2010	.	.	.	.	.	.	.	.r	.	.	.r	.r
Spergulario rupicolae-Halimionetum portulacoidis Bioret & Géhu 2008	.	o1	.	Or	.	.	.	.	.	.	.	.
Spergulario rupicolae-Limonietum dodartii Géhu, Franck & Scoppola 1984	.	.	.	.	.	.	.	.	.	.	.	o+
Spergulario rupicolae-Limonietum occidentalis Bioret & Géhu 2008	o+	.r	.	.	.	.	.	.r	.	.	.+	.+
Umbilico rupestris-Asplenietum billoti de Foucault 1979	.	.	.	.	.	.	.+	.	.	.r	.	.
Soncho arvensis-Rumicetum rupestris Bioret & Géhu 2002	.	.	.	.r	.	.	.r	.	.	.	.r	.
Gpt à *Carex extensa* et *Armeria maritima*	/o+	.	.	.	.	.	.r	.	.r	.	.	.
Maritime grassland												
Spergulario rupicolae-Armerietum maritimae Géhu 2008	.	/+	.	/+	/+	/+	.	/+	/+	/+	/+	/+
Spergulario rupicolae-Silenetum maritimae Bioret & Géhu 2000	.	/+	.	/+	/+	/+	/+	.	/+	/+	/+	.r
Armerio maritimae-Agrostietum maritimae Bioret & Géhu 2008	.	.	.	.	.	.	.r	.r	o+	.	.	.
Armerio maritimae-Festucetum pruinosae Géhu 2008	O3	O2	O4	O3	O3	O4	O4	O3	O3	O4	O3	O3
Festuco pruinosae-Elymetum atherici Géhu 2008	01	o1	o1	.	.	.	.	o+	.	.	.	.
Armerio maritimae-Festucetum pruinosae à *Anthyllis vulneraria*	.	.	.	.	O2	O2	O2	O1	O2	O1	O3	.
Dactylo oceanicae-Daucetum gummiferi Géhu 2008	.	.	.	.	or	o+	or	.	.	.	o+	.
Festuco pruinosi-Osmundetum regalis Géhu & Bioret 2000	.	.	.	.	.	.	o+	.	o+	o+	.	O+
Flayed grass												
Armerio maritimae-Plantaginetum coronopodis (Vanden Berghen 1979) Géhu & Bioret 2008	o+	.+	o1	o+	o+	.	o+	o+	.r	.+	o+	o+
Bromo ferronii-Trifolietum arvensis Géhu 2008	.	.	.	.+	.r	.r	.	.	.	.	o+	.
Dactylo oceanicae-Sedetum anglici J.-M. & J. Géhu & Caron 1978 *corr.* Géhu 2008	/+	.+	.+	o+	o+	.+	.	.	.	.	.	.
Desmazerio marinae-Brometum ferronii Bioret 1994	.r	.	.	.	.	.	.r	.+	.	.	.	.r
Sagino maritimae-Catapodietum marini R. Tx. 1963	.	.+	.	.+	.	.	.r	.+	.	.	.+	.+
Sedo anglici-Festucetum huonii Bioret *et al.* 2014	.	.	.	.+	o+	o+	.	.	.	.	.r	.
Sedo anglici-Scilletum vernae Bioret 1994	.	.	.+	.	.r	.	.	.	.	.	.	.
Pre-mantle												
Hedero helicis-Silenetum maritimae Géhu 2008	.	O3	.	.	.	.	.	.	.	.	.	.
Orobancho hederae-Hederetum helicis Géhu (1961) 2006	.	.	.	or	o+	.	.	.	.	.	.	.
Anthropogenic vegetation												
Gpt à *Carpobrotus edulis*	O1	.	.	Or	.	.	.	.	.	.	o+	.
Other												
Rochers et lichens	O3	O3	01	.	02	01	O2	03	02	02	02	02

Location : 1,2 : Plougonvelin ; 3 : Locmaria-Plouzané ; 4, 11 : Plogoff ; 5,6 : Cléden-Cap-Sizun ; 7 : Beuzec-Cap-Sizun ; 8 : Plonévez-Porzay ; 9 : Goulien ; 10 : Poullan-sur-Mer ; 12 : Plomodiern

The distribution of coastal phytocoenoses, and by extension of coastal geopermaseries, depends on bioclimatic, geomorphologic and pedological factors. In 1984 Géhu and colleagues presented a map of French bioclimatic stages, proposing creation of a thermo-Atlantic plain level and the subdivision of the coastal hill level into two types: hyperAtlantic and Atlantic [31]. These great units have been divided more precisely based on bioclimatic divisions, recent meteorological data (23 coastal stations studied from 1981 until 2010), and the chorology of plant associations. Eleven geopermaseries have been distinguished on the French Channel-Atlantic coast (Table 2). This division completes the Géhu publications, in which he described four sigmassociations on cliffs of Nord Pas-de-Calais, Normandy and Brittany [20–22].

Table 2 Summary table of géosynrelevés from French Atlantic cliffs

	Cliffs geopermaseries of sedimentary Parisian area			Cliffs geopermaseries of Massif armoricain				Cliffs geopermaseries of Aquitain area			
	1	2	3	4	5	6	7	8	9	10	11
Department	62	62	76	22 - 35 - 50	22 - 29	56 - 44 - 85	56	85	17	64	64
Location	Cap Gris-Nez / Cap d'Alprech	Cap Blanc-Nez	Pays de Caux (Etretat)	Barfleur to Perros-Guirec	Perros-Guirec to Plomeur	Ploemeur to St-Gilles-Croix-de-Vie	Groix - Belle-Ile - Yeu	Jard-sur-Mer	St-Palais & Vaux-sur-Mer / Suzac	Biarritz	Corniche basque
Number of géosynrelevés	2	2	5	39	120	51	17	4	3	2	7
Number of syntaxa	12	4	5	37	48	44	37	14	12	4	3
Chasmo-halophytic vegetation											
Spergulario rupicolae-Crithmetum maritimi (Roux & lahondère 1960) Géhu 1962 *nom. inv.*	2	.	.	V	V	V	V	3	1	.	.
Armerio maritimae-Asplenietum marini Géhu & Géhu-Franck 1984	.	.	.	II	III	I	I	.	.	.	.
Crithmo maritimi-Betetum maritimae Géhu 2008	.	.	.	II	I	I	+	.	2	.	.
Armerio maritimae-Inuletum crithmoidis Géhu 2006	.	.	.	I	I	+	I	.	.	.	.
Crithmo maritimi-Juncetum acuti Bioret 2008	.	.	.	I	r	.	+	.	.	.	.
Crithmo maritimi-Juncetum maritimi Géhu & Duquef 2009	.	.	.	+	r	I	I	.	.	.	.
Gpt à *Carex extensa* et *Armeria maritima*	.	.	.	+	+	I	.	.	.	.	.
Umbilico rupestris-Asplenietum billoti de Foucault 1979	.	.	.	II	I	r	.	.	.	.	.
Crithmo maritimi-Limonietum normannici Lahondère & Bioret 1995	.	.	.	r	.	.	.	.	.	.	.
Umbilico rupestris-Asplenietum obovati Géhu & Biondi 1994	.	.	.	.	+	.	.	.	.	.	.
Armerio maritimae-Cochlearietum officinalis Géhu & Géhu-Franck 1984	.	.	.	.	IV	.	.	.	.	.	.
Soncho arvensis-Rumicetum rupestris Bioret & Géhu 2002	.	.	.	.	r	.	.	.	.	.	.
Apio graveolens-Rumicetum rupestris Bioret & Géhu 2002	.	.	.	.	.	I	.	.	.	.	.
Crithmo maritimi-Limonietum ovalifolii (Kuhnholtz-Lordat 1926) Lahondère, Bioret & Botineau 1991	.	.	.	.	.	I	+	.	2	.	.
Crithmo maritimi-Limonietum occidentalis Pavillard 1928	.	.	.	.	.	.	.	.	.	2	.
Crithmo maritimi-Plantaginetum maritimae (Pavillard 1928) Guinea 1949	.	.	.	.	.	.	.	.	.	.	V
Gpt à *Asplenium marinum*	.	.	I	.	.	.	.	.	.	.	.
Gpt à *Asplenium marinum* et *Sonchus maritimus*	.	.	.	.	.	.	.	.	1	.	.
Vegetation of *Riccardio pinguis-Eucladion verticillatii* Bardat *in* Bardat *et al.* 2004 *prov.*	1	2	.	.	.	.	.	.	.	.	.
Halo-chomophytic vegetation											
Spergulario rupicolae-Halimionetum portulacoidis Bioret & Géhu 2008	.	.	.	+	I	IV	IV	1	.	.	.
Spergulario rupicolae-Limonietum occidentalis Bioret & Géhu 2008	.	.	.	I	I	II	III	.	.	.	.
Armerio maritimae-Plantaginetum maritimae Géhu 2000	.	.	.	+	+	I	.	.	.	.	.
Spergulario rupicolae-Frankenietum laevis Bioret & Lahondère 2010	.	.	.	.	I	III	IV	1	.	.	.
Spergulario rupicolae-Limonietum dodartii Géhu, Franck & Scoppola 1984	.	.	.	.	r	II	I	1	.	.	.
Gpt à *Phragmites australis* (on the plane or at the foot of the cliff)	.	.	.	r	+	+	I	1	.	.	.
Tuberario guttatae-Romuleetum columnae Provost 1978	.	.	.	+	.	r	+	.	.	.	.
Jasiono maritimae-Anthyllidetum maritimae Géhu & Duquef 2009	.	.	.	+	r	.	.	.	.	.	.
Anthemido nobilis-Agrostietum capillaris (Allorge 1922) de Foucault *in* Royer *et al.* 2006	.	.	.	.	+	.	.	.	.	.	.
Ophioglosso lusitanici-Isoetetum histricis de Foucault 1988	.	.	.	.	r	.	.	.	.	.	.
Gpt à *Juncus maritimus* et *Inula crithmoides*	.	.	.	.	.	r	I	.	.	.	.
Crithmo maritimi-Senecionetum cinerariae Géhu 2005	.	.	.	.	.	+	.	.	.	.	.
Junco maritimi-Caricetum extensae (Corillion 1953) Géhu 1976	.	.	.	.	.	r	.	.	.	.	.
Spergulario rupicolae-Puccinellietum maritimae Bioret & Gallet 2014	.	.	.	.	.	+	I	.	.	.	.
Brassicetum oleraceae Géhu 1962	.	2	III	.	.	.	.	.	.	.	.
Gpt à *Brassica oleracea* et *Diplotaxis tenuifolia*	.	2	.	.	.	.	.	.	.	.	.
Samolo valerandi-Caricetum vikingensis Géhu 1982	2	.	.	.	.	.	I	1	.	.	.
Agrostio stoloniferae-Caricetum vikingensis Géhu 1979	.	.	.	.	.	.	+	.	.	.	.
Soncho maritimi-Schoenetum nigricantis (Lahondère 1979) de Foucault 1984	.	.	.	.	.	.	+	.	1	.	.
Sagino maritimae-Spergularietum marinae Bioret 2006	.	.	.	.	.	.	.	.	1	.	.
Azonal vegetation along freshwater seeps											
Scirpetum compacti (Van Langendonck 1931) Beeft. 1957	.	.	.	.	r	+	.	.	.	.	.
Pulicario dysentericae-Juncetum inflexi de Foucault *in* Royer *et al.* 2006	1	.	.	.	.	.	.	.	.	.	.
Gpt à *Epilobium hirsutum* et *Pulicaria dysenterica*	2	.	.	.	.	.	.	.	.	.	.
Aerohaline grass											
Armerio maritimae-Festucetum pruinosae Géhu 2008	.	.	.	V	V	V	V	.	.	.	.
Carici arenarii-Festucetum pruinosae Bioret & Géhu 2008	.	.	.	+	I	II	+	.	.	.	.
Spergulario rupicolae-Silenetum maritimae Bioret & Géhu 2000	.	.	.	I	III	+	III	.	.	.	.
Armerio maritimae-Juncetum gerardii Bioret *et al.* 2014	.	.	.	+	r	II	II	.	.	.	.
Spergulario rupicolae-Armerietum maritimae Géhu 2008	.	.	.	III	IV	I	.	.	.	.	.
Festuco pruinosae-Elymetum atherici Géhu 2008	.	.	.	I	II	I	.	.	.	.	.
Gpt à *Silene maritima* et *Festuca pruinosa*	.	.	.	II	I	r	.	.	.	.	.
Festuco pruinosi-Osmundetum regalis Géhu & Bioret 2000	.	.	.	+	+	.	.	.	.	.	.
Dactylo oceanicae-Daucetum gummiferi Géhu 2008	.	.	.	II	+	.	.	.	.	.	.
Sileno zetlandicae - Festucetum pruinosae Bioret & Géhu 2008	.	.	.	r	r	.	.	.	.	.	.
Armerio maritimae-Agrostietum maritimae Bioret & Géhu 2008	.	.	.	.	I	II	+	.	.	.	.
Festuco pruinosae-Ononidetum maritimae Bioret & Géhu 2008	.	.	.	.	+	II	III	.	.	.	.
Dauco gadeceaui-Festucetum pruinosae Bioret & Géhu 2008	.	.	.	.	.	r	.	.	.	.	.
Sileno zetlandicae-Festucetum pruinosae Bioret & Géhu 2008	.	.	.	.	+	.	.	.	.	.	.
Festuco huonii-Plantaginetum littoralis (J.-M & J. Géhu 1976) Bioret, Bouzillé, de Foucault, Géhu & Godeau 1988	.	.	.	.	.	.	III	.	.	.	.
Dauco intermedii-Festucetum pruinosae (Liger 1956) Géhu & de Foucault 1978	.	.	V	.	.	.	.	.	.	.	.
Cirsio acaulis-Festucetum pruinosae Bioret & Géhu 2008	2	.	.	.	.	.	.	.	.	.	.
Apio graveolentis-Tussilaginetum farfarae Géhu *et al.* 1983 *ex* Bioret & Géhu 2008	2	.	.	.	.	.	.	.	.	.	.
Tuberario guttatae-Corynephoretum canescentis Géhu 1964	.	.	.	.	.	+	+	4	.	.	.
Leucanthemo crassifolii-Festucetum pruinosae Géhu & Géhu-Franck 1980 *corr.* F. Pietro & Loidi 1984	.	.	.	.	.	.	.	.	.	.	V
Gpt à *Carex distans* et *Festuca pruinosa*	.	.	.	.	.	.	.	.	.	2	.
Gpt à *Festuca pruinosa* et *Spartina versicolor*	.	.	.	.	.	.	.	.	.	1	.
Crithmo maritimi-Elytrigietum acutae Bioret, Demartini, Géhu & Glemarec 2016	.	.	.	.	.	II	.	3	1	.	.

	Cliffs geopermaseries of sedimentary Parisian area			Cliffs geopermaseries of Massif armoricain				Cliffs geopermaseries of Aquitain area			
	1	2	3	4	5	6	7	8	9	10	11
Department	62	62	76	22 - 35 - 50	22 - 29	56 - 44 - 85	56	85	17	64	64
Location	Cap Gris-Nez / Cap d'Alprech	Cap Blanc-Nez	Pays de Caux (Etretat)	Barfleur to Perros-Guirec	Perros-Guirec to Plomeur	Ploemeur to St-Gilles-Croix-de-Vie	Groix - Belle-île - Yeu	Jard-sur-Mer	St-Palais & Vaux-sur-Mer / Suzac	Biarritz	Corniche basque
Number of géosynrelevés	2	2	5	39	120	51	17	4	3	2	7
Number of syntaxa	12	4	5	37	48	44	37	14	12	4	3
Flayed grass											
Armerio maritimae-Plantaginetum coronopodis (Vanden Berghen 1979) Géhu & Bioret 2008	1	.	.	II	III	IV	V	.	.	.	.
Sagino maritimae-Catapodietum marini R. Tx. 1963	2	.	I	II	II	IV	V	1	.	2	.
Dactylo oceanicae-Sedetum anglici J.-M. & J. Géhu & Caron 1978 *corr.* Géhu 2008	.	.	.	IV	III	II	III	.	.	.	.
Catapodio marini-Trifolietum scabri Géhu & de Foucault 1978 *nom. inv.*	1	.	.	I	r	III	II	1	.	.	.
Desmazerio marinae-Brometum ferronii Bioret 1994	1	.	.	II	I	II	+	1	.	.	.
Soncho oleracei-Cochlearietum danicae Géhu & de Foucault 1978	.	.	.	I	r	II	II	1	.	.	.
Bromo ferronii-Trifolietum arvensis Géhu 2008	.	.	.	I	I	r	.	1	.	.	.
Catapodio marini-Parapholisetum incurvae Géhu & de Foucault 1978	1	.	.	.	r	II	I	1	1	.	.
Sedo anglici-Festucetum huonii Bioret 2014	.	.	.	r	I	.	+	.	.	.	.
Sedo anglici-Festucetum armoricanae Bioret *et al.* 2014	.	.	.	+	.	.	.	.	.	.	.
Sedo anglici-Scilletum vernae Bioret 1994	.	.	.	.	I	.	.	.	.	.	.
Bromo ferronii-Anthoxanthetum aristati Bioret, Bouzillé, de Foucault, Géhu & Godeau 1988	.	.	.	.	.	.	II	.	.	.	.
Spergulario rupicolae-Salicornietum ramosissimae Bioret, Lahondère & Khelifi 2001	.	.	.	.	.	.	+	.	.	.	.
Spergulario rupicolae-Plantaginetum coronopodis Bioret & Géhu 2013	.	.	.	.	.	.	+	.	.	.	.
Dactylo oceanicae-Limonietum dodartii (Lahondère 1979) Géhu, Franck & Scoppola 1984	.	.	.	.	.	.	.	.	2	.	.
Dactylo hispanicae-Helichrysetum stoechadis Géhu, Franck & Scoppola 1984	.	.	.	.	.	.	.	.	1	.	.
Leucanthemo crassifolii-Helichrysetum staechadis (Allorge & Jovet 1941) Géhu & Géhu-Franck 1980	.	.	.	.	.	.	.	.	.	.	IV
Gpt à *Helichrysum stoechas* et *Helianthemum apenninum*	.	.	.	.	.	.	.	.	1	.	.
Pre-mantle											
Orobancho hederae-Hederetum helicis Géhu (1961) 2006	.	.	.	II	II	I	.	.	.	.	.
Hedero helicis-Silenetum maritimae Géhu 2008	.	.	.	I	+	I	.	.	.	.	.
Hedero helicis-Ruscetum aculeati Géhu 2007	.	.	.	r	r	.	.	.	.	.	.
Nitrophilous grass											
Atriplici hastatae-Betetum maritimae (Arènes 1933) Géhu 1968	.	.	II	.	r	r	.	.	.	.	.

1 – Sandstone-marly cliff geopermaseries of Boulonnais (Fig. 4)

This geopermaseries includes vegetation growing on marly cliffs, cut through by Portlandian sandstone (Gris-Nez Cape, Alprech Cape). Meteorological data come from the Calais station. The 10.8 °C mean annual temperature and 722.6 mm average annual rainfall are characteristic of a subhumid Atlantic bioclimate. The substrate is quite soft, with fresh water running slowly down from the top of the cliffs, favorable to hygrophilous permaseries (*Samolo valerandi-Caricetum vikingensis* Géhu et al. 1982, *Apio graveolentis-Tussilaginetum farfarae* Géhu et al. 1983 *ex* Bioret et Géhu 2008…).

2 – Marly and chalky-cliff geopermaseries of Boulonnais

This geopermaseries occurs on limestone cliffs of Cape Blanc-Nez (Pas-de-Calais). The bioclimate is subhumid Atlantic. Only a few permaseries occur on the steep slopes of Cape Blanc-Nez. The Cape Blanc-Nez cliffs are phytosociologically and geomorphologically similar to those of Haute-Normandie, and involve in particular the *Dauco intermedii-Festucetum pruinosae* (Liger 1956) Géhu and de Foucault [32]. Maritime grassland is missing.

3 – Chalky-cliff geopermaseries of Pays de Caux (Fig. 5)

This geopermaseries occurs on the steep cliffs of Etretat. The 11.4 °C mean annual temperature and 785 mm average annual rainfall (Cape Hève meteorological station) are characteristic of a subhumid Atlantic bioclimate. A few permaseries grow

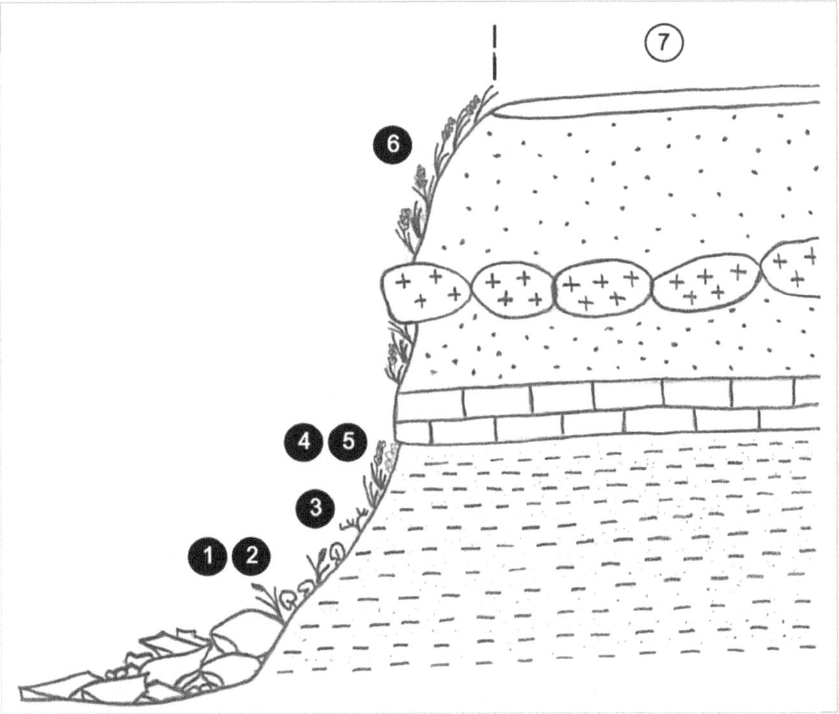

Fig. 4 Theoritical zonation of the Boulonnais coastal cliffs permaseries
(1) *Samolo valerandi-Caricetum vikingensis*; (2) *Apio graveolentis-Tussilaginetum farfarae*; (3) *Spergulario rupicolae-Crithmetum maritimi*; (4) *Pulicario dysentericae-Juncetum inflexi*; (5) *Ricardio pinguis-Eucladion verticillatii*; (6) *Cirsio acaulis-Festucetum pruinosae*; (7) Contact: agricultural crops.

in vertical side crevices (*Brassicetum oleraceae* [16]) or in less steep parts of the cliff, where the substrate is deeper (*Dauco intermedii-Festucetum pruinosae* Liger 1956). Géhu and de Foucault [32].

The following three geopermaseries, growing on the ancient bedrock of the Armorican massif, share the *Armerio maritimae-Festucetum pruinosae* Géhu [28] maritime grassland. They are distinguished by the presence of differential associations, which indicate chorological or ecological variations.

4 – Crystalline-cliff geopermaseries of the Golfe Normand-Breton
This geopermaseries is characterized by a hyper-oceanic Atlantic subhumid bioclimate. At the four meteorological stations the average annual temperature varies between 11.9 °C and 12.2 °C, while the average annual rainfall varies between 615.5 mm (Granville) and 754.2 mm (Bréhat). The geopermaseries are differentiated by the chasmo-halophytic vegetation of the *Crithmo maritimi-Limonietum normannici* Lahondère & Bioret 1995 and the flayed grass of the *Sedo anglici-Festucetum armoricanae* Bioret et al. 2014.

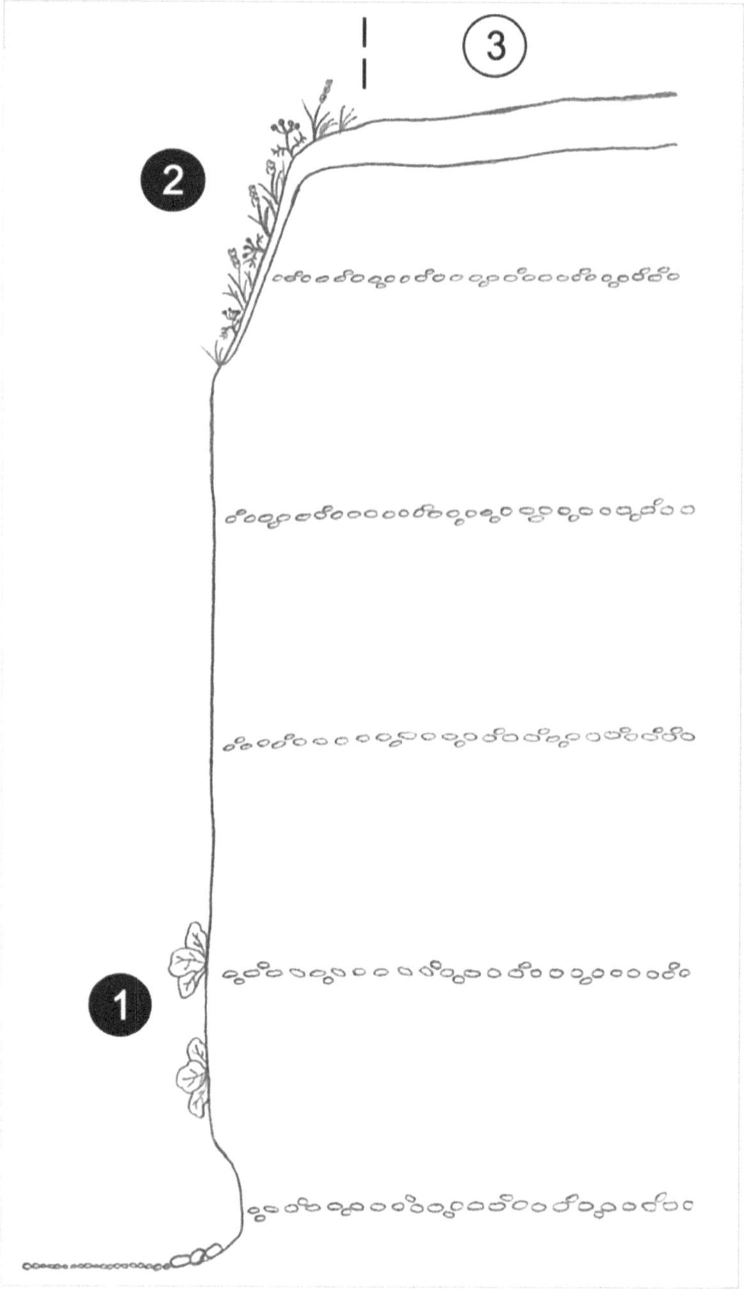

Fig. 5 Theoritical zonation of the Pays-de-Caux coastal cliffs permaseries
(1) *Brassicetum oleraceae*; (2) *Dauco intermedii-Festucetum pruinosae*; (3) Contact: agricultural crops.

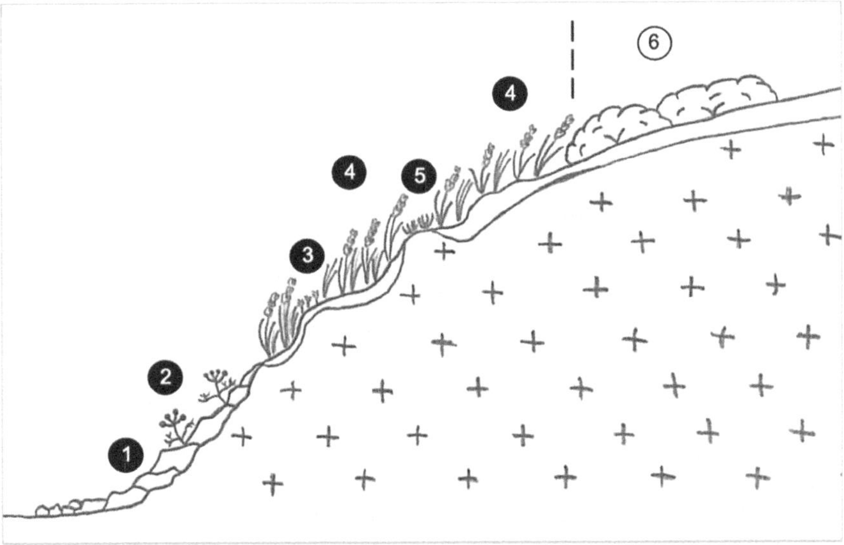

Fig. 6 Theoritical zonation of the Western part of Massif armoricain coastal cliffs permaseries (1) *Armerio maritimae-Cochlearietum officinalis*; (2) *Spergulario rupicolae- Crithmetum maritimi*; (3) *Desmazerio marinae-Brometum ferronii*; (4) *Armerio maritimae-Festucetum pruinosae*; (5) *Dactylo oceanicae-Sedetum anglici*; (6) Contact: *Ulici maritimi-Ericetum cinereae.*

5 – Crystalline-cliff geopermaseries of the western Massif Armoricain, under a hyper-Atlantic climate (Fig. 6)

This geopermaseries, confined to extreme western France, is characterized by hyper-Atlantic associations such as the *Armerio maritimae-Cochlearietum anglicae* Géhu & Géhu-Franck 1984. Except for the stations at Brest and Penmarc'h, mean annual temperatures range from 11.9 to 12.2 °C, and average annual rainfall is between 819 and 1006 mm. Brest is slightly cooler (11.5 °C) and has more annual precipitation (1210 mm). On the contrary, conditions are milder at Penmarc'h: 12.5 °C on average, and 664.4 mm of rainfall. The bioclimate is hyper-Atlantic subhumid inferior or humid inferior.

6 – Crystalline-cliff geopermaseries of the southern Massif Armoricain, under a thermo-Atlantic climate

More thermophilous than the previous, this geopermaseries prefers a mild climate, less subject to storms. The temperatures are cooler at the northern limit of geopermaseries (12 °C in Lorient) and higher at its southern limit (12.9 °C in Château d'Olonne). Rainfall decreases progressively from north to south (950.9 mm in Lorient to 761.7 mm in Château d'Olonne). The bioclimate is thermo-Atlantic subhumid superior to humid inferior. The geopermaseries is defined mainly by the chasmo-halophytic vegetation of the *Apio graveolens-Rumicetum rupestris* Bioret & Géhu 2002 and the *Crithmo maritimi-Limonietum ovalifolii* (Kuhnholtz-Lordat 1926) Lahondère, Bioret & Botineau 1991.

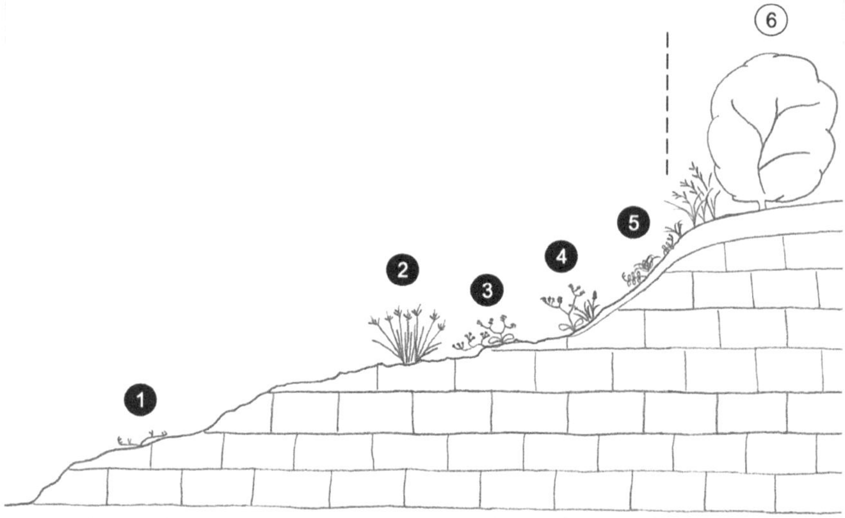

Fig. 7 Theoritical zonation of the right bank of Gironde estuary cliffs permaseries (1) *Sagino maritimae-Spergularietum marinae*; (2) *Soncho maritimi-Schoenetum nigricantis*; (3) *Crithmo maritimi-Limonietum ovalifolii*; (4) *Dactylo oceanicae-Limonietum dodartii*; (5) *Astragalus monspessulanus* and *Helianthemum apenninum community*; (6) Contact: *Phillyreo latifoliae-Quercetum ilicis*.

7 – Crystalline-cliff geopermaseries of islands of the southern Massif Armoricain, under a thermo-Atlantic climate

In general, islands are characterized by original vegetation, often linked to endemic plant species and syntaxa, justifying the creation of new geopermaseries. The South Armorican islands (Groix, Belle-Île, Yeu) are characterized by the *Festuco huonii-Plantaginetum littoralis* (J.-M & J. Géhu 1976) Bioret, Bouzillé, de Foucault, Géhu & Godeau 1988 and by the *Bromo ferronii-Anthoxanthetum aristati* Bioret, Bouzillé, de Foucault, Géhu & Godeau 1988. Rainfall is lower than on the mainland (from 701.4 to 769.8 mm per year). Temperatures are similar to those of geopermaseries of the southern Armorican Massif (11.9 °C to 13 °C). The bioclimate is still thermo-Atlantic subhumid inferior or subhumid superior.

8 – Cliff-top dune geopermaseries on decalcified sand substrates

The distinction of this geopermaseries is more related to its ecology than to its geographical distribution. It develops on low cliffs covered with a veneer of decalcified sands. The characteristic association is the *Tuberario guttatae-Corynephoretum canescentis* Géhu 1964. The bioclimate is thermo-Atlantic.

9 – Calcareous-cliff geopermaseries of the right-bank of the Gironde estuary, under a thermo-Atlantic climate (Fig. 7)

Besides a chalky substrate, there is no other similarity between the geopermaseries of the small limestone cliffs around Royan (Vaux-sur-Mer, Saint-Palais-sur-Mer) and the Normando-Picardian geopermaseries. Located much farther south in

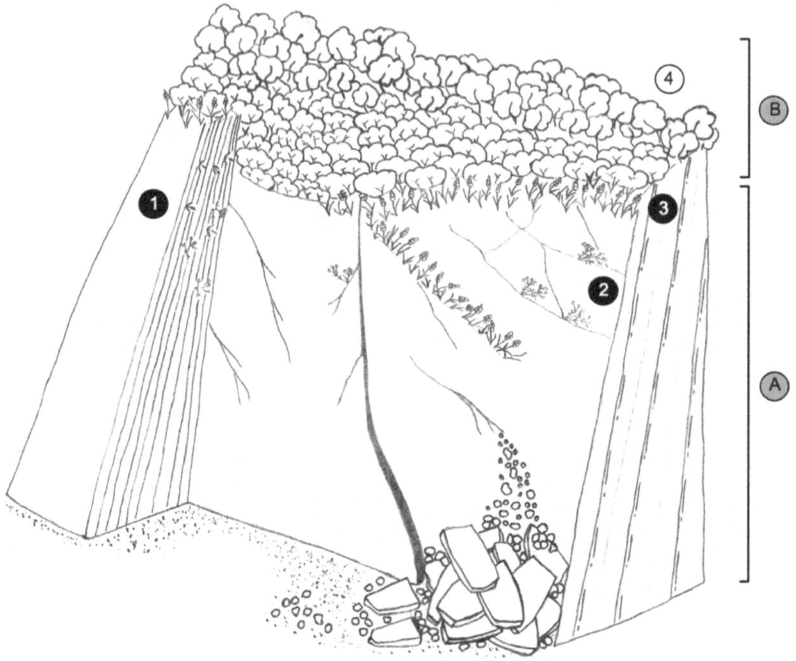

Fig. 8 Theoritical zonation of the Basque Corniche coastal cliffs permaseries
(1) *Crithmo maritimi-Plantaginetum maritimae*; (2) *Leucanthemo crassifolii-Helichrysetum staechadis*; (3) *Leucanthemo crassifolii-Festucetum pruinosae*; (4) Contact: *Leucanthemo crassifolii-Ericetum vagantis, Rubo ulmifolii-Tametum communis.*
(A) *Leucanthemo crassifoliae-Festuco pruinosae geopermasigmetum*; (B) Contact: *Rubo ulmifolii-Tamo communis geominorisigmetum.*

France, under a much milder climate, these vegetation types are more thermophilous. Unfortunately, frequent tourist activity has led to the destruction of most places where maritime grasslands can grow. The geopermaseries is characterized by its chasmo-halophytic vegetation: the *Sagino marinae-Spergularietum marinae* Bioret 2006 and the *Soncho maritimi-Schoenetum nigricantis* (Lahondère 1979) de Foucault 1984. The bioclimate is thermo-Atlantic.

10 - Chalky and sandstone-cliff geopermaseries around Biarritz, under a thermo-Atlantic climate

After a gap of 230 km, where the landscape of Gironde and Landes is dominated by coastal dunes, cliffs reappear south of the Adour (Pyrénées-Atlantiques). The mean annual temperature is higher (14.3 °C in Biarritz), and average annual precipitation is important (1450.9 mm in Biarritz). The rocky coast of Biarritz is composed of calcareous sandstone and Nummulites calcareous sandstone. The geopermaseries is very localized and characterized by *Crithmo maritimi-Limonietum occidentalis* Pavillard 1928, strictly limited to this sector. The maritime grassland that grows on the slopes is quite different from those found in other coastal geopermaseries. It has not yet been characterized phytosociologically.

11 - Marly-chalky or sandstone cliff geopermaseries of the Corniche Basque, under a thermo-Atlantic climate (Fig. 8)
At the extreme south of the Aquitaine Basin the rocky coasts of the Corniche Basque stand out for their geologic and phytosociologic originality. These are high cliffs, almost vertical, carved in the Cretaceous flysch. The chasmo-halophytic vegetation and maritime grassland are synendemic to this coast: *Crithmo maritimi-Plantaginetum maritimae* (Pavillard 1928) Guinea 1949, *Leucanthemo crassifolii-Festucetum pruinosae* Géhu & Géhu-Franck 1980 *corr.* F. Pietro & Loidi 1984, *Leucanthemo crassifolii-Helichrysetum staechadis* (Allorge & Jovet 1941) Géhu & Géhu-Franck 1980. As at Biarritz, temperature and precipitation are much higher than in the north: 14.7 °C and 1483 mm annually at Socoa station.

Synsystem of Rocky Coastal cliffs of the French Channel-Atlantic Coastline
The 75 syntaxa are organized into the following synsystem:

AGROSTIETEA STOLONIFERAE Th. Müll & Gors 1969
- *Potentillo anserinae-Polygonetalia avicularis* Tüxen 1947
  - *Loto tenuis-Festucenalia arundinaceae palustris* Julve *ex* de Foucault, Catteau & Julve *in* de Foucault & Catteau 2012
    - *Loto tenuis-Trifolion* fragiferi (Westhoff, Van Leeuwen & Adriani 1962) de Foucault 2008
      - *Agrostio stoloniferae-Caricetum vikingensis* Géhu 1979
      - *Apio graveolens-Rumicetum rupestris* Bioret & Géhu 2002
      - *Samolo valerandi-Caricetum vikingensis* Géhu 1982
      - *Soncho arvensis-Rumicetum rupestris* Bioret & Géhu 2002
  - *Loto pedunculati-Cardaminenalia pratensis* Julve *ex* de Foucault, Catteau & Julve *in* de Foucault & Catteau 2012
    - *Mentho longifoliae-Juncion inflexi* Th. Müller & Görs *ex* de Foucault 2008
      - *Pulicario dysentericae-Juncetum inflexi* de Foucault *in* Royer *et al.* 2006

ARMERIO MARITIMAE-FESTUCETEA PRUINOSAE Bioret & Géhu 2008
- *Crithmo maritimi-Armerietalia maritimae* Géhu 1964
  - *Crithmo maritimi-Armerion maritimae* Géhu 1968
    - *Armerio maritimae-Asplenienion marini* Géhu 2008
      - *Armerio maritimae-Asplenietum marini* Géhu & Géhu-Franck 1984
    - *Crithmo maritimi-Limonienion binervosi* Géhu & Géhu-Franck 1984
      - *Armerio maritimae-Cochlearietum officinalis* Géhu & Géhu-Franck 1984
      - *Armerio maritimae-Inuletum crithmoidis* Géhu 2006
      - *Armerio maritimae-Plantaginetum coronopodis* (Vanden Berghen 1979) Géhu & Bioret 2008
      - *Armerio maritimae-Plantaginetum maritimae* Géhu 2000
      - *Crithmo maritimi-Betetum maritimae* Géhu 2008
      - *Crithmo maritimi-Juncetum acuti* Bioret 2008
      - *Crithmo maritimi-Juncetum maritimi* Géhu & Duquef 2009
      - *Crithmo maritimi-Limonietum normannici* Lahondère & Bioret 1995
      - *Crithmo maritimi-Limonietum occidentalis* Pavillard 1928
      - *Crithmo maritimi-Limonietum ovalifolii* (Kuhnholtz-Lordat 1926) Lahondère, Bioret & Botineau 1991

Crithmo maritimi-Plantaginetum maritimae Guinea 1949
Crithmo maritimi-Senecionetum cinerariae Géhu 2005
Dactylo oceanicae-Limonietum dodartii (Lahondère 79) Géhu, Franck & Scoppola 1984
Plantagini coronopodis-Spergularietum rupicolae Bioret & Géhu 2011
Spergulario rupicolae-Crithmetum maritimi (Roux & lahondère 1960) Géhu 1962 *nom. inv.*
Spergulario rupicolae-Frankenietum laevis Bioret & Lahondère 2010
Spergulario rupicolae-Halimionetum portulacoidis Bioret & Géhu 2008
Spergulario rupicolae-Limonietum dodartii Géhu, Franck & Scoppola 1984
Spergulario rupicolae-Limonietum occidentalis Bioret & Géhu 2008
Spergulario rupicolae-Puccinellietum maritimae Bioret & Gallet 2014
Sileno maritimae-Festucenion pruinosae (Géhu & Géhu-Franck 1984) Bioret & Géhu 2008
Apio graveolentis-Tussilaginetum farfarae Géhu *et al.* 1983 *ex* Bioret & Géhu 2008
Armerio maritimae-Agrostietum maritimae Bioret & Géhu 2008
Armerio maritimae-Festucetum pruinosae Géhu 2008
Armerio maritimae-Juncetum gerardii Bioret, Demartini, Géhu & Glémarec 2014
Carici arenarii-Festucetum pruinosae Bioret & Géhu 2008
Cirsio acaulis-Festucetum pruinosae Bioret & Géhu 2008
Dactylo oceanicae-Daucetum gummiferi Géhu 2008
Dauco gadeceaui-Festucetum pruinosae Bioret & Géhu 2008
Dauco intermedii-Festucetum pruinosae (Liger 1956) Géhu & de Foucault 1978
Festuco huonii-Plantaginetum littoralis (J.-M & J. Géhu 1976) Bioret, Bouzillé, de Foucault, Géhu & Godeau 1988
Festuco pruinosae-Elymetum atherici Géhu 2008
Festuco pruinosae-Ononidetum maritimae Bioret & Géhu 2008
Festuco pruinosae-Osmundetum regalis Géhu & Bioret 2000
Hedero helicis-Silenetum maritimae Géhu 2008
Jasiono maritimae-Anthyllidetum maritimae Géhu & Duquef 2009
Leucanthemo crassifolii-Festucetum pruinosae Géhu & Géhu-Franck 1980 *corr.* F. Pietro & Loidi 1984
Leucanthemo crassifolii-Helichrysetum staechadis (Allorge & Jovet 1941) Géhu & Géhu-Franck 1980
Orobancho hederae-Hederetum helicis Géhu (1961) 2006
Sileno zetlandicae - Festucetum pruinosae Bioret & Géhu 2008
Spergulario rupicolae-Armerietum maritimae Géhu 2008
Spergulario rupicolae-Silenetum maritimae Bioret & Géhu 2000
Gpt à *Carex cf. repens* et *Festuca pruinosa*
Gpt à *Festuca pruinosa* et *Heracleum sphondylium* subsp. *trifoliolatum*
Gpt à *Festuca pruinosa* et *Spartina versicolor*
Brassicion oleracei Rivas-Martínez, Fernández González & Loidi 1999
Brassicetum oleraceae Géhu 1962

ARRHENATHERETEA ELATIORIS Braun-Blanq.1949 *nom. nud.*

Plantaginetalia majoris Tüxen *ex* Von Rochow 1951
Lolio perennis-Plantaginion majoris Sissingh 1969
Anthemido nobilis-Agrostietum capillaris (Allorge 1922) de Foucault *in* Royer *et al.* 2006

ASPLENIETEA TRICHOMANIS (Braun-Blanq. *in* Meier & Braun-Blanq. 1934) Oberd. 1977
Androsacetalia vandellii Braun-Blanquet *in* Meier & Braun-Blanquet 1934 *nom. corr. in* Bardat *et al.* 2004
Asplenio billotii-Umbilicion rupestris de Foucaut 1988
Umbilico rupestris-Asplenietum billoti de Foucault 1979
Umbilico rupestris-Asplenietum obovati Géhu & Biondi 1994

ASTERETEA TRIPOLII Westhoff & Beeftink in Beeftink 1962
Glauco maritimae-Puccinellietalia maritimae Beeftink & Westhoff *in* Beeftink 1962
Limonio vulgaris-Plantaginenion maritimae Géhu & Géhu-Franck 1984 *nom. nud.*
Glauco maritimae-Juncion maritimi Géhu & Géhu-Franck *ex* Géhu *in* Bardat *et al.* 2004
Junco maritimi-Caricetum extensae (Corillion 1953) Géhu 1976

CAKILETEA MARITIMAE Tüxen & Preising *in* Tüxen 1950
Cakiletalia integrifoliae Tüxen *ex* Oberdorfer 1950 *corr.* Rivas-Martínez, Costa & Loidi 1992
Atriplicion littoralis Nordhagen 1940
Atriplici hastatae-Betetum maritimae (Arènes 1933) Géhu 1968

CRATAEGO MONOGYNAE-PRUNETEA SPINOSAE Tüxen 1962
Prunetalia spinosae Tüxen 1952
Lonicerion periclymeni Géhu, de Foucault & Delelis 1983
Hedero helicis-Ruscetum aculeati Géhu 2007

HELIANTHEMETEA GUTTATI (Braun-Blanq. *ex* Rivas Goday 1958) Rivas Goday & Rivas Mart. 1963 *em.* Boullet 2004
Helianthemetalia guttati Braun-Blanquet *in* Braun-Blanquet, Molinier & Wagner 1940
Thero-Airion Tüxen *ex* Oberdorfer 1957
Bromo ferronii-Anthoxanthetum aristati Bioret, Bouzillé, de Foucault, Géhu & Godeau 1988
Tuberario guttatae-Romuleetum columnae Provost 1978

ISOETO DURIEUI-JUNCETEA BUFONII Braun-Blanq. & Tüxen *ex* V.West., Dijk & Paschier 1946
Isoetalia velatae (Braun-Blanquet 1936) de Foucault 1988
Ophioglosso lusitanici-Isoetion histricis (Braun-Blanquet 1936) de Foucault 1988
Ophioglosso lusitanici-Isoetetum histricis de Foucault 1988

KOELERIO GLAUCAE-CORYNEPHORETEA CANESCENTIS Klika *in* Klika & V.Novak 1941
Corynephoretalia canescentis Klika 1934
Miboro minimae-Corynephorion canescentis Loiseau & Felzines 2007
Tuberario guttatae-Corynephoretum canescentis Géhu 1964

PHRAGMITI AUSTRALIS-MAGNOCARICETEA ELATAE Klika *in* Klika & V.Novák 1941
Scirpetalia compacti Hejny *in* Holub, Hejny, Moravec & Neuhausl 1967 *corr.* Rivas-Martínez, J.C. Costa, Castroviejo & Valdes Berm. 1980
Scirpion compacti A.E.Dahl & Hadac 1941 *corr.* Rivas- Martínez, J.C. Costa, Castroviejo & Valdes Berm. 1980
Scirpetum compacti (Van Langendonck 1931) Beeft. 1957

SAGINETEA MARITIMAE V.Westh., C.Leeuwen & Adriani 1962
Saginetalia maritimae Westhoff, C.Leeuwen & Adriani 1962
Saginion maritimae Westhoff, C.Leeuwen & Adriani 1962
Bromo ferronii-Trifolietum arvensis Géhu 2008
Catapodio marini-Parapholisetum incurvae Géhu & de Foucault 1978
Catapodio marini-Trifolietum scabri Géhu & de Foucault 1978 *nom. inv.*
Desmazerio marinae-Brometum ferronii Bioret 1994
Sagino maritimae-Catapodietum marini R. Tx. 1963
Soncho oleracei-Cochlearietum danicae Géhu & de Foucault 1978
Frankenietalia pulverulentae Rivas-Martínez *ex* Castroviejo & Porta 1976
Frankenion pulverulentae Rivas-Martínez *ex* Castroviejo & Porta 1976
Sagino maritimae-Spergularietum marinae Bioret 2006

SCHEUCHZERIO PALUSTRIS-CARICETEA FUSCAE Tüxen 1937
Caricetalia davallianae Braun-Blanquet 1949
Hydrocotylo vulgaris-Schoenion nigricantis de Foucault 2008
Caricenion pulchello-trinervis Julve *ex* de Foucault 2008
Soncho maritimi-Schoenetum nigricantis (Lahondère 1979) de Foucault 1984

SEDO ALBI-SCLERANTHETEA BIENNIS Braun-Blanquet 1955
Sedo albi-Scleranthetalia biennis Braun-Blanquet 1955
Sedion anglici Braun-Blanquet *in* Braun-Blanquet & Tüxen 1952
Dactylo oceanicae-Sedetum anglici J.-M. & J. Géhu et Caron 1978 *corr.* Géhu 2008
Sedo anglici-Scilletum vernae Bioret 1994
Festucenion huonii Bioret *et al.* 2014
Sedo anglici-Festucetum armoricanae Bioret *et al.* 2014
Sedo anglici-Festucetum huonii Bioret *et al.* 2014

THERO-SUAEDETEA SPLENDENTIS Rivas Mart. 1972 *em.* Rivas Mart., Fern.Gonz. & Loidi 1998
Thero-Salicornietalia dolichostachyae Tüxen *ex* Boullet & Géhu *in* Bardat *et al.* 2004
Salicornion europaeo-ramosissimae Géhu & Géhu-Franck *ex* Rivas-Martínez 1990
Spergulario rupicolae-Salicornietum ramosissimae Bioret, Lahondère & Khelifi 2001

5 Conclusion

By highlighting 11 geopermaseries, the geoserial analysis revealed the diversity of plant landscapes of French Channel-Atlantic rocky coastline.

By studying catenas at the regional scale, geosynphytosociology provides an overview of the layout of coastal vegetation, at the national scale.

This study shows that geopermaseries of the supra-littoral zone are characterized by particular geological, geomorphological, bioclimatic, and phytoecological contexts.

For conservation, analysis of geopermasynrelevés can identify the most diverse or the most vulnerable coastal areas and vegetation (incomplete geopermaseries, presence of indicator associations of anthropogenic disturbance...) in order to implement appropriate protection measures.

References

1. Bardat J, Bioret F, Botineau M, Boullet V, Delpech R, Géhu J-M, Haury J, Lacoste A, Rameau J-C, Roux G, Royer J-M, Touffet J (2004) Prodrome des végétations de France. Collection Patrimoines naturels, 61. Muséum national d'histoire naturelle, Paris, p 171
2. Béguin C, Géhu J-M, Hegg O (1979) La synphytosociologie: une approche nouvelle des paysages végétaux. Doc Phytosociol N.S. IV:49–69
3. Bioret F (1989) Contribution à l'étude de la flore et de la végétation de quelques îles et archipels ouest et sud armoricains. Ph. D. thesis, Nantes University, p 480
4. Biondi E, Calandra R, Gigante D, Pignatelli S, Rampiconi E, Venanzoni R (2002) Il paesaggio vegetale della provincia di Terni. Arti grafiche Iezzi, Terni, p 104
5. Bioret F, Géhu J-M (2002) Révision phytosociologique des végétations à Rumex rupestris sur les littoraux atlantiques européens. J Botanique de la Société Botanique de France 20:45–54
6. Bioret F, Géhu J-M (2008) Révision phytosociologique des végétations halophiles des falaises littorales atlantiques françaises Fitosociologia 45(1):75–116
7. Bioret F, Lahondère C (2010) Les végétations à Frankenia laevis et à Limonium à nervures parallèles des hauts schorres sablonneux et des falaises du littoral atlantique français. Braun-Blanquetia 46:129–137
8. Bioret F, Lazare J-J, Géhu J-M (2011) Evaluation patrimoniale et vulnérabilité des associations végétales du littoral atlantique français. J Botanique de la Société Botanique de France 56:39–67
9. Braun-Blanquet J (1928) Pflanzensoziologie. Grundzüge der Vegetationkunde. Biol. Studientbücher 7:1–330
10. Braun-Blanquet J (1932) Plant sociology - The study of plant communities (translation from Pflanzensoziologie by Fuller & Conard), p 439
11. Braun-Blanquet J, Pavillard J (1922) Vocabulaire de sociologie végétale, vol 1, p 16. Montpellier
12. De Foucault B, Bioret F (2010) Contribution au prodrome des végétations de France: les Saginetea maritimae Westoff et al. 1962. J de Botanique de la Société Botanique de France 50:59–83
13. De Foucault B, Catteau E (2012) Contribution au prodrome des végétations de Botanique de France 59:5–131
14. Demartini C, Bioret F, Lazare J-J (2017) Typologie et cartographie des géopermaséries et des géocurtaséries des végétations littorales des côtes Manche-Atlantique françaises. Doc Phytosociol, Série 3, 6:316–332
15. Dupias G, Rey P (1985) Document pour un zonage des régions phyto-écologiques. CNRS, p 39

16. Géhu J-M (1962) Quelques observations sur la falaise crétacée du Cap Blanc-Nez (P. de C.) et étude de la végétation de la paroi abrupte: Brassicetum oleraceae ass. nov. Bulletin de la Société Royale de Botanique de Belgique 95(1):109–129
17. Géhu J-M (1963) L'excursion dans le nord et l'ouest de la France de la Société internationale de phytosociologie. Bulletin de la Société Botanique du Nord de la France 16(3):105–189
18. Géhu J-M (1974) Sur l'emploi de la méthode phytosociologique sigmatiste dans l'analyse, la définition et la cartographie des paysages. Comptes Rendus de l'Académie des Sciences Paris 279(Série D):1167–1170
19. Géhu J-M (1975) Essai systématique et chorologique des principales associations du littoral atlantique français. Los anales de la real academia de farmacia XLI(2):207–227
20. Géhu J-M (1977) Le concept de sigmassociation et son application à l'étude du paysage végétal des falaises atlantiques françaises. Vegetatio 34:117–125
21. Géhu J-M (1979) La phytocoenologie comme outil de définition des secteurs côtiers à préserver. Publications du CNEXO: Actes de Colloques, no 9, pp 154–164
22. Géhu J-M (1986) Des complexes de groupements végétaux à la phytosociologie paysagère contemporaine. Informatore Botanico Italiano 18:53–83
23. Géhu J-M (1988) Schéma synsytématique des principales classes de végétations littorales sédimentaires européennees avec références à d'autres territoires holarctiques. Annali di Botanica LVI(1):5–52
24. Géhu J-M (1991) L'analyse synphytosociologique et géosynphytosociologique de l'espace. Théorie et méthodologie. Colloques Phytosociologiques, Phytosociologie et Paysages, Versailles, XVII:11–45
25. Géhu J-M (1994) Schéma synsystématique et typologique des milieux littoraux français atlantiques et méditerranéens. Colloques Phytosociologiques, Typologie phytosociologique des habitats, Bailleul, XXII:183–212
26. Géhu J-M (1995) Inventaire typologique des milieux terrestres littoraux menacés à des fins conservatoires. Vol I-VI. Fondation de France, les côtes de France (Corse inclue), p 608
27. Géhu J-M (1998) Schéma synsystématique des principales classes de végétations littorales sédimentaires européennes avec références à d'autres territoires holarctiques. Annali di Botanica 61(1):5–52
28. Géhu J-M (2008) Étude des associations végétales des sentiers littoraux de Dinard et Saint-Énogat (France, Ille-et-Vilaine) suivie d'un guide itinéraire. J Botanique de la Société Botanique de France, 41:47–80
29. Géhu J-M (2009) Signification phytosociologique de Brassica oleracea ssp. oleracea sur les falaises atlantiques européennes. Fitosociologia 46(1):3–10
30. Géhu J-M, Bioret F (2000) Les végétations à Osmunda regalis des falaises littorales armoricaines. Colloques Phytosociologiques, Données de la Phytosociologie sigmatiste: Structure, gestion, utilisation, Bailleul, XXVII:203–210
31. Géhu J-M, Géhu-Franck J, Bournique C (1984) Sur les étages bioclimatiques de la région eurosibérienne française. Doc Phytosociol N.S., VIII:29–43
32. Géhu J-M, De Foucault B (1978) Phytosociologie de la pelouse aérohaline des falaises de craie de Haute-Normandie (France). Doc Phytosociol N.S. 3:289–294
33. Géhu J-M, Duquef H (2009) Observations phytosociologiques complémentaires sur les falaises de la Côte d'émeraude (France, Côtes-d'Armor et Ille-et-Vilaine). Bulletin de la Société Botanique du Centre-Ouest, N.S. 40:231–280
34. Géhu J-M, Franck J, Bournique C (1982) La végétation du littoral Nord-Pas-de-Calais. Synthèse phytocoenologique. Université de Paris V, p 361
35. Géhu J-M, Franck J, Scoppola A (1984) Observations sur la végétation aérohaline des falaises maritimes du Centre-Ouest français. Doc Phytosociol N.S. VIII:147–164
36. Géhu J-M, Géhu J (coll. M. Bigot) (1980) Aperçu phytosociologique sur les falaises d'Hendaye et de Saint-Jean-de-Luz (Pays basque). Doc Phytosociol N.S. 5:363–374
37. Géhu J-M, Rivas-Martínez S (1981) Notions fondamentales de phytosociologie. In: Dierschke H (ed) Syntaxonomie. Berichte der Internationalen Symposien der Internationalen Vereinigung für Vegetationkunde, pp 5–33

38. Lahondère C (1980) La végétation des falaises autour de Biarritz. Bulletin de la Société Botanique du Centre-Ouest, N.S. 10:37–44
39. Lahondère C (1986) La végétation des falaises des côtes charentaises. Bulletin de la Société Botanique du Centre-Ouest, N.S. 17:33–53
40. Lahondère C (1990) Les pelouses sèches littorales autour de Royan (Charente-Maritime). Bulletin de la Société Botanique du Centre-Ouest, N.S. 21:29–39
41. Lahondère C (2005) Les formations sèches de la partie charentaise de l'estuaire de la Gironde, de Saint-Palais-sur-Mer à Mortagne-sur-Gironde. Compte rendu de la session phytosociologique 2004. Bulletin de la Société Botanique du Centre-Ouest, N.S. 36:481–512
42. Lalanne A (2017) Le programme de cartographie nationale des végétations naturelles et semi-naturelles de France: CarHAB, Doc Phytosociol Série 3, 6:236–244
43. Lazare J-J (2009) Phytosociologie dynamico-caténale et gestion de la biodiversité. Acta Botanica Gallica 156(1):49–61
44. Lazare J-J, Lanniel K (2003) Une sous-association nouvelle de fourrés du Rubo ulmifolii-Tametum communis du littoral basque. J Botanique de la Société Botanique de France 21:33–35
45. Lazare J-J, Dartiguelongue S, Dicharry C, Lartigue A (2001) Érosion des communautés végétales dunaires du littoral aquitain. Poster, XVIIIe Jornadas de fitosociología, "Condiciones extremas y vegetación", León, 19–22 IX 2001
46. Rivas-Martínez S (1976) Sinfitosociologia, una nueva metodologia para el estudio del paisaje vegetal. Anales del instuto Botanico Cavanilles 33:179–188
47. Rivas-Martínez S (2005) Notions on dynamic-catenal phytosociology as a basis of landscape science. Plant Biosystems 139(2):135–144
48. Rivas-Martínez S (2007) Mapa de series, geoseries y geopermaseries de vegetación de España (Mémoria del mapa de vegetacion potential de España) parte I. Itinera Geobotanica, 17(new series):1–435
49. Rivas-Martínez S (2011) Mapa de series, geoseries y geopermaseries de vegetación de España (Mémoria del mapa de vegetacion potential de España) parte II. Itinera Geobotanica, 18 (2):425–800
50. Tüxen R (1973) Vorschlag zur Aufnahme von Gesellschaftskomplexen in potentiell natürlichen Vegetationgebieten. Acta Botanica Academiae Scientiarum Hungaricae 19(1–4):379–384

Symphytosociology, a Tool for Landscape Monitoring: A Case Study from the Swiss Alps

Jean-Paul Theurillat, Michele Di Musciano, Olivier Duckert, and Claude Béguin

Abstract A comparison of vegetation using a symphytosociological approach was carried out in the Aletsch region, a UNESCO World Heritage Centre located around the Aletsch Glacier in the Valais Alps (Switzerland). For this, relevés performed in 1982 and 2014 in 61 operational geographic units (OGUs) located near the treeline, with a mean elevation above 2000 m, were analysed. During the 32-year interval between the two surveys, several environmental changes occurred, mainly a decrease in traditional pasturing, a 1.5 °C increase in the mean annual temperature, and an increase in deer population. Numerical analyses of the relevés showed that the landscape remained globally stable. Nevertheless, in 56% of the OGUs, there was a change in cover greater than 10% either for trees (> 2.5 m), dwarf shrubs (< 0.5 m) or the herbaceous vegetation. In particular, *Picea abies* increased at lower elevation, and *Pinus cembra* dominated the regeneration of young trees (0.5–2.5 m) outside the forested area. Signs of thermophilisation were found in 74% of the OGUs. Deer were present in all the OGUs, and in some places, they had a strong impact on the vegetation, such as in the Aletschwald Nature Reserve. The resurvey confirms that symphytosociological relevés are a promising tool for landscape monitoring.

J.-P. Theurillat (✉)
Centre Alpien of Phytogeography, Foundation J.-M. Aubert, CP 71, 1938 Champex-Lac, Switzerland
e-mail: jean-paul.theurillat@unige.ch

J.-P. Theurillat
Department of Botany and Plant Biology, Section of Biology, University of Geneva, CP 71, 1292 Chambésy, Switzerland

M. Di Musciano
Department of Life, Health and Environmental Sciences, University of L'Aquila, 67100 Coppito, Italy
e-mail: michele.dimusciano@univaq.it

O. Duckert
Grenat Sarl, 1955 Chamoson, Switzerland
e-mail: olivier.duckert@grenat-sarl.ch

C. Béguin
2067 Chaumont, Switzerland
e-mail: clfbeguin@bluewin.ch

F. Pedrotti and E. O. Box (eds.), *Tools for Landscape-Scale Geobotany and Conservation*, Geobotany Studies, https://doi.org/10.1007/978-3-030-74950-7_14

Keywords Symphytosociology · Vegetation · Landscape · Deer · Climate change · Thermophilisation · Monitoring · Switzerland · European Alps · Valais

1 Introduction

Symphytosociology (also termed symphytocoenology) is a branch of phytosociology aiming at the analysis of vegetation specifically at landscape scale. It is based on comparison of quantitative relevés of all the vegetation units occurring within a given portion of a landscape or operational geographic unit (OGU). The particular assembly of vegetation units forms a so-called vegetation complex, and abstract types of such complexes are called sigmassociations, a term meaning "a sum of associations" (from the Greek letter *sigma,* $\sum$, meaning sum). Consequently, it is possible to build a typology to describe the recurring vegetation complexes within a landscape. According to the complexity of the landscape itself, and the size of the OGU, different organisational levels of sigmassociations have been distinguished. The theoretical ground and the methodological approach were developed in the 1970s, mainly by Tüxen [14], Béguin and Hegg [2], Rivas-Martínez [9] and Géhu [6], see also Theurillat [12, 13] and Béguin et al. [3]. Several approaches have been developed to define sigmassociations. Among them, the "Swiss School" [5, 10, 11] proceeds with a preliminary delimitation of the study area in uniform OGUs (landscape units), in which only one kind of potential vegetation would likely occur. This exhaustive approach allows a simultaneous mapping of the vegetation landscape of the study area.

The great advantage of the symphytosociological relevé is that it can include vegetation units at any level, from very broad units at the level of the phytosociological class to very detailed units, corresponding to fine variations of a given association (variant, facies), fragmentary aspect, or basal unit in the sense of Kopecký and Hejný [12]. In this respect, such relevés can distinguish particularities within the landscape and small differences in its use. Compared to detailed mapping of the vegetation, there is a trade-off between precision at the unit level and that of the location, a kind of Heisenberg's uncertainty principle. The symphytosociological relevé is not limited regarding the precision of the vegetation units but cannot offer precision in their location within the area of the relevé, in as much as the surface occupied by the vegetation unit is very small. Conversely, fine-scale mapping allows great precision of location but only down to the resolution permitted by the scale; hence, not enabling the mapping of very tiny patches.

In this contribution, we are interested in comparison over a 32-year interval (1982 versus 2014) of the vegetation landscape of a section of the Aletsch region, which is a UNESCO World Heritage Centre located around the Aletsch Glacier in the Valais Alps (Switzerland). In the years 1981/82 (hereafter 1982), the vegetation landscape was studied along an elevation transect from 700 m up to 2230 m. Since then, several environmental changes co-occurred in this region:

(1) The traditional land use with pasturing decreased strongly in the 1980s, especially in the upper part of the study area.
(2) The mean annual temperature increased by 1.5 °C during the 32-year interval.
(3) Deer population, which was scarce at the beginning of the 1980s, increased drastically in the 1990s, especially on the north side of the Aletschwald Nature Reserve, due to decreased human activity. The deer population fluctuates between 100 and >200 individuals on less than 10 km^2.

In order to see if these environmental changes had already produced a diachronic change at the landscape level, especially around the tree line, the upper part of the transect was resurveyed during the years 2013/14 (hereafter 2014), and the landscape relevés of the two surveys were compared.

2 Methods

2.1 Location

The study area is located in the upper part of the right-hand side of the Rhone River valley (46°22′ N, 8°02′ E; Fig. 1). The bedrock is siliceous and pertains to the southern border of the Aar massive. An elevation transect of 8.5 km^2 was investigated in 1981/82, from the bottom of the Rhone River at 700 m up to the crest at the tree line at 2230 m (southeast side) and then down to 1600 m, to reach the Great Aletsch Glacier (northwest side). The crest can be divided into a western part, with the Riederhorn reaching 2230 m, a central part at Riederfurka (2065 m) and an eastern part up to Hohflue (2227 m).

Until the 1950s, the southeast side of the study area was used intensively in traditional farming, with cultivation up to the 1500 m, hay up to 1900 m, pasturing at a higher elevation, and wood-cutting up to the tree line. Pasturing in the upper part, above 2000 m, was maintained quite intensively until the end of the 1970s and then extensively until the 2000s. Tourism developed in the 1960s, mainly skiing activities during winter and hiking during summer. On the northwest side, there has been intensive pasturing in the upper part of the eastern part together with wood-cutting in the past. However, since 1933, the Aletschwald in the eastern part was made a nature reserve from the glacier to the crest, and so human activity has subsided, except hiking along strictly limited paths.

2.2 Symphytosociological Relevés

For the 1981/82 study [4, 12, 13], the area was delimited exhaustively in 1–8 ha OGUs, with the aid of tatopographical map (1:10,000) and aerial views, based on three main criteria of more or less homogenous (a) slope, (b) aspect and

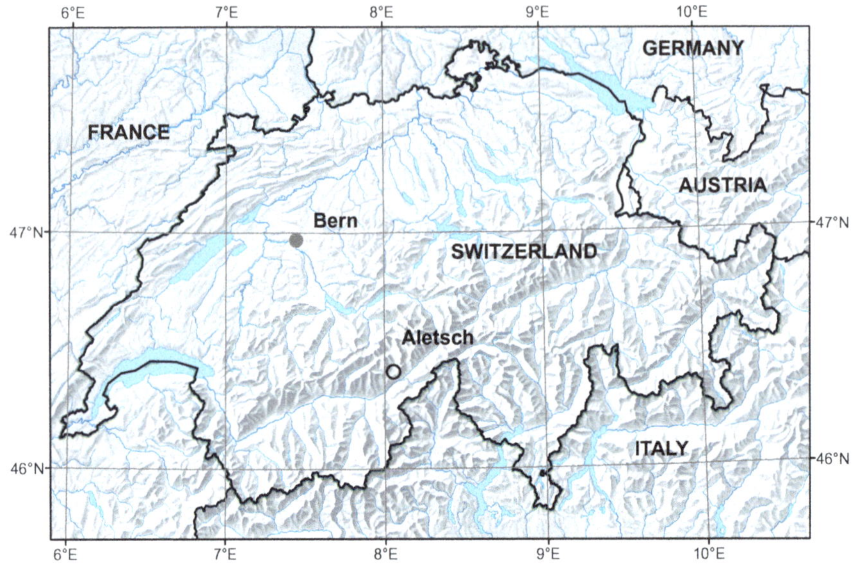

Fig. 1 Location of the study area (Aletsch) on the right-hand side of the Rhone river valley (Valais, Switzerland)

(c) vegetation structure. Along the whole elevation transect from the Rhone River to the glacier, 289 OGUs were delimited according to the above criteria. The symphytosociological relevés of the OGUs were performed with the help of the topographic map and a Thommen altimeter. In each relevé, the following elements were recorded and quantified (see Table 1 as an example):

(a) The vegetation units, according to a list of as detailed as possible units. The units were established with the assistance of vegetation relevés, performed during a preliminary survey. These are working units where physiognomy is emphasised; thus, corresponding to subassociations, variants or facies of already, formally described vegetation units and, otherwise, to undetermined physiognomic, fragmentary or basal units (see [12]). In addition, some of the vegetation units, usually the dominant ones, were subdivided further according to the state of their development and visible disturbance (e.g., vegetation normally developed, vegetation poorly developed). If needed, these stages could be combined during data analysis. Each unit bears an identifier number that is indicated in the relevé, which is used in the numerical analyses.
Examples of working units for mesothermophilous dwarf scrub dominated by *Juniperus communis* ssp. *alpina* are as follows:

775 *Vaccinio-Juniperetum* typical ø
776 *Vaccinio-Juniperetum* typical °
777 *Vaccinio-Juniperetum 'sphagnetosum'*
778 *Vaccinio-Juniperetum 'rhododendretosum'* ø

779 *Vaccinio-Juniperetum 'saxifragetosum aizoon'*
780 *Vaccinio-Juniperetum 'dryopteridetosum spinulosae'*
782 *Vaccinio-Juniperetum 'festucetosum variae'*
RH1 *Vaccinio-Juniperetum 'callunetosum'*
RH37 *Vaccinio-Juniperetum* with *Rubus idaeus*
RH46 *Vaccinio-Juniperetum* degraded by herbicide
RH49 *Vaccinio-Juniperetum* with *Juniperus sabina*
RH52 *Vaccinio-Juniperetum* typical var. *Epilobium angustifolium* (somehow ruderal)
RH53 *Vaccinio-Juniperetum* typical var. *Urtica dioica*

The numbers 775–782 correspond to units established during the 1982 survey; numbers RH1–RH53 represent units added during the resurvey in 2014; ø designates normally developed, and ° denotes poorly developed.

Table 1 Relevés in 1982 and 2014 of the operational geographic units (OGUs) dominated by the mesophilous cembran pine forest. Manual classification: see text. The changes mainly attributable to deer between 1982 and 2014 are highlighted in light grey. Data related to geomorphology, land use, human activities, phanerophytic structure and isolated trees have been omitted

		Mesophilous cembran pine forests, Aletschwald (Eb)									
	Classification 1992	typical						cool		cold	
	OGU	276	276	277	277	278	278	275	275	266	266
	Year of the relevé	2014	1982	2014	1982	2014	1982	2014	1982	2014	1982
	Vegetation structure										
012	Trees (> 2.5 m)	8B	7B	7B	7B	7B	7B	7	7B	6B	6B
013	Shrubs (< 2.5 m)	2B	2C	2C	2C	1C	2C	2B	3G	2C	4D
014	Dwarf shrubs (< 0.5 m)	4B	5B	5C	6B	5B	6B	4C	5C	6C	7B
015	Swards, grasslands	4B	4B	5C	4B	5B	3B	5C	5C	5C	4B
016	Open vegetation	4B	4B	3C	3C	3F	3B	2D	2D	2C	2G
	Land use										
S00	Paths of deer	3F		2F		3F		2F		2F	
31	Dead trees on the soil	2B	2B	2	1B	2	1B	2	2B	2	2B
	Vegetation units										
332	Urtico-Chenopodietum boni-henrici°					1B					
333	Sileno dioicae-Urticetum dioicae										
360	Plantagini-Poetum supinae typique ø			2F		2F		2F	1A		
361	Plantagini-Poetum supinae typique°			1G				1F		1F	
522	Alchemillo-Festucetum rubrae typique			3F		3F		3C		4F	1G
757	Agrostietum schraderianae	2B°	2C		1B		2B		2G	3C°	3C
741	Adenostylo-Cicerbitetum alpinae typique°		1A								
743	Adenostylo-Cicerbitetum var. Peucedanum ostruthium		1B								
745	Calamagrostietum villosae peucedanetosum								3G		2G
744	Calamagrostietum villosae gentianetosum		3C	4C	3B	5B°	3B	5C	5C	5C	3C
730	Alnetum viridis typique ø				1A				2A		2B
731	Alnetum viridis typique°	2F								1B	
732	Alnetum viridis salicetosum helveticae										2B
736	Alnetum viridis rhododendretosum ø		2G	1A			1A		2B		2B
737	Alnetum viridis rhododendretosum°							2A		2B	
760	Salicetum helveticae							1A	3G		4C
778	Vaccinio Juniperetum rhododendretosum ø		3B		5B		6C		3A		7C
784	Vaccinio-Rhododendretum typique ø	4B	5C	4C	5B	3B	5C		5C	5B	2B
785	Vaccinio-Rhododendretum typique°			4C		5B		4C		5C	
788	Vaccinio-Rhododendretum salicetosum helveticae ø							2A	2D		5C
804	Vaccinio-Rhododendretum salicetosum helveticae°									2B	
806	Vaccinio-Rhododendretum juniperetosum					2A		2A		3	
824	Vaccinio-Pinetum cembrae typique ø	7B	7B	6B	6B	6B	6B	3C	4C		3C
825	Vaccinio-Pinetum cembrae typique°							4C		6B	
826	Vaccinio-Pinetum cembrae calamagrostietosum villosae ø		5C	5B	3B	5B	3B	7B	7C		5C
827	Vaccinio-Pinetum cembrae calamagrostietosum villosae°	4B									
828	Vaccinio-Pinetum cembrae juniperetosum ø				3B		5B		3D	2B	5C
829	Vaccinio-Pinetum cembrae juniperetosum°					2A		2A			

(b) The main vegetation structure (trees, shrubs, dwarf shrubs, herbs – both closed and open herbaceous vegetation) was evaluated separately and quantified in the field. This approach helps to control the quantification of each single vegetation unit; the sum of the cover of all the vegetation units of a given structure must correspond to the cover of the main vegetation structure to which they belong. Isolated trees were recorded separately and placed in two categories of height (0.5–2.5 m and > 2.5 m).
(c) The phanerophytic vegetation (> 0.5 m height) was divided further into < 2.5 m and > 2.5 m isolated broadleaf trees, < 2.5 m and > 2.5 m isolated coniferous trees, coniferous patches, coniferous forests and native shrubs.
(d) The predominant land use (mowing, pasturing, cultivation, fallow, manure).
(e) Geomorphological features (e.g., terrain roughness, screes, aquatic elements).
(f) All anthropogenic activities (e.g., types of constructions, types of roads and other kinds of transport, ski runs).

Except for surfaces up to 10 m^2, the presence of all the above features was quantified in the OGUs in proportion of their occurrence, according to height classes of cover (1 = $\leq$ 10 m^2, 2 = $\leq$ 1%, 3 = 1–5%, 4 = 5–10%, 5 = 10–25%, 6 = 25–50%, 7 = 50–75%, 8 = 75–90%, 9 = > 90%). A manual classification (tables of relevés) produced seven main landscape units.

2.3 *Resurvey*

For the 2014 study, 61 OGUs with mean elevation above 2000 m, covering 1,736 km^2, were re-sampled with the help of a Garmin 60 GPS and a 1:10,000 orthophoto. The delimitations of the OGUs in 1980 were previously reported on the orthophoto (Fig. 2). Where needed, the delimitations were corrected in the field and were fixed with additional GPS points. The same vegetation units as those for the 1982 survey were used, with new units added when needed.

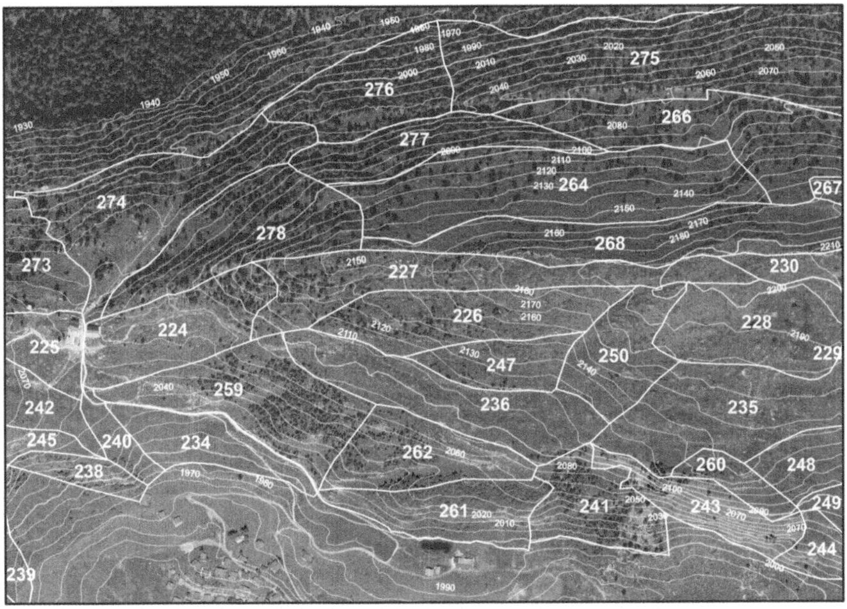

Fig. 2 Example of operational geographic units (OGUs) reported on the orthophoto for the 2014 survey. OGUs were delimited in 1980 on a 1:10,000 topographic map, according to slope, aspect and vegetation structure

2.4 Numerical Analyses

Data from both the 1982 and 2014 studies were analysed numerically in performing Ward D's cluster analysis, McQuitty's principal component analysis (PCA), and canonical correspondence analysis (CCA) with the VEGAN package in R (https://github.com/vegandevs/vegan), using a rank transformation of 1–9 for the nine classes of the cover data. The analyses were made by lumping the states "vegetation normally developed" and "vegetation poorly developed" into a single unit. Lumping is a conservative approach that minimises the differences between the OGUs, to get closer to the manual classification.

2.5 Nomenclature

If not otherwise mentioned, the nomenclature of the species followed Aeschimann et al. [1]. Trees and shrubs: cembran pine (*Pinus cembra*; larch (*Larix decidua*); spruce (*Picea abies* L.); green alder (*Alnus viridis*); Swiss willow (*Salix helvetica*); *Juniperus nana* Willd. = *J. communis* ssp. *alpina* Celak.

3 Results

3.1 Classification of the 1982 Relevés

The symphytosociological relevés of the 1982 survey were analysed manually [12]. For the upper part of the transect, above 2000 m, seven main vegetation landscape units and three main sub-groups were identified within the 61 OGUs. These seven landscapes (between brackets the more importantly as diagnostic associations determining the landscape type when they are dominant in the OGU) were classified as follows:

A: landscape of cliffs (*Asplenio-Primuletum hirsutae, Festucetum variae* s.l.)
B: landscape of boulders (*Cryptogrammetum crispae*)
C: landscape of thermophilous cembran pine forests (*Cotoneastro integerrimae-Pinetum cembrae*)
D: landscape of spruce forests (*Homogyno alpinae-Piceetum abietis*)
E: landscape of mesophilous cembran pine forests (*Larici deciduae-Pinetum cembrae*) with a rocky variant (Ea, Riederhorn) and a gentler variant (Eb, Aletschwald)
F: landscape of thermophilous dwarf heaths (*Junipero nanae-Arctostaphyletum uvae-ursi*) with a warm and a cold variant
G: landscape of mesophilous heaths s.l. (*Vaccinio-Juniperetum nanae, Vaccinio-Rhododendretum ferruginei*) with the sub-groups

  G1: mesothermophilous dwarf heaths (*Vaccinio-Juniperetum nanae*, *Geo montani-Nardetum strictae*) with a warm, a crest, and a cold variant
  G2: dwarf heaths with *Calamagrostis villosa* (*Vaccinio-Juniperetum nanae, Calamagrostietum villosae* s.l.)
  G3: cryophilous dwarf heaths (*Vaccinio-Rhododendretum, Empetro-Vaccinietum*).

The manual classification was globally validated by Ward's cluster analysis, performed on the 1982 relevés (Fig. 3). The main divisions of the cluster analysis were also reinforced with PCA (not shown).

The principal division in the cluster analysis separated the OGUs of the forested landscape where no pasturing occurs (cluster 1) from the OGUs of the non-forested landscape submitted to pasturing (cluster 2). The group with OGUs not submitted to pasturing (cluster 1) was divided further into two geographic-ecological clusters, the first (cluster 1.1) corresponding to the western part of the study area with rockier slopes (Riederhorn), and the second (cluster 1.2) associated with the less rocky northern aspect of the eastern part of the study area (Riederfurka, Aletschwald). The western cluster 1.1 was subdivided further into spruce forests (D) (cluster 1.1.1) and cembran pine forests (cluster 1.1.2), with a division occurring in cluster 1.1.2 between two OGUs of rocky thermophilous cembran pine forests (C) (cluster 1.1.2.1) and less rocky OGUs with thermophilous cembran pine forests and mesophilous cembran pine forests (E) (cluster 1.1.2.2). The eastern cluster 1.2

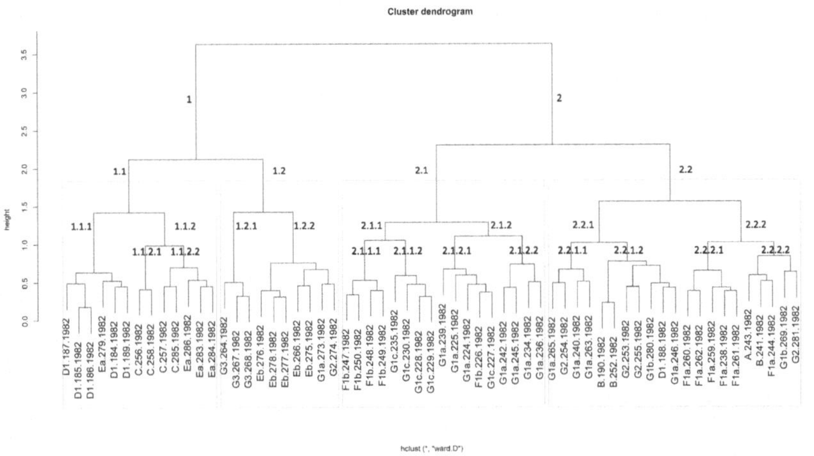

Fig. 3 Ward's cluster analysis (Ward D) of the 1982 relevés. A–G: former manual classification (see text). Cluster 1: forested operational geographic units (OGUs) with no pasturing. Cluster 2: non-forested OGUs submitted to pasturing

mainly contained OGUs of the landscape of the cryophilous dwarf heaths (G3) (cluster 1.2.1) and mesophilous cembran pine forests (E) (cluster 1.2.2).

Cluster 2, where pasturing occurs, was divided into two parts. Cluster 2.1 corresponded to heaths that were still more or less pastured, and cluster 2.2 resembled OGUs with a relatively lower pasturing pressure. Cluster 2.1 was then divided into four geographic-ecological sub-clusters. On the one hand, the south of the eastern part of the study area (2.1.1), with a further subdivision into thermophilous dwarf heaths (F) (cluster 2.1.1.1) and mesophilous dwarf heaths (G1) (cluster 2.1.1.2); and, on the other hand, the mesophilous dwarf heaths (G1) of the south side of the central part of the study area that were further subdivided into an upper (cluster 2.1.2.1) and lower part (cluster 2.1.2.2). Cluster 2.2 contained OGUs with a comparatively lower pasturing pressure. This was a mixture of mesophilous heaths (G1), heaths with *Calamagrostis villosa* (G2), thermophilous dwarf heaths (F), boulders (B) and a cliff (A). This was divided further, on the one hand, into a sub-cluster of the eastern part of the Riederhorn (cluster 2.2.1.1) and a sub-cluster of the southwest part of the Riederhorn (cluster 2.2.1.2); and, on the other hand, into a sub-cluster of thermophilous dwarf heaths in the eastern part (cluster 2.2.2.1) plus a mixed sub-cluster 2.2.2.2, containing the unique OGUs occupied with a cliff.

The most important units of the vegetation landscape obtained by the manual classification could be recognised as sub-groups in the numerical classification, although the divisions are not always clear-cut. For the landscape units recognised by manual classification, cluster analysis showed the following:

(1) The OGUs containing mesophilous cembran pine forests (E) were separated geographically. This finding can be explained because, on the rocky slopes on

the north side of the Riederhorn, a mixture of mesophilous and thermophilous cembran pine forests existed (Fig. 3, cluster 1.1.2).

(2) The OGUs of the cryophilous dwarf heaths (G3) were grouped with the OGUs of the mesophilous cembran pine forests located on the north side (Fig. 3, cluster 1.2) rather than with other OGUs of dwarf heaths because they did not contain swards with *Nardus stricta*.

(3) The OGUs of the cold variant of the thermophilous heaths (Fb) formed a sub-cluster (Fig. 3, cluster 2.1.1.1) within the mesothermophilous heaths (G1) (Fig. 3, cluster 2.1) because they were located at higher elevation and thus also contain mesophilous heaths.

(4) The unique OGU of cliffs (A) was classified with the warm variant of the thermophilous heaths (F) (Fig. 3, cluster 2.2.2) because the OGU of the cliffs contained thermophilous heaths.

(5) The three OGUs of the boulders (B) did not form a cluster because their delimitation was not restricted solely to boulders.

(6) The sub-group G2 was not recognised as a group because the *Calamagrostietum villosae* s.l. is a complex unit and because several other kinds of heaths also occur in the five OGUs assigned manually to this landscape. Three out of five of these OGUs were part of a heterogeneous cluster (Fig. 3, cluster 2.2.1) containing OGUs of the Riederhorn, mainly with mesophilous heaths and some boulders.

3.2 Changes Between 1982 and 2014

3.2.1 Cluster Analysis

Ward's cluster analysis on the relevés from both 1982 and 2014 (Fig. 4) showed the same clustering structure as the cluster analysis of the relevés in 1982. In a first step, the forested OGUs (cluster 1) were separated from the non-forested OGUs (cluster 2). Instead of having a geographical subdivision in cluster 1, however, the OGUs with spruce forest (D) now formed a separate cluster (1.1), and all the OGUs with cembran pine forests (C, E) were in the same cluster (1.2.2), except the cryophilous heaths (G3) (cluster 1.2.1). In the non-forested cluster 2, the subdivision had not changed, with a cluster (2.1) for the more or less pastured OGUs and another cluster (2.2) for the non-pastured OGUs with cliffs (A), boulders (B), thermophilous heaths (F) and mesophilous heaths with *Calamagrostis villosa* (G2).

There was a change between clusters 1 and 2 for two OGUs (D.188, G1.269), and in one OGU (F.238) between clusters 2.1 and 2.2. Changes between the higher sub-clusters occurred for OGU G3.264 in cluster 1.2, and OGU G1.236 in cluster 2.2.1. At a lower level, a change occurred for 17 OGUs, of which seven with thermophilous heaths (F) were placed in clusters 2.1.2 and 2.2.2 and three with boulders (B) placed in cluster 2.2.

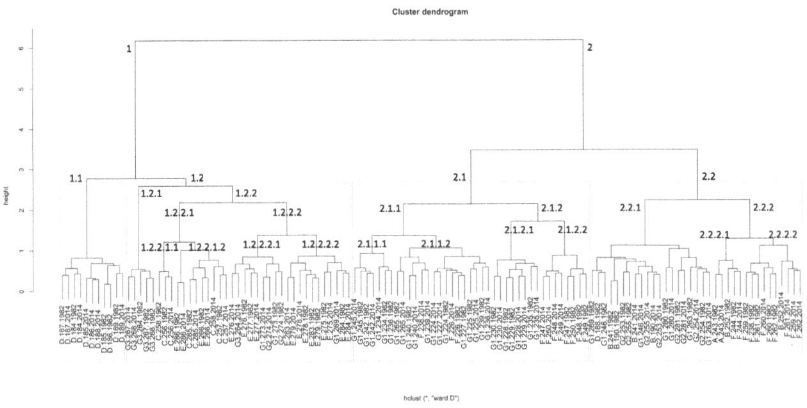

Fig. 4 Ward's cluster analysis (Ward D) of the relevés from both surveys (1982, 2014). A–G: former manual classification (see text). Cluster 1: forested operational geographic units (OGUs). Cluster 2: non-forested OGUs

On the other hand, when taking only the relevés from 2014 (Fig. 5), there was a definite change in the main clustering, with a first division between the more or less pastured OGUs (cluster 1) and the non-pastured OGUs (cluster 2), instead of between the forested and non-forested OGUs. This behaviour shows a convergence between the forested OGUs and the non-forested OGUs that are no longer pastured. In addition, the non-pastured OGUs of cluster 2 were not subdivided into forested and non-forested OGUs. There was one cluster (2.1) with the OGUs of the Aletschwald Reserve, plus two adjacent OGUs with a recolonisation by mesophilous cembran pine forest; and another cluster (2.2) with a sub-cluster 2.2.1 that was about the same as cluster 2.2 in Fig. 3, for the relevés in 1982. Nonetheless, sub-cluster 2.2.1 was without most of the OGUs of the thermophilous dwarf heaths (F) (Fig. 3, cluster 2.2.2.1) and included, instead, most of the OGUs of the thermophilous cembran pine forest (C). The other sub-cluster 2.2.2 contained mainly the OGUs of the spruce forests (D) and the remaining OGUs of the mesophilous cembran pine forests (E).

3.2.2 Canonical Correspondence Analysis (CCA)

The five classes of vegetation structure (trees > 2.5 m, shrubs 0.5–2.5 m, dwarf shrubs < 2.5 m, closed herbaceous vegetation, and open herbaceous vegetation) were seen as driving factors for change in the OGUs between 1982 and 2014 (Fig. 6). The modification in each category between 1982 and 2014 was statistically significant. Changes in the cover of trees, dwarf shrubs and herbs were the main factors of change. Globally, when looking at the position of the group centroids, a centripetal convergence trend of the relevés appeared in 2014, relative to the relevés in 1982, except for the boulders (B) and the mesophilous heaths with

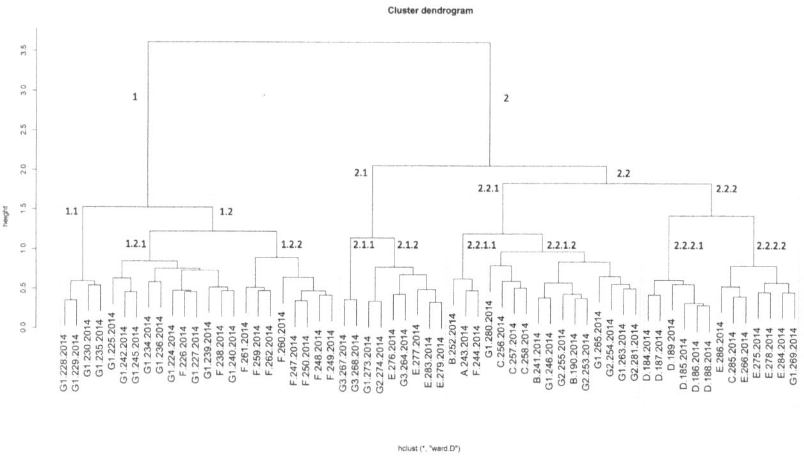

Fig. 5 Ward's cluster analysis (Ward D) of the 2014 relevés. A–G: former manual classification (see text). Cluster 1: operational geographic units (OGUs) with more or less pasturing. Cluster 2: OGUs with no pasturing

Calamagrostis villosa (G2). The convergence was also confirmed by cluster analysis of the relevés in 2014 at the highest level (Fig. 5).

There was a marked increase in the tree cover of the non-forested OGUs (A, B, F, G1, G2), as exemplified by the change in the average centroid. A decrease occurred in the cover of the dwarf shrubs in the spruce forests (D), the mesophilous cembran pine forests (E), the warm variant of the thermophilous dwarf heaths (F) and the mesophilous heaths with *Calamagrostis villosa* (G2). The cover of the herbaceous vegetation increased in the spruce forests (D) and mesophilous cembran pine forests (E) but decreased in the warm variant of the mesothermophilous dwarf heaths (G1). The cryophilous dwarf heaths (G3) appeared stable, besides an increase in the herbaceous vegetation, in one OGU.

The convergence was substantiated by comparing the CCA performed separately on the relevés from 1982 and those from 2014 (not shown). For the 1982 relevés, the four main subdivisions resulting from the cluster analysis (Fig. 3) could be individualised on the CCA, with forested and cryophilous OGUs on one side (clusters 1.1 and 1.2; C, D, E, G3) and more or less pastured OGUs on the other side (clusters 2.1 and 2.2; A, B, F, G1, G2). For the 2014 relevés, only the two principal subdivisions, pastured OGUs (cluster 1; F, G1) versus non-pastured OGUs (cluster 2; A, B, C, D, G2, G3; Fig. 5), could be identified.

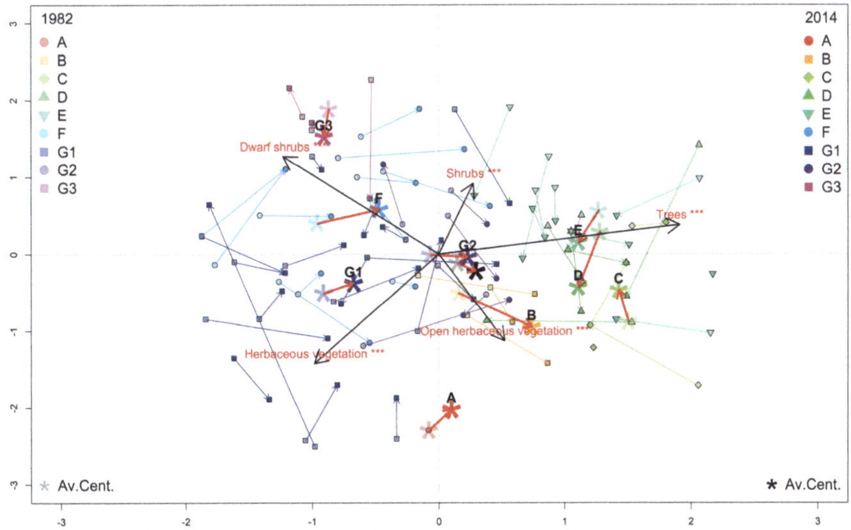

Fig. 6 Canonical correspondence analysis (CCA) of the relevés from both surveys (1982, 2014). The driving factors are the five classes of vegetation structure (trees > 2.5 m, shrubs 0.5–2.5 m, dwarf shrubs < 2.5 m, closed herbaceous vegetation, and open herbaceous vegetation)

3.3 *Landscape Change*

Comparing the change of land cover between 1982 and 2014, one can see that 34 OGUs (56%) showed a cover change greater than 10%, either in tree cover (> 2.5 m), dwarf shrubs (< 0.5 m) or herbaceous vegetation s.l. Proportionally, the forest-dominated OGUs were the more stable, with 9 out of 19 (47%) showing no change between 1982 and 2014 (Fig. 4). The modification in tree cover corresponded mainly to an increase in patches of spruce and spruce forests at the lowest elevations, in five OGUs on the southeast side; as well as an increase in mesophilous cembran pine patches and forests in four OGUs on the northwest side, outside the nature reserve. The change in dwarf-shrub cover did not show any clear pattern for any given vegetation units. However, at the lowest elevation, there was mainly a decrease in thermophilous heaths and an increase in mesophilous heaths. At thc highest elevation, there was mainly a decrease in mesothermophilous heaths and an increase in mesophilous heaths. There was no pattern in the changes of the herbaceous vegetation. Most of the modifications occurred with the *Calamagrostis villosa* meadows and the *Nardus stricta* swards, with either an increase or a decrease of their cover.

Cembran pine dominated the regeneration of young trees (0.5–2.5 m) outside the forested area. A regeneration density of more than 10 individuals/ ha was observed in 43 OGUs (70%), of which 20 OGUs had a density between 10 and 29 (33%), 17 between 30 and 59 (28%), and 7 OGUs with more than 60 individuals, reaching up to 154 individuals (data not shown), respectively. Regeneration occurred at all elevations, with 19 OGUs located at mean elevation below 2100 m. By comparison, a density of more than 10 individuals/ ha of spruce occurred in only 18 OGUs (29.5%); two thirds of these were located at mean elevation higher than 2100 m. Young spruce reached a maximum of 45 individuals/ ha.

3.4 Thermophilisation

Among the differences in the vegetation units, an increase in some thermophilous units was noticeable, particularly for the OGUs at the lowest elevation. These units corresponded mainly (1) to a new presence or an increase of shrubs from warmer conditions (plant communities with *Juniperus sabina*, shrubby *J. communis* ssp. *alpina* (> 0.5 m), *J. communis* var. *intermedia, Rosa spp., Amelanchier ovalis, Cotoneaster integerrima* and shrubby *Populus tremula*); (2) to increase of thermophilous swards (*Festucetum variae* s.l., *Carici sempervirentis-Brachypodietum rupestris*); and (3) to increase in the thermophilous dwarf heath (*Junipero-Arctostaphyletum uvae-ursi* s.l.). A sign of thermophilisation also appeared in a few cases, with a distinct decrease or even the disappearance of some cryophilous units, mostly small patches of snowbed vegetation (*Alchemillo pentaphyllae-Salicetum herbaceae* s.l.). Evident signs of thermophilisation (positive and negative) emerged in 18 OGUs, which were also visible, but less importantly, in 27 other OGUs. In total, 74% of the OGUs displayed some sign of thermophilisation. No thermophilisation was apparent in most of the OGUs containing the mesophilous cembran pine forests, except two OGUs on steep, rocky slopes.

3.5 Impact of the Deer

The presence of deer was evident in all OGUs, with frequent incidence of droppings, primarily in 26 OGUs with paths. Strong impact by deer on the vegetation existed in two zones. One was the Aletschwald Nature Reserve on the north side, with the OGUs of mesophilous cembran pine (Table 1) and where human activity is limited to restricted hiking; the other location was on the southeast side, in the OGUs of spruce forests where there is hardly any human activity. The main impact was from trampling, affecting the dwarf-shrub vegetation, mostly *Rhododendron ferrugineum* and *Juniperus communis* ssp. *alpina* communities. The result was a reduction in the mesophilous and mesothermophilous heaths (G1), and the occurrence of more or less highly degraded patches of these vegetations.

In the nature reserve, especially in the OGUs with mesophilous cembran pine forests (Table 1), the degradation and destruction of *Rhododendron ferrugineum* and *Juniperus communis* ssp. *alpina* dwarf shrubs led to local increase in the clonal grass *Calamagrostis villosa*, both as a substituting meadow and as the understory of mesophilous cembran pine forest. In some resting places, even ruderal vegetation with *Urtica dioica* had developed. There was also trampled vegetation, with *Poa supina* along the paths. The browsing had strongly decreased or, in some OGUs, even eliminated the small patches of green alder, Swiss willow and tall herbs, with *Adenostyles alliariae* and *Peucedanum osthruthium*. In the spruce forests, trampling and browsing also led to the degradation and destruction of *Rhododendron ferrugineum* and *Juniperus communis* ssp. *alpina* in the openings and an increase in herbaceous vegetation, either with *Festuca rubra* in the mesophilous parts or with *Nardus stricta* swards. In contrast, on the southeast part of the study area, the durability of some extensive swards with *Nardus stricta*, despite cessation of pasturing, could have resulted from browsing by deer.

4 Discussion

Comparison of the symphytosociological relevés made in 2014 with those made in 1982 verified that the landscape around the tree line remained globally the same at the level of the plant communities. In the conservative cluster analysis (Fig. 4), only five OGUs (8%) changed their classification at the level of the higher clusters, and 17 (28%) at the level of the lower clusters (Fig. 4). However, when comparing photos of the southeast side taken in the 1980s versus those of today, one can see distinctly that the tree cover has increased, notably for spruce at the lowest elevations. The importance of the structural transformation of the landscape was revealed by comparing the cover of the five main categories of vegetation, with 56% of the OGUs showing a change greater than 10%. Since vegetation in the non-forested units was changed more, we are inclined to think that the main driver of the observed differences is the abandonment of the traditional pasturing that had shaped the landscape near tree line for centuries. The noticeable decrease of thermophilous heaths and increase in mesophilous heaths at lowest elevation, albeit counterintuitive under climatic warming of 1.5 °C, are consistent with a decrease in pasturing pressure, increased shade produced by increasing spruce, and the cessation of the manual *Rhododendron ferrugineum* and *Juniperus nana* dwarf shrubs, which can now re-grow on the southeast part of the study area. The same reasoning applies to the decrease in mesothermophilous and increase in mesophilous heaths in the upper OGUs in the eastern part. Therefore, pasturing was a strong structuring element of the landscape, and maintaining it would be important for the stabilisation of the landscape in the face of climate change [7]. In the absence of pasturing, it is known that deer can substitute for cattle to maintain some kinds of swards, with browsing by 100 deer being about equivalent to that of 30 cattle.

On the other hand, the importance and rapidity of the modifications observed near tree line cannot be explained only by the abandonment of pasturing. A simultaneous acceleration factor is also needed. The increase in the mean annual temperature by 1.5 °C since the beginning of the 1980s is probably the explanation, especially for the development of the spruce forests in the lowest OGUs. Such a temperature increase corresponds to a theoretical rise by the tree line of 270 m. Therefore, the non-forested OGUs around 2000 m are now distant from a climax equilibrium that would correspond to the spruce forest, namely, the speed at which spruce forest can now develop. A warmer vegetation period is also reflected by signs of thermophilisation in 74% of the OGUs, with increase or emergence of shrubs and swards at relatively lower elevation in warmer conditions, and disappearance of cryophilous vegetation in the fragmentary snowbeds in the OGUs of the cryophilous heaths (G3) that occurred on the northern side, in the upper part of the nature reserve.

The nature reserve, in which no human activity has occurred since its creation in 1933, would have been an opportunity to observe whether the climax vegetation of the mesophilous cembran pine forests (E) would show some resistance to the temperature increase of 1.5 °C. However, the strong presence of deer degraded the dwarf heaths by trampling; caused the disappearance of some vegetation types, such as green alder, by browsing; and caused ruderalisation along paths and in resting places. Deer presence also favoured the expansion of the grass *Calamagrostis villosa*, both by trampling and browsing (Table 1). Although there was no sign of thermophilisation in the corresponding OGUs, which could be considered a sign of stability against the current warming, the long-term impact of deer could be a local accelerator of a deeper change in these OGUs. Ironically, in the absence of top predators, like wolf, the large, concentrated deer population, hardly regulated by selective shooting, could drive future vegetation of the nature reserve back to a state resembling the situation that existed 100 years ago, when the emblematic mesophilous cembran pine forest was pastured by cattle, sheep and goat, and submitted to wood-cutting [8].

5 Conclusion

Symphytosociological relevés are an effective and potent tool for detailed landscape monitoring, especially with the existing facilities offered by geo-referencing that allow selecting precise standard plots in a given area. Comparison of symphytosociological relevés is more powerful than comparison of vegetation maps, even at very fine scale, which cannot reach the level of detail afforded by symphytosociological relevés because of scale limits.

Although the plant communities have not been modified profoundly at the landscape level during the last 30 years, the symphytosociological resurvey in 2014 revealed a steady trend towards increases of dwarf shrubs and tree cover. Therefore, there is a need for some management, in the face of the ongoing climate change and increasing deer population, to maintain the attractiveness of a traditional landscape shaped by millennia of human activity in places like the Aletsch World Heritage site.

Acknowledgements This research has been supported by the Swiss National Science Foundation within the Man and Biosphere project (FNRS 4.331.2.79.55 to Claude Béguin) and the Service des forêts, des cours d'eau et du paysage of the canton Valais (support 2012 to Jean-Paul Theurillat).

References

1. Aeschimann D, Lauber K, Moser DM, Theurillat JP (2004) Flora alpina. Paris, Berlin
2. Béguin C, Hegg O (1975) Quelques associations d'associations (sigmassociations) sur les anticlinaux jurassiens recouverts d'une végétation naturelle potentielle (essai d'analyse scientifique du paysage). Doc Phytosociol 9–14:9–18
3. Béguin C, Hegg O, Camenisch M, Theurillat JP (2016) Jean-Marie Géhu et ses liens avec la Suisse: la symphytosociologie s.l. et le sauvetage des documents de la Station Internationale de Géobotanique Méditerranéenne et Alpine. Doc Phytosociol (Numéro Spécial) 8:109–119
4. Béguin C, Theurillat JP (1984) Landschaftsökologische Studie in der Region Aletsch (MAB 6) nach einer modifizierten symphytosoziologischen Methode. Verh Ges Ökol 12:149–157
5. Chalumeau A, Bioret F (2013) Méthodologie de cartographie phytosociologique en Europe: approches symphytosociologique et géosymphytosociologique. Synthèse bibliographique. Institut de Géoarchitecture, Université de Bretagne Occidentale, Brest
6. Géhu JM (1977) Le concept de sigmassociation et son application à l'étude du paysage végétal des falaises atlantiques françaises. Vegetatio 34:117–125
7. Henne PD, Bigalke M, Büntgen U, Colombaroli D, Conedera M, Feller U, Frank D, Fuhrer J, Grosjean M, Heiri O, Luterbacher J, Mestrot A, Rigling A, Rössler O, Rohr C, Rutishauser T, Schwikowski M, Stampfli A, Szidat S, Theurillat JP, Weingartner R, Wilcke W, Tinner W (2018) An empirical perspective for understanding climate change impacts in Switzerland. Reg Environ Change 18:205–221
8. Hess E (1933) La Forêt d'Aletsch, monument naturel. Bull Murith Soc Valais Sci Nat 51:78–111
9. Rivas-Martínez S (1976) Sinfitosociologia, una nueva metodologia para el estudio del vegetal. Anal Inst Bot Cavanilles 33:179–188
10. Schwabe A (1991) Perspectives of vegetation complex research and bibliographic review of vegetation complexes in vegetation science and landscape ecology. Excerpta Bot Sect B Sociol 28:223–243
11. Schwabe A (1997) Sigmachorology as a subject of phytosociological research: a review. Phytocoenologia 27:463–507
12. Theurillat JP (1992) Etude et cartographie du paysage végétal (symphytocoenologie) dans la région d'Aletsch (Valais, Suisse). Développement historique et conceptuel de la symphytocénologie, niveaux de perception, méthodologie, applications. Beitr Geobot Landesaufn Schweiz 68:1–384 & annexe
13. Theurillat JP (1992) Abgrenzungen von Vegetationskomplexen bei komplizierten Reliefverhältnissen, gezeigt an Beispielen aus dem Aletschgebiet (Wallis, Schweiz). Ber Reinhold-Tüxen-Ges 4:147–166
14. Tüxen R (1973) Vorschlag zur Aufnahme von Gesellschaftskomplexen in potentiell natürlichen Vegetationsgebieten. Acta Bot Acad Sci Hungar 19:379–384

Ecology

Why Does Mediterranean Vegetation Seem So Diverse?

Elgene O. Box

Abstract Worldwide patterns of biodiversity have been catalogued and mapped at least since the 1990s, based partly on studies from much earlier. For plant biodiversity, construed most easily as species richness, theories on controlling factors have also been summarized and analyzed by numerous authors, with particular emphasis on relationships between plant species richness and ecosystem function in general, especially primary productivity. A less studied aspect of plant diversity and vegetation structure involves the basic ecological plant types and their morphological characters that make up the vegetation – and may also be related to function. In the absence of complete data worldwide, models are required for global coverage. The purpose of this study is to use an existing model of the climatic limits of major plant growth forms to predict potential geographic patterns of richness in basic plant forms and form characters, in order to understand why Mediterranean-type regions, and southern Italy in particular, appear so rich. Highest levels of total richness in potential plant forms and pheno-physiognomic characters are predicted for tropical and some subtropical regions, but richness in coexisting main structural elements is suggested to be high in the more open vegetation of regions with Mediterranean-type climates, perhaps the Mediterranean region proper. With global warming some of this structural richness may be lost.

Keywords Climatic envelopes · Form richness of vegetation · Main structural complex · Mediterranean-type regions · Physical vegetation structure · Plant form characters · Plant life forms · PFORMS model · Southern Italy

1 Introduction

It has been observed by many that biodiversity, in many taxonomic groups, including vascular plants, increases from high latitudes toward the tropics – albeit not monotonically on continental west sides (e.g. [19, 34, 37, 48, 55, 60, 66, 67, 75]).

E. O. Box (✉)
Department of Geography, University of Georgia, Athens, GA 30602, USA

F. Pedrotti and E. O. Box (eds.), *Tools for Landscape-Scale Geobotany and Conservation*, Geobotany Studies, https://doi.org/10.1007/978-3-030-74950-7_15

Worldwide statistics on biodiversity have been catalogued, among others, by Groombridge [31], Mittermeier et al. [49], and Olson and Dinerstein [52], and patterns of plant species richness have been mapped by Barthlott et al. [4]. Theories on the factors that control plant species richness have also been suggested by many authors (e.g. [2, 32, 38, 42, 43, 77], cf [24]) and have been summarized, among others, by Shmida and Wilson [62], Diamond [22], Palmer [53], Pausas and Austin [54], and Evans et al. [25]. Relationships between plant species richness and ecosystem function have been explored by, among others, Brophy et al. [18], Hooper et al. [35] and Midgley [46], with an especially intense discussion regarding the relationship between plant species richness and primary productivity (e.g. [47, 73, 74]). A novel and perhaps quite useful idea that plant species richness may be controlled by light rather than productivity or other factors was provided by Gilbert [28].

Tropical regions appear to have the greatest diversity in species and perhaps other biological aspects (e.g. taxonomic groups, trophic webs, ecological niches). Transitional areas may also be quite diverse, however, to the extent that they contain elements from both adjacent regions. Regions with Mediterranean-type climates are transitional between the temperate and tropical zones, and their floras contain elements of both (e.g. [5, 33, 56]). In North America, a third component, the so-called Madro-Tertiary geoelement, arose in the developing region of Mediterranean-type climate (s.l.) between the Arcto-Tertiary and the Neotropical-Tertiary geofloras, and has been ascribed to mid-continental drying [3]. Hotspots of diversity in the mediterranean regions have been noted by Médail and Quézel [44, 45] and Skeels and Cardillo [63] (cf [21, 59]). The tropical region may also be richest in different basic plant types. An early world map of the potential co-existence of different plant growth forms as major structural elements in vegetation stands, however, suggested high numbers in regions with Mediterranean-type and other warm, subhumid climates [7].

In mediterranean climates, the usual response of vegetation along gradients from wetter to drier conditions starts, at the wetter end, with low sclerophyll forests (*Quercus* species around the Northern Hemisphere), with moderate-sized leaves and deep root systems that permit the trees to remain active during the dry summer (e.g. [71]). Intermediate levels of precipitation involve broad areas of sclerophyll scrub (*maquis, macchia, matorral*, chaparral, kwongan, etc.), also with deep roots but with smaller, hard but still evergreen leaves. But this response to increasing summer dryness may be discontinuous toward the dry end, with the increasingly smaller, harder sclerophylls replaced suddenly, at least partly, by plants with larger, softer but deciduous leaves, where soil water is insufficient for continued plant activity in summer. These more drought-adapted plants are largely dwarf shrubs, with shallow roots and soft summer-deciduous leaves that permit only a short springtime growing season. These different responses to varying degrees of summer dryness lead to the coexistence of great variety in plant structures, foliage types and seasonality patterns in relatively open vegetation stands, in which dominance through greater shade tolerance becomes less relevant.

Major reasons for the historical importance of the Mediterranean region include not only its crossroads location between Europe, Africa and Asia, but also the diversity of its landscapes and their resources. Already the ancient Greeks saw vegetation as the basis for recognizing and delimiting different regions ([70], cf [69]). Diversity, in this sense, means not only species richness but also the variety of different landscapes, plant types and their products. Anyone can appreciate the diversity of sizes, shapes and seasonal patterns of plant in Mediterranean cultural landscapes. But a closer look also reveals a high apparent degree of structural complexity, i.e. diversity, in natural vegetation, based on the large number of plant forms coexisting in the mostly open vegetation not dominated by closed forest overstoreys. Diversity in plants can be viewed in many ways: taxa, morphological attributes, functional "traits", more generalized ecological plant types, etc. Global coverage and scrutiny may only be possible for those aspects that can be generalized for modeling, which means mainly generalized basic plant types, their structural attributes, and the overall structure (complexity) of the vegetation stands that they build. Such basic plant types can be related to whole, recognizable organisms (as opposed, for example, to functional diversity) and are perhaps the only aspects that can be modeled adequately to provide global [terrestrial] coverage. As a result, only richness in plant forms and characters will be treated in this study.

'Diversity', as used herein, will refer to the main physical structure of vegetation stands, as represented by the number and mix of the different larger plant forms that build the vegetation structure, i.e. the 'main structural elements'. The Russian literature called these main structural contributors 'edificators' (Эдификаторы, e.g. [1]), and the word has been used occasionally in Western ecology. More commonly, these have been called the 'main structural elements' (e.g. [27]; [17] p. 112; [51] p. 30; [14] p. xxv; [39] p. 613). Open vegetation stands appear more diverse because more different shapes and sizes of plants can be seen. Consider, for example, what a bird seeking shelter or a temporary perch might see when approaching from the outside.

High diversity of coexisting plant types in Mediterranean vegetation was first suggested by a world map of form diversity predicted from a global-scale climatic envelope model of plant types [7]. The high diversity in Mediterranean areas was emphasized again by a recent assessment of potential diversity in plant form characters, based on an updated version of the world envelope model and a larger number of sites [16]. Both studies found high form diversity in areas that are subhumid to only moderately humid, generally warm (i.e. with little or no frost), somewhat seasonal, and have wet and dry seasons. Such conditions often produce more open vegetation, as in deserts and grasslands. In these situations, and in polar climates, the vegetation is usually dominated by only a few plant forms, although some deserts (e.g. the Sonoran) but may appear surprisingly form-rich (but still not have a large number of forms in total). This leaves scrub vegetation, perhaps especially that of the drier Mediterranean-type climates, as potentially the most form-diverse. But why does the Mediterranean region proper seem more diverse than other regions of Mediterranean-type climate?

The purpose of this study is to use an existing model (PFORMS) of the climatic limits of major plant growth forms to predict potential geographic patterns of richness in basic plant forms and form characters, in order to understand why Mediterranean-type regions, and southern Italy in particular, appear so rich. In keeping with the title of this book, there is a tool involved, namely a climate-based general model of basic plant types and their environmental relationships.

2 Modeling Plant Forms and Characters

The idea of basic ecological plant types is not new and grew out of a need to understand general ecological relationships without having to treat large numbers of different species individually. Plant types defined entirely by structure, such as deciduous broad-leaved trees or stem-succulents, had been called 'growth forms' (*Wuchsformen*) by Drude, based on the *Hauptformen* of Humboldt [36], the 54 types of Grisebach [30], and his own 55 postulated forms [23]. Structural types interpreted as ecological adaptations to environmental conditions were called 'life forms' (*Lebensformen*) by Warming [72] and could be interpreted as basic ecological types, grouping taxa with similar ecological requirements and morphological responses to similar environmental conditions. The first modern system of plant forms was designed for temperate zones [58], but the approach was later extended to the tropics (e.g. [41, 61]). A comprehensive, global classification of plant life forms is given as Appendix A in Mueller-Dombois and Ellenberg [50]. The possibility of form change over the course of a lifespan was incorporated in the concept 'living form' (жизная форма) in the Russian literature (see summaries by [26, 68]). The currently popular term 'plant functional type' (PFT) seems to have originated just before the large IIASA workshop on 'global' vegetation change" (April 1988; see preface in [65], cf [78]). Initially the concept of PFTs developed along quite different pathways (see [12]), including the "function follows form" approach of Box [7] but also approaches that made no reference to form whatsoever (see also review by [64]). Most simply, PFTs have been defined as "functionally similar plant types" ([12], cf [29]) or as "groups of species that use the same resources and respond to the environment in a similar way" [54].

Since plant metabolism, transpiration and other processes operate through the size, texture and seasonality of plant photosynthetic (and some other) surfaces, basic plant types must be described by morphological (pheno-physiognomic) attributes. The most important of these seem to be:

- Permanence – mainly woody vs non-woody, plus plant architecture (branching, etc.)
- Size – relative to other plants
- Seasonality patterns – but not just evergreen vs deciduous
- Texture of the photosynthetic surface: "hard" or "soft" (gas exchange), shade tolerance, etc.

Consequently, the original model focused on these attributes to describe basic plant types [7, 9].

The basic approach consists of three submodels [7, 16]. The first is a set of structurally defined "basic plant types" (growth forms) defined by form characters. The six characters used to represent the basic plant forms are structural type (e.g. tree, shrub, forb), relative plant size, type of leaf or other photosynthetic organ (e.g. phyllode, green woody stem), relative leaf size, seasonality habit, and consistency of the photosynthetic surface (e.g. sclerophyllous, malacophylous, ligneous). A sample of the plant types and their characters, as coded in the model, is given in Table 1.

The second submodel is a set of climatic variables hypothesized to represent the effective climatic environment (and for which data can be obtained readily worldwide). These variables include monthly, seasonal-extreme, and annual temperature and precipitation, plus an annual moisture index (precipitation divided by potential evapotranspiration). Each site also includes absolute minimum temperature, either recorded (if over enough years to be reasonable) or estimated (often better) based on the difference between mean monthly minimum and absolute minimum at nearby sites where it has been recorded reliably (cf [16]).

The plant forms and climatic variables are linked by a third submodel, which consists of a climatic "envelope" for each plant type, defined by apparent tolerance limits for that plant type relative to each of the climatic variables. The limiting values in these envelopes involve commonly accepted values where possible, such as cardinal temperatures (e.g. [40], cf [76]), as well as empirical values obtained from worldwide comparisons. Envelopes for a few plant types prominent in mediterranean vegetation are shown in Table 2, which also defines the climatic variables used. Other details of the model have been described elsewhere [7, 9, 16] and require too much space to repeat here.

The model, now called PFORMS, has been updated and improved from time to time, in particular by 1) by improving the estimator for potential evapotranspiration (cf [8]); 2) by adding absolute minimum temperature as a limiting factor (cf [11]); and 3) by recognizing shade tolerance as an important attribute of leaf structure (cf [13, 17]). Each of these improved model predictions significantly and made the results more reliable globally. The model has also been improved by adding more plant forms and by making the whole set more geographically balanced, especially by including major understorey forms in all regions. There are now 115 such plant forms in the model (see list in Box and Fujiwara [17]). Finally, the climatic envelopes have been re-calibrated repeatedly, based largely on field experience, for better fit to the actual and potential geographic ranges of the various plant types. A sample PFORMS result, for Palermo (Sicily), is shown in Table 3, along with naturally occurring local species corresponding to the forms predicted.

Table 1. Sample of plant forms and characters in the PFORMS model

Form	**Struc**	Size	**Ltype**	Lsize	**Surf**	**Habit**
Tropical Rainforest Trees	T	T	B	L	L	EG
Xeric Raingreen Trees	T	N	B	S	M	RG
Mediterr. Sclerophyll Trees	T	N	B	S	S	EG
Submediterr. Needle Trees	Tp	N	N	N	S	EG
Temperate Needle Trees	Tp	N	N	N	Cs	EG
.						
Palmiform Tuft-Trees	RT	N	F	L	S	EG
.						
Mediterranean Shrubs	S	N	B	S	S	EG
Mediterranean Dwarf Shrubs	S	S	B	V	P	SE
.						
Compact Stem-Succulents	SS	S	-	-	F	EG
.						
Short Bunch Grasses	G	S	G	N	Mr	M
.						
Summergreen Forbs	F	N	B	N	M	SG
.						
Xeric Thallophytes	Th	S	B	S	S	G

Each plant type is a structurally defined growth form, defined by the six characters shown but also interpreted tacitly as an adaptation to environmental conditions, i.e. each form is also a 'life form' *sensu* Drude and others (see main text). For example, Mediterranean Shrubs are shrubs (multiple main stems) of "normal" size, with broad but small leaves which are evergreen but drought adapted sclerophylls, e.g. *Rosmarinus officinalis* or *Phillyrea* spp. Among the 115 plant types included, almost all combinations of the six form characters (most important in boldface) are unique. Abbreviations:

Struc (structural type): T = tree, Tp = pole tree, RT = tuft tree, S = shrub, SS = stem-succulent, G = graminoid, F = forb, Th = cryptogam (not shown: arborescents, semi-shrubs, ferns, vines, epiphytes)
Size (relative plant size): T = tall/large, N = normal (medium), S = small, D = dwarf
Ltype (green organ): B = broad leaf, N = needle, F = frond, L = linear leaf, G = graminoid leaf (not shown: green woody stem)
Lsize (relative leaf size): L = large, N = normal (medium), S = small, V = very small
Surf (green surface): S = sclerophyllous, L = laurophyllous (shade tolerant), C = coriaceous, M = malacophyllous (Mr = reinforced), P = pubescent (mainly soft), F = fleshy (succulent) (not shown: ligneous)
Habit (seasonality): EG = evergreen, SE = semi-evergreen, RG = raingreen, SG = summergreen (not shown: ephemeral)

Table 2. Climatic envelopes for selected Mediterranean-climate plant forms, with climatic variables used and their limiting values

	BT	Tmax	Tmin	Tabmin	MIy	Pmax	Pmin	PmTmx
Mediterr. Sclerophyll Trees								
Maximum	25.0	32.0	15.0	15.0	2.50	****	75	75
Minimum	12.0	20.0	2.0	- 15.0	0.50	60	0	0
Xeric Summergreen Shrubs								
Maximum	28.0	35.0	12.0	12.0	1.00	150	30	150
Minimum	4.5	14.0	- 30.0	- 50.0	0.25	30	0	10
Mediterr. Dwarf Shrubs								
Maximum	28.0	35.0	15.0	15.0	1.40	220	40	50
Minimum	10.0	16.0	3.0	- 10.0	0.15	30	0	0
Summergreen Forbs								
Maximum	25.0	35.0	18.0	2.0	****	****	****	****
Minimum	3.0	9.0	- 50.0	- 80.0	0.20	20	0	0

These sample envelopes are from the PFORMS model used herein. A climatic envelope consists of a set of upper and lower limiting values of the [climatic] envelope variables, of which there are eight:

BT = annual biotemperature (sum of above-freezing monthly means, divided by 12 (°C)
Tmax = mean temperature of the warmest month (°C)
Tmin = mean temperature of the coldest month (°C)
Tabmin = absolute minimum temperature (°C)
MIy = annual moisture index (precipitation/potential evapotranspiration)
Pmax = average precipitation of the wettest month (mm)
Pmin = average precipitation of the driest month (mm)
PmTmx = average precipitation of the warmest month (mm)
Asterisks (****) indicate unspecified (open-ended) and presumably unimportant limits. Occurrence of a plant form is hypothesized to be possible only between the limiting values for all envelope variables. See also Box [8-10, 16, 17].

3 Methodology

Estimation and mapping of plant form and character diversity, in Mediterranean-type regions and worldwide, followed essentially the same steps as in the global study ([16]: if commercially blocked, please write to the author). This methodology consisted of five steps:

1. Develop an appropriate, large, geographically balanced world data-base of site climatic values
2. Predict the potential plant-form occurrences, using the climatic-envelope model, at each climatic site
3. Place the predicted plant forms at each site into the standard layers recognized in vegetation stands

Table 3. Predicted PFORMS model results for Palermo (Sicily)

Palermo, Sicilia 38.12N, 13.35E, 71m Biotemperature = 17.4°C, MIy (P/PET) = 0.55
Prediction: 35 forms, 62% cover by woody forms, 93% cover (spring maximum) by herbaceous forms

T1:	Tall Sclerophyll Trees	MIy	.05	(*Eucalyptus camaldulensis Euc. globulus*)
	Mediterr. Sclerophyll Trees	MIy	.05	*Quercus ilex*, *Ceratonia siliqua*, *Olea europaea*
	Xeric Raingreen Trees	Tmin	.01	(*Jacaranda ovalifolia*; *Acacia* spp.?)
T2:	Dwarf-Needle Small Trees	MIy	.15	*Juniperus phoenicea*, *J. oxycedrus*; *Erica arborea* (*Cupressus semperv.*)
	Raingreen Small Trees	Tmin	.11	(*Albizia jul.*, *Acacia karroo*, *Erythrina viarum*)
	Xeric Stemgreen Arborescents	MIy	.23	*Calycotome spinosa*, *Cytisus scoparius*, *Spartium junceum*, *Genista*
	Raingreen Arborescents	Tmin	.11	(*Euphorbia dendroides*)
	Xeric EG Tuft-Arborescents	Tmin	.11	*Dracaena draco* (*Yucca rostrata*)
	Big-Sclerophyll Arborescents	Tmin	.11	???
	Tall-Columnar Stem-Succulents	Tmin	.06	- - - - -
	Arborescent Stem-Succulents	Tabmin	.42	(*Opuntia tomentosa*, *O. ficus-indica*)
	[Raingreen Vines]	Tmin	.01u	*Vicia* spp. (*Fallopia*?); [*Bougainvillea*?]
	[Wintergreen Bush Epiphytes]	MIy	.28	*Viscum album*, *Loranthus europaeus*
S1:	Needle-Leaved EG Shrubs	MIy	.14u	*Asparagus acutifolius*; *Erica multiflora*
	Xeric Summergreen Shrubs	Tmin	.10u	*Pistacia terebinthus*, *Vitex agnus-castus*
	Mediterranean EG Shrubs	MIy	.09u	*Rosmarinus*, *Phyllirea*, *Cistus*, *Halimium*, *Rhamnus alaternus*, *Lavandula*
	Xeric Rosette-Shrubs	Tmin	.36	*Chamaerops humilis* (*Agave ameri*, *Aloë vera*)
	Hot-Desert Evergreen Shrubs	MIy	.18	*Atriplex halimus*, *Salsola oppositifolia*
	Frutescent Stem-Succulents	MIy	.46	(*Opuntia robusta*, trunkless *O. ficus-indica*)
S2:	Mediterranean Dwarf-Shrubs	Tmin	.25u	*Thymus*, *Thymelaea hirsuta*, *Micromeria graeca*
	Xeric Dwarf Shrubs	MIy	.23u	*Ephedra* spp.
	Xeric Cushion Shrubs	MIy	.32	*Sarcopterium spinosum*; *Capparis spinosa*, *C. orientalis*
	Xeric Semi-Shrubs	MIy	.32u	*Salvia officinalis*, *S. fruticosa*; *Helichrysum italicum*, *Euphorbia bivonae*
	Xylopodial Semi-Shrubs	Tabmin	.26	???
	Compact Stem-Succulents	MIy	.04	[*Ferocactus*, *Echinocactus*]
H:	Short Bunch Grasses	MIy	.33	*Brachypodium phoenicoides*, *B. ramosum*; *Stipa barbata*, *S. sicula*
	Desert Grasses	MIy	.14	*Festuca caerulescens* (*Aristida adscen.* ssp. *coer.*, *Cenchrus ciliaris*)
	Sclerophyllous Grasses	Tmin	.01	*Lygeum spartum* (*Stipa tenacissima*?)
	Xeric Cushion Forbs	MIy	.33u	*Sedum acre*
	Succulent Forbs	Tabmin	.17	*Sedum stellatum*, *Portulaca sicula* (*Pilea microphylla*, *Kalanchoë daigr.*)
	Raingreen Forbs	Tmin	.16	*Ferula*, *Foeniculum*, *Thapsia*, thistles; *Asphodelus* (geophyte)
	Summergreen Forbs	Tabmin	.12	*Urtica pilulifera*, *Acanthus mollis* (semi-ever.)
	Raingreen Ferns	Tabmin	.26	*Selaginella denticulata*; *Cheilanthes*, *Ceterach* (poikilohydrous)
	[Cryptic Stem-Succulents]	MIy	.41u	*Sedum dasyphyllum*
M:	Xeric Thallophytes	MIy	.50	crustose lichens

The plant forms predicted for Palermo (northwestern Sicily) by the PFORMS climatic-envelope model are shown in the left-hand column, organized by traditional stand layers: T1 = canopy trees, T2 = understorey tree layer, S1 = upper shrub layer, S2 = lower shrub layer, H = ground layer, M = moss layer. Vines and epiphytes are placed in the T2 layer. Topo-specific forms not used in the richness analysis are shown in brackets. The closest limiting factor in the envelope design is shown in the second column (e.g. MIy for moisture index, see Table 2 for abbreviations), along with the proximity to closest envelope limit (0–1 scale; u = upper value limiting) (see also Box [8, 9, 16]). With predicted maximum woody cover of 62%, this result can be interpreted as open vegetation dominated by Mediterranean-type shrubs and dwarf shrubs (sclerophyll and summergreen), with scattered smaller trees and arborescents – plus exotics such as *Eucalyptus* and cacti. Note that, in Mediterranean-type climates, summer-deciduous forms may be considered raingreen.

Actual plant taxa occurring in and near Palermo are shown at right, juxtaposed with the plant forms that they represent. Non-native but perhaps well naturalized taxa are shown in parentheses; probably not naturalized but widely planted ornamental taxa (including some taxa from the famous local botanical garden) are shown in brackets. Question marks indicate missing information or, for particular taxa, unknown occurrence or naturalization status. Unpredicted but widely occurring taxa (mainly on special microsites) include *Pinus* spp., plus species that are hard to classify, such as the tall, tussocky cane grass *Ampelodesmus mauritanicus* and the forb *Eryngium campestre*, which though sclerophyllous could also be considered raingreen, since it does generally dry out rather completely in late summer. Some of these unpredicted forms were rejected by only one envelope variable; see discussion of these "near misses" later in the main text. Matches of plant taxa to particular plant forms were made based on field experience (four visits to Palermo) and field notes, supplemented by images and verbal descriptions from numerous websites, including Wikipedia. Species names and verification of native or naturalized status were based largely on the checklist of the flora of Sicily by Raimondo et al.[57] and supplemented by field notes on naturally occurring species.

Palermo is representative of model results and potential accuracy at many if not most sites studied. It is used here as an example because: 1) it is representative; 2) it is where the form-richness results in the latter part of this paper were first presented; and 3) it illustrates a basic premise of this study, namely that subhumid warm-temperate climates, such as the Mediterranean type in southern Italy, have unusually high richness in co-existing plant types, co-occurring in open vegetation stands (cf Figure 1, from Box 1981[8], page 100).

4. Count and synthesize occurrences of forms and the individual form characters
5. Map the resulting patterns.

Each site in the data-base has long-term average (pre-1990) values of the monthly, seasonal, and annual variables, plus absolute minimum temperature (see Table 2). The envelope model PFORMS was run initially on 1866 such sites worldwide, but more sites were added for mapping, deleting some coastal and island sites and adding many more sites inland, for a total of 2889 sites.

Forms predicted by the model are normally displayed by synusiae but were re-assigned (program PFdiv) to the usual standard layers recognized in vegetation stands, i.e. the emergrent, canopy, subcanopy, shrub, low-shrub, herb (ground) and moss-lichen layers, as shown in Table 3. Each plant form has a single character value for relative plant size (e.g. large, medium, small), representing a maximum potential that is not realized everywhere. Placement of forms into stand layers was thus adjusted for the particular site, based on the climatic moisture balance as a suggestion of actual stand height and strata. For example, 'Tall Sclerophyll Trees' (potentially in a super-tree layer) would only occur as shorter trees if a site is too dry. As a result, one can distinguish closed versus open stands and understand stand structure more easily.

The numbers of plant forms predicted to occur, and the numbers of different values for the individual form characters (i.e. structural type, seasonality, etc.), were then counted for each site and summed for world totals (also by program PFdiv). The individual sites richest in plant forms and characters were also determined and listed (by PFdiv). The site counts were mapped, over the world's land areas, by a distance-squared-decay contour algorithm (program PIXMAP) and colored (program PIXPAINT) by assigning color codes from Adobe Photoshop, which did the final mapping (see details in [16]). All programs except Photoshop were written in FORTRAN by the author.

Finally, for the current study, the site and map results were analyzed regionally, focusing especially on richness within the Mediterranean region, in comparison with worldwide patterns and with other regions of Mediterranean-type climate, i.e. California, central Chile, the Cape region of South Africa, and the west-facing coastal areas of southwestern and southeastern Australia. In order to understand the apparent diversity of Mediterranean vegetation, special attention is paid to the groups of main plant forms (main structural elements) that provide the main physical structure of vegetation stands. This 'main structural complex' consists essentially of those forms that receive full sunlight and are not in a shady understorey. For closed forests this means only the canopy trees; for open forests, both the canopy and the tree understorey (T2) layer are involved. For woodlands and scrub, all woody forms are included, plus larger stem succulents, forbs and grasses. In Table 3, for example, because the vegetation predicted for Palermo is somewhat open (estimated cover 62%), the main structural complex includes the T1, T2, S1 and S2 layers, for a total of 23 forms out of the total of 35 predicted to occur. Vines, epiphytes and cryptophytes (s.l.) are always excluded from this calculus.

4 Results Worldwide

Results are presented as tallies of plant-form richness for particular sites and regions (based on 1866 sites) and as world maps (based on 2889 sites). These are first results, so there are some obvious errors, especially the bad spatial interpolation in some parts of the maps, such as northern Scandinavia, southwestern Arabia, the Mackenzie Valley of northwestern Canada, and Chukotka, where too few climate sites were available.

The total number of plant forms predicted for sites worldwide, and interpolated for mapping, is shown in Fig. 1 (from [16]). Total potential form richness appears highest in much of non-Amazon Brazil and other humid warm-temperate to subtropical parts of South America; southern Mexico and northern Central America; southern China; northeastern Australia; small equatorial and larger humid-subtropical parts of Africa; and northernmost New Zealand. These areas involve mostly tropical regions with wet and dry seasons, smaller equatorial areas (mainly in Africa), and more subtropical to warm-temperate humid areas. A bit lower but still relatively high form richness is also predicted for areas with Mediterranean-type climates, including Italy, submediterranean Galicia (northwestern Iberia), central Chile and southern Australia – but not California. High potential plant-form richness is also predicted in southern Britain, but this was precluded by the Pleistocene glaciations.

These worldwide results are summarized by region and richness levels in Table 4. Overall, the two areas potentially richest in plant forms seem to be non-Amazon tropical and subtropical Brazil and subtropical southern China. Africa

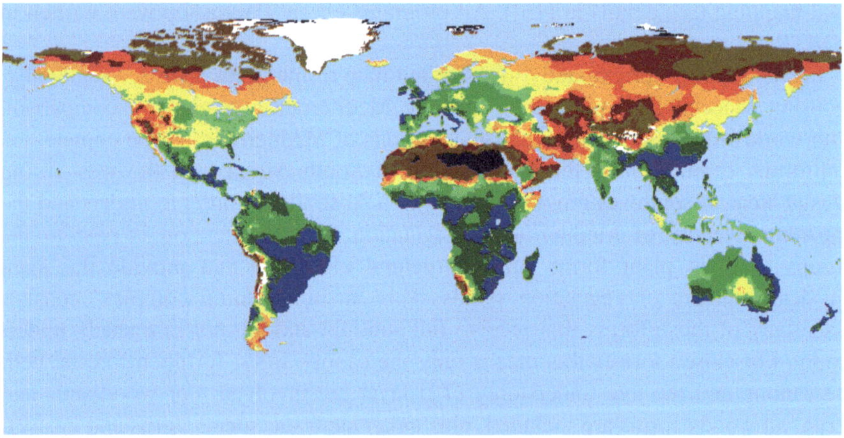

Fig. 1. Total Number of Plant Forms Predicted by the PFORMS Model.
The most forms predicted at any site was 56-61 (see Table 5).
Legend: Dark brown = 1-4; Light brown = 5-7; Dark red = 8-9; Red = 10-11; Light orange = 12-14; Yellow = 15-17; Light green = 18-19; Medium green = 20-24; Dark green = 25-29; Blue = 30-39; Purple = > = 40; Black = no forms predicted

Table 4. Predicted plant-form richness in world regions (excluding Mediterranean)

Region	60s	50s	40s	30s	Richest
Brazil (mainly not Amazon)	2	6	14	24	Rio J. area: Nova Friburgo (61), Resende
Africa (excl. South Africa)	1	10	35	62	Zimbabwe (Chipenge, 61)
- Western (incl. Nigeria)		2	10	13	Ghana (51)
- Central (incl. Uganda)		4	18	25	Uganda (52)
- Eastern (incl. Eth., Sudan)		3	3	15	Nairobi area (upland Kenya, 55)
- S-central (incl. Mozam.)	1	1	2	9	Zimbabwe; Beira (Mozamb., 51)
South Africa		3	6	15	KwaZulu-Natal (e.g. Pietermaritzburg, 57)
Madagascar			2	4	Mandritsara (44), Antsirabe (41)
China (incl. Taiwan)		17	14	11	Yunnan (59-52), Guangdong, Guangxi
Southeast Asia		2	8	8	Hanoi (54); Bhamo (Myanmar, 53)
- Philippines				2	Luzon (39-35)
- Indonesia				6	N Sumatra (38), Denpasar (Bali, 38)
Australia (all eastern)		4	14	9	Queensland: Atherton tableland (54)
Meso-America		1	7	5	
- Mexico		1	5	3	Tuxpan (55), Xilitla (48)
Argentina-Paraguay-Uruguay		1	11	4	Tucuman (51)
South America (other)			5	7	Mérida (Venezuelan Andes, 43)
- Pacific			2	5	S Chile: Temuco (42), Valdivia (40)
South Asia		1	8	23	
- India		1	7	16	Dibrugarh (50)
New Zealand			6	11	Northland (49), Auckland area (47)
Europe (excl. Mediterr.)			5	11	submediterranean: Genova (43), Oviedo
Southeastern USA			2	11	
- Florida			2	6	south Florida (43-40)
Turkey			1	2	Trabzon (42)
Japan				15	Pacific: Shikoku (39), Kii peninsula (38)
Arabia				1	Hejaz mountains (30)
Pacific islands				1	Fiji (32)

Shown by region are the numbers of modeling sites (out of 1866) at which the total number of plant forms predicted fell into the respective deciles. For example, Brazil had 2 sites with 60 or more forms predicted, 6 with 50–59 forms, 14 with 40–49 forms, and 24 with 30–39 forms predicted. The regions richest in plant-form potential seem to be subtropical Brazil and subtropical China.

as a whole is also relatively rich in potential forms, with Australia and especially tropical Asia and Meso-America further down the list. The everwet equatorial regions of South America and the East Indies are predicted to be less form-rich because both light-demanding forms and forms adapted to shorter growing seasons are largely precluded in the shady environments of tropical rainforests, which are strongly dominated by evergreen shade-tolerant trees and understorey forms. Remaining areas of Europe, most of the USA, and much of temperate East Asia are predicted to be still less form-rich, dominated largely by summergreen forms. Form richness is predicted to be lowest in the driest and coldest areas but also low in some cool-temperate humid areas (e.g. Canada, Manchuria).

Predicted form richness in different stand layers was treated in Box [16], with overstorey form richness predicted to be highest in warm subhumid to quite dry areas with no forest canopy. Overstorey form richness was also predicted to be high in (sub)tropical wet-dry areas but lower in areas of closed forest, especially tropical rainforests and Northern Hemisphere forests, where summergreen deciduous trees or boreal conifers dominate strongly. Shrub-layer form richness was generally predicted to be higher, throughout both the everwet and wet-dry tropics, and in both humid and summer-dry warm-temperate areas. Less diverse shrub layers are suggested for quite dry areas, and no shrub layer at all was predicted [correctly] for the vast permafrost areas of Siberia, where *Larix* pole trees have only lichen-moss ground layers between, with occasional dwarf shrubs. Species richness is generally highest in herb layers, but forest areas herein were generally predicted to have relatively few forms, with significantly more only in the tropics and subtropical southern China, and in relatively steep transitions to arid areas. Large areas of more or less uniform understorey form richness, at 8–12 forms, were predicted over much of temperate to subboreal North America, eastern Europe and northern China. Similarly large areas of 15–20 predicted plant understorey forms are predicted for tropical wet-dry and warm-temperate to subtropical humid areas. The areas of greatest understorey form richness are predicted to occur in humid subtropical to warm-temperate areas of southern China, and to a lesser extent in southeastern Africa and eastern to southeastern Brazil.

Geographic patterns are probably more important than site results, since individual sites may involve artifacts such as predicted occurrences just outside or inside envelope boundaries. Predicted form richness at the richest sites worldwide is compared with that of the world's regions of Mediterranean-type climate in Table 5. Three sites, two in Brazil and one in Zimbabwe, had 60 or 61 plant forms predicted (out of 115). Southern China, however, had the most sites with at least 50 forms predicted. Almost all these locations are tropical or subtropical, but not equatorial. Indeed the map (Fig. 1) also suggests a general reduction of form richness in equatorial areas, where the lack of a dry season precludes opportunities for plant forms adapted to shorter growing seasons. On the other hand, the predicted form-rich sites in Table 5 all occur in climates with driest-month precipitation below 30 mm (except one site in southern Brazil). The actual climate types involved are mostly tropical wet-dry in Brazil, Mesoamerica and southern Africa; and humid subtropical to warm-temperate, with colder winters, in southern China

Table 5. Predicted plant-form richness at the richest sites versus richer sites in Mediterranean climates

Richest		Lat	Elev	Tmax	Tmin	Py	Pmin	MIy	Forms	Climate
Chipinge (SE Zimb.)	Zimbabwe	-20.2	1131	20.9	14.6	1119	19	1.13	**61**	Tropical wet-dry
Nova Friburgo (Rio J)	Brazil	-22.3	850	21.0	13.2	1506	20	1.61	**61**	Humid subtropical
Resende (Rio de J.)	Brazil	-22.5	404	23.7	16.7	1591	23	1.42	60	Humid subtropical
Rio Blanco (Veracruz)	Mexico	18.8	1260	22.1	16.5	1877	25	1.78	60	Tropical wet-dry
Hlabisa (KwaZulu)	S Africa	-28.1	512	23.1	16.4	1073	16	1.00	60	Subtropical wet-dry
Jinghong (Yunnan)	China	21.2	552	25.7	15.6	1196	11	1.00	59	Humid subtropical
Hong Kong Obsv.	China	22.3	32	28.8	16.1	2212	24	1.75	58	Humid subtropical
Pietermaritzburg (KZ)	S Africa	-29.6	684	22.1	13.3	906	11	**0.91**	57	Subtropical wet-dry
Zhanjiang (Guangd.)	China	21.2	26	28.9	15.8	1495	21	1.18	57	Humid subtropical
Nanning (Guangxi)	China	22.8	80	28.5	13.0	1354	27	1.14	56	Humid subtropical
Hekou (Yunnan)	China	22.5	137	27.5	15.3	1784	25	1.45	56	Humid subtropical
Cachoeira do C. (MG)	Brazil	-20.3	1105	20.6	14.8	1538	13	1.56	56	Tropical wet-dry upland
Jonkershoek (W Cape)	S Africa	-33.8	274	21.1	11.4	1068	29	1.23	56	Submediterranean
Camboriu (Sta.Clara)	Brazil	-27.0	10	23.4	14.8	1538	63	1.49	55	Humid subtropical
Tuxpan de RC (Ver.)	Mexico	20.9	28	27.2	19.1	1314	16	1.00	55	Tropical wet-dry
Xiamen (Fujian)	China	24.5	139	28.6	12.9	1180	28	1.03	55	Humid subtropical
Guangzhou (+similar)	China	23.1	18	28.5	13.5	1672	30	1.40	55	Humid subtropical
Dagoretti (Nairobi)	Kenya	-1.3	1798	18.8	14.8	1066	16	1.18	55	Bimodal wet-dry montane
Mediterranean										
Montpellier	France	43.6	80	23.1	**6.0**	732	**27**	1.00	39	Submediterranean
Napoli	Italy	41.0	88	24.2	**7.1**	928	**23**	**1.17**	**41**	Mediterranean
Valencia	Spain	39.5	13	25.0	11.0	468	9	0.51	33	Mediterranean
Lisboa	Portugal	38.7	95	22.0	10.6	703	4	0.83	27	Atlantic-Mediterranean
Messina (E Sicily)	Italy	38.2	59	26.2	11.8	836	**15**	**0.86**	**53**	Mediterranean
Palermo (NW Sicily)	Italy	38.2	71	25.2	10.1	512	2	0.55	35	Mediterranean
Pasadena (LA)	California	34.1	263	23.4	12.1	518	1	0.56	25	Mediterranean
Perth (SW Austr.)	Australia	-31.9	19	24.1	13.2	865	8	0.88	**40**	Mediterranean
Santiago	Chile	-33.4	520	**20.2**	8.0	360	2	0.50	24	Cool-Mediterranean
Cape Town (W Cape)	S Africa	-33.9	12	**21.1**	12.4	612	**15**	0.69	33	Cool-Mediterranean
Albany (SW Austr.)	Australia	-34.9	71	**19.8**	11.7	806	**22**	0.99	**49**	Cool-submediterranean
Temuco	Chile	-38.7	110	**16.1**	**7.5**	1153	**40**	**1.99**	**42**	Cool-submediterranean

Predicted plant-form richness is shown for the richest sites in the world data-base, ordered by decreasing richness; and (bottom) for richer representative sites in regions with Mediterranean-type climate (ordered by latitude). Boldface type indicates relatively high or low values in each column. None of the richest sites has a Mediterranean-type climate (see last column), although Jonkershoek is perhaps submediterranean. Among the Mediterranean-type sites, less summer-dry sites are often richer in forms, including some submediterranean sites. Abbreviations: Lat = latitude; Elev = elevation (m); Tmax, Tmin = mean temperature of warmest/coldest month respectively; Py = annual precipitation, Pmin = precipitation of driest month; MIy = annual Py/potential evapotranspiration, Forms = number of plant forms predicted by the PFORMS model

(plus Hlabisa in southeastern South Africa). Only one location (Pietermaritzburg) has annual precipitation below 1000 mm, which is less than annual PET. The only Mediterranean-type sites in the list are submediterranean Jonkershoek (55 forms, Western Cape, South Africa) and truly Mediterranean Messina (53 forms, on the strait between peninsular Italy and Sicily).

What do these potentially richest areas have in common? Initially, it seems that they all:

- are relatively **warm** but with a cooler season, i.e. tropical or subtropical but not equatorial;
- have mild winter (coolest month < 19 °C, permitting occurrence of some temperate forms);
- are in generally moist climates;
- have **seasonal** precipitation, but with at least 10 mm (on average) even in the driest month; and.
- may be at a range of elevations (not > 1800 m).

Overall, this suggests relatively long growing seasons, with evergreen or seasonally green closed-canopy forests (sometimes with many epiphytes). Elevation (and higher latitude) result in cooler but still mild winters up to a point, and this may permit some temperate forms to occur without excluding frost-sensitive tropical forms. These very richest sites are not an exact match to the mapped geographic patterns, which show wide areas of high predicted form richness in more wet-dry to generally subhumid warm areas.

Predicted numbers of structural types, leaf types, seasonality types and surface types among the plant forms predicted are shown as separate world maps in Fig. 2 (each with a different color scheme!). The most structural types (upper left of Fig. 2) are predicted for both everwet and wet-dry tropical areas, as well as summer-dry areas, before dropping off fairly abruptly in the transition to the temperate zone. Again, California seems to be the outlier among the world's mediterranean areas. Fewest structural types are predicted in non-wooded arid and polar areas, but numbers are also low in forest areas throughout the higher-latitude continental areas of the Northern Hemisphere. The only broad areas of predicted intermediate richness in structural types are in areas of temperate deciduous forest.

The most types of leaves or other photosynthetic organs (upper right of Fig. 2) are predicted in tropical wet-dry areas, areas of Mediterranean-type climate (except California), and some more humid warm-temperate areas, including eastern Australia, and southeastern Africa and South America. Again, erroneous high richness is predicted in southern Britain. Richness in photosynthetic organs is predicted to be low in Northern high latitudes and drier continental areas, and lowest in arid and polar areas. Questionable results include the lower values in the southeastern Amazon (yellow) and northwestern Australia (light green).

The potential number of different seasonal foliation patterns represented by the plant forms predicted is shown in the lower left of Fig. 2. High-latitude Northern areas involve mostly just two potential patterns: evergreenness among conifers and tundra dwarf shrubs, and summergreen deciduousness among the broad-leaved trees, larger tundra shrubs (e.g. *Salix*), and graminoids. The highest predicted number of seasonality patterns, unlike the top-row panels, is in drier rather than wetter tropical and subtropical areas, including areas of Mediterranean-type climate (this time including California) but also more humid warm-temperate to subtropical climates. Fewer seasonal habits are suggested in the equatorial and wetter wet-dry areas, where most forms are evergreen or raingreen.

Different photosynthetic surface types (lower right of Fig. 2) include sclerophylls, shade-tolerant laurophylls and soft (deciduous) malacophylls, as well as the "harder" surfaces of succulents, phyllodes and green woody stems. Fewer types are predicted throughout almost all of the Northern Hemisphere, except for the Mediterranean area and humid subtropical to tropical (wet-dry) southern Asia. The only Southern Hemisphere areas predicted to have few surface types are in extreme southern South America and the cool-subtropical Atacama and Namib desert areas, plus some of extreme southern Australia. The most co-occurring surface types are predicted for tropical wet-dry, Mediterranean-type, and humid warm-temperate to subtropical areas. Fewer surface types are predicted in temperate grasslands but

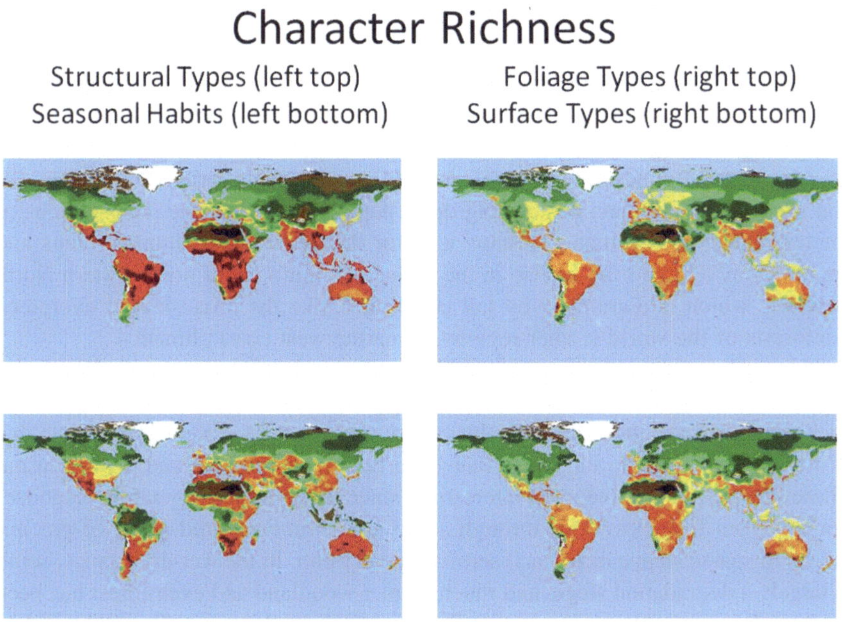

Fig. 2. Number of Plant Form Characters represented among the Plant Forms Predicted The four panels represent:
a) plant structural types (upper left), such as trees, shrubs, stem-succulents, graminoids;
b) foliage types (upper right), including broad leaves, needles, and leafless (e.g. succulents);
c) seasonal foliation habits (lower right), including evergreen, summergreen, and ephemeral; and
d) types of the photosynthetic surfaces (lower right), such as sclerophyllous, coriaceous, soft (malacophyllous), and woody (photosynthetic stems).
Structural Types (upper left): maximum of 13 at any site
Dark red = 12-13; Red = 11; Orange = 10; Light orange = 9; Yellow = 8; Light green = 7; Medium green = 6;
Dark green = 5; Light brown = 4; Dark brown = 1-3; Black = no forms predicted;
Foliage Types (upper right): maximum of 6 at any site
Red = 6; Orange = 5; Yellow = 4; Green = 3; Dark green = 2;
Light brown = 1; Black = no forms predicted Intervening colors result from interpolation;
Seasonal Habits (lower left): maximum of 8 at any site
Dark red = 8; Red = 7; Orange = 6; Yellow = 5; Green = 4;
Dark green = 3; Light brown = 2; Dark brown = 1; Intervening colors result from interpolation Black = no forms predicted;
Surface Types (lower right): maximum of 7 at any site
Dark red = 7; Red = 6; Orange = 5; Yellow = 4; Green = 3;
Light brown = 2 Dark brown = 1; Intervening colors result from interpolation Black = no forms predicted

only slightly lower than in Northern forest areas, both of which are dominated by malacophylls.

Two apparent anomalies are of particular interest, since their results, which occur on multiple maps, were not expected within the context of generally understandable patterns. These two areas are California (mediterranean) and submediterranean Atlantic Europe, from (southern) Britain to Galicia. Although present-day representation is low by climatically potential evergreen forms in Atlantic Europe, one can find palms, numerous conifers, other evergreens, and even eucalypts in local botanical gardens. California is different in that it covers a moderately large area on a large continent, and is thus more "subcontinental", with perhaps more frequent and severe droughts (and desiccating downslope winds), than the other regions of Mediterranean-type climate. Perhaps it is for this reason that summergreens and conifers play a large role, as also in the temperate rainforest of northwestern North America, which is dominated by tall conifers, unlike the broad-leaved evergreen rainforests of the world's other regions of marine west-coast climate.

If one takes a zonal view of world bioclimates and vegetation, *sensu* Walter for example (cf [15]), one can sort the zones into those that are everwet (no dry season, such as equatorial); those that are always dry (no wet season, i.e. deserts); one that is always cold (polar); and those that have both a wet season and a dry season. These last are perhaps the most interesting, since these permit vegetation responses ranging from forest (wettest) through scrub and shrubland (mid-range of precipitation) to sparse steppe or (dwarf) scrub at the dry end. In the wet-dry tropics, scrub is largely a degradation stage, and much former woodland and even forest has been converted permanently to savanna. This leaves only the generally subhumid to slightly humid Mediterranean-climate regions with large areas of climatically more or less stable, open *maquis*, chaparral, kwongan or other scrub vegetation. Scrub is by definition a mix of woody plant forms, in which no single form is dominant and form diversity should thus be relatively high. In addition, the transitional geographic positions of the various Mediterranean-type regions permit potential mixing of tropical, subtropical and warm-temperate evergreen elements with temperate summergreen and some tropical raingreen elements.

5 Why Does the Mediterranean Seem So Rich?

The maps of predicted form and character richness (Figs. 1 and 2), the overview of predicted richness in regions (Table 4) and the brief overview of sites worldwide with greatest predicted form richness (Table 4) all suggested the greatest numbers of potential plant forms in tropical (but not equatorial) to subtropical regions with significant precipitation seasonality. One major advantage leading to high form (and species) richness in the frost-free tropics involves the appearance of numerous epiphyte and other frost-sensitive forms, even if some of these "additional" forms are only in understoreys and do not contribute greatly to overall vegetation structure. The areas suggested as richest in plant forms generally have the climatic potential for closed forest, even if only seasonally closed (i.e. deciduous). These results, however, also suggested fairly high potential form richness in at least some regions with Mediterranean-type climate.

Table 6. Predicted plant-form richness for representative sites in Southern Italy, plus Montpellier

	Elev	Py	**Pmin**	P/PET	**Total Forms**	Near Misses	**Main Mix**	**Main/ Total**(%)
Montpellier	5m	772	24	**1.01**	**39**	**31**	15	48
Roma (città)	45m	830	16	**1.03**	39	29	23	59
Napoli	88m	928	**23**	**1.17**	**41**	28	23	56
Salerno	40m	1333	14	**1.37**	**46**	28	12	26
Reggio-Calabria	15m	568	6	0.58	27	17	16	60
Sassari	224m	600	8	0.74	31	23	21	**68**
Cagliari	4m	432	3	0.50	25	16	14	56
Messina	59m	836	**15**	0.86	**53**	25	**34**	64
Palermo	71m	512	2	0.55	**35**	18	23	**66**
Floresta	**1250m**	1467	23	**2.72**	30	21	11	37
Siracusa	23m	529	3	0.56	27	18	16	60

Numbers of plant forms predicted to occur, as well as near misses, are shown for sites from central to southern Italy, plus submediterranean Montpellier for comparison. The "main mix" (right) is the number of woody and other larger forms predicted, i.e. the 'main structural elements' (see main text), those that provide the main structure of the vegetation. Their fraction of the total number of forms is shown in the last column. Highest values in most columns are shown in boldface. Highest form richness generally occurs at less dry sites, but only Salerno and montane Floresta are truly wet for Mediterranean conditions and should have closed forest overstoreys. For the other sites, all with potentially more open vegetation, more than half of the forms predicted would contribute significantly to the overall stand structure. This is what makes the vegetation of Mediterranean-climate regions so structurally diverse. Messina, with the most total and main-mix forms predicted, is subhumid (P/PET < 1.0) but does have relatively high summer humidity and precipitation. With 53 forms, Messina is the only Mediterranean site that approaches the richness level of the world's form-richest sites (see Table 5). The large number of near misses at Montpellier reflects its submediterranean position, where many forms may be precluded only barely by one aspect of winter temperature. Abbreviations: Elev = elevation; Py, P = annual precipitation (mm); Pmin = precipitation of driest summer month (mm); PET = annual potential evapotranspiration (mm).

In order to understand the apparent structural complexity (form richness) of mediterraanean vegetation, we must look at what I will call the 'main structural complex' at each location, i.e. those larger forms (main structural elements, *vide supra*), that receive full sunlight and are not submerged in shady understoreys. This main structural complex is what builds the physical structure of a vegetation stand and gives mediterranean vegetation its characteristic appearance of diversity. Even non-forest vegetation may develop a fairly dense overstorey, as in some chaparral, *maquis*, *matorral*, *macchia*, etc.; but with disturbance, the more diverse mix of potential plant forms is revealed. This can perhaps be understood best by looking at the Palermo example (Table 3). For Palermo, 35 plant forms were predicted to be possible in the climatic-potential natural vegetation. Of these 35 (but omitting the vines, epiphytes and cryptic succulents), 23 of the remaining 32 forms can be considered part of the main structure (i.e. layers T1, T2, S1 and S2). Even without the three predicted forms that do not occur at Palermo (no species examples shown in the righthand column), this is a high percentage. This is why Mediterranean vegetation seems so diverse.

Corresponding values for total forms predicted and those in the main structure ("main mix") are shown in Table 6 for various sites in southern Italy. Here also one sees high percentages for the main-mix forms, except at the much wetter sites, Salerno and Floresta, which potentially should have closed forest. The results in Table 6 also suggest that potential form richness is generally greater if sites are **sub-humid** (i.e. potentially have open vegetation) and have more **summer rainfall** (as represented by Pmin). Up to the point of potential forest canopy closure (i.e. P/PET around 1.0), there is also a less clear suggestion of increasing total form richness (but not a higher main-mix percentage). Overall annual water balance (annual P and P/PET) appears to be less important. At global scale there was some suggestion that diversity in general decreases along gradients of increasing stress. In Mediterranean climates, the main stress factor is summer dryness, except on the poleward margins where significant winter frost reduces vitality and growing seasons.

Another aspect of the model results, shown also in Table 6, involves the number of "near misses", i.e. those plant forms that were not predicted to occur because they were rejected by only one of the screening variables. Many of these near-miss forms, especially those rejected by a moisture variable, may nevertheless occur on special microsites in natural situations but especially in cultural landscapes. The near misses at Palermo included the following plant forms:

- Limited by winter cold (Tmin):	Bottle Trees, Bottle Palms (with water-storing trunks)
	Leaf-Succulent Shrubs (e.g. *Crassula*)
- Limited by insufficient cold (Tmin):	Cold-Suffrutescent Shrubs (e.g. *Artemisia*)
- Limited by overall dryness (P/PET):	Tropical Evergreen Microphyll Trees (e.g. Meliaceae)
	Tropical Xeric Needle Trees (e.g. *Casuarina*)
	Sub-Mediterranean Needle Trees (e.g. *Cedrus*)
	Temperate Needle Trees (pines, e.g. *Pinus sylvestris*)
	Laurophyll Small Trees (e.g. *Laurus nobilis*)
	Temperate Evergreen Small Trees (e.g. *Ilex aquifolium*)
	Palmiform Tuft-Trees and Treelets (i.e. palms)
	Mesic Stemgreen Arborescents (e.g. *Spartium junceum*)
	Tropical Evergreen Shrubs (perhaps facultatively deciduous)
	Mesic Semi-Shrubs
- Limited by lack of aridity (P/PET):	Ephemeral Desert Forbs
	Mat-Forming Thallophytes (i.e. mosses)
- Limited by summer dryness (Pmin):	Summergreen Arborescents

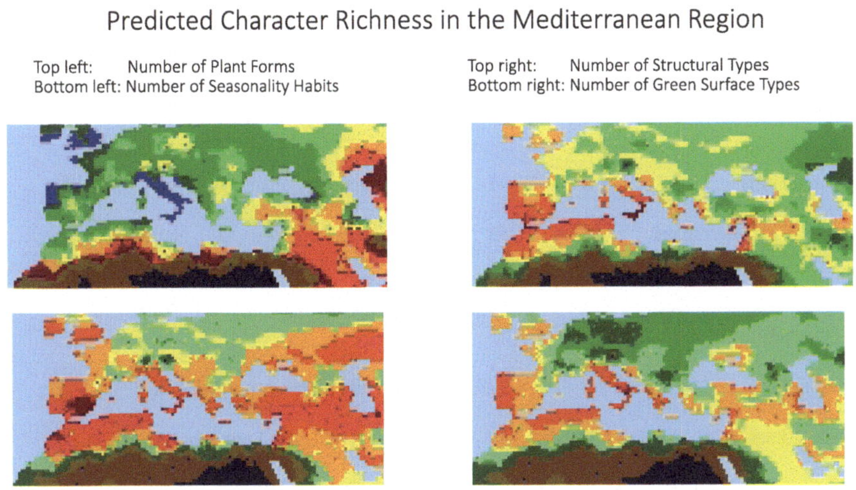

Fig. 3. Mediterranean Sections of the Form and Character Maps of Figures 1 and 2. The four panels represent: a) total numbers of plant forms predicted (upper left); b) number of plant structural types (upper right); c). seasonal foliation habits (lower left); and d) textures of the photosynthetic surfaces (lower right). See Figures 1 and 2 for color legends.

The forms limited by low winter temperatures can often be found near buildings and are certainly represented in the Palermo Botanical Garden, except for cold-winter *Artemisia*. Most of the forms limited by overall (annual) or summer dryness also occur on more favorable microsites, especially the palms, needle trees, and leafless stemgreen *Spartium junceum*. Only the desert ephemerals, which generally require arid climates with only short wet periods, are probably not represented in the Palermo Botanical Garden. For Palermo there were 18 such near misses predicted, and the numbers are even higher for most of the other sites in Table 6. Indeed, these numbers are generally a bit higher than for sites in most world regions with other climate types.

Structural richness results not only from the number of coexisting plant forms at a place but also from the number of different plant structural types, types of photosynthetic organ, and seasonality patterns found among the different plant forms possible. World patterns for such 'character richness' were shown in Fig. 2, but it is difficult to see details of individual regions on such world maps. Albeit with coarse resolution, the Mediterranean portions of the maps in Figs. 1 and 2 are reproduced in Fig. 3. From these regional sections it is possible to see more clearly the suggestion of especially high plant character richness in central and southern Italy (all four maps in Fig. 3) and some other regions for particular characters.

The total number of plant structural types (trees, shrubs, forbs, etc.) predicted to be possible in the potential vegetation is especially high in southern Italy and still relatively high in the southern Iberian Peninsula, the near-coastal Levant, and some

Table 7. Predicted plant-form richness in Mediterranean regions of the World

Region	n	NF	Richest Sites
California (incl. Baja)	5	18-25	Pasadena (25), San Fernando (24); Ensenada (Mexico, 24)
Central Chile	2	24-41	Temuco (42), Santiago (24)
Western Mediterranean (Europe)	22	21-53	Messina (53); Porto (43), Napoli (41), Montpellier (39), Palermo (35), Valencia (33)
Eastern Mediterranean (Europe)	8	21-44	Split (Croatian coast, 44); Tirane (36)
Maghreb	5	18-37	Al-Kala (northeastern Algeria, 37)
Middle East	9	15-40	Rasht (Iran, 40); Lenkoran (Azerbaijan, 29)
Cape region (SW Africa)	1	33	Cape Town (33)
Southwestern Australia	7	26-49	Albany (49), Perth (40)

The range of the total numbers of plant forms predicted at representative sites in each region are shown in the NF column. For each region, the site with the most forms predicted is listed first, followed by other form-rich sites. For example, Messina (eastern Sicily) was predicted to be the richest site in the Western Mediterranean, with 53 plant forms predicted, followed by Porto with 43 forms. With 53 forms, Messina is the only truly Mediterranean site in the world data-base (n=2889) that approaches the potential richness of the richest sites in Table 5. The Mediterranean region proper appears to be potentially the richest of the world's five regions of Mediterranean-type climate. California seems to have fewer plant forms than the other regions, although the core mediterranean region in central Chile is also low. (Temuco and Albany are both more humid submediterranean.)

Abbreviations: n = the total number of sites in each region; NF = number of forms predicted (range).

coastal areas around the Aegean Sea. These are all coastal or near-coastal areas that receive somewhat more precipitation and humidity in the otherwise quite dry Mediterranean summer. Predicted richness in seasonality habits is generally highest in the same areas, although here the largest area of highest predicted richness seems to have shifted to the [drier] southeastern quadrant of Spain. For richness in types of leaf or other photosynthetic surfaces (sclerophyll, malacophyll, etc.), southern Italy is again predicted to have the highest levels, followed this time, though, by the windward, more humid Atlantic side of the Iberian Peninsula. Taken together, these patterns suggest a special "Garden of Eden" position for southern Italy, as was suggested to this author by the very first predictive results in 1981 (cf [7], p. 100; or see [16], Fig. 1).

The Mediterranean region is compared with the world's other regions of Mediterranean-type climate in Table 7, which shows the range of total plant forms predicted at a representative set of sites in each region. These results (and the maps) suggest high form richness in the Mediterranean region proper, perhaps especially in the western Mediterranean but only slightly less in the eastern Mediterranean. The predicted form richness then decreases more significantly in the drier Levant and on into the more continental inland Middle East (e.g. Iraq, Iran) – with the notable exception of Rasht (40 forms), in the relatively wet climatic strip just south of the Caspian Sea. The Mediterranean region may be more form-rich than the other four comparable regions because it is not restricted to a narrow ocean-coastal strip on a large continent, which may result in drier summers. The Mediterranean Sea is in just the right latitude to provide precipitation throughout much of its inland extent. The Levant and inland Middle East are drier, and apparently thus less diverse, because they are farther from the winds that bring moisture from the sea. North Africa is less form-rich because it is nearer the core zone of subtropical high pressure and is simply more arid overall.

The Australian and South African mediterranean regions are especially species-rich and are predicted here also to have high form richness, but both are in reality dominated by shrub forms, with few trees and almost no deciduous elements. California (including the northern part of Baja California) is predicted to be less form-rich than the other regions, perhaps because it is generally drier in summer (see Table 5). This is curious since California is usually regarded as one of the world's main hotspots of species richness. California is the only one of the world's mediterranean regions that extends far inland without a large source of water (such as the Mediterranean Sea). This greater continentality means colder winter temperatures (at least inland) as well as less and more irregular precipitation. This may explain partly the greater importance of summergreen forms and conifers in western North America. The dry Middle East is somewhat comparable in this respect.

Finally, we might also look at submediterranean regions, which already have mild enough winter temperatures to permit sclerophyllous and other evergreen broad-leaved plants but are not yet so summer-dry as to preclude summergreens. This suggests high potential form richness – but not necessarily. For the main vegetation structure (cf Table 6), the summer-drier true mediterranean climate produces more open vegetation with more forms contributing to stand structure, even if some water-demanding summergreen forms are lost. And even though winters are relatively mild in the submediterranean zone, they do not permit the truly subtropical or tropical forms, such as larger stem-succulents, most palms, hot-desert shrubs, many tropical deciduous (raingreen) plants, and most xeric arborescents. So, the true mediterranean zone, which does pick up these forms, may indeed have greater potential form richness than the submediterranean.

6 Conclusion

To begin with, these are only results from a necessarily imperfect model – some will say playing with numbers and abstractions. At best such models can do no more than suggest potentials, which may or may not be worth looking into further. At least, though, the model herein does provide greater resolution of vegetation into constituent plant types than do other global models, and it is based as far as possible on accepted climatic limits, such as cardinal temperatures. This model is based on the general logic that form follows function (and *vice versa*), i.e. that there are general reasons why drier climates tend to have smaller plants with smaller, often harder leaves: i.e. that there is a strong relationship between plant pheno-physiognomy and plant function.

One might normally expect highest values of both plant-form and character richness in the tropics, lowest values in polar and arid areas, perhaps low values also in boreal areas, and perhaps quite variable results in some oceanic areas (cf [20]). Taken together, the results shown herein suggest:

- greater potential richness (at least for plant forms) in areas with mild or no real winter;
- greatest potential richness in climates with both a wet season and a dry season; and
- depressed potential richness over large high-latitude land masses and in the everwet tropics.

The geographic regions predicted to be richest in different plant forms seem to have two features in common: they are mostly warm, with marginal frost (i.e. not equatorial), and have some degree of dryness, either a wet-dry seasonality pattern or an overall subhumid character (see fuller list in [16], Fig. 3).

The results also suggest the potential for many co-existing forms in the main vegetation structure, and fairly high potential richness overall, in climates that are:

- warm-temperate to subtropical (with little or no frost), permitting evergreen-summergreen overlap;
- subhumid to moderately humid, permitting somewhat open vegetation, not closed forests;
- seasonal (dry and wet seasons), fostering the co-occurrence of evergreen and raingreen forms; and
- not too arid, where richness decreases.

Richness also seems enhanced in such situations by some amount of precipitation in the dry season. This description fits Mediterranean-type climates well, and the result is what gives the vegetation of Mediterranean-type climates its characteristic appearance of diversity.

Along the main gradients of increasingly favorable climatic conditions (increasing simultaneous warmth and wetness), there seems to be a general pattern in the model results, as perhaps also in nature, of increasing potential richness in plant forms. This suggests that new forms appear before old forms drop out, and that the greatest potential richness should thus appear in transitional zones. As the climate permits new forms, though, these usually appear first as ornamentals or in other cultivations before they appear in natural vegetation, for example the widely recognized 'laurophyllization' of the vegetation south of the Alps, by which evergreen broad-leaved garden species have escaped and become naturalized. This also contributes to the overall impression of high diversity in Mediterranean landscapes, which are of course largely cultural landscapes now.

Along broad-scale geographic gradients, increasing stress may lead to more ecological niches for plant species, with less chance for dominance by individual species. As the stress becomes harsher, though, species diversity usually decreases. In mediterranean climates, the most significant stress factor is often summer dryness. But before this point is reached, do the increases in stress, perhaps species richness, and species niches lead also to more plant forms? The model results suggest perhaps the opposite. Is there some "intermediate stress hypothesis" for plant forms, and do these results reflect patterns in floristic richness or other patterns that seem worth pursuing?

Apparent "hotspots" of plant diversity have been mapped by Groombridge et al. ([31], Fig. 15.1) and by [4]. Hotspots in general have been catalogued in more detail by Groombridge et al. ([31], Tables 15.1, 15.6), including the 241 Centres of Plant Diversity identified at that time by the Plant Conservation Programme of the International Union for the Conservation of Nature. The greatest concentrations of overall diversity hotspots are in Southeast Asia, Meso-America (and south to Ecuador), and to a slightly lesser extent from the eastern Mediterranean eastward into Iran, and in various parts of central to southern Africa. The table of endemism hotspots for higher plants (Groombridge et al. [31], Table 15.1) lists 18 hotspots, most of which are also in the tropics, including everwet regions (e.g. western Amazonia, northern Borneo, the Colombian Chocó, and peninsular Malaysia). Four of the world's five regions of Mediterranean-type climate are included, but not the Mediterranean region itself. Some areas with higher levels of predicted form richness on multiple maps seem to coincide with both these compilations, including subtropical to tropical southern China and Southeast Asia, areas of Mediterranean-type climate (except California), central to southern Mexico, parts of New Zealand, and various parts of Australia. Southern China and Meso-America have long been suggested as areas of great diversity and speciation (e.g. Barthlott et al. [4]). High levels of taxonomic radiation also occurred in Australia and New Zealand, but these have been attributed partly to small taxonomic stocks radiating on long isolated but climatically diversifying land masses.

Possession of a seasonal but overall humid climate has already been suggested as a common factor among the sites with the very most total plant forms predicted. A relatively mild winter (or none at all) seems important, but the existence of drier periods during the year seems even more important for the development of high potential plant form and character richness. Some drier areas with greater richness may be seen as "marginal" climates, i.e. transitions between wet and dry, which immediately suggests these as natural cradles for adaptation and evolution.

With global warming, the Mediterranean region is expected to become generally drier, as the subtropical high-pressure belt moves northward into southern Europe. A similar fate is expected in the world's other regions of Mediterranean-type climate, and California in particular has already seen historic drought in recent years. Since mediterranean vegetation is already largely open vegetation, greater dryness would suggest less apparent diversity in vegetation stands rather than any gain from the more open vegetation that would accompany greater dryness. Water-demanding summergreen forms may be lost the most, and there would be only small gains among desert forms. Mediterranean zones are expected to advance poleward, upward, and perhaps inland – but perhaps not in southern Australia and South Africa, where there is no more land to advance into. The relatively large, inland submediterranean zone in Italy may be reduced but may also move northward across the Alps. For that portion of the Middle East that receives significant winter rainfall and can be called Mediterranean, the prognosis is more difficult. Indeed the inland climate of Iran, in particular, is somewhat unusual already, with winter-rain and summer-rain floristic elements mixing, without a large representation of the sclerophyll element (e.g. [6]).

Of course, with environmental change, plant characters may also change as the plant species adapt to the new conditions. The greatest potential for change in form characters probably involves seasonal habits, since these can be quite "plastic", (facultative rather than obligatory) in many taxa, even under current conditions, especially in the tropics and subtropics. Leaf consistency and size may also change in some taxa, with leaves perhaps becoming "harder" or smaller or both with increasing dryness. Least changeable is probably the overall plant architecture, i.e. the structural type.

References

1. Alekseenko MI (1968) Edifictors and dominants of the plant communities of chalk outcrops in the basin of the Northern Donets River. Biol. Nauka v Univers. I Ped. Inst. Ukrainy za 50 Let, pp 84–85. Kharkov University (in Russian)
2. Auerbach M, Shmida A (1987) Spatial scale and the determinants of plant species richness. Trends Ecol Evol 2:238–242
3. Axelrod DI (1958) Evolution of the Madro-Tertiary geoflora. Bot Rev 24(7):433–509
4. Barthlott W, Lauer W, Placke A (1996) Global distribution of species diversity in vascular plants: towards a world map of phytodiversity. Erdkunde 50:317–327+map
5. Blondell J, Aronson J, Bodiou J-Y, Boeuf G (2010) The Mediterranean region: biological diversity in space and time, 2nd edn. Oxford University Press, Oxford
6. Blumler MA, Plummer JC (2015) Deciduous woodlands in the near eastern fertile crescent, and a comparison with California. In: Box EO, Fujiwara K (eds) Warm-temperate deciduous forests around the northern hemisphere, pp 267–276
7. Box EO (1981) Macroclimate and plant forms: an introduction to predictive modeling in phytogeography. Tasks for vegetation science, vol 1, p 258. Dr. W. Junk BV, The Hague
8. Box EO (1986) Some climatic relations of the vegetation of Argentina, in global perspective, pp 181–216. Veröff. Geobot. Inst. ETH, Stiftung Rübel (Zürich)
9. Box EO (1987) Plant life forms in Mediterranean environments. Annali Di Botanica 45(2):7–42
10. Box EO (1995) Factors determining distributions of tree species and plant functional types. Vegetatio 121:101–116
11. Box EO (1995) Climatic relationships of the forests of East and South-East Asia. In: Box EO et al (eds) Vegetation science in forestry: global perspective based on forests ecosystems of East and Southeast Asia. Kluwer, Dordrecht, pp 23–55
12. Box EO (1996) Plant functional types and climate at the global scale. J. Veg. Sci. 7:309–320

13. Box EO (1997) Bioclimatic position of evergreen broad-leaved forests. In: Island and high-mountain vegetation: biodiversity, bioclimate and conservation, pp 17–38. Proceedings, annual IAVS meeting, Tenerife, April 1993. Universidad de La Laguna, Servicio de Publicaciones, Tenerife (Canary Islands, Spain).
14. Box EO (ed) (2016) Vegetation structure and function at multiple spatial, temporal and conceptual scales, p 578. Springer, Cham
15. Box EO (2016b) Global bioclimatic zonation. In: Box EO (ed) Vegetation structure and function at multiple spatial, temporal and conceptual scales, pp 3–52. Springer
16. Box EO (2019) Form and character diversity of potential world vegetation. Flora 254. https://doi.org/10.1016/j.flora.2019.03.001
17. Box EO, Fujiwara K (2005, 2013) Vegetation types and their broad-scale distribution. In: van der Maarel E (ed) Vegetation ecology, pp 106–128, van der Maarel E, Franklin J (eds) 2nd edn., pp 455–485+plate 15.1 (color). Wiley, London
18. Brophy C, Dooley A, Kirwan L, Finn JA, McDonnell J, Bell Th, Cadotte MW, Connolly J (2017) Biodiversity and ecosystem function: making sense of numerous species interactions in multi-species communities. Ecology 98(7):1771–1778
19. Crame JA (2001) Taxonomic diversity gradients through geological time. Divers Distrib 7:175–189
20. Crawford RMM (2008) Plants at the margin. Cambridge University Press, Cambridge, p 478
21. Davis SD, Heywood VH, Hamilton AC (1994) Centres of plant diversity: a guide and strategy for their conservation. Vol. 1: Europe, Africa, Southwest Asia and the Middle East. World Wildlife Fund and World Conservation Union, Cambridge
22. Diamond J (1988) Factors controlling species diversity: overview and synthesis. Ann. Missouri Bot. Gard. 75:117–129
23. Drude O (1896) Deutschlands Pflanzengeographie. Handbuch Dt. Land-. u. Volkskunde, Engelhorn, Stuttgart, p 502
24. Erwin DH (2009) Climate as a driver of evolutionary change. Curr Biol 19:R575–R583
25. Evans KL, Warren PH, Gaston KJ (2005) Species-energy relationships at the macroecological scale: a review of the mechanisms. Biol. Rev. 80:1–25
26. Fekete G, Szuko Lacza J (1970, 1971) A survey of plant life-form systems and the respective research approaches. Annales Hist Nat Musei Nation Hungarici Pars Botanica 62:115–127, 63:37–50
27. Gavilan RG (2003) Does global warming pose a threat to Mediterranean biodiversity? Bocconea 16(1):379–395
28. Gilbert N (2011) Plant biodiversity theory debunked. Nature. https://doi.org/10.1038/news.2011.553. Accessed 22 Sept 2011
29. Gillison AN (2013) Plant functional types and traits at the community, ecosystem and world level. In: van der Maarel E, Franklin J (eds) Vegetation ecology, pp 347–386
30. Grisebach ARH (1872) Die Vegetation der Erde nach ihrer klimatischen Anordnung, Leipzig
31. Groombridge B (ed) (1992) World conservation monitoring centre. Global biodiversity: status of the earth's living resources, p 585. Chapman & Hall, London
32. Hawkins BA, Field R, Cornell HV, Currie DJ, Guégan J-F et al (2003) Energy, water, and broad-scale geographic patterns of species richness. Ecology 84(12):3105–3117
33. Heywood VH (1995) The Mediterranean flora in the context of world biodiversity. Ecol Mediterr 20:11–18
34. Hillebrand H (2004) On the generality of the latitudinal diversity gradient. Am Nat 163:192–211
35. Hooper DU, Chapin FS, Ewel JJ et al (2005) Effects of biodiversity on ecosystem functioning: a consensus of current knowledge. Ecol Monogr 75(1):3–35
36. von Humboldt A (1806) Ideen zu einer Physiognomik der Gewächse, p 28. F. G. Cotta, Tübingen. Reprinted 1957 by Akad, p 19. Verlagsgesellschaft, Leipzig
37. von Humboldt A (1808) Ansichten der Natur mit wissenschaftlichen Erlaeuterungen. J. G. Cotta, Tübingen
38. Huston M (1979) A general hypothesis of species diversity. Am Nat 113:81–101

39. Keith DA (ed) (2017) Australian vegetation, 3rd edn. Cambridge University Press, Cambridge
40. Larcher W (1973) Limiting temperatures for life functions in plants. In: Precht H, et al. (eds) Temperature and life. Springer-Verlag, Berlin
41. Lebrun J (1966) Les formes biologiques dans la végétation tropicale. Mém Soc Fr 1966:166–177
42. MacArthur RH, Levins R (1967) The limiting similarity, convergence and divergence of coexisting species. Am Nat 101:377–385
43. MacArthur RH, Wilson EO (1967) The theory of island biogeography. Princeton University Press, Princeton
44. Médail F, Quézel P (1997) Hot-spots analysis for conservation of plant biodiversity in the Mediterranean basin. Ann Mo Bot Gar 84:112–127
45. Médail F, Quézel P (1999) Biodiversity hotspots in the Mediterranean basin: setting global conservation priorities. Consv. Biol. 13(6):1510–1513
46. Midgley GF (2012) Biodiversity and ecosystem function. Science 335:174–175
47. Mittelbach GG, Steiner CF, Scheiner SM, Gross KL, Reynolds HL et al (2001) What is the observed relationship between species richness and productivity? Ecology 82:2381–2396
48. Mittelbach GG, Schemske DW, Cornell HV, Allen AP, Brown JM et al (2007) Evolution and the latitudinal diversity gradient: speciation, extinction and biogeography. Ecol Lett 10:315–331
49. Mittermeier RA, Gil PR, Mittermeier CG (1997) Megadiversity: earth's biologically richest nations. Conservation International, Washington
50. Mueller-Dombois D, Ellenberg H (1974) Aims and methods of vegetation ecology. Wiley, New York, p 547
51. Nakhutsrishvii G (2013) The vegetation of Georgia (South Caucasus). Springer, Heidelberg
52. Olson DM, Dinerstein E (1998) The global 200: a representation approach to conserving earth's most biologically valuable ecoregions. Conserv Biol 12:502–515. World Wildlife Fund, Washington
53. Palmer MW (1994) Variation in species richness: towards a unification of hypotheses. Folia Geobotan Phytotax 29:511–530
54. Pausas JG, Austin MS (2001) Patterns of plant species richness in relation to different environments: an appraisal. J Veg Sci 12:153–166
55. Pianka ER (1966) Latitudinal gradients in species diversity: a review of concepts. Am Nat 100:33–46
56. Quézel P (1985) Definition of the mediterranean region and origin of its flora. Dr Junk W, Dordrecht
57. Raimondo FM, Domina G, Spader V (2010) Checklist of the vascular flora of Sicily. Quad Bot Amb Appl 21:189–252
58. Raunkiaer C (1934) The life forms of plants and statistical plant geography. Clarendon, Oxford
59. Reid WV (1998) Biodiversity hotspots. Trends Ecol Evol 13:275–280
60. Rohde K (1992) Latitudinal gradients in species diversity – the search for primary cause. Oikos 65:514–527
61. Schnell R (1970–1977) Introduction à la Phytogéographie des Pays Tropicaux, vols 4 (see vol II: chapter 3). Gauthier-Villars, Paris
62. Shmida A, Wilson MV (1985) Biological determinants of species diversity. J Biogeogr 12:1–21
63. Skeels A, Cardillo M (2017) Environmental niche conservatism explains the accumulation of species richness in Mediterranean-hotspot plant genera. Evolution 71(3):582–594
64. Smith TM, Shugart HH, Woodward FI, Burton PJ (1993) Plant functional types. In: Solomon AM, Shugart HH (eds) Vegetation dynamics and global change. Chapman & Hall, New York, pp 272–292
65. Solomon AM, Shugart HH (eds) (1993) Vegetation dynamics and global change. Chapman & Hall, New York

66. Stehil FG, Douglas RG, Newell ND (1969) Generation and maintenance of gradients in taxonomic diversity. Science 164:947–949
67. Stevens GC (1989) The latitudinal gradient in geographical range: how many species coexist in the tropics. Am Nat 133:240–258
68. Szuko-Lacza JS, Fekete G (1969, 1972) A survey of plant life-form systems and the respective research approaches. Annales Hist Natur Musei Nat Hungarici Pars Botanica 61:129–139, 64:53–62
69. Thanos CA (2005) The geography of Theophrastus' life and of his botanical writings. In: Karamanos, AJ, Thanos, CA (eds) Biodiversity and natural heritage in the Aegean, pp 113–131
70. Theophrastus (1916, 1926) Enquiry into plants (history of plants), vols I–II. In: Hort AF (ed) Ancient Greek with English translation. Harvard University Press
71. Walter H (1968, 1973) Die Vegetation der Erde in öko-physiologischer Betrachtung, 3rd edn., vol 1, p 744 and vol II, p 1002. Fischer-Verlag, Stuttgart
72. Warming E (1895) Plantesamfund: Grunträk af den ökologiske Plantgeografi. København. English version 1909: oecology of plants, p 422. Humphrey Milford & Oxford University Press, Oxford
73. Whittaker RJ (2010) Meta-analyses and mega-mistakes: calling time on meta-analysis of the species richness-productivity relationship. Ecology 91(9):2522–2533
74. Whittaker RJ, Willis KJ, Field R (2001) Scale and species richness: towards a general, hierarchical theory of species diversity. J Biogeogr 28:453–470
75. Willig MR, Kaufman DM, Stevens RD (2003) Latitudinal gradients of biodiversity: pattern, process, scale and synthesis. Ann Rev Ecol Evol Syst 34:273–309
76. Woodward, I. 1987. *Climate and Plant Distribution*. Cambridge University Press. 174 pp.
77. Wright DH (1983) Species-energy theory: an extension of species-area theory. Oikos 41:496–506
78. Wullschleger S, Epstein HE, Box EO, Euskirchen ES et al (2014) Plant functional types in earth system models: past experiences and future directions for application of dynamic vegetation models in high-latitude ecosystems. Ann Bot 114:1–16. https://doi.org/10.1093/aob/mcu077

Coastal Dune Vegetation Zonation in Italy: Squeezed Between Environmental Drivers and Threats

Alicia T. R. Acosta

Abstract The present work provides an overview of recent studies in coastal dune systems performed by the Plant Ecology Laboratory, Roma 3 University (Italy), highlighting major findings. In this review major habitat types along the sea-inland environmental gradient are described, emphasizing their ecological value in terms of biodiversity and landscape heterogeneity. A synthesis of major threats is also presented. Occupying transition zones between terrestrial and marine ecosystems, coastal dune systems are characterized by strong environmental gradients which determine the coexistence of different plant communities in a relatively small area. Despite the presence of numerous habitats which are considered priorities for international conservation goals, coastal dune ecosystems have undergone consistent transformations and are currently considered to be highly endangered. The understanding of major threats is a crucial step toward developing effective management strategies to safeguard coastal dune biodiversity, in the hope of preserving the functionality and uniqueness of these fragile ecosystems.

Keywords Coastal vegetation · Coastal dune systems · Dune species · Threatened ecosystems

1 Introduction

By occupying the transition between terrestrial and marine ecosystems, coastal dunes represent particularly interesting environments both from an ecological and a landscape perspective. The complex, compressed zonation of plant communities along the sea-inland gradient is, in fact, one of the most interesting features of Mediterranean sandy shores [1, 10, 24, 25, 43, 46, 48], giving rise to a unique biodiversity in terms of both number of different habitats and species composition [49]. Mediterranean coastal areas have been the scene of a wide variety of human

A. T. R. Acosta (✉)
Department of Sciences, Rome 3 University, V. le Marconi 446, 00146 Rome, Italy
e-mail: acosta@uniroma3.it

F. Pedrotti and E. O. Box (eds.), *Tools for Landscape-Scale Geobotany and Conservation*, Geobotany Studies, https://doi.org/10.1007/978-3-030-74950-7_16

activities for millennia. However, anthropic pressure intensified greatly over the course of the twentieth century, with remarkable land-use changes for urbanization, harbor and industrial centers, roadways, and increased tourist activities, just to cite a few examples. Particularly after World War II, various related human pressures caused severe losses in the extent of dune systems, the same factors still threatening these communities to this day. It is widely acknowledged that coastal dune habitats are currently in a critical conservation condition in the whole Mediterranean Basin. In fact, coastal sandy ecosystems are currently pinpointed as being vulnerable to high, widespread biodiversity loss, making them among the most endangered habitats at European level [25, 26]. The continuous degradation of coastal dune systems is clearly connected to the increasing intensification of some human activities, such as the recreational use of beaches. Especially during the warm, sunny summer months, many coastal areas attract a great number of local visitors and international tourists. Thus, human activities may also represent a serious threat, even though at the same time, they provide many coastal areas with their main source of income. The understanding of these threats is a crucial step toward developing effective management strategies to safeguard coastal dune biodiversity. If these ecosystems are managed according to a thorough, sustainable conservation strategy, human activities could also represent a long-lasting economic income for local populations.

The present work brings together recent published studies on coastal dune systems performed by the Plant Ecology laboratory, Roma 3 University (Italy). Thus, here I present an outline of the 'state of the art' of our studies in coastal dunes by summarizing major findings. This review is not intended as a comprehensive synthesis of the studies conducted in coastal dunes in Italy, but an overview of our previous works.

2 Coastal Dune Vegetation Zonation: Environmental Gradients, Habitats and Species Diversity

On sandy coastal habitats, factors related to substrate and to wind action vary along the sea-land ecotone, forming a marked directional stress and natural disturbance gradient. Environmental factors shift along this gradient, concomitantly with plant communities. It is interesting to note that the gradient is similar both in the Mediterranean and in oceanic coastal ecosystems [18], with analogous trends along the dune profile [21]. However, we could also highlight some peculiarities of Mediterranean systems, such as the relatively minor importance of wind-related variables as compared to soil properties. In particular, organic matter and grain-size variability are closely correlated with the distribution of plant communities along the coastal zonation [22].

In Mediterranean ecosystems, vegetation is organized along a well-known zonation ranging from pioneer annual communities on the drift line to

Fig. 1 Simplified representation of the vegetation zonation in Mediterranean coastal dune ecosystems. Vegetation is organized along a well-known zonation ranging from pioneer annual communities on the drift line to Mediterranean scrubs on fixed dunes. EUNIS Habitats (see Table 1) were identified by different letters: **A** B1.1 Sand beach driftlines, **B** B1.3 Shifting coastal dunes, **C** B1.4 Coastal stable dune grassland (grey dunes), **D** B1.6 Coastal dune scrub.

Mediterranean scrub on fixed dunes. This zonation is observed not only in vascular plants but also in other taxonomic groups, such as bryophytes [32]. Habitats of conservation interest can be identified based on the presence of characteristic, diagnostic plant species, a list of which has been standardized across Europe in the EUNIS classification of habitats (EUNIS B1; European Environment Agency 2008). Moreover, several dune plant assemblages have been included in the Annex I of the Habitats Directive [9] (Table 1, Fig. 1).

Acosta et al. [2] analyzed coastal dune vegetation zonation in relation to dune morphology. Later, Bazzichetto et al. [7] described the morphological profile of each coastal dune habitat along the zonation, using morphological variables such as elevation, slope, curvature and sea distance. By integrating field-collected vegetation data and remotely sensed imagery, these authors found that habitats occur differentially, and that dune elevation and sea distance are key factors in shaping habitat occurrence along local gradients. Moreover, species richness varied along the sea-to-land gradient, reaching its peak in the intermediate sector of the vegetation zonation. Nevertheless, species richness seems to differ significantly only between the upper beach and all the other habitat types [4]. The upper beach is a very selective environment with extreme maritime conditions. In fact, very few, highly specialized vascular plant species can survive here, with specific morphological and physiological traits usually interpreted as a response to salt, drought and unstable substrate. Outside the drift line, coastal dune plant communities tend to be perennial, although annual grasslands are an interesting exception. Pioneer psammophilous plants have evolved a variety of adaptations that allow survival, growth and reproduction under relatively harsh conditions. Proceeding inland, vegetation is

Table 1. EUNIS, Nature 2000 and main phytosociological Alliances.

EUNIS Classification	*EU Habitats Directive Annex I* 92/43/EEC	*Description*	*Prodromo della Vegetazione d'Italia* http://www.prodromo-vegetazione-italia.org/habitat
B1.1 Sand beach driftlines	1210 Annual vegetation of drift lines	Just above the normal tide limit, where drift material accumulates and the sand may be rich in nitrogenous organic matter. Annual, nitrophilous and ephemeral vegetation community is exposed to wind disturbance and flooding.	*Euphorbion peplis* Tüxen 1950
B1.3 Shifting coastal dunes	2110 Embryonic shifting dunes 2120 Shifting dunes along the shoreline with Ammophila arenaria (white dunes)	Mobile sands with low vegetation cover and poor substrate. The pioneer vegetation community is dominated by dune-forming rhizomatous plants. In the Mediterranean the habitat is limited to a fairly flat upper beach, still subject in part to inundation.	• *Agropyrion juncei* (Tüxen in Br.-Bl. & Tüxen 1952) Géhu, Rivas-Martínez & Tüxen 1972 in Géhu, Costa, Scoppola, Biondi, Marchiori, Peris, Franck, Caniglia & Veri 1984 • *Ammophilion australis* Br.-Bl. 1933 Em. Géhu & Géhu-Franck 1988
B1.4 Coastal stable dune grassland (grey dunes)	2130 Fixed coastal dunes with herbaceous vegetation (grey dunes) 2210 Crucianellion maritimae fixed beach dunes 2230 Malcolmietalia dune grasslands 2240 Brachypodietalia dune grasslands with annuals	Fixed or semifixed dunes with perennial-chamaephyte grasslands community partially sheltered from winds. Ephemeral therophytic vegetation of deep sands in dry interdunal depressions of the coasts. Pseudo-steppe and short-grass perennial grasslands rich in therophytes.	• *Crucianellion maritimae* Rivas Goday & Rivas-Martínez 1958 • *Alkanno-Maresion nanae* Rivas Goday ex Rivas Goday & Rivas-Martínez 1963 corr. Díez Garretas, Asensi & Rivas-Martínez 2001 • *Laguro ovati-Vulpion fasciculatae* Géhu & Biondi 1994 • *Thero-Brachypodion* ramosi Br.-Bl. 1925 • *Tuberarion guttatae* Br.-Bl. in Br.-Bl., Molinier & Wagner 1940 nom. mut. propos. Rivas-Martínez, Diaz, Fernández-González, Izco, Loidi, Lousa & Penas 2002
B1.6 Coastal dune scrub	2160 Dunes with *Hippophae rhamnoides* 2250 Coastal dunes with *Juniperus* spp 2260 *Cisto-Lavenduletalia* dune sclerophyllous scrubs	Scrub with junipers on coastal sand dunes. Junipers are often relatively low and prostrate, sometimes erect. Evergreen maquis dominated by shrub species with high cover values and less exposed to the harsh coastal conditions.	• *Pruno spinosae-Rubion ulmifolii* O. Bolòs 1954 • *Juniperion turbinatae* Rivas-Mart. 1975 corr. 1987 • *Teucrion mari* Gamisans et Muracciole 1984 • *Cisto eriocephali-Ericion multiflorae* Biondi 2000

progressively less exposed to the harsh maritime environment and gradually less tolerant to salt spray, strong winds and sand burial [1]. A significant variation in the degree of specialization was also observed along the vegetation zonation: habitats closer to the sea and under greater stress held higher proportions of specialized species and had higher average specialization levels [13]. Regarding fine-scale community assembly, Santoro et al. [40] highlighted how community assemblage patterns were strongly linked to the sea-land environmental gradient. In pioneer foredune communities, an aggregated assemblage was observed. In contrast, progressing along the zonation, assemblage patterns first shifted to random and then became segregated in the fixed dunes. These results are consistent with the "Stress Gradient Hypothesis" [8], which predicts that along an environmental gradient, facilitation drives the assembly of plant communities where conditions are harsh, whereas competition is predominant where conditions are less severe. In addition, Conti et al. [17] highlighted the importance not only of the environmental conditions along the already well-known sea-land gradient, but also of environmental heterogeneity at fine local scale, suggesting that micro-abiotic filtering might play a more important role than previously anticipated.

3 Major Threats

3.1 Land Use Changes and Landscape Fragmentation

Most studies on coastal dune vegetation changes have been conducted using remotely sensed data, generally agreeing that the composition and spatial patterns of coastal landscapes have been drastically modified by human activities over the last 50 years. In particular, the expansion of artificial areas is an important driver of this transformation, together with afforestation and the loss of land to new agricultural activities [28]. As on other Mediterranean coasts, agricultural and afforestation lands tend to be concentrated in areas of the inner sectors of the coastal zonation, mostly affecting fixed dunes. Therefore, it seems that the typical environmental gradient of coastal ecosystems mentioned before, which constrains the natural vegetation zonation, also shapes human activities such as agriculture and afforestation. In contrast, urban expansion involves the whole landscape and seems less constrained by this gradient, as no preferences for any particular natural dune cover type have been observed [28].

It seems that landscape patterns and transformations affect taxonomic diversity more moderately than do local abiotic and biotic site conditions, although a slight decline on highly disturbed coastal landscapes was observed, mainly in patches sharing borders with artificial areas [30]. Surprisingly, the effect of the landscape spatial pattern on functional diversity was stronger: a generalized increase in functional diversity was seen in human-dominated contexts, but it decreased in those areas that have undergone high urbanization rates in the last 60 years. In this way, even if a short-term enrichment in functional traits related to human presence was evident, functional diversity could be negatively affected in the long term.

For floristic changes, Del Vecchio et al. [19] analyzed changes in plant species composition of coastal dune habitats over a 20-year period in a sector of the Adriatic coast. Subsequently, Prisco et al. [35] assessed major changes between the late 1960s and the present day at national level, taking advantage of the extensive spatial and temporal coverage of a coastal dune vegetation data-base [34]. Unexpectedly, results in protected areas showed positive trends in species cover but without clear evidence of climatic preferences, at least when using Ellenberg's indicator values [35]. A possible explanation is that the recent incorporation of Natura 2000 sites into the protected-area network may have limited the effects of urbanization and beach exploitation activities on coastal habitats, thus promoting species cover. However, it is also worth noting that large stretches of Italy's coastlines remain unprotected and are at risk of further biodiversity declines. Regarding recent erosion/accretion processes, positive trends in vegetation cover were also observed by using permanent vegetation transects, but only at stable or accreting sites [36]. In particular, the cover of the key species *Ammophila arenaria* increased at sites where the shoreline is accreting, so underlining that highly relevant plant species of mobile dunes can develop effectively without erosion processes and with relatively low anthropic disturbance. As could have been anticipated, negative trends were observed at sites with coastal erosion [36].

3.2 Trampling

Tourism is a growing activity with great economic potential, in terms of both employment and income. The greatest growth in tourism is occurring in the sub-sector of coastal and marine tourism, while the Mediterranean region plays host to around 33% of the world's tourism industry [5]. Among direct threats, increasing numbers of visitors may also mean increased trampling and the consequent degradation of coastal dune ecosystems [5]. A four-year experiment highlighted how fencing can be an effective method for promoting the passive recovery of coastal dune vegetation, in terms of both species cover and species richness, as well as their spatial organization [41]. This capacity for rapid recovery is, in fact, a typical feature of many coastal dune plants, which are often well adapted to high levels of stress and disturbance [12, 51] and are thus able to recover quickly once a source of disturbance has been limited. Therefore, fencing demonstrates a high potential as a passive recovery strategy for improving the conservation status of degraded dune habitats.

3.3 Alien Species

Invasive species constitute a major threat to biodiversity, and their impact on natural and semi-natural habitats has been the focus of many ecological studies. In Europe, coastal dunes are among the habitats most invaded by alien species [14]. Expansion of alien species has characterized many coastal dune systems of the whole Mediterranean basin, leading to drastic changes in floristic composition. Indeed, native species of coastal areas are in a fragile equilibrium with their natural environment, and as such they are extremely sensitive to human-driven modifications [13].

At regional scale, human population density was the most important factor negatively related to native species richness in central Italy. On the other hand, alien species richness was strongly related to climatic factors, especially to precipitation regimes (more aliens in areas with more precipitation). Secondarily, alien introductions appear to be related to recent urban sprawl and associated gardening [11]. In addition, fragmentation of natural habitats in the landscape (cf. [23]) may directly enhance the invasibility of ecological communities, thus reducing the extent of natural habitats and increasing isolation and the decline of ecological quality of the remaining fragments. The resulting fragmented pattern has the potential to boost habitat colonization by alien species from the surrounding matrix. Malavasi et al. [29] found evidence that fragmentation of natural coastal habitats enhances the number of aliens present in each vegetation patch. This increase in patch invasion was observed in correspondence with the recent expansion of artificial areas closer to natural habitats, confirming the existence of consistent links between human-driven landscape changes and the presence and distribution of alien species on coastal dunes. As in other human-modified landscapes, recent construction directly on the dunes, road development and urban sprawl all seem to have promoted alien invasion. In addition, human structures probably increase the propagule pressure of alien species, although adjacency to other natural dune patches may also facilitate their spread. In fact, patches closer to other natural patches tend to host a higher average number of aliens, most likely because other invaded natural patches may facilitate alien establishment elsewhere in the landscape. Some aliens widely present on Mediterranean coasts, such as *Carpobrotus* spp. and *Agave americana*, with extensive vegetative growth, lead to the formation of extensive mats that may facilitate their further expansion into other neighboring vegetation patches.

At local scale, results support the hypothesis that variation in abiotic conditions can explain differences in invasibility, although propagule pressure seems to be the main determinant of alien species occurrence and spread [12]. Thus, the major role played by propagule pressure cannot be ignored, as it remains the main determinant of alien species occurrence and spread even along a strong natural environmental gradient [12].

In central Italy 8.5% (57 species) of the registered total vascular flora was alien, but only few such species were considered invasive, such as the genus *Carpobrotus* (*C. acinaciformis*/*C. edulis*), *Agave americana, Erigeron canadensis* (=*Conyza canadensis*), *Oenothera biennis* and *Ambrosia coronopifolia* [3]. It is worth emphasizing that most alien species came from America, but there are some significant differences in alien distribution along the Adriatic and Tyrrhenian coasts. Actually, significantly higher rates of alien species of African origin (mainly tropical or from the Mediterranean fynbos) were found on the Tyrrhenian side, while higher rates of extra-tropical American and Eurasian species were found on the Adriatic. In fact, the genus *Carpobrotus* (from the Cape Region of South Africa) and *Agave americana* (from Mexico), probably the most diffused invasive aliens along the Tyrrhenian coast, were quite rare on the Adriatic. On the other hand, *Oenothera biennis* and *Ambrosia coronopifolia* (from extra-tropical North America) were the most diffused invasive aliens on the Adriatic coast.

The Genus Carpobrotus

Currently, invasion by *Carpobrotus* spp. constitutes one of the most severe threats to the Mediterranean basin's coastal plant communities [44, 50]. In Italy two different species have been identified, *Carpobrotus edulis* and *C. acinaciformis*, although important taxonomical problems still remain unsolved [45]. This succulent perennial taxon reaches high cover values and is able to form monospecific populations which seriously endanger the survival of many dune habitats, from shifting dunes with *Ammophila arenaria* to fixed dunes with *Juniperus* spp.

Adult *Carpobrotus* plants have been recognized to have an undesirable direct impact on native plants in several areas of its secondary distribution range, with negative effects on germination, survival, growth and reproduction of the victimized native species [6, 15, 33, 50]. Several studies have shown that *Carpobrotus* invasion ultimately affects patterns of native species diversity [27, 42], confirming that its successful establishment probably operates through the replacement and exclusion of native species, rather than coexistence. In conjunction with the decline in taxonomic diversity, Jucker et al. [27] reported a parallel loss in functional diversity as *Carpobrotus* abundance increased, this suggests that the invasive species is somehow acting as a filter in the process of native community assembly by preferentially excluding species with specific life-history traits. In particular, on the basis of a plant-trait analysis, *Carpobrotus* appears to exclude closely related and ecologically similar taxa preferentially [27], and ruderal and nitrophilous species may substitute for the typical native dune species [31]. Moreover, it has been shown how this alien invasion influences the fine-scale assembly of native coastal communities, by leading to a shift to randomness in community structure [40].

Carpobrotus has also been involved in many indirect negative effects on the invaded native coastal ecosystems. Actually, it has been recognized as a major driver of shifts in soil conditions and disruption of soil geochemical processes [33, 39]. In particular, it seems that in coastal dune pioneer habitats, *Carpobrotus* invasion is likely to affect soil physico-chemical and biological processes [39]. These soil modifications could ultimately inhibit germination and affect the survival of the specialized native dune species.

Beach Litter

Among various threats, marine litter seems to be one with a global distribution and probably one of the most recognized pollution problems in the world's oceans. Litter washed ashore is called "beach litter" and is one of the most obvious signs of marine litter pollution [37]. Sandy shores are generally considered important sinks for floating debris, which, after stranding, generally becomes trapped in sand or might be blown farther inland. In central Italy, plastic and polystyrene are the most common litter items, in line with findings which confirm plastic as the most widespread marine litter worldwide [37]. Moreover, there are differences in the spatial distribution of the beach litter along the coastal zonation. A relationship between beach-litter distribution and the sea-land gradient was also found: coastal habitats on the upper beach, shifting dunes and the transition to the fixed dunes were similarly threatened by beach litter, with the highest levels of litter accumulation. On the other hand, fixed dune habitats were less influenced by beach litter. It seems that fixed-dune habitats may act as a natural barrier protecting the inner part of the dune system from marine litter dispersion [37]. Results on the various potential sources of litter (urban areas, rivers and beach users) suggested that solid pollution along coastal ecosystems was significantly related to the nearby presence of urban areas. In fact, urban areas seem to be the main potential source of beach litter, and rivers were probably the main, but not the only, carriers of litter items from their origin to the marine environment [38].

Overall, accumulation of large amounts of beach litter is considered an aesthetic loss to the value of the coastal environment, with potential adverse effects on tourist activities. For this reason, cleaning beaches is one of the most important management issues for coastal planning and is usually carried out by mechanical equipment. Unfortunately, mechanical techniques allow no distinction between beach litter and biological resources such as plants, animals, and organic debris naturally deposited on shorelines and often removes the latter completely [37]. Therefore, the presence of beach litter should be considered an indirect threat to biotic communities, promoting changes in the dune morphology and erosion processes. In particular, this threat is mostly relevant in those coastal areas characterized by intense recreational use of beaches.

4 Conclusions

Although the great diversity of Italian coastal dune habitats is now widely recognized, many threats should also be emphasized. The consequences of these threats are habitat and species loss, fragmentation, the final outcome sometimes being the complete destruction of coastal dune systems. The remarkable "artificialization" of coastal areas is probably one of the most serious concerns in Mediterranean countries. Coastal dune plant species are, to a large degree, highly specialized in coping with the severe constraints imposed by the coastal environment along the sea-land gradient: wind erosion, sand burial, aridity and incoherent soils. However, they are not able to cope with the combination of human-induced environmental changes outlined in this review. The high sensitivity of coastal plant species to such changes leads to increasing concern about imminent biodiversity loss (both taxonomical and functional) in coastal dune habitats. In particular, human impact is likely to affect the most specialized coastal plants more strongly, while more generalist and alien plants might be able to survive, leading to an overall biotic homogenization process among communities. The critically poor conservation conditions make these habitats those with the highest risk level; as such they require further research and special, and even urgent, management measures in the hope of preserving the functionality and uniqueness of these fragile ecosystems for future generations.

Acknowledgements This review is the outcome of the intensive work of many people. There are so many that I could fill a long list. I would not like to forget anyone, but will just mention some of them: Manuele Bazzichetto, Marta Carboni, Luisa Conti, Maria Laura Carranza, Silvia Del Vecchio, Stefania Ercole, Francesca Izzi, Tommaso Jucker, Marco Malavasi, Irene Prisco, Gianluca Poeta, Riccardo Santoro, Marta Gaia Sperandii and Angela Stanisci.

References

1. Acosta ATR, Stanisci A, Ercole S, Blasi C (2003) Sandy coastal landscape of the Lazio region (Central Italy). Phytocoenologia 33(4):715–726
2. Acosta ATR, Ercole S, Stanisci A, Pillar VDP, Blasi C (2007) Coastal vegetation zonation and dune morphology in some Mediterranean ecosystems. J Coast Res 236:1518–1524
3. Acosta ATR, Carranza ML, Di Martino L, Frattaroli A, Izzi CF, Stanisci A (2008) Patterns of native and alien plant species occurrence on coastal dunes in Central Italy. In: Tokarska-Guzik B, Brock JH, Brundu G, Child L, Daehler CC, Pyšek P (eds) Plant Invasions: human perception, ecological impacts and management. Backhuys Publishers, Leiden, pp 235–248
4. Acosta ATR, Carranza ML, Izzi CF (2009) Which habitat contributes best to plant species diversity in coastal dunes? Biodivers Conserv 18:1087–1098
5. Acosta ATR, Jucker T, Prisco I, Santoro R (2013) Passive recovery of Mediterranean coastal dunes following limitations to human trampling. In: Martinez LM, Gallego-Fernandez JB, Hesp PA. Restoration of coastal dunes, pp 187–198. Springer, Cham
6. Affre L, Suehs CM, Charpentier S, Vilà M, Brundu G, Lambdon P, Traveset A, Hulme PE (2010) Consistency on the habitat degree of invasion for three invasive plant species across Mediterranean islands. Biol Invasions 12(8):2537–2548

7. Bazzichetto M, Malavasi M, Barták V, Acosta ATR, Moudrý V, Carranza ML (2018) Modeling plant invasion on Mediterranean coastal landscapes: an integrative approach using remotely sensed data. Landscape Urban Plan 171:98–106
8. Bertness M, Callaway RM (1994) Positive interactions in communities. Trends Ecol Evol 9:191–193
9. Biondi E, Blasi C, Burrascano S, Casavecchia S, Copiz R, Del Vico E, Galdenzi D, Gigante D, Lasen C, Spampinato G, Venanzoni R, Zivkovic L (2009) Italian Interpretation Manual of the 92/43/EEC Directive Habitats, https://vnr.unipg.it/habitat/index.jsp
10. Biondi E, Brugiapaglia E, Allegrezza M, Ballelli M (1989) La vegetazione del litorale marchigiano (Adriatico centro-settentrionale). Colloq Phytosociol 19:429–460
11. Carboni M, Thuiller W, Izzi F, Acosta ATR (2010) Disentangling the relative effects of environmental versus human factors on the abundance of native and alien plant species in Mediterranean sandy shores. Divers Distrib 16(4):537–546
12. Carboni M, Santoro R, Acosta ATR (2011) Dealing with scarce data to understand how environmental gradients and propagule pressure shape fine-scale alien distribution patterns on coastal dunes. J Veg Sci 22(5):751–765
13. Carboni M, Zelený D, Acosta ATR (2016) Measuring ecological specialization along a natural stress gradient using a set of complementary niche breadth indices. J Veg Sci 27 (5):892–903
14. Chytrý M, Pyšek P, Wild J, Pino J, Maskell LC, Vilà M (2009) European map of alien plant invasions based on the quantitative assessment across habitats. Divers Distribs 15(1):98–107
15. Conser C, Connor EF (2009) Assessing the residual effects of *Carpobrotus edulis* invasion, implications for restoration. Biol Invasions 11(2):349–358
16. Ciccarelli D (2015) Mediterranean coastal dune vegetation: are disturbance and stress the key selective forces that drive the psammophilous succession? Estuar Coast Shelf S 165(5):247–253
17. Conti L, de Bello F, Lepš J, Acosta ATR, Carboni M (2017) Environmental gradients and micro-heterogeneity shape fine-scale plant community assembly on coastal dunes. J Veg Sci 28:762–773
18. Doing H (1985) Coastal foredune zonation and succession in various parts of the world. Vegetatio 61:65–75
19 Del Vecchio S, Prisco I, Acosta ATR, Stanisci A (2015) Changes in plant species composition of coastal dune habitats over a 20-year period. AoB PLANTS 7:plv018
20. European Commission (2013) Interpretation Manual of European Union Habitats. EUR28
21. Feola S, Carranza ML, Schaminée JHJ, Janssen JAM, Acosta ATR (2011) EU habitats of interest: an insight into Atlantic and Mediterranean beach and foredunes. Biod Conser 20 (7):1457–1468
22. Fenu G, Carboni M, Acosta ATR, Bacchetta G (2013) Environmental factors influencing coastal vegetation pattern: new insights from the Mediterranean Basin. Folia Geobot 48 (4):493–508
23. Forman RTT, Godron M (1986) Landscape Ecology. Wiley, New York
24. Gerdol R, Piccoli F (1984) Sand dune vegetation in the Po Delta (N Italy). Ecologia Mediterranea 10(3–4):119–131
25. Janssen JAM, Rodwell JS, Criado MG, Gubbay S, Arts GHP, Haynes T, Nieto A, Sanders N, Landucci F, Loidi J, Ssymank A, Tahvanainen T, Valderrabano M, Acosta ATR, Aronsson M, Arts G, Attorre F, Bergmeier E, Bijlsma RJ, Bioret F, Biţă-Nicolae C, Biurrun I, Calix M, Capelo J, Čarni A, Chytrý M, Dengler J, Dimopoulos P, Essl F, Gardfjell, Gigante D, Giusso del Galdo G, Hájek M, Jansen F, Jansen J, Kapfer J, Mickolajczak A, Molina JA, Molnár Z. Paternoster D, Piernik A, Poulin B, Renaux B, Schaminée JHJ, Šumberová K, Toivonen H, Tonteri T, Tsiripidis I, Tzonev R, and Valachovič M (2016) European Red List of Habitats. Part 2. Terrestrial and freshwater habitats. European Union. https://doi.org/10.2779/091372.

26. Genovesi P, Angelini P, Bianchi E, Dupré E, Ercole S, Giacanelli V, Ronchi F, Stoch F (2014) Specie e habitat di interesse comunitario in Italia: distribuzione, stato di conservazione e trend. ISPRA, Serie Rapporti, 194/2014
27. Jucker T, Carboni M, Acosta ATR (2013) Going beyond taxonomic diversity: deconstructing biodiversity patterns reveals the true cost of iceplant invasion. Divers Distrib 19:1566–1577
28. Malavasi M, Santoro R, Cutini M, Acosta ATR, Carranza ML (2013) What has happened to coastal dunes in the last half century? A multitemporal coastal landscape analysis in Central Italy. Landscape Urban Plan 119:54–63
29. Malavasi M, Carboni M, Cutini M, Carranza ML, Acosta ATR (2014) Landscape fragmentation, land-use legacy and propagule pressure promote plant invasion on coastal dunes: a patch-based approach. Landscape Ecol 29(9):1541–1550
30. Malavasi M, Conti L, Carboni M, Cutini M, Acosta ATR (2016) Multifaceted analysis of patch-level plant diversity in response to landscape spatial pattern and history on Mediterranean dunes. Ecosystems 19:850–864
31. Malavasi M, Santoro R, Cutini M, Acosta ATR, Carranza ML (2016) The impact of human pressure on landscape patterns and plant species richness in Mediterranean coastal dunes. Plant Biosyst 150(1):73–82
32. Murru V, Marignani M, Acosta ATR, Cogoni A (2018) Bryophytes in Mediterranean coastal dunes: ecological strategies and distribution along the vegetation zonation. Plant Biosyst 152 (5):1141–1148
33. Novoa A, González L, Moravcová L, Pyšek P (2013) Constraints to native plant species establishment in coastal dune communities invaded by *Carpobrotus edulis*: implications for restoration. Biol Conserv 164:1–9
34. Prisco I, Carboni M, Acosta ATR (2012) VegDunes – a coastal dune vegetation database for the analysis of Italian EU habitats. Biodiv Ecol 4:191–200
35. Prisco I, Carboni M, Jucker T, Acosta ATR (2016) Temporal changes in the vegetation of Italian coastal dunes: identifying winners and losers through the lens of functional traits. J Appl Ecol 53(5):1533–1542
36. Prisco I, Stanisci A, Acosta ATR (2016) Mediterranean dunes on the go: Evidence from a short-term study on coastal herbaceous vegetation. Estuar Coast Shelf S 182:40–46
37. Poeta G, Battisti C, Acosta ATR (2014) Marine litter in Mediterranean sandy littorals: spatial distribution patterns along central Italy coastaldunes. Mar Poll Bull 89:168–173
38. Poeta G, Conti L, Malavasi M, Battisti C, Acosta ATR (2016) Beach litter occurrence in sandy littorals: the potential role of urban areas, rivers and beach users in central Italy. Estuar Coast Shelf S 181:231–237
39. Santoro R, Jucker T, Carranza ML, Acosta ATR (2011) Assessing the effects of *Carpobrotus* invasion on coastal dune soils. Does the nature of the invaded habitat matter? Community Ecol 12(2):234–240
40. Santoro R, Jucker T, Carboni M, Acosta ATR (2012) Patterns of plant community assembly in invaded and non-invaded communities along a natural environmental gradient. J Veg Sci 23(3):483–494
41. Santoro R, Jucker T, Prisco I, Carboni M, Battisti C, Acosta ATR (2012) Effects of trampling limitation on coastal dune plant communities. Envir Manag 49:534–542
42. Santoro R, Carboni M, Carranza ML, Acosta ATR (2012) Focal species diversity patterns can provide diagnostic information on plant invasions. J Nat Conserv 20(2):85–91
43. Sburlino G, Buffa G, Filesi L, Gamper U, Ghirelli L (2013) Phytocoenotic diversity of the N-Adriatic coastal sand dunes – the herbaceous communities of the fixed dunes and the vegetation of the interdunal wetlands. Plant Sociol 50(2):57–77
44. Sintes T, Moragues E, Traveset A, Rita J (2007) Clonal growth dynamics of the invasive *Carpobrotus* affine *acinaciformis* in Mediterranean coastal systems: a non-linear model. Ecol Model 206(1):110–118
45 Suehs CM, Affre L, Médail F (2004) Invasion dynamics of two alien Carpobrotus (Aizoaceae) taxa on a Mediterranean island: I. Genetic diversity and introgression. Heredity 92(1):31–40

46. Stanisci A, Acosta ATR, Ercole S, Blasi C (2004) Plant communities on coastal dunes in Lazio (Italy). Annali Di Botanica 4:115–128
47. Suehs CM, Affre L, Médail F (2004) Invasion dynamics of two alien *Carpobrotus* (Aizoaceae) taxa on a Mediterranean island: II reproductive strategies. Heredity 92(6):550–556
48. Vagge I, Biondi E (1999) La vegetazione delle coste sabbiose del Tirreno settentrionale italiano. Fitosociologia 36(2):61–95
49. van der Maarel E (2003) Some remarks on the functions of European coastal ecosystems. Phytocoenologia 33:187–202
50. Vilà M, Tessier M, Suehs CM, Brundu G, Carta L, Galanidis A, Lambdon P, Manca M, Médail F, Moragues E, Traveset A, Troumbis AY, Hulme PE (2006) Local and regional assessments of the impacts of plant invaders on vegetation structure and soil properties of Mediterranean islands. J Biog 33(5):853–861
51. Wilson B, Sykes MT (1999) Is zonation on coastal sand dunes determined primarily by sand burial or salt spray? A test in New Zealand dunes. Ecol Lett 2:233–236

Constraints on Evergreen Broad-Leaved Forests in the Southeastern United States

Elgene O. Box and Kazue Fujiwara

Abstract In a humid, mild-winter warm-temperate climate, as occurs across most of the southeastern US coastal plain, one would normally expect to find "laurel forests" dominated by temperate-zone evergreen tree taxa, as in other warm-temperate regions. Instead, on the sandy uplands of the geologically young coastal plain, one finds a topogenic, landscape-scale mosaic of wetlands, forests, open woodlands and scrub, some of it maintained by recurring fires. As a result, the potential over the coastal plain is not extensive forest but rather, where soil, topography and fire permit, a slow progression to woods dominated by coriaceous or harder-leaved evergreen broad-leaved trees, with somewhat open canopies and a greater role for deciduous trees, even at maturity. Colder winters to the north constrain most evergreen forest types, but there are also subtler, non-climatic constraints. The purpose of this paper is to describe briefly the main evergreen broad-leaved forest types and their dynamics, and to evaluate the climatic and non-climatic factors that constrain their distributions, especially within the context of local climatic warming and drying.

Keywords Warm-temperate climate · Humic sand · Topogenic vegetation · Limiting temperatures · Laurophyll · Sclerophyll · Evergreen forest succession · Fire cycle

1 Introduction

Spreading live oaks (*Quercus virginiana*), dripping with Spanish moss (*Tillandsia usneoides*) and complemented by fragrant but primitive-looking Southern magnolias (*Magnolia grandiflora*), all evergreen, constitute the character forests and settlement vegetation of the original American South, i.e. the US southeastern

E. O. Box (✉)
Department of Geography, University of Georgia, 210 Field Street, Athens, GA 30602, USA

K. Fujiwara
Yokohama City University, Yokohama, Japan

F. Pedrotti and E. O. Box (eds.), *Tools for Landscape-Scale Geobotany and Conservation*, Geobotany Studies, https://doi.org/10.1007/978-3-030-74950-7_17

coastal plain, which extends from southern New Jersey to south Florida and west to coastal Texas. Although other forests may be evergreen and broad-leaved, the term 'evergreen broad-leaved forest' is commonly used for forests dominated by extra-tropical evergreen broad-leaved trees, which constitute the "zonal" vegetation of humid warm-temperate climates, generally understood as having no regular dry season and mild but not frost-free winters [12, 15, 16], cf [49, 58, 66, 71, 74, 121, 128, 136, 160–162]. Such forests are found in humid warm-temperate to subtropical parts of East Asia, southern Brazil, eastern Australia, and New Zealand. In the southeastern US, on the other hand, the potential for characteristic evergreen broad-leaved forests is restricted to the warm-temperate climate of the coastal plain and is clearly separated from the subtropical forests of southernmost Florida by intervening non-forest areas and by the cessation of frost [32].

That part of the world most similar physiographically to eastern North America is East Asia [6, 53], cf [54], where humid warm-temperate areas originally had "laurel forests" with mesomorphic, glossy, shade-tolerant laurophylls, called "lucidophylls" by Kira [75], cf [150]. These warm-temperate forests extended southward from the Tokyo area, southernmost Korea and the Chang Jiang (Yangtze) estuary near Shanghai, through subtropical southeastern China [142–144, 174]; the Okinawa islands [102]; and northern Taiwan [26, 65, cf 145]. Most of this forest was lost long before ecological studies began, and the secondary deciduous forests were interpreted until recently as the potential natural vegetation. Laurel forests were finally recognized as the true potential natural vegetation, based on small remnants of original or very similar forest that remained around temples, shrines, and even some farmhouses [101, 104, 106]). Such laurel forests are dominated by *Persea* (= *Machilus*, Lauraceae) and laurophyllous *Quercus* and *Castanopsis* (Fagaceae), along with laurophylls from families such as Theaceae and Myrsinaceae [44, 142, 144, 174], [46], cf [118, 119]. It was also recognized by Kira [72], however, that deciduous forests could persist within the warm-temperate zone in some interior parts of Japan (with colder winters). Such forests were called 'warm-temperate deciduous forests', occur also in China, and were treated in detail by Box and Fujiwara [16].

The US southeastern coastal plain is one of the world's largest regions in which the main substrate, geologically young sand, does not match the prevailing climate, which would normally produce zonal red-yellow podzolic clay soils. The coastal plain is delimited inland by the Fall Line, a slight escarpment (at which European explorers met the first river rapids) which still presents areas of nutrient-poor sandhills left from old oceanic shorelines. Most of the coastal plain is covered by relatively fine sand with a significant humic content (over limestone in Florida). This is not a bad substrate for plant growth but, in combination with the low-lying topography and warm climate, produces sharp vertical gradients between dry conditions above and wet, sometimes saturated conditions not far below the surface, where groundwater may be high, at least seasonally. This translates also into steep horizontal gradients, between distinctly hydromorphic vegetation in lower-lying areas and distinctly xeromorphic vegetation on areas only a few meters or only decimeters higher [122, 168, 169].

As in East Asia, the original forests of the US South were lost early, to plantation agriculture and to Northern shipbuilding [173]. In the US there was no tradition of preserving original forest, but some large evergreen broad-leaved trees, usually live oaks, have remained where agriculture was not possible, suggesting that such trees may have been important in the original forest. Due to recurring fire, both natural and man-induced (e.g. [3, 79]), forests probably never covered the coastal plain even where topography permitted. As a result of topography, substrate and fire, the "natural" vegetation of the southeastern coastal plain is a mosaic of deciduous forest, pinelands, sclerophyll live-oak woods and scrub, various kinds of wetlands, and agricultural land covers—albeit with increasing wooded cover throughout the 1900s due to land abandonment, except in Florida [127, cf 141, 60]. Some areas of evergreen broad-leaved forest were recognized fairly early (e.g. [48, 59, 84]), but most current forests on the coastal plain involve a variable mix of evergreen and deciduous broad-leaved trees called 'Southern Mixed Hardwoods' [126], which was eventually accepted as the potential climax forest type on the coastal plain (cf [80, 108]).

With global warming, this and more completely evergreen forest types could perhaps expand northward. What factors would constrain this expansion? Critical factors include not only the role of fire, nutrient requirements, and substrate limitations or affinities but also attributes of potential main structural tree species. Much information on evergreen broad-leaved taxa and woods is well known to local naturalists but needs to be published and examined within the context of geographic zonation, environmental constraints, and global warming. The purpose of this paper is to describe briefly the structure and dynamics of the main evergreen broad-leaved forest types of the southeastern US, and to examine constraints on their distributions and their prospects under global warming.

2 Data and Methods

The basic methodology of this study is to evaluate environmental conditions near the geographic limits of native evergreen broad-leaved forests and woodlands, and to explore the potential dynamics of these forests under environmental change, especially global warming and perhaps drying. Constraints on these forests are both climatic and substrate-related, so data are needed for both aspects, as well as for representative vegetation stands.

For vegetation, sites were sought to represent the full variation in evergreen broad-leaved forest types. This involved fieldwork in Florida, Georgia, and the Carolinas but extending west to the Hill Country of central Texas [14]. Forest stands were described by the relevé method of Braun-Blanquet [19], see [45], cf [112]. This method provides complete three-dimensional descriptions of stand composition and structure, and species cover and clustering (sociability), by layer. A relevé quadrat in a southeastern forest is typically about 25 × 25 m and requires about one hour to execute. The data-base for evergreen forests was supplemented, beginning in 1988, by relevés obtained during a three-year Japan-funded project to inventory the vegetation of eastern North America [105]. The resulting combined data-base involves about 100 relevés for evergreen broad-leaved or related forests.

Climatic data include long-term mean monthly temperature and precipitation at sites across the Southeast, but also extreme temperature minima. In particular, the 'absolute minimum temperature', i.e. the coldest temperature that can occur, was suggested quite early as an important limiting factor, especially for evergreen broad-leaved trees [20, 131, 132]. Even where measured, though, records of absolute minima are often not long enough to capture the true absolute minimum. The global geography of absolute minima is quite different from that of cold-month means, due to land-sea configurations, monsoon systems, etc. (Box 1995a). Thus one cannot simply assume a predetermined offset from cold-month mean temperature (as done by [125], cf [89]). On the other hand, reasonable estimates of absolute minima can be made based on the local difference between long-term absolute minima and the cold-month mean (program TXTRAP, cf [13,9]). The site values of absolute minimum temperature used herein are the lower of the measured value or the estimate.

Topography and substrate properties come from existing maps (e.g. [40, 41, 156, 67, 120]) but were corroborated by data recorded at the relevé sampling sites. The physical properties recorded at data sites are qualitative but include topographic position, general soil texture, and soil humic and moisture content (at sampling time).

3 The Coastal Plain Environment

Winter temperatures on the coastal plain involve January means near or only a bit below 10°C (higher in Florida), usually light frost, but absolute minima generally not below about −15°C (see Table 1). This makes the coastal plain a warm-temperate climate, not "subtropical" as erroneously claimed by various unfortunate variants of Köppen's terminology (cf [12], p. 11). In addition to mean temperatures, Table 1 also shows, for sites across the Southeast, the absolute minimum temperature, the number of consecutive warm months (mean at least 10°C); average precipitation for the year and the driest month; and the climatic moisture index (i.e. annual precipitation divided by potential evapotranspiration). The sites in Table 1 are listed from north to south, in five sub-regions. The Interior South includes inland sites on the piedmont or uppermost coastal plain, with winters that permit only deciduous forests. The Warm-Temperate Transition includes sites with absolute minima only a bit below −18°C. Sites in the true Warm-Temperate region generally have cold-month means of at least 8°C and absolute minima above or not far below −15°C. Most of Florida is also warm-temperate but shows higher mean minima and all 12 months above 10°C. Finally, data are also shown for subtropical Miami, where the cold-month mean is above 18°C and the absolute minimum is −2.8°C, which perhaps occurred only once.

Table 1 Climatic data for representative sites across the US South

	Lat	Elev	Tmax	Tmin	Tabmin	Lwarm	Py	Pmin	MIy
Interior South									
Lexington, Kentucky	38.0	298	24.6	0.5	-29.4	7	1128	63	1.49
Charlottesville, Virg.	38.0	265	24.7	2.0	-23.3	7	1160	74	1.45
Nashville, Tennessee	36.1	180	26.7	4.3	-27.2	8	1182	63	1.28
Little Rock, Arkansas	34.7	78	27.7	3.9	-25.0	9	1274	78	1.31
Oxford, Mississippi	34.4	116	26.7	5.2	-25.0	9	1346	70	1.39
Dallas, N Texas	32.8	134	30.0	7.2	-22.2	9	917	47	.82
Warm-Temperate Transition									
Baltimore, Maryland	39.3	4	25.7	1.5	-21.7	7	1076	76	1.35
Norfolk (SE Va coast)	36.9	8	25.7	4.9	-19.4	8	1186	72	1.32
Raleigh-Durham, N Car.	35.9	132	25.6	4.7	-22.8	9	1156	68	1.28
Columbia, S Carolina	34.0	74	27.1	7.9	-18.9	9	1128	58	1.08
Athens, Georgia	33.9	246	26.5	5.7	-20.0	9	1230	76	1.27
Montgomery, Alabama	32.3	67	27.6	8.1	-20.6	10	1326	61	1.23
Waco, central Texas	31.6	157	30.0	7.8	-20.6	10	800	42	.70
Austin, central Texas	30.3	193	29.3	9.7	-18.9	11	812	46	.68
Warm-Temperate									
Cape Hatteras (NC islands)	35.3	2	25.6	7.7	-14.4	9	1416	87	1.44
Wilmington (NC coast)	34.3	9	26.5	8.4	-17.8	9	1288	66	1.25
Charleston (SC coast)	32.8	3	27.7	10.1	-12.2	12	1221	61	1.09
Savannah (Georgia coast)	32.1	15	27.6	10.5	-16.1	12	1227	50	1.08
Mobile (Alabama coast)	30.7	66	27.3	10.6	-18.3	12	1607	84	1.40
Biloxi (Mississippi coast)	30.4	5	27.8	10.9	-17.2	12	1528	80	1.31
Baton Rouge, Louisiana	30.5	20	27.8	11.3	-16.7	12	1477	83	1.25
New Orleans, Louisiana	30.0	1	28.2	12.2	-13.9	12	1518	84	1.25
Houston, SE Texas	30.0	30	28.7	11.2	-15.0	12	1188	72	.98
Florida									
Pensacola (W Fla coast)	30.5	36	27.5	11.4	-15.0	12	1521	93	1.30
Tallahassee (NW Florida)	30.4	25	27.3	11.5	-18.9	12	1504	71	1.29
Jacksonville (NE Fla coast)	30.4	9	27.9	12.8	-13.9	12	1309	50	1.07
Gainesville (N Florida)	29.6	29	27.3	12.7	-14.4	12	1310	44	1.08
Tampa (mid-Gulf Florida)	28.0	8	27.9	15.9	-7.8	12	1254	43	.96
Sarasota (S-Gulf coast, Fla)	27.3	2	27.7	16.8	-5.6	12	1317	32	1.05
Subtropical									
Miami (SE Florida coast)	25.8	3	28.2	19.8	-2.8	12	1472	48	1.03

Abbreviations are as follow: Lat = latitude, Tmax = mean temperature of the warmest month, Py = average annual precipitation, Elev = elevation (m), Tmin = mean temperature of the coldest month, Pmin = avg. precip. of driest month, Tabmin= absolute minimum temperature, Lwarm = number of months above 10°C (mean), MIy = Py/PETy, where PETy = avg. annual potential evapotranspiration. All temperatures are in °C and all precipitation amounts in millimeters.

Temperatures in the Southeast are compared in Table 2 with those at comparable sites in East Asia and Europe. Milder winters extend farther north in Europe, as on all continental west sides. For east sides, at comparable latitudes and geographic position, mean winter temperatures are generally higher in eastern North America but absolute minima are lower than in East Asia [9]. In East Asia the monsoon system insures consistently low winter temperatures, but the east-west orientation of most mountain ranges blocks the extremely cold air masses from further north. North America has much less winter-monsoon effect, since it has the Gulf of Mexico to moderate winter cold; and it has no east-west mountains that affect the east. So winter means are higher and extremes are lower, especially in the southeastern US. It is for this reason that evergreen broad-leaved forests extend further north in East Asia (see map in Fujiwara and Box [46], p. 305; cf [137]).

Table 2 Minimum temperatures at comparable humid sites around the Northern temperate zone

Latitude	Southeastern US		East Asia		Europe	
43°	Durham (N.Hamp.)	-5.5 -34.4	Sapporo (Hokkaidō)	-5.4 -28.5	Marseille (France)	6.5 -16.8
42°	Boston (Mass.)	-2.6 -27.8	Hakodate (Hokkaidō)	-3.6 -21.7	Roma (Italia)	7.2 -8.7
41°	New York	-0.4 -26.1	Aomori (Honshū)	-2.6 -24.0	Porto (Portugal)	9.4 -4.1
40°	Wilmington (Delaware)	-0.3 -26.1	Beijing*	-4.5 -27.4	Madrid* (600m)	5.4 -14.8
39°	Baltimore (Maryland)	1.5 -21.7	Dalian (Liaoning)	-5.0 -21.1	Valencia (España)	11.0 -8.0
38°	Fredericksbg. (Virginia)	1.9 -29.4	Cangzhou (Hebei)	-2.9 -20.0	Palermo (Sicilia)	12.4 -1.7
37°	Norfolk (Virginia)	4.9 -19.6	Nagano* (Honshū)	-1.2 -17.0	Sevilla (España)	10.7 -5.0
36°	Raleigh* (NC)	4.7 -22.8	Tōkyō (Honshū)	3.8 -9.2	Tarifa (España)	13.2 -3.3
35°	Hatteras (NC)	7.7 -14.4	Mokpo (Korea)	1.2 -14.2	Iraklion (Kriti)	12.1 -0.5
34°	Wilmington (NC)	8.4 -17.8	Hiroshima (Honshū)	4.0 -8.6	Rabat (Maroc)	12.2 -3.2
33°	Charleston (SC)	10.1 -12.2	Fukuoka (Kyūshū)	5.2 -8.1	Hefa (Israel)	13.8 -2.8
32°	Savannah (Georgia)	10.5 -16.1	Nanjing (Jiangsu)	2.3 -13.8	Tel Aviv (Israel)	12.3 -2.0
31°	Brunswick (Georgia)	11.5 -15.0	Shanghai	3.4 -12.1	Agadir (Maroc)	13.9 -2.8
30°	New Orleans* (Louisiana)	12.2 -13.9	Chongqing* (Sichuan)	8.1 -2.5		
29°	Daytona B. (Florida)	14.5 -9.4	Quxian (Zhejiang)	5.2 -10.0		
28°	Tampa (Florida)	16.1 -7.8	Wenzhou (Zhejiang)	7.8 -7.5		
27°	Sarasota (Florida)	16.8 -5.6	Nanping (Fujian)	9.4 -6.1		
26°	Miami (Florida)	19.8 -2.8	Fuzhou (Fujian)	11.1 -3.0		

The sites are listed from north to south, are mostly near-coastal (except with asterisk), and are paired as far as possible by comparable geographic position (i.e. distance from a coastline). The top number at each site is the coldest-month mean temperature; the number directly under this is the absolute minimum temperature, i.e. the lowest ever measured (or expected, indicated by 'e', see Data and Methods). Temperatures for some additional US sites are given in Table 1.

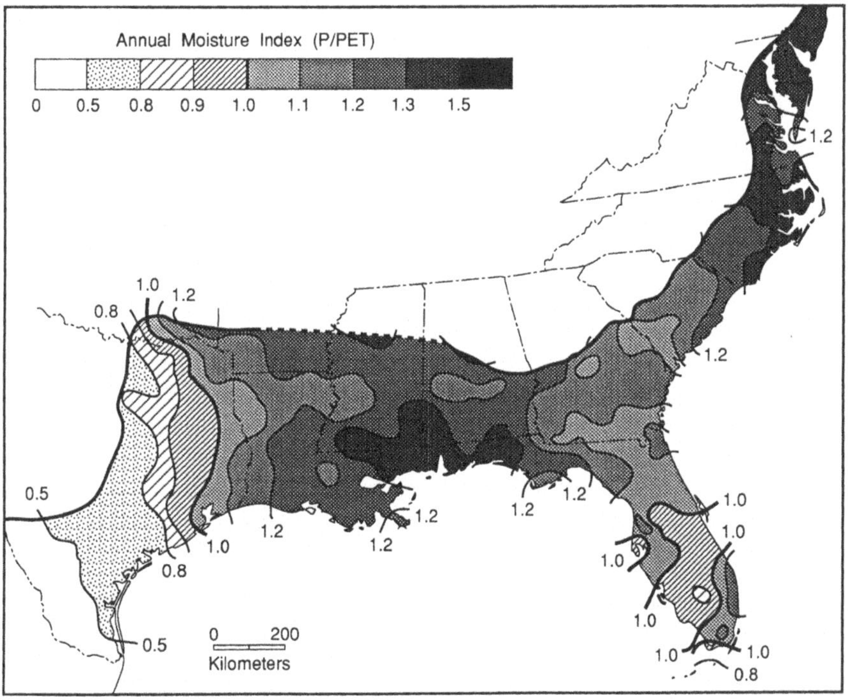

Fig. 1 Climatic Moisture Balance in the Southeastern USA. This moisture balance is estimated as average annual precipitation divided by annual potential evapotranspiration (PET). By this measure, the wettest areas on the southeastern coastal plain are along the northern coast of the Gulf of Mexico and the mid-Atlantic coast (New Jersey to North Carolina). The driest areas are in central to south-central Florida and in Texas, where precipitation decreases rapidly westward as Gulf moisture becomes less available to onshore winds. Note that most of South Carolina, southern Georgia, and northern Florida is also relatively dry, as is the area of the sharp gradient from Louisiana into east Texas. Annual PET is estimated in the southeastern USA by the method of Holdridge [64], which appears to be relatively accurate at warm-temperate latitudes [8]
Climatic Moisture Balance in the Southeastern USA. This moisture balance is estimated as average annual precipitation divided by annual potential evapotranspiration (PET). By this measure, the wettest areas on the southeastern coastal plain are along the northern coast of the Gulf of Mexico and the mid-Atlantic coast (New Jersey to North Carolina). The driest areas are in central to south-central Florida and in Texas, where precipitation decreases rapidly westward as Gulf moisture becomes less available to onshore winds. Note that most of South Carolina, southern Georgia, and northern Florida is also relatively dry, as is the area of the sharp gradient from Louisiana into east Texas. Annual PET is estimated in the southeastern USA by the method of Holdridge [64], which appears to be relatively accurate at warm-temperate latitudes [8]

The other major climatic factor for vegetation is available moisture, depicted here by the climatic moisture index (see above), as mapped in Fig. 1. In Louisiana and Arkansas, and westward into eastern Texas and Oklahoma, there is a sharp westward reduction in rainfall as less moisture becomes available to prevailing winds coming over land rather than the Gulf of Mexico. The pattern in Fig. 1 also suggests greater moisture stress in southern Georgia, South Carolina and especially central to south-central Florida, where much of the landscape is indeed more open. Forests in Florida are often topographically circumscribed and referred to as hammocks. Precipitation in the Southeast is highly variable from year to year, and relative to long-term seasonal averages. This is due partly to the inherent variability in the shape and position of the jet stream over North America, but perhaps even more due to El Niño events.

The southeastern climate is conditioned greatly by the surrounding ocean and Gulf of Mexico, which moderate winter temperatures (Fig. 2). As a result, the warm-temperate climate does not reach far inland, beyond the coastal plain, except through some of the broader river valleys, especially the Mississippi. The warm-temperate climate also does not extend to subtropical south Florida, where frost essentially disappears and forests are dominated rather suddenly by subtropical taxa [32], cf [93, 115]. Compared with other regions of humid warm-temperate climate, that of the southeastern US is more *subcontinental*, i.e. it shows relatively high temporal variability in both temperature and precipitation. The effect of this variability on vegetation is exacerbated by topography and substrate.

The substrate across the southeastern US coastal plain is largely humic fine sand, dotted with deep sterile sandhills representing old dune systems; moist depressions, often with shallow peat accumulations; and broad moist bottomlands along

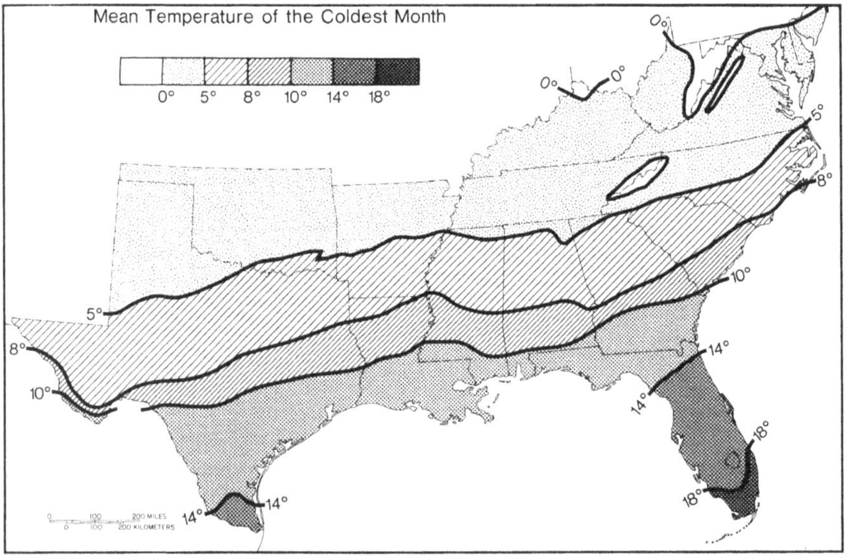

Fig. 2 Winter Temperatures in the Southeastern USA. The map shows the mean temperature of the coldest month (Tmin, in January), in degrees Celsius. Greller [56] also mapped the 5°C isotherm for mean nighttime minimum in January (running between the 10°C and 14°C Tmin isotherms) and related this to the northern boundary of his evergreen broad-leaved forest zone. At global scale, even short-term exposure to temperatures below about −15°C has been identified as potentially lethal to temperate-zone evergreen broad-leaved tree species (e.g. Sakai and Larcher [134]), and this isotherm generally follows the poleward limits of warm-temperate evergreen broad-leaved forests worldwide (e.g. Woodward [173]). In the Southeast, this −15°C isotherm for absolute minimum temperature runs near the 10°C isotherm for Tmin. Harder-leaved trees, such as live oaks and southern magnolia, may tolerate short-term exposure to slightly lower temperatures, perhaps as low as about −18°C, a value which may extend even to areas near the Gulf of Mexico coastline (e.g. Tallahassee)

tributaries of larger rivers (e.g. [27, 67, 169, 96, 120]). The lower-lying areas may have swamp forests (*Taxodium*, *Nyssa*, *Chamaecyparis*, etc.), fire-maintained 'pocosin' shrublands [140], and other wetlands or near-wetlands. Mesomorphic laurophyll "bay" forests may occur in moist depressions with peat accumulations, while xeromorphic scrub (with *Pinus*, deciduous and evergreen *Quercus*, etc.) dominates the highest, driest (and nutrient-poorest) deep-sand areas (e.g. [27]). A few areas do have relatively extensive richer soils. One is the loess hills of Louisiana, with silt-loam soils over loess [146], cf [35]. Another is the Apalachicola bluff area of western Florida, where limestone is near the surface and nutrient conditions may be dramatically better [59]. Such areas can support more mesomorphic laurophyll trees as well as more nutrient-demanding deciduous forests [36], cf [108]. All of this means that one must attempt to distinguish immediately between 'zonal' vegetation, i.e. that corresponding to the humid warm-temperate climate, on the one hand, and 'azonal' vegetation determined by unusual edaphic, topographic, or pyrogenic factors. On the southeastern coastal plain, azonal vegetation may cover a larger area than does the potential zonal vegetation. Such topographic contrasts are especially apparent in the climatically drier parts of the region (central Florida through the Carolinas) but occur even in the wettest areas as well (western Florida through southern Louisiana).

Fire, ignited by lightning strikes or by humans, is the joker in all this and is fostered by the often dry conditions above the surface [86, 117, 25, 78]. Climate determines the amount of fuel available and the length and severity of dry periods, but weather regulates the moisture content of dead fuels and the ignition and spread of wildfires. Wildfires burn off remaining or incipient broad-leaved forests on the uplands and maintain their replacing pine-palmetto savannas on lower flats [27]. Pollen analyses suggest that fire and the fire-adapted pine vegetation, including the characteristic *Pinus palustris-Aristida stricta* open woodlands, expanded rapidly to cover large areas of the coastal plain about 5000–6000 years ago [166, 167].

4 Evergreen Broad-Leaved Tree Taxa

The evergreen broad-leaved woody flora of the southeastern US is not large, as compared with comparable parts of East Asia (cf [46]). This may be due to biogeographic history and the small size, throughout evolutionary time, of the near-coastal, milder-winter region in the southeastern US. The evergreen broad-leaved trees native to the southeastern US are listed in Table 3 and fall into two distinctly different groups, namely trees with: 1) hygromorphic, shade-tolerant dark green laurophylls (*Lorbeerblätter*, [128], cf [73, 74, 157], as in laurel forests elsewhere), or 2) distinctly harder, xeromorphic leaves with a wide range of shade tolerance (but less than true laurophylls). The main laurophylls are the three "bay" species: *Persea borbonia* (Lauraceae, "red bay", plus almost identical *P. palustris*), *Magnolia virginiana* ("sweet bay"), and *Gordonia lasianthus* (Theaceae, "loblolly bay"), so called because they are slightly aromatic and were used by early settlers as

Table 3 Main evergreen broad-leaved woody taxa in the temperate US South

Laurophylls			
Quercus laurifolia	Fagaceae	canopy tree	floodplains
Gordonia lasianthus	Theaceae	tree	moist peaty soils
Magnolia virginiana	Magnoliaceae	tree	mesic to moist soils
Persea borbonia	Lauraceae	tree	mesic to moist soils
Persea palustris	Lauraceae	small tree	moist peaty soils
Vaccinium arboreum	Ericaceae	small tree	humic sand to mesic
Osmanthus americanus	Oleaceae	arborescent	mesic
Ardisia escallonioides	Myrsinaceae	arborescent	mesic (subtropical)
Rapanea (=*Myrsine*) *guianensis*	Myrsinaceae	arborescent	mesic (subtropical)
Illicium floridanum	Illiaciaceae	large shrub	rich moist soils
Lauro-sclerophylls			
Magnolia grandiflora	Magnoliaceae	tree	mesic to moist soils
Ilex opaca	Aquifoliaceae	tree	mesic to moist soils
Coriaceous			
Quercus hemisphaerica	Fagaceae	pioneer tree	humid sand to mesic
Ilex cassine	Aquifoliaceae	small tree	humic sand to mesic
Ilex vomitoria	Aquifoliaceae	small tree	mainly humic sand
Ilex myrtifolia	Aquifoliaceae	smaller tree	mesic to moist
Lyonia ferruginea	Ericaceae	small tree	mesic bottomlands
Leucothoe axillaris	Ericaceae	large shrub	mesic bottomlands
Symplocos tinctoria (semi-EG)	Symplocaceae	arborescent	mesic
Cyrilla racemiflora (late-decid.)	Cyrillaceae	arborescent	wet depressions
Cliftonia monophylla	Cyrillaceae	arborescent	wet depressions
Myrica cerifera	Myricceae	shrub	mesic to wet depressions
Sclerophylls			
Quercus virginiana	Fagaceae	canopy tree	humid sand to mesic
Quercus fusiformis	Fagaceae	tree	dry to mesic, incl. limestone
Quercus geminata	Fagaceae	small tree	deep, nutrient-poor sand
Quercus myrtifolia	Fagaceae	shrub	nutrient-poor sand

Laurophylls are darker, glossy green and shade-tolerant, as in the "laurel forests" of East Asia, southern Brazil—and tropical rainforests. Coriaceous leaves are pliable but usually thicker, and are usually somewhat less shade-tolerant. Sclerophylls are hard (rigid), resist water loss better, and are usualy not shade-tolerant, although *Q. virginiana* does regenerate well under its own canopy. Lauro-sclerophylls are an unusual combination of hardness and shade-tolerance, both of which lead to slower growth rates.

cooking substitutes for 'bay leaves' (*Laurus nobilis*). There are also two prominent but somewhat puzzling lauro-sclerophylls, *Magnolia grandiflora* and *Ilex opaca* (much like European *I. aquifolium*), which have quite hard but dark, shade-tolerant leaves, thus grow slowly, but can reach canopy height with time.

The most important and widely occurring tree is *Quercus virginiana* (live oak), whose spreading crown, long lifespan (to 300 years or more), and intermediate shade tolerance permit it to grow up and dominate, where topography and history permit, across most of the coastal plain. *Q. virginiana* is evergreen, not semi-evergreen, and dominates forests that are warm-temperate, not subtropical or only maritime, as characterized erroneously by Thyroff et al. [153] in an otherwise useful documentation of live oak's phenotypic plasticity (e.g. sclerophyllous to semi-sclerophyllous; cf [111]). *Q. virginiana* may dominate from maritime forests ([7, 17, 170], pp 185) to broad coastal-plain uplands (e.g. [115, 170], pp 188) and west to the more continental but still warm-temperate Hill Country of central Texas, where it hybridizes with very similar *Q. fusiformis* [158], cf [113, 152]. *Q. virginiana* also extends into south Florida, where it has long been seen as a "successional" tree in subtropical hammocks (e.g. [1], and even into western Cuba. This may be an example of the contention by Loehle [92] that trees often grow well at their southern range limits and are limited there by competition rather than physiology.

Two other important oaks may deserve the descriptor 'semi-evergreen', namely *Q. laurifolia*, a bottomland species that does lose many leaves in the northern part of its range (near the Fall Line); and *Q. nigra*, a widespread "tardily deciduous" tree that may become essentially evergreen at the southern end of its range in central Florida [68, 113]. The other important evergreen oak is *Quercus hemisphaerica*, a fast-growing, mainly upland secondary species (red-oak group), which was long confused with *Q. laurifolia* (white-oak group) [42], cf [154, 155]. The most sclerophyllous oaks are small-tree *Q. geminata* and shrubby *Q. myrtifolia*, which may dominate clearly on deeper, nutrient-poor sandhills and sandy flats. Other arborescent evergreen hollies include coriaceous *Ilex vomitoria* and *I. cassine*, both of which can be quite abundant locally. Somewhat similar ecologically are the coriaceous-leaved understorey trees *Vaccinium arboreum* and *Lyonia ferruginea*. These evergreen broad-leaved species and their ecology are described in various local identification manuals, the best of which are by Godfrey [52], Kurz and Godfrey [82], Hunt [68] and Muller [113] (but see also [29, 38], and Lakele and Wunderlin [87]).

The occurrence of distinctly hygromorphic and xeromorphic evergreens results from the strong moisture gradients in the sandy landscape. The relative shade tolerance and growth rates of laurophyll and sclerophyll evergreens, and deciduous trees, are significant for vegetation dynamics on sufficiently mesic sites, where all three forms can occur. The laurophylls are the true counterparts of the "laurel forest" taxa of East Asia and elsewhere, and are found on the US southeastern coastal on sites that remain moist or at least mesic throughout the summer, such as shallow depressions (see [170], p 74). These laurophyll species are more shade-tolerant than deciduous species or other evergreen species, and thus can form permanent "bay forests" where permanent moisture availability permits. The oaks and other broad-leaved evergreens grow more slowly (except *Q. hemisphaerica*, e.g. [51]) and are potential dominants on the humic sands across most of the coastal plain. The most sclerophyllous oaks (*Q. geminata* and shrubby *Q. myrtifolia*) dominate on deep, nutrient-poor sandhills (e.g. [27]).

5 Evergreen Broad-Leaved Forest Types

As abandoned upland areas grow back in humid warm-temperate climates, in the absence of fire, they pass through a succession from initial colonizers, to tall grasses, then light-demanding pines, then more shade-tolerant deciduous trees, which can invade once the pines have provided a less harsh, shadier, more humid microclimate. Where winters are not too cold, the deciduous trees can then be replaced at least partly by broad-leaved evergreen trees. The basic hypothesis is that, eventually, shade-tolerant laurophyll trees or other evergreen broad-leaved trees such as oaks, with intermediate shade tolerance, will replace deciduous trees with higher light requirements, as in East Asia [11, 47]. The logic of shade tolerance as a mechanism of succession in temperate forests was suggested early by Cowles [31] and explained further by Bazzaz [4], but without reference to evergreen broad-leaved trees, which do not occur in their northern areas of the USA. The shade-tolerance theory of succession was extended to include laurophyll trees through experience in East Asia, especially the southern half of Japan, where the replacement of deciduous forests (long thought to be the potential) by evergreen laurel forests was documented by Miyawaki [100], albeit without explaining shade-tolerance as the mechanism (cf [103, 107]). Evergreen trees have several advantages over deciduous trees: earlier shoot growth and photosynthesis in spring (e.g. [24]), other competitive advantages in forest canopies (e.g. [37, 62, 116], Cavender-Bares et al. [23], [124]), greater shade tolerance, even in evergreen oaks (e.g. Thyroff et al. [152]); the ability to regenerate under deciduous canopies as well as under their own shade; and permanent occupation of canopy space. This is true to some extent also for more xeromorphic *Q. virginiana*, which does regenerate under its own, less dense canopy (many personal observations).

Although the coastal-plain climate suggests evergreen broad-leaved forests, this potential was not recognized until recently by most American botanists and ecologists. Syntheses describing the biomes and other vegetation regions of North America have treated the southeastern coastal plain variously as "longleaf-pine forest" [135], "southeastern pine forest" [139], "southeastern evergreen forest" [18], "southern mixed forest" [80], "pine forest" (e.g. [39]), "warm-temperate mixed forest" [91] or "Vegetation of the Southeastern Coastal Plain" [27]. The term "Southern mixed hardwoods" was enshrined by Quarterman and Keever [126] to describe potential dominance by a varying mix of deciduous and evergreen broad-leaved tree species, plus conifers (cf [108]). Only a few writers, mainly from outside the US, recognized the region as potential "evergreen broad-leaved forest" (e.g. [137]). Probably the first American to describe the evergreen broad-leaved forest zone as such was Greller [55, 56, 57], who expanded Braun's [18] regionalization of the eastern US forests to include a semi-evergreen transition and a zone of potential evergreen broad-leaved forest.

As a first step in clarifying the position of evergreen broad-leaved forests in the Southeast, two attempts were made by the present authors to identify basic types of such forests and related vegetation. The first attempt (see Table 4, from [14]) recognized five types of evergreen broad-leaved forest on humic sand, as well as laurophyll "bay" forests in scattered moist depressions, oak-dominated floodplain forests, and more open *Q. geminata* woodlands on deep sterile sand. The second attempt (see Table 5, from Fujiwara and Box [46]) was based on phytosociological tablework and recognized types by actual or potential dominant species, mainly *Q. virginiana* but also successional *Q. hemisphaerica*, harder-leaved *Q. geminata* on sterile sand, laurophylls in moist depressions, and *Q. laurifolia* in bottomlands and floodplains. More floristic detail and consideration of zonal and pedogenic position for mostly the same types was provided later by Greller [57], who also related his types to those of other authors. It is not possible to treat all these types here, so we take six main types, each of which is described briefly below, with some consideration of dynamic status: zonal (climatic climax), topogenic, edaphic, or pyrogenic. Examples are given in the Appendix (Relevé Tables A.1, A.2, A.3, A.4, A.5

Table 4 Evergreen broad-leaved forest types identified by Box and Fujiwara [14]

Forest Type	Habitat	Canopy species	Understorey species
Evergreen Oak Forests			
Maritime oak forest	dunes	*Quercus virginiana*, *Q. geminata*, (with *Magnolia grandiflora*)	*Serenoa repens*, *Myrica cerifera*, *Ilex vomitoria*, *Juniperus*
Coastal oak forest	coastal sands (flat)	*Quercus virginiana*, *Q. laurifolia* *Q. hemisphaerica* (*Magnolia grandiflora*, *Carya*)	*Persea borbonia*, *Lyonia ferruginea*, *Vaccinium arboreum*, *Serenoa repens*
Mesophytic forest	moist depressions	*Quercus laurifolia*, *Q. virginiana*, *Magnolia grandiflora* (*Carya glabra*)	*Ilex opaca*, *Persea borbonia*, *Lyonia*, *Vaccinium arboreum*, *Osmanthus*, *Serenoa repens*
Oak-palm forest (transition to subtropical)	mesic /wet	*Quercus virginiana*, *Q. laurifolia*, *Q. hemisphaerica*, *Sabal palmetto*	*Sabal palmetto*, *Serenoa repens*, *Persea* (*Ulmus*, *Acer*, *Carya*)
Upland oak forest	mesic/ subhumid	*Quercus virginiana*, *Q. hemisphaerica*, (*Carya*? *Magnolia*?)	*Ilex vomitoria*, *Persea borbonia*, *Magnolia virginiana*, *Carya*
Xeric oak woodlands	subhumid	*Quercus virginiana*, *Q. geminata*, (*Q. fusiformis*) *Juniperus*	*Ilex vomitoria*, *Rhus* spp., *Juniperus*, Rosaceae, Rhamnaceae
Laurophyll "bay" forests	wet depressions	*Magnolia virginiana*, *Persea* spp., *Gordonia lasianthus*, (*Nyssa biflora*, *Taxodium*)	*Persea*, *Gordonia*, *Ilex*, *Lyonia*, *Nyssa*, *Myrica*, *Rhododendron*, *Acer rubrum*
Floodplain forests	briefly flooded	*Quercus laurifolia*, *Q. michauxii*, *Magnolia virginiana*, *Acer rubrum*, *Liquidambar*, *Pinus*, *Liriodendron*	*Ilex opaca*, *Acer rubrum*, *Magnolia* , *virginiana*, *Persea*, *Liquidambar*, *Ulmus* (*Sabal*)
Swamp forests (deciduous canopy)	hydric	*Taxodium* spp., *Nyssa biflora*, *N. aquatica*	*Persea*, *Magnolia virginiana*, *Acer rubrum*, *Fraxinus*, *Myrica*

All taxa listed are evergreen except: *Carya*, *Liquidambar*, *Liriodendron*, *Nyssa*, *Q. michauxii* and coniferous *Taxodium*, plus (mainly in understoreys) *Acer*, *Fraxinus*, *Ulmus*, Rosaceae and Rhamnaceae. The *Carya* species is mainly *C. glabra*, but others may occur on the more deciduous uplands. The main *Nyssa* species is *N. biflora* (plus *N. sylvatica* in drier uplands), but *N. aquatica* may dominate in quite hydric situations. *Taxodium* is mainly *T. distichum*, but shorter *T. ascendens* also occurs in some ponds.

Table 5 Evergreen broad-leaved woody communities identified by Fujiwara and Box [46]

Forest/Woodland Communities	Habitat	Canopy species	Understorey species
***Quercus virginiana* communities**			
Q. virginiana-Magnolia grandiflora	humic sand	***Q. virginiana*, *Magnolia grandiflora***, plus ep. *Tillandsia usneoides*, *Polypodium polypod.*	*Ilex opaca*, *Persea borbonia*, *Vaccinium arboreum*
Q. virginiana-Sabal palmetto (central Florida)	moist	***Q. virginiana*, *Sabal palmetto*** (T2), plus epiph. *Tillandsia* spp, *Nephrolepis*, *Polypodium pol.*	*Psychotria nervosa*, *Eugenia*, spp, *Ardisia*, *Oplismenus*
Q. virginiana-Serenoa repens	coastal dunes	***Q. virginiana***, plus epiphytic *Tillandsia usneoides*, *Polypodium polipod.*	***Serenoa repens***, *Persea*, *Ilex* spp
Q. virginiana-Pinus taeda (coastal NC-Virg)	sand	***Q. virginiana*, *Pinus taeda***	*Vaccinium* spp, *Myrica cerifera*, *Acer rubrum*
Q. virginiana-Juniperus ashei (central Texas)	limestone	***Q. virginiana*** (cf, *Q. fusiformis*) (*Q. texensis*, *Q. sinuata* var. *breviloba*)	***Juniperus ashei***, *Mahonia trifoliata*, *Yucca rupicola*
***Quercus hemisphaerica* communities**			
Typical type (to 36m)	humic sand	***Q. hemisphaerica***, *Q. virginiana*, *Magnolia grandiflora*, ep. *Tillandsia usneoides*	*Ilex opaca*, *Vaccinium arboreum*,
Q. hemisph.-Serenoa repens	humic sand	***Q. hemisphaerica***, *Magnolia grandiflora*, *Q. virginiana*, ep. *T. recurvata*, *T. usneoides*	***Serenoa repens***, *Ilex vomitoria*, *Persea*, *Vitis rotundifolia*
Q. hemisph.-Juniperus silicicola	coastal dunes	***Q. hemisphaerica***, *Pinus taeda*	*Juniperus silicicola*, *Ilex opaca*, *Persea*, *Vitis rotundifolia*
***Quercus laurifolia* communities**			
Q. laurifolia-deciduous	bottomlands	***Q. laurifolia***, *Q. nigra*, *Magn. virginiana* *Acer rubrum*, *Liquidambar* (*Taxodium*)	*Ilex opaca*
Q. laurifolia-Sabal palmetto (central-south Florida)	floodplains	***Q. laurifolia*, *Sabal palmetto*** (T2) *Tillandsia usneoides* (*Taxodium distichum*)	*Persea*, *Acer rubrum*
Q. laurifolia-Sabal minor (north from Florida)	seasonally flooded	***Q. laurifolia***	*Sabal minor*, *Ilex opaca*, *Magnolia virg.*, *Acer rubrum*
***Quercus geminata* communities**			
Q. geminata-Serenoa repens (sometimes under sparse pine)	sand	(*Pinus clausa*) *Tillandsia recurvata*, *T. usneoides*	***Q. geminata*, *Serenoa***, *Q. myrtifolia*, *Q. chapmanii*, *Q. laevis*, *Lyonia lucida*, *Aristida stricta*
Pinus palustris-Q. geminata (savanna-woods)	sand	*Pinus palustris*	***Q. geminata***, *Persea* spp, *Aristida stricta*
Q. hemisphaerica-Q. geminata	sand	***Q. hemisphaerica*, *Q. geminata***, (*Q. virginiana*, *Q. laevis*, *Q. margaretta*)	***Serenoa***, *Persea*, *Vaccinium* spp, *Aristida stricta*
Bay Forest (laurophyll) communities			
Magnolia virginiana-Persea palustris	humic depressions	***Magnolia virginiana***, sometimes *Nyssa biflora*, epiphytic *Tillandsia usneoides*	***Persea*** spp, *Ilex* spp, *Acer rubrum*, various ferns
Gordonia lasianthus-Lyonia lucida	humic depressions	***Gordonia lasianthus***, sometimes*Pinus serotina*, with *Magnolia virginiana*, *Persea*, plus *Nyssa biflora*, *Liquidambar*, *Acer rubrum*	***Lyonia lucida***, *Ilex coriacea*, *I. glabra*, *L. mariana*, *Cyrilla*
***Bursera simaruba* communities** (subtropical)			
Bursera-Coccoloba Hammocks	sand over limestone	***Bursera simaruba***, *Coccoloba diversifolia*, *Simarouba glauca*, *Mastichodendron*, *Ficus aurea*, *Q. virginiana*; successional canopy of *Lysiloma latisiliqua*	*Nectandra coriacea*, *Eugenia axillaris*, *Ardisia escallon.*, *Psychotria nervosa*, sometimes *Krugiodendron*, *Sabal palmetto*
Thickets with *Metopium toxiferum* (Florida Keys)	limestone	***Bursera simaruba***, *Maniklara bahamensis*	***Metopium toxiferum***, *Thrinax*, *Guapira*, *Capparis*, *Eugenia*

All taxa listed are evergreen except: *Quercus nigra*, *Q. michauxii*, small tree *Q. laevis*; *Q. texensis* and *Q. sinuata* var. *breviloba* in Texas; some *Vaccinium* spp., *Acer rubrum, Carya glabra, Liquidambar, Nyssa biflora, Lyonia mariana*, subtropical *Lysiloma latisiiqua,* and coniferous *Taxodium*. Relevés in central Texas were all near Austin and show *Q. virginiana*, but almost indistinguishable *Q. fusiformis* occurs near there and further west. The communities are described in more detail in [46].

and A.6) and illustrate most, if not all, of the common characteristics of these main evergreen broad-leaved forest types.

Bay forests (also "bayheads" or "baygalls"; see Relevé Table A.1) are dominated by shade-tolerant laurophyll trees and thus represent the only true counterparts to the "laurel forests" of East Asia. On the US coastal plain, though, they occur only in small, moist depressions scattered very widely over the otherwise sandy substrate. Such depressions often have a thin accumulation of peat, which keeps them permanently moist through the hot summers (see [170], p 74). The main tree taxa are a small sample from the main laurophyll genera of East Asia: *Persea borbonia*, *P. palustris*, *Magnolia virginiana*, and *Gordonia lasianthus*. Of these, *M. virginiana* is the most common and occurs also in floodplains and some more mesic upland forests and woody savannas. The combined canopy and tree understorey of bay forests is usually fairly dense and may also contain evergreen *Ilex opaca*, *I. cassine*, and occasional remnant pines or deciduous trees. As a result, the lower understorey is quite variable and consists largely of juveniles, other shade-tolerant woody species such as laurophyllous *Ilex* spp. and *Lyonia lucida*, and sometimes abundant ferns. Vines and epiphytes may occur but are not abundant. Bay forests are most common from the Carolinas to central Florida. The examples are from the southern part of this range, where *Gordonia* becomes more prominent.

Beech-Magnolia forests (see Relevé Table A.2) are co-dominated by deciduous beech (*Fagus grandifolia*) but are included here because they, plus a variant with co-dominant [evergreen] *Ilex opaca*, present the most luxurious development of evergreen southern magnolia (*Magnolia grandiflora*). These forests were recognized early [48, 123] as a potential climatic-climax forest type at least on certain mesic upland sites on the southeastern coastal plain (see also [80]). Due to the generally droughty sand substrate and the prevalence of fire, however, they are found only in small, topographically protected areas with better soils. Based on old land records from 1821, widespread forest cover by "primeval" beech-magnolia forest was reconstructed by Delcourt and Delcourt [34] on the loess hills of Louisiana; similar beech-magnolia forests have been described on mesic sites in the Apalachicola bluffs area of western Florida [5, 36, 83]. The "purest" old-growth stands are not species-rich, may have few epiphytes and little understorey, vines (e.g. *Smilax*) only in lower layers, and may have only stately beech and magnolia in canopies that can reach 40 m (e.g. [172]). Other forests co-dominated by *Magnolia grandiflora* include some bluff forests on seepage slopes in southern Georgia ([170], pp 169) and the *Magnolia grandiflora-Ilex opaca* type of Veno [159] in northern Florida.

The stability of beech-magnolia associations was questioned due to poor regeneration by magnolia under its own dense canopy, although it does regenerate with an admixture of other canopy trees [81], cf [43]. As a result, beech-magnolia is now often considered as a variant of the highly variable but self-perpetuating Southern Mixed Hardwoods, now accepted as the potential forest type because continuing human disturbance insures greater variability in canopy and overall diversity. Relevé Table A.2 shows an old-growth beech-magnolia forest in Louisiana, protected from fire by its location in the bend of a river [172]. The other two examples are from species-rich ‘southern mixed hardwoods’ forests in western Florida, with beech-magnolia co-dominance. The first (Apalachicola Bluff) is on a gentle “steephead” slope (cf [83]) and the other in a slight depression.

Live Oak-Magnolia forests (see Relevé Table A.3), i.e. a combination of *Quercus virginiana* (live oak) and *Magnolia grandiflora*, were first described as a consistent association by Daubenmire [33], cf [85], and are potentially the most widely occurring evergreen broad-leaved forest type across the somewhat drier uplands as well as on old coastal dunes and other drier (i.e. sandier) lowland areas. Although live oak is the potential long-term dominant, slow-growing magnolia remains in the canopy and understorey. The canopy may also include occasional deciduous companions, especially *Carya glabra* and *Q. nigra*. Palmettos (*Sabal* or *Serenoa*) may occur in understoreys, along with evergreen shrubs, but the overall number of species may not be large.

The maritime version of this forest type, called ‘maritime forest’ since Bourdeau and Oosting [7], is illustrated by the first example (Naval Live Oaks) in Relevé Table A.3, which is from coastal dunes near Pensacola (western Florida). This forest actually contains more magnolia than shown in the relevé. The best inland example in Relevé Table A.3 is the last (Dudley Hammock, see photo in [170], p 188), well protected on a military reservation in southeastern Georgia. The middle example is from a sandy levee along the Mississippi River near its delta south of New Orleans. In this variant there is little magnolia, perhaps due to the high groundwater. Very species-poor live oak-magnolia forests were also found by the authors on very low-lying areas along the Texas coast north of Corpus Christi.

Upland Oak forests (cf Relevé Table A.4) are mainly secondary and differ from the previous type by having less live oak and magnolia, at least in their current composition. They are usually dominated currently by *Quercus hemisphaerica*, a pioneer evergreen oak that colonizes wide areas after forest cutting or other wide-area disturbance (e.g. [51]). Such woods are found especially in various state parks in Florida, where *Q. hemisphaerica* grew up quickly on abandoned land acquired by the state in the first half of last century. If live-oak propagules are available, and there is no fire, these forests may succeed to the live oak-magnolia type described above and could cover wide upland areas; if not, due to ongoing disturbance, they may remain for decades as a mix of co-dominant evergreen (*Q. hemisphaerica*) and deciduous oaks (mainly *Q. nigra* and *Q. falcata*). *Q. hemisphaerica* lives only about 100 years, but in some cases its dominance in relatively mature stands may continue for decades, as in the Kolomoki Mounds (southwestern Georgia) example in Relevé Table A.4, which has no live oak at all (cf [51]). The other examples, both from northern Florida, have live oaks but only in the canopy. These forests continue west to southeastern Texas and can be fairly species-rich, with a relatively consistent woody composition (as suggested by the three examples).

Oak-Palm forests (cf Relevé Table A.5) combine live-oak (*Q. virginiana*) canopies with often dense palmetto (*Sabal palmetto*) subcanopies, plus other oaks and some subtropical understorey species (cf [147]). These forests are often circumscribed by topography into discrete hammocks, and occur mainly in central to south-central Florida, in the transition from the temperate-zone flora to the north and the subtropical flora to the south [32]. Live oak occurs mainly in the canopy, while *Sabal* shows better regeneration. *Tillandsia* species are especially common, as well as vines and other epiphytes, both warm-temperate and subtropical. Subtropical species increase southward, but there may also be some admixture of *Q. laurifolia* or *Q. hemisphaerica* (not both together), depending on site conditions. The first and last examples in Relevé Table A.5 are from central Florida; the middle example (Little Corkscrew, southwestern Florida) illustrates the southward increase in subtropical epiphytes.

Floodplain and Bottomland forests in the South (cf Relevé Table A.6) occur in many forms, some entirely deciduous, many semi-evergreen, and some mainly evergreen. The most common evergreen broad-leaved dominant of floodplain forests is *Q. laurifolia*, but *Magnolia virginiana* may be co-dominant or at least abundant in many situations, along with various deciduous canopy and understorey trees. *Q. virginiana* and *Magnolia grandiflora* are much less common in floodplains. Warm-temperate vines and epiphytes are common, while composition of the herb layer depends strongly on topography and hydrology. The first example (Pearl River), with strong dominance by *Q. laurifolia*, is from southeastern Louisiana (cf [171]). The last two examples are from bottomlands surrounded by sandy uplands: Tiger Creek in central Florida and the Beidler Forest near the Fall Line sandhill area of central South Carolina, below Columbia. Despite the distance between these examples, their tree composition is remarkably consistent.

Other evergreen broad-leaved woods on the coastal plain include, in particular, the xeric upland oakwoods on the deep, sterile sand of old dunes. These oakwoods are dominated by *Q. geminata*, a shorter, often visibly scrubby oak tree with sclerophylls harder than those of *Q. virginiana*. There are also oak-pine sandhill savannas with deciduous *Q. laevis*, and shorter oak scrub co-dominated by evergreen *Q. geminata*, *Q. myrtifolia* and *Q. chapmanii*; these two sandhill types are distinguished by fire regime [114].

6 Constraints and Vegetation Dynamics

Forests may be constrained by limits on their main tree species or on forest development itself. The main evergreen broad-leaved trees are constrained first of all by climatic conditions, in particular winter cold and increasing dryness westward (and southward in Florida). For temperature, at least three possible limitations have been suggested. Mean temperature of the coldest month was used by Köppen [76] to delimit temperate climates, and the 0°C isotherm for mean January temperature was shown by Hocker [63] to fall close geographically to the northern limit of "southern pines"; also, the northern limit of laurel-forest trees along the northern Pacific coast of Japan falls near the 0.5°C mean January isotherm. Due to variations in continentality, though, there is no consistent relationship worldwide between winter mean temperatures and the boundaries of extra-tropical vegetation types (except in the case of permafrost). On the other hand, Greller [56] showed that the 5°C isotherm for mean *minimum* temperature in the coldest month does match the northern limit of southeastern evergreen broad-leaved tree species fairly well. Unfortunately, this relationship cannot be tested globally because this minimum gets defined differently in different data sources.

More important may be the clear physiological limitation by ice formation within plant cells, which can occur during even short-term exposure [129, 20, 50], cf [134], and desiccation damage due to frozen soil (e.g. Sakai [130], [151]). For temperate-zone evergreen broad-leaved tree species, absolute minimum temperature appears to be a limiting factor globally [174]. Estimates of this limiting 'cardinal temperature' have converged at about −15°C [20, 88, 131–133, 165]. This would appear to limit the region of potential evergreen broad-leaved forests and woodlands in the Southeast to most of Florida, the southern part of Georgia, and narrow strips along the Carolina and Gulf coasts. Harder-leaved evergreen broad-leaved trees (i.e. *Quercus virginiana* and *Magnolia grandiflora*) survive somewhat lower absolute minima (to about −18°C; cf [132]) but are outside their natural range further north, where planted or escaped trees do survive near settlements and sometimes in closed forests (numerous personal observations).

The other main climatic constraint, of course, is available moisture. Topography and substrate aside, the geographic pattern of climatic moisture (see Fig. 1) suggests forest potential all across the Southeast, but decreasing rapidly west of the Mississippi River and less rapidly southward in Florida. There is a corresponding general gradient in vegetation, with forests declining in stature, then opening up into open woodlands, and finally giving way to grasslands or savannas. Within this climatic framework, vegetation varies over much shorter distances in the landscape-scale mosaic of wetter and drier areas dictated by topography, substrate and recurring fire. The constraints are obvious, but understanding their effects is complicated because, even on drier surfaces, groundwater may not be far below the surface, at least in winter and spring. This groundwater effect is involved in the concept of 'flatwoods' (well illustrated by [170], pp 194) and can be seen well at the University of Florida preserve near Welaka, which preserves examples of 'xeric hammock' (*Q. virginiana*) with lower groundwater; 'mesic hammock' (*Magnolia grandiflora-Ilex opaca*) with higher groundwater; and 'hydric hammock' (evergreen *Quercus* spp.-*Sabal palmetto-Liquidambar*) with groundwater near or at the surface, at least seasonally (see initial descriptions by [84], cf [57]).

The potential water holding capacity of upland sandy soils may be influenced more subtly by another factor, namely their humic content, which depends on the rate at which plant litter decomposes and is made available for incorporation into the soil. In the early 1980s, Meentemeyer [99] made a globally applicable climatic model for potential litter decomposition rate (percent annual litter disappearance), and Box mapped the balance between annual primary production and the amount of litter decomposed. (This was done using a world mapping system which was lost completely when science was shoved off mainframe computers and their high-speed page printers, onto do-it-all-yourself desktop machines). The resulting map was thus never published but divided the world into regions where litter could potentially accumulate (be produced faster than it could decompose) or decompose completely, with no accumulation (decomposition potentially faster than production). The dividing line between these regions of potential litter accumulation and disappearance ran through the US South, roughly along the Fall Line. Of course, not all litter is the same: evergreen leaves with more lignin generally decompose more slowly than do deciduous leaves [98]. Nevertheless, there is the suggestion that microclimatic conditions may play a role in limiting the amount of humus available to developing soils. Humus may not be incorporated rapidly into clay per se, but it can be held on the soil surface if there is a more or less permanent layer of dead leaves and other fine litter. If the litter decomposes too fast, as in the warmer winters of the coastal plain, less humus would be available for eventual incorporation into the soil. Of course, a humus layer also provides a more humid microclimate for the establishment of species of the middle and latter stages of succession.

The particular problem in the US South is that there are four potential limitation mechanisms for evergreen broad-leaved forests, and the critical values for all four fall close together geographically, near or along the Fall Line boundary between the coastal plain and the piedmont. The two non-climatic boundaries occur at the Fall Line: 1) the major substrate discontinuity between coastal-plain sand and piedmont clay; and 2) the more hypothetical boundary between climates of potential litter accumulation versus disappearance. The two thermal boundaries represent infrequent low temperatures: 3) the 5°C isotherm for mean minimum winter temperatures [56], and 4) the −15°C isotherm for absolute minimum temperature (although in reality, unusual extremes occur in spots, with some outliers still further south, e.g. −18°C in Tallahassee; cf Table 1). These isotherms fall across the middle of the coastal plain, near the Florida-Georgia border a bit further south (and closer to the coast in the Carolinas and westward to Texas). This location approximates the transition between mainly evergreen live-oak forests and semi-evergreen "southern mixed hardwoods" forests, with their substantial deciduous components.

Numerous species range boundaries fall in or near this narrow region of multiple hypothetical limitations, but there are also constraints on forest development in general. The first of these is fire, which may be caused by lightning strikes or set by man [86, 25, 78]. Florida, in particular, is one of the most lightning-prone areas in the whole world. Fire may not always kill mature trees, but it does remove seedlings, saplings and young trees quite effectively. Bay forests, beech-magnolia forests, floodplain forests, maritime live-oak forests, and even the transitional oak-palm forests are usually circumscribed topographically and thus relatively well protected from fire. The live oak-magnolia and upland-oak types, though, can be quite vulnerable to fire and thus to retrogression to the fire-maintained (pyroclimax) pine savannas that have historically covered large expanses on the coastal plain (cf [27]). If the fire cycle could be broken, the broad uplands might become live oak forest – but there is always a natural source of new fires: dry years and lightning. The interplay of fire, topography, substrate and potential vegetation is illustrated especially well by Wharton [170, p 178].

Another constraint on the development of evergreen broad-leaved forests involves competition from deciduous trees, which generally grow faster (e.g. [138])

and have an advantage where they cannot be shaded out by more shade-tolerant evergreens. Due to more rapid growth and response to disturbance, deciduous trees may also have an advantage near the northern limit of the evergreen broad-leaved trees (cf [69, 92, 116]). Sensitivity to winter cold may also limit broad-leaved evergreens northward even before lethal cold is reached ([174], p 69). Most of the native evergreen broad-leaved tree species seem quite able to grow on piedmont substrates, but they may not be as able as deciduous trees to grow densely and rapidly enough to form and dominate forests. Secondary deciduous forests may succeed to evergreen live-oak forests only if live oaks can create sufficiently dense canopies. Drier climatic conditions may result in less dense evergreen canopies (as well as more fires) and thus a reduced advantage of shade tolerance. This too suggests a greater role for deciduous trees, especially with continuing disruption due to warming and drying.

Finally, in today's world there is another major factor that influences forests, namely break up of continuous woody vegetation cover into ever smaller fragments by man's activities. This reduction of extensive forest canopies permits more light penetration from the sides—as well as population fragmentation and various edge effects. Landscape fragmentation and other disturbance thus favor deciduous trees, as depicted in Fig. 3 (from [11]). As a result, much of the currently remaining forest

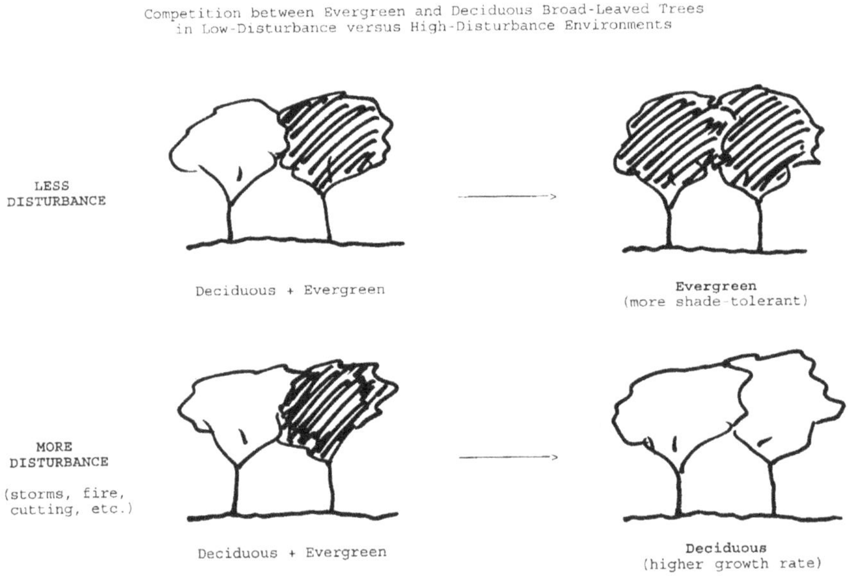

Fig. 3 Results of Competition between Evergreen and Deciduous Forest Canopies in Humid Undisturbed versus Disturbed Landscapes (from [11]). In humid climates, evergreen canopy trees are generally more shade-tolerant than are deciduous canopy trees. Thus, in undisturbed landscapes, where forest canopies may be relatively closed over large areas, the more shade-tolerant evergreen trees generally shade out and replace deciduous trees. On the other hand, in disturbed landscapes, where forests are much more fragmented and more light is available laterally, faster-growing deciduous trees are usually able to retain their canopy dominance

and woodland is deciduous or at best only semi-evergreen, depending not only on physical factors but even more on recent land-use history and the resulting local availability or lack of live-oak propagules.

7 Conclusion

Unlike warm-temperate climatic regions elsewhere, the southeastern US is climatically drier and the coastal plain provides a harsh, immature, nutrient-poor substrate in which water and nutrient conditions vary drastically over short distances, from deep, xeric sands to permanently moist depressions, seasonally inundated flats, or even more hydric bottomland forests and swamps. As a result, the original landscape was a forest-scrub-wetland mosaic, and its current potential is not for "laurel forests" over large areas but rather, where soil, topography and fire permit, for a perhaps slow progression to domination by coriaceous-leaved, even sclerophyll forests and woods with somewhat more open canopies and a greater role for deciduous trees and dense understoreys, even at maturity. Where limestone underlies shallow sand, as in northern and central Florida, the greater availability of nutrients seems to favor deciduous trees (cf [94, 95, 109]). Even so, large trees of *Q. virginiana* or *Q. hemisphaerica* may also occur, suggesting that these evergreen species may have been dominant or co-dominant before the original forests were cut. As in East Asia, secondary forests on the coastal plain are overwhelmingly deciduous. The long periods of time required for evergreen broad-leaved trees to return to dominance, perhaps 100 years, explain partly why the regional forest potential has been described as mainly deciduous "Southern mixed hardwoods."

Taken together, these environmental conditions suggest a complicated pattern of vegetation limitation across the coastal plain and raise many questions concerning vegetation dynamics and status. Zonal warm-temperate laurophyll forests occur on the coastal plain only in moist depressions that compensate for drier overall conditions, as suggested by the 'law of relative habitat constancy' ([163], pp 32–33). The term 'climax' has been criticized frequently, especially in Europe, but does have the advantage of permitting useful, widely used terminology, such as climatic climax, edaphic climax, and fire climax [21, 28], cf [97]. These terms can reflect the vegetation status more completely than the single term 'potential natural vegetation' – which still has not resolved the role of fire in vegetation. (Can a fire-climax vegetation type be seen as potential natural vegetation, *sensu* Tüxen or Miyawaki?) Pyrogenic vegetation types, such as pine woodlands/savannas, represent arrested development of vegetation which might, in the absence of fire, succeed to zonal vegetation on "zonal" sites or to topo-climax or edaphic-climax vegetation on corresponding azonal sites.

Since most currently mature evergreen broad-leaved forest stands were not observed (or identified as important) during earlier successional stages, it is difficult to match secondary forests and mature evergreen broad-leaved forests in characteristic successional sequences. Many studies document changes in species composition, invoking "succession" but without identifying any mechanism except fire suppression (e.g. [61, 85, 110, 159]). Evidence that evergreen broad-leaved forests can replace deciduous forests must come from remaining large trees (e.g. *Q. virginiana*) and from good regeneration of their seedlings and saplings in shady understoreys.

How might these limiting factors, and vegetation, change with global warming and other environmental changes? Isotherms for mean and mean-minimum winter temperatures will move northward with general warming, but it is much more difficult to anticipate absolute minima, which may be the best temperature delimiter for temperate-zone evergreen broad-leaved tree species and for a warm-temperate climate in general (see global systematization in [12], cf [174]). In Florida, for example, many new absolute minima were reached during the 1980s, when winters elsewhere in eastern North America were returning to normal after the unusually cold winters of the latter 1970s. This may be due to weakening of the polar vortex with overall global warming, which lets the coldest polar air masses escape southward more easily [77]. Even if the [potential] absolute minima remain the same, though, they may become less frequent. In addition, most global and regional models have suggested that global warming will bring generally drier conditions in the US South (e.g. [70, 90, 164]). Ever since their beginnings in the 1980s, global climate models have consistently suggested that, in a warmer world, the subtropical high-pressure belts would move farther away from the equator, in both the northern and southern hemispheres ([148], cf [2]). The recent drier conditions and more frequent, larger wildfires in the western USA, Australia, and some Mediterranean areas, suggest that this is already happening, at least on continental west sides.

Warming will mean two things: the northern climatic limit for evergreen broad-leaved trees would move northward, onto the clay substrates of the piedmont; and potential evapotranspiration rates would be higher. If this latter is not compensated by sufficient increases in precipitation, there would be net drying. Net drying would mean that:

- current areas of open woodland and pine savanna (e.g. central Florida) would become even more open;
- current areas of upland forest may become too dry to support closed forest;
- some bottomland and floodplain forests, and some wetlands, could contract;
- some moist flats (e.g. pocosins) and depressions (bay forests) could also dry out, at least partly; and
- .the overall potential for evergreen broad-leaved forests, and other forests, would be threatened.

Some forest types are somewhat protected by topography, against drying as well as fire, but drying may make them more susceptible to encroachments by fire from surrounding landscapes. Upland forests, already in the driest situations, will become threatened even more by fire than now. Although changing physical conditions will influence species range boundaries, these will depend also on properties of the species themselves: phenology and abilities to reproduce, germinate, colonize and become established, and to compete with other species.

Warming and drying may also affect some non-climatic factors. Topography and substrate type will not be affected by global warming, except perhaps along the immediate coast, due to sea-level rise. Soil humic content may not be affected quickly, but soil moisture and near-surface temperatures and humidity will be affected (warmer and drier), with consequent effects on litter decomposition, infiltration of water into the soil, seedling establishment and thus succession. These consequences may be difficult to anticipate, and the high variability between wet and dry years, as well as between warmer and colder winters, will make interpretation of constraints and future potentials even more difficult. Wildfires can be expected to become more frequent. Meanwhile, landscape fragmentation continues, causing further drying and local warming.

The main purpose of this study was thus to identify the types of evergreen broad-leaved forest that do occur and try to understand their constraints. From this information it may be possible to infer forest dynamics and ultimate potential, based on understorey composition, remaining large individual trees in similar situations, and observed patterns of species replacement over sufficient periods of time. As a final prognosis, though, we must unfortunately expect a reduced potential for warm-temperate evergreen broad-leaved forests on the coastal plain, with more open landscapes and a greater role for faster-growing deciduous species, including subtropicals. The area in which winters permit evergreen broad-leaved trees may well expand northward, in cities as well as countryside. Even in the absence of land conversion, other disturbance by humans, and fires, however, the area over which evergreen broad-leaved trees can form more or less permanent evergreen broad-leaved forests will probably be reduced. There seem to be two main unknowns:

1. Milder winters may not change absolute temperature minima significantly; in some places, new record lows continue to occur, in spite of more general global warming.
2. Can the coastal-plain canopy dominants, especially live oak, colonize and capture areas currently covered by deciduous forest on piedmont red clay? Live oaks can grow on clay, as in piedmont cities, but can they colonize these areas and outcompete the existing forest?

Mangrove species are being seen further north in Florida (e.g. [22]), but otherwise there has been very little published evidence of subtropical species moving northward or of evergreens moving from the coastal plain onto the piedmont. This is partly because individual observations and anecdotes are not publishable units, which require well documented multi-year studies. One exception to this lack is the *Cinnamomum-Ardisia* forest near Tallahassee documented by Clewell and Tobe [30]. At any rate, species are probably migrating northward without being recognized, but changes in forest structure will be slow to follow [92].

Appendix: Composition and Structure of the Main Evergreen Broad-Leaved Forest Types

The six main types of evergreen broad-leaved forests in the southeastern USA, as identified in the main text, are described here by means of Braun-Blanquet relevés, which show species composition and abundance in individual forest layers and thus the physical structure of the forests themselves. Each of the following six tables (identified as Relevé Tables A.1–A.6, as opposed to the Tables in the main text) includes three relevés, from different areas. Each relevé has four columns, for the T1, T2, S and H layers of the respective forests. The numbers in these layer columns represent species cover and sociability on the normal Braun-Blanquet scale of 1-5, with a plus sign (+) for sporadic occurrence.

Table A.1 Laurophyll "Bay" forest examples from the Southeastern US coastal plain

	Persea Bay Lake Louise, Ga				Mag.virg-Gordonia Lake Kissimmee, Fl				Mag.virg-Gordonia Tiger Creek, Fl			
	18m	9m	3m	.5m	18m	8m	4m	1m	24m	13m	5m	1m
	75%	10%	70%	15%	70%	60%	30%	50%	85%	20%	20%	60%
Magnolia virginiana	2.2	+.2	1.2		4.4	3.3	+	1.2	3.3	2.2	2.3	1.2
Gordonia lasianthus		1.2			1.1	3.3	2.2	+	3.3	1.1	2.3	+.2
Persea palustris	3.3	1.2		2.2		+	2.2	+				
Persea borbonia											1.2	+.2
Nyssa biflora	1.1	+.2			1.1	+	+		2.2			
Acer rubrum									2.2	2.2		
Pinus elliottii					1.1			+				
Pinus serotina	+											
Tillandsia usneoides						+.2	+.2		+.2	+.2		
Ilex opaca						1.2		+				
Ilex cassine										2.2		
Myrica inodora						1.1	1.2					
Ilex coriacea			3.3	+.2								
Lyonia lucida			2.2	1.2							1.1	
Rhododendron viscosum			2.2									
Clethra alnifolia v. tom.			1.2									
Itea virginica											2.2	
Viburnum cassinoides			+	+								
Smilax laurifolia			1.2			+	+					+.2
Vitis rotundifolia						+	+	+	+.2			1.1
Rhus radicans			+	+								
Blechnum occidentale												3.3
Osmunda cinnamomea				+.2				3.4				3.3
Woodwardia areolata				1.2								+.2

All three sites were in slight depressions, some with peat, in the otherwise sandy coastal plain topography. Lake Louise is in south-central Georgia and has a boardwalk-accessible "bay" forest on one side. Lake Kissimmee is in central Florida, where ***Gordonia lasianthus*** becomes more prominent. Tiger Creek is a very diverse preserve, also in central Florida. Evergreen species are shown in boldface. Numbers under the headings represent the height and cover percentage of each stand layer (T1, T2, S and H). The following companion species occurred only infrequently (+ or +0.2) in a lower layer (unless epiphytic) of only one relevé.

At Lake Louise, S: ***Myrica cerifera***, ***Leucothoe axillaris***, *Vaccinium stamineum*, *V. atrococcum*. At Lake Kissimmee, T2: ***Tillandsia fasciculata***, ***T. setacea***; S: ***Vaccinium arboreum***, *Decumaria barbara*; H: ***Quercus hemisphaerica***, *Osmunda regalis*, *Vaccinium* "hairy", Araceae sp., and Rosaceae "maloid". At Tiger Creek, S: *Vaccinium* sp.; H: ***Tillandsia recurvata***.

Table A.2 Beech-Magnolia forest examples from the US Southeastern coastal plain

	Fag-Mag + Illicium Zemurray, La				Rich Fagus-Magnolia Apalachicola Bluff				Rich So. Mixed Hwds. Wakulla Springs			
	40m 80%	20m 20%	5m 60%	.2m 10%	40m 80%	20m 20%	3m 15%	.6m 20%	32m 80%	13m 20%	6m 30%	.8m 35%
Fagus grandifolia	4.4		1.1		3.3				3.3		+	+
Magnolia grandiflora	2.2	1.1	1.2		2.3	1.2	+.2		2.2	2.2	+	+
Liquidambar styraciflua	2.2				2.2		+		2.2		+	+
Quercus alba	2.1				1.1				2.2			
Quercus michauxii	1.1	1.1			1.1			+	2.2		+.2	+.2
Quercus shumardii									2.2	1.1	+	
Quercus nigra	+						+			1.1	+.2	
Carya glabra					2.2		+		1.1	1.1	1.2	
Carya cordiformis									2.2		+	
Fraxinus americana									1.1		+.2	
Tilia americana									2.2		1.1	
Liriodendron tulipifera					1.1							
Pinus glabra	+			+								
Quercus muhlenbergii					+							
Platanus occidentalis					+							
Ilex opaca						1.1	+	+		2.2	3.3	1.2
Persea borbonia							+.2			1.1	1.1	
Prunus caroliniana			1.1				+.2	+.2		1.2		
Carpinus caroliniana		1.2	1.2	+		1.2		+.2			+	
Cornus florida						1.1	1.2	+			1.1	
Ostrya virginiana						2.2	2.2					
Halesia diptera		1.2	1.2									
Symplocos tinctoria		1.2	2.2	+.2								
Nyssa sylvatica		+				+	+					
Illicium floridanum			3.4	2.2			+.3					
Ilex vomitoria			+.2								1.2	
Quercus hemisphaerica							1.1	+			2.3	
Hamamelis virginiana			+				+				+.2	+.2
Carex "narrow"				2.2				2.2				
Chasmanthium sessiliflorum				1.2								
Hexastylis arifolia				1.2				+.2				
Mitchella repens				+				+				1.2
Anisostichus capreolata				+.2				+				+
Rhus radicans				+				+		+.2		1.1
Vitis cinerea									+			
Vitis rotundifolia				+						+		
Tillandsia usneoides						+	+					
Polypodium polypodioides						+	+					
Quercus austrina							+.2	+				
Juniperus virginiana							+	+				
Cercis canadensis							+	+			1.1	
Callicarpa americana							+	+			1.1	
Cornus foemina							+				+.2	
Asimina parviflora											2.2	
Osmanthus americanus											1.2	
Smilax smallii			+	+								
Smilax laurifolia											1.1	
Smilax bona-nox												3.4
Smilax pumila												2.2
Euonymus americanus												2.2
Polystichum acrostichoides								1.2				
Thelypteris kunthii								1.3				
Tradescantia subaspera								1.1				
Carex baltzellii								1.2				
Arisaema dracontium								+				+
Dioscorea quaternata								+			+	
Oplismenus setarius								+.2				1.2
Asplenium platyneuron												1.2
Viola walteri												1.2
Panicum sp.												1.2

The Zemurray Estate is in eastern Louisiana, north of Lake Ponchartrain, and contains classic examples of tall, mature beech-magnolia forest, sometimes with relatively little understorey. Apalachicola Bluff is a TNC preserve in western Florida and contains especially species-rich examples of beech-magnolia forest with other canopy trees and rich understoreys. Wakulla Springs is a state park in northern Florida, near Tallahassee, and also contains a species-rich beech-magnolia forest with other canopy trees. These last two stands could also be considered especially species-rich examples of "Southern Mixed Hardwoods" (sensu Quarterman & Keever 1962). Evergreen species are shown in boldface. (*Polypodium polypodioides* is poikilohydrous and perhaps best described as semi-evergreen.) The numbers under the headings represent the height and cover percentage of each stand layer (T1, T2, S and H). The following companion species occurred only infrequently (+ or +0.2) in one lower layer of only one relevé:

At Zemurray, S: ***Magnolia virginiana***, *Trichomanes petersii*; H: *Erythrina herbacea*, *Epifagus virginiana*, *Vaccinium tenellum*, ***Ligustrum sinense***, *Elephantopus tomentosus*, *Smilax rotundifolia*, ***Smilax glauca***, *Smilax* sp., *Vitis* sp., Liliaceae sp., *Panicum* sp.

At Apalachicola Bluff, T2: *Halesia caroliniana*, *Planera aquatica*, *Prunus* (decid.); S: ***Vaccinium arboreum***, ***Pinus taeda***, *Hydrangea quercifolia*, *Aesculus pavia*, ***Smilax auriculata***; H: *Ampelopsis arborea*, *Passiflora lutea*, *Parthenocissus quinquefolia*, *Agrimonia microcarpa*, *Cnidoscolus stimulosus*, *Croomia pauciflora*, *Sabaletonia*, *Chasmanthium latifolium*, *Aster* sp., *Elephantopus* sp., *Senecio* sp., *Asplenium* sp., *Viola* sp., *Panicum* "narrow", *Panicum* "wide'
At Wakulla Springs, S: *Aralia spinosa*, *Bumelia lanuginosa*, *Chionanthus virginicus*, *Crataegus marshallii*, *C. viridis*, *Ilexambigua*, *Morus rubra*, *Prunus serotina*, ***Sabal palmetto***, *Ulmus alata*, *Vaccinium stamineum*, *Viburnum nudum*, *Menispermum canadense*; H: *Botrychium dissectum*, *Desmodium galactia*, *Rubus flagellaris*, *Ruellia caroliniensis*, ***Sabal minor***, *Solanum americanum*, *Uniola laxa*, *Desmodium* "big", *Sanicula* sp., *Carex* "short", *Carex* "big".

Table A.3 Live Oak-Magnolia forest examples from the Southeastern US coastal plain.

	Maritime Live Oak Naval Oaks				Live Oak Levee Mississippi Delta				Mature Oak-Magnolia Dudley's Hammock			
	18m	8m	3m	.4m	22m	12m	4m	1m	26m	16m	7m	.3m
	75%	30%	50%	30%	85%	15%	20%	85%	85%	30%	60%	2%
Quercus virginiana	4.4	2.2	2.3	2.2	5.5				4.4			1.2
Magnolia grandiflora	+	1.1		+						3.3		
Pinus glabra			(+)						2.2	1.2	2.2	+
Carya glabra	1.1		+						1.1	1.2		
Quercus nigra					1.1	1.1		+	3.3	2.2	1.2	1.2
Quercus hemisphaerica		1.2									+	+.2
Vitis rotundifolia	1.2	1.1			1.2						+.2	+.2
Tillandsia usneoides		+							+.2	1.2	1.1	
Quercus alba										1.1		
Nyssa biflora										1.1	+	
Asimina parviflora		+.2	+									
Serenoa repens			2.3								2.3	
Vaccinium arboreum			2.2	1.2							2.2	
Persea borbonia			1.1			2.2	1.2	+				
Lyonia ferruginea											3.3	
Vaccinium myrsinites			1.1									
Vaccinium stamineum			+.2								+.2	
Vaccinium elliottii											3.3	
Vaccinium corymbosum											1.2	
Gaylussacia frondosa											1.2	
Anisostichus capreolata			1.2	1.2								
Rhus copallina			+	+								
Campsis radicans					2.3			1.2				
Rhus radicans					1.2	1.1	+	+				
Berchemia scandens					1.1		+	+				
Ulmus americana						2.2	1.2					
Acer rubrum v. drummondii						+						
Sabal minor							2.2	+				
Liquidambar styraciflua							+				+	
Osmanthus americanus											1.1	
Ilex opaca											+	+
Thelypteris kunthii								5.4				
Polypodium polypodioides									+.2	1.2	+.2	
Mitchella repens												1.2
Carex sp.				1.2								

The Naval Live Oaks is a preserve of maritime live-oak forest on sand only a few meters above the nearby sea level, near Pensacola in western Florida. The Mississippi Delta site is on a levee (sand) of the Mississippi river in Plaquemines Parish, southeast of New Orleans, at elevation of only a very few meters above sea level (high groundwater). Dudley's Hammock is an especially good example of a mature oak-magnolia hammock, located and well preserved within Moody Air Force Base in southeastern Georgia. Numbers under the headings represent the height and cover percentage of each stand layer (T1, T2, S and H). Evergreen species are shown in boldface. (*Polypodium polypodioides* is semi-evergreen, and *Quercus nigra* is "tardily deciduous".) The following companion species occurred only infrequently (+ or + 0.2) in only one lower layer (S or H) of only one relevé:

At Naval Oaks, S: ***Smilax laurifolia***, ***Smilax smallii***; H: ***Ilex vomitoria***, *Hypericum reductum*, *Solidago* sp. on the Mississippi levee, S: *Cornus foemina*, *Sambucus canadensis*; H: ***Smilax rotundifolia***, *Parthenocissus quinquefolia*, *Tradescantia* sp.
At Dudley's Hammock, S: *Quercus michauxii*, *Carya tomentosa*, *Acer rubrum*; H: ***Persea palustris***, ***Lyonia lucida***, ***Gelsemium sempervirens***, ***Smilax bona-nox***, *Pteridium aquilinum*, *Uniola laxa*, *Scleria triglomerata*.

Table A.4 Upland Oak forest examples from the Southeastern US coastal plain.

	Q.hemisphaerica Kolomki Mounds				Upland Live Oak Suwannee River				Live Oak Hammock UF Welaka Preserve			
	25m	12m	4m	.6m	24m	12m	6m	.4m	30m	15m	6m	.6m
	80%	20%	45%	20%	85%	20%	40%	25%	85%	20%	70%	5%
Quercus virginiana					5.4				4.4			
Quercus hemisphaerica	5.5	2.2	3.1	2.3	1.2	2.2	2.3	2.2	3.3	+.		+.3
Magnolia grandiflora		2.2	3.3	1.2			+			1.1		
Pinus glabra	2.2		+				+					
Carya glabra					2.2		+					
Quercus nigra				+			1.2	2.2	2.2			+.2
Sabal palmetto								1.1		2.2	4.4	1.2
Pinus taeda					1.1							
Vitis rotundifolia	2.3	1.2	1.1	1.2		1.1	+		1.2		1.1	+
Rhus radicans		1.1		2.2				1.1				
Gelsemium sempervirens		+	+	1.2				+				
Smilax bona-nox				1.2			1.1	2.2				
Parthenocissus quinquefolia				+								+
Anisostichus capreolata								1.1				
Tillandsia usneoides		1.2	+.2		+.2	+.2	+.2		1.2	1.2		
Polypodium polypodioides					+.2	1.2		+	1.2	+.2		
Prunus serotina		1.1	+.2	+.2			+	+.2		+		
Morus rubra		1.2										
Ilex opaca			1.1			1.1	+	3.3				
Vaccinium arboreum			+				1.2	+.2				
Liquidambar styraciflua						1.1	+					
Celtis laevigata			1.2				+					
Fraxinus americana			+				+					
Callicarpa americana			2.2				+.2				+	+.2
Viburnum prunifolium			1.2									
Cornus florida			+				+					
Amelanchier arborea			+				+					
Bumelia lanuginosa			+				+.2					
Vaccinium elliottii			+				+.2					
Vaccinium stamineum							1.2				+	
Ilex ambigua							1.1					
Nyssa sylvatica							+	+				
Mitchella repens				1.3				2.3				
Elephantopus tomentosus								1.1				
Panicum "big"								1.1				1.2

Kolomoki Mounds is a state park in southwestern Georgia and contained this especially good example of an upland ***Quercus hemisphaerica*** forest that may succeed to ***Q. virginiana*** dominance. In the 1980s, many state parks in Florida were in this stage of succession. The Suwanee River site, already dominated by live oak, is in a state park in northern Florida, on a humic-sand upland near the river. The Welaka site is a preserve owned by the University of Florida, in north-central Florida (southeast from Gainesville); it contains good examples of this "moist" live-oak hammock as well as several other forest types. Numbers under the headings represent the height and cover percentage of each stand layer. Evergreen species are shown in boldface. (*Polypodium polypodioides* is semi-evergreen.) The following companion species occurred only infrequently (+ or +0.2) in only one lower layer (S or H) of only one relevé.

At Kolomoki Mounds, S: *Fagus grandifolia*, *Carya tomentosa*, *Nyssa biflora*, *Diospyros virginiana*, *Acer leucoderme*, ***Euonymus americanus***, *Rhamnus caroliniana*, *Aralia spinosa*, ***Nandina domestica***, ***Smilax laurifolia***, *Ribes* sp.; H: *Sassafras albidum*, ***Yucca filamentosa***, ***Sabal minor***, ***Asplenium platyneuron***, *Smilax herbacea*.

At Suwanee River, S: *Asimina triloba*, *Chionanthus virginicus*, ***Ilex vomitoria***, ***Juniperus virginiana***, *Viburnum nudum*, *Campsis radicans*; H: *Smilax pumila*, *Cocculus carolinus*, *Cnidoscolus stimulosus*.

At the Welaka Preserve, S: ***Persea borbonia***, *Malus* sp.; H: *Smilax rotundifolia*, *Galium* sp.

Table A.5 Oak-Palm forest examples from the Southeastern US coastal plain

	Live Oak-Sabal Lake Kissimmee				Rich Q.virg-Q.laur Little Corkscrew				Live Oak-Sabal Merritt Island			
	30m	12m	4m	.7m	22m	12m	4m	1m	20m	14m	4m	.5m
	70%	35%	60%	15%	70%	40%	30%	20%	80%	40%	40%	20%
Quercus virginiana	4.3				3.3			+.3	5.4			1.2
Sabal palmetto	2.2	3.3	4.3	1.1	1.2	3.3	2.2	2.2	+	3.3	+.2	1.1
Quercus laurifolia					3.3		+.2	+.2	2.1			
Quercus hemisphaerica		1.3		1.1								
Ulmus americana									1.1		+	
Tillandsia usneoides	1.2	1.1	+		+.2	1.2			2.2	2.2	+.2	
Celtis laevigata		1.2	2.2						+	+	+.2	+
Citrus sinensis		+.2	1.2							+.2	+.2	
Carya glabra		+										
Quercus nigra			+	+								
Diospyros virginiana			+			+	+	+				+
Callicarpa americana			1.2	+							+	
Erythrina herbacea			+.2								+	
Parthenocissus quinquef.		1.1		1.1			+	+		+		+.2
Rhus radicans		+.2		1.3		1.2	+			+.2	+	1.1
Vitis aestivalis			+						1.3			
Vitis rotundifolia			+.2	+								
Gelsemium sempervirens			+.2	+.2								
Smilax auriculata			+.2	+							+	
Smilax bona-nox				+				+				
Smilax hispida								1.1				
Berchemia scandens									+		+	
Campsis radicans								+				+
Polypodium polypodioides		+.2				1.2	+	+.2		+		
Tillandsia recurvata		+.2			+	1.2				+.2		
Tillandsia setacea					+.2	+				+.2		
Tillandsia utriculata					+.2							
Tillandsia fasciculata						+				+		
Tillandsia polystachya						+.2						
Tillandsia "short yellow-green"						+.2						
Nephrolepis exaltata						+.2						
Encyclia tampensis						+						
Catopsis floribunda						+.2						
Polypodium aureum						+.2						
Morus rubra										3.3	+	
Eugenia sp.										2.2	3.3	2.2
Myrsine floridana						+	2.3	1.1			1.2	
Psychotria nervosa											3.3	+.2
Ardisia escallonioides											+.2	+
Blechnum serrulatum				1.3				2.3				
Woodwardia sp.				1.3								
Paspalum sp.								1.2				

Lake Kissimmee is in central Florida, surrounded by low flat terrain with high groundwater. Little Corkscrew Island is in the Corkscrew Swamp sanctuary and is the farthest south of all relevés shown; this represents the transition region from temperate to subtropical flora, as suggested by the numerous epiphytes. Merritt Island is on the central east coast of Florida, near Cape Canaveral. Numbers under the headings represent the height and cover percentage of the vegetation layers (T1, T2, S and H). Evergreen species are shown in boldface. (*Polypodium polypodioides* is semi-evergreen.) The following companion species occurred only infrequently (+ or + 0.2) in only one lower layer of only one relevé:

At Lake Kissimmee, S: ***Cinnamomum camphora***, *Cephalanthus occidentalis*, *Myrica inodora*, *Asimina obovata*,
Apios americana, ***Eugenia*** sp., *Desmodium* sp., *Polypodium* "big", Malvaceae sp.; H: ***Ilex opaca***, ***Citrus paradisi***, *Oplismenus setarius*, *Psoralea* sp., *Hypericum* sp., *Galium* sp., *Carex* "big", *Panicum* sp., Compositae sp.
At Little Corkscrew, T2: ***Ficus aurea***; S: ***Persea borbonia***, *Bumelia reclinata*, ***Myrica cerifera***, ***Serenoa repens***, ***Schinus terebinthifolius***, ***Vittaria lineata***; H: ***Baccharis halimifolia***, *Boehmeria cylindrica*, *Rubus argutus*,
Mikania scandens, ***Smilax*** sp., *Viola* sp., *Erechtites* (?) sp., *Arundinella* (?) sp., Rubiaceae sp.
At Merritt Island, T2: *Persea palustris*; S: *Acer rubrum*, ***Illicium floridanum***, ***Psychotria sulzneri***, Convolvulaceae sp., ***Ligustrum sinense***; H: *Mikania cordifolia*, *Ampelopsis arborea*.

Table A.6 Bottomland evergreen oak forests on the Southeastern US coastal plain

Bottomland Evergreen Oak Forests on the Southeastern US Coastal Plain

	Quercus laurifolia Pearl River Basin				Mag.virg-Q.laur Tiger Creek				Qu.laurifolia Beidler Forest			
	28m	18m	6m	.6m	27m	14m	4m	.8m	20m	13m	4m	.9m
	80%	20%	20%	30%	80%	15%	25%	30%	75%	40%	20%	80%
Quercus laurifolia	5.4	2.3		2.3	2.2	1.2		+.2	4.4	2.2		1.2
Magnolia virginiana					2.2		2.2				1.2	+.2
Liquidambar styraciflua	1.2								1.2		+	
Quercus nigra	1.1			+.2	2.3	1.1						
Quercus michauxii	1.1									1.1	+	
Acer rubrum		1.2	+.2		2.3		+.2.	+	2.2	1.1		+.2
Nyssa biflora					2.2				1.1	+		
Ulmus americana		1.1										
Fraxinus americana		1.1										
Fraxinus caroliniana										1.1		+
Liriodendron tulipifera									1.1	+		
Carya glabra									1.1			
Pinus glabra										+	+	+.2
Sabal palmetto						2.2	2.2	1.2				
Sabal minor			+									+.2
Ilex cassine						1.2	+					
Carpinus caroliniana			1.2							3.3		1.2
Quercus lyrata			+	+								
Halesia diptera		+	+									
Itea virginica			+				1.2				+	+.2
Cephalanthus occidentalis			+				1.1	+				
Berchemia scandens				1.2						1.1	+.2	
Campsis radicans				1.2				+.2				
Gelsemium sempervirens				+.2				+.2				+
Smilax laurifolia				+								+
Smilax rotundifolia				+						1.1	+.2	+
Rhus radicans				+		+.2		1.2				1.2
Boehmeria cylindrica				+				+				+
Saururus cernuus				+.2								+
Vitis rotundifolia						1.2		+.2		+	+	
Tillandsia usneoides						1.2	+					
Tillandsia fasciculata						+.2						
Persea palustris							1.2	+.2		1.1	1.2	+.2
Ilex opaca										1.1	+	
Myrica cerifera							1.2				1.2	+
Diospyros virginiana							+.2				+	
Sambucus canadensis							+	+				
Smilax bona-nox							+					+
Blechnum serrulatum								2.2				
Osmunda regalis								2.3				+.2
Osmunda cinnamomea								2.2				
Houstonia sp.								2.3				
Rhynchospora sp.								2.3				
Cyperaceae sp.								2.3				
Panicum sp.								1.2				
Symplocos tinctoria											1.2	
Vaccinium tenellum											2.2	
Celtis laevigata											+	+
Decumaria barbara											+.2	1.2
Carex sp.												3.3
Mitchella repens												2.2
Hypericum hypericoides												1.2
Lyonia lucida												1.2
Vaccinium myrsinites												1.2
Elephantopus nudatus												1.2
Solidago arguta												1.2
Carex "very fine"												1.2

All relevés represent floodplain forests, with evergreen species shown in boldface. Each stand was dominated by ***Quercus laurifolia***, with some deciduous canopy trees, and often with evergreen ***Magnolia virginiana*** or ***M. grandiflora***. The Pearl River basin is in southeastern Louisiana and shows a classic ***Q. laurifolia*** floodplain forest. Tiger Creek is a diverse preserve in central Florida; the example shown here had a mixed canopy of ***Q. laurifolia*** and ***M. virginiana***, plus deciduous, typically bottomland trees. The Biedler Forest stand is in the Fall Line sandhills of central South Carolina, is slightly more open, and is dominated by ***Q. laurifolia*** plus deciduous trees. Numbers under the headings represent the height and cover percentage of each stand layer (T1, T2, S and H). The following companion species occurred infrequently (+ or + 0.2, or 1.1 if not completely identified) in a non-canopy layer (T2, S or H) of not more than two relevés:

At the Pearl River basin, T2: *Nyssa sylvatica*; S: *Betula nigra*, *Crataegus* sp., *Ilex decidua*; H: ***Crinum americanum***,
Hypericum sp., *Leersia* sp., *Euphorbia* sp., *Commelina* sp., Acanthaceae sp., *Carex* sp., *Panicum* "big wide".
At Tiger Creek, S: ***Baccharis halimifolia***, *Callicarpa americana*, *Cornus foemina*, *Psilotum nudum*; H: *Phytolacca americana*, *Rubus argutus*, *Parthenocissus quinquefolia*, *Thelypteris* sp., *Galactia* sp., *Sorghum* sp., Gramineae sp.
At Biedler Forest, S: ***Juniperus virginiana***, *Ilex decidua*, *Cornus amomum*; H: ***Euonymus americanus***, *Rubus trivialis*,
Ulmus alata, *Viburnum dentatum*, *Ampelopsis arborea*, *Anisostichus capreolata*, *Clethra alnifolia*, ***Lonicera japonica***,
Mikania scandens, *Smilax glauca*, *Cornus amomum*, *Desmodium rotundifolium*, *Hexastylis arifolia*,
Hypoxis micrantha, *Lobelia elongata*, *Polygonum hydropiperoides*, *Rhododendron canescens*, *Thelypteris kunthii*,
Triadenum virginicum, *Woodwardia areolata*, *Uniola laxa*, *Aster* sp., *Cirsium* sp., *Galium* sp., *Solidago* "small lvs",
***Panicum* "small", *Panicum* "big", *Carex* sp.**

References

1. Alexander TR (1953) Plant succession on Key Largo, Florida, involving Pinus caribaea and Quercus virginiana. Q J Fla Acad Sci 16(3):133–138
2. Angell JK (2006) Changes in the 300-mb north circumpolar vortex, 1963–2001. J Climate 19:2984–2994
3. Bartram W (1791) (facs. 1973). Travels through North and South Carolina, Georgia, East and West Florida. Beehive Press, Savannah. 142 pp (facs. of London eds.)
4. Bazzaz FA (1979) The physiological ecology of plant succession. Ann Rev Ecol Syst 10:351–371
5. Blaisdell RS, Wooten J, Godfrey RK (1974) The role of magnolia and beech in forest processes in the Tallahassee, Florida, Thomasville, Georgia, area. Procs Tall Timbers Fire Ecol Conf 13:363–397
6. Boufford DE, Spongberg SA (1983) Eastern Asian-Eastern North American phytogeographical relationships a history from the time of linnaeus to the twentieth century. Annals Missouri Bot Garden 70(3):423–439
7. Bourdeau PF, Oosting HJ (1959) The maritime live oak forest in North Carolina. Ecology 40:148–152
8. Box EO (1986) Some climatic relations of the vegetation of Argentina. global perspective. Veröff. Geobot. Inst. ETH, Stiftung Rübel (Zürich), pp 181–216
9. Box EO (1995a) Climatic relations of the forests of east and Southeast Asia. In: Box EO, et al. (eds) Vegetation Science in Forestry, Handbook of Vegetation Science, 12/1. Kluwer, Dordrecht, pp 23–55

10. Box EO (1995b) Global potential natural vegetation: dynamic benchmark in the era of disruption. In: Murai Sh (ed) Toward Global Planning of Sustainable Use of the Earth – Development of Global Eco-engineering. Elsevier, Amsterdam, pp 77–95
11. Box EO (1997) Bioclimatic position of evergreen broad-leaved forests. In: Island and high-mountain vegetation: biodiversity, bioclimate and conservation. Procs., annual IAVS meeting, Tenerife, April 1993. Universidad de La Laguna, Servicio de Publicac., Tenerife (Canary Islands), pp 17–38
12. Box EO (2016) Global bioclimatic zonation. In: Box EO (ed) Vegetation Structure and Function at Multiple Spatial, Temporal and Conceptual Scales. Springer, pp 3–52
13. Box EO, Crumpacker DW, Hardin ED (1993) A climatic model for plant species locations in Florida. J Biogeography 20:629–644
14. Box EO, Fujiwara K (1988) Evergreen broad-leaved forests of the southeastern United States: preliminary description. Bull. Inst Envl Sci Techn Yokahama Nat Univ 15:71–93
15. Box EO, Fujiwara K (2013) Vegetation types and their broad-scale distribution. In: van der Maarel E (ed) Vegetation Ecology, 2nd edn. Blackwell Scientific, Oxford, , pp 455–485
16. Box EO, Fujiwara K (2015) Warm-temperate deciduous forests around the northern hemisphere Springer. Cham, Heidelberg, p 292
17. Bratton SP (1993) Survivorship of evergreen hardwoods after wildfire in maritime forest, Cumberland Island National Seashore, Georgia. Castanea 58:34–44
18. Braun EL (1950) Deciduous forests of eastern North America. Blakiston, Philadelphia, p 596
19. Braun-Blanquet J (1964) (1928, 1951). Pflanzensoziologie: Grundzüge der Vegetationskunde, 3rd edn. Springer, Berlin, p 330 (English 3rd ed. 1965: Plant Sociology, Hafner, New York)
20. Burke MJ, Gusta LV, Quamme HA, Weiser CJ, Li PH (1976) Freezing and injury in plants. Annual Review Plant Physiol 27:507–528
21. Cain SA (1939) The climax and its complexities. Amer Midland Naturalist 21:146–181
22. Cavanaugh K, et al (2019) Climate-driven regime shifts in a mangrove-salt marsh ecotone over the past 250 years. Procs Natl Acad Sci 116(43):21602–21608
23. Cavender-Bares J, Apostol S, Moya I, Briantais J-M, Bazzaz FA (1999) Chilling-induced photoinhibition in two oak species: are evergreen leaves inherently better protected than deciduous leaves? Photosynthetica 36(4):587–596
24. Chabot BF, Hicks DJ (1982) The ecology of leaf life spans. Ann Rev Ecol Syst. 13:229–259
25. Chen E, Gerber JF (1990) Climate. In: Myers RL, Ewel JJ (eds) Ecosystems of Florida, pp 11–34
26. Chen Y-F (1995–98) Taiwan Zhibei [Vegetation of Taiwan]. (in Chinese). Yushan, Chenxing, vol 3. Qianwe Publishers, Taipei, pp 303 + 621 + 632
27. Christensen NL (2000) (1988) Vegetation of the Southeastern coastal plain. In: Barbour MG, Billings WD (eds) North American Terrestrial Vegetation, 2nd edn. Cambridge University Press, pp 398–448
28. Clements FE (1936) Nature and structure of the climax. J Ecol 24:253–284
29. Clewell, A. F. 1985. Guide to the Vascular Plants of the Florida Panhandle. 605 pp.
30. Clewell AF, Tobe JD (2011) Cinnamomum-Ardisia forest in Northern Florida. Castanea 76 (3):245–254
31. Cowles HC (1911) The causes of vegetational cycles. Ann Assoc Amer Geographers 1:3–20
32. Crumpacker DW, Box EO, Hardin ED (2001) Temperate-subtropical transition areas for native trees and shrubs in Florida: present locations, predicted changes under climatic warming, and implications for conservation. Nat Areas J 21(2):136–148
33. Daubenmire RF (1990) The Magnolia grandiflora - Quercus virginiana forest of Florida. Amer Midland Naturalist 123:331–347
34. Delcourt HR, Delcourt PA (1974) Primeval magnolia-holly-beech climax in Louisiana. Ecology 55:638–644
35. Delcourt HR, Delcourt PA (1975) The Blufflands: Pleistocene pathways into the Tunica Hills. Amer Midland Naturalist 94:385–400

36. Delcourt HR, Delcourt PA (1977) Pre-settlement magnolia-beech climax of the Gulf Coastal Plain: quantitative evidence from the Apalachicola River Bluffs, north-central Florida. Ecology 58:1085–1093
37. Dhaila S, Singh SP, Negi GCS, Rawat YS (1995) Shoot-growth phenology of co-existing evergreen and deciduous species in an oak forest. Ecol Res 10:151–159
38. Duncan WH, Duncan MB (1988) Trees of the Southeastern United States. Univ, Georgia Press, Athens, p 322
39. Eyre SR (1968) Vegetation and Soils: A World Picture, 2nd edn. Arnold Co., London, p 314
40. Fenneman NM (1931) Physiography of Eastern United States. McGraw-Hill, New York
41. Fenneman NM (1946) Physical Divisions of the United States. Map at scale 1:7,000,000. US Geol. Survey, Reston (Virginia)
42. Fernald ML (1946) Identification and re-identification of North American plants: the identity of Quercus laurifolia. Rhodora 48:137–145
43. Fowells HA (1965) Silvics of Forest Trees of the United States. Dept. Agric. Handbook 271, Washington, 762pp
44. Fujiwara K (1981–86) 日本の常緑広葉森林の群落体系(Nihon-no jōryoku kōyō shinrin-no gunraku taikei) [Phytosociological investigation of the evergreen broad-leaved forests of Japan]. Bull Inst Env Sci Tech Yokohama Nat Univ 7:67–133, 8:121–150, 9:139–160, 13:99–149 (in Japanese)
45. Fujiwara K (1987) Aims and methods of phytosociology or "vegetation science." In: Takeda Y (ed) Papers on Plant Ecology and Taxonomy to the Memory of Dr. Satoshi Nakanishi, pp 607–628. Kobe Geobotan. Society
46. Fujiwara K, Box EO (1994) Evergreen broad-leaved forest region of the Southeastern United States. In: Miyawaki A, et al. (eds) Vegetation in Eastern North America, pp 273–312. Tokyo Univ. Press
47. Fujiwara K, Box EO (1999) Evergreen broad-leaved forests in Japan and Eastern North America. In: Klötzli F, Walther G-R (eds) Conference on Recent Shifts in Vegetation Boundaries of Deciduous Forests, especially Due to General Global Warming, pp 273–300. Birkhäuser
48. Gano L (1917) A study in physiographic ecology in northern Florida. Bot Gaz 63:337–372
49. Gaussen H (1955) Les climates analogues à l'echelle du monde. Compt Rend Acad Agr, France, p 41
50. George MF, Burke MJ, Pellett HM, Johnson AG (1974) Low-temperature exotherms and woody-plant distribution. HortScience 9:519–522
51. Gibson DJ (1992) Vegetation-environment relationships in a Southern Mixed Hardwood forest. Castanea 57:174–189
52. Godfrey RK (1989) Trees, Shrubs, and Woody Vines of Northern Florida and Adjacent Georgia and Alabama. Univ. of Georgia Press, Athens
53. Graham A (ed) (1972) Floristics and Paleofloristics of Asia and Eastern North America. Elsevier, Amsterdam, p 278
54. Gray A (1846) Analogy between the flora of Japan and that of the United States. Amer J Sci Arts 2(2):135–136
55. Greller AM (1980) Correlation of some climatic statistics with distribution of broad-leaved forest zones in Florida, USA. Bull Torrey Botan Club 107:189–219
56. Greller AM (1989) Correlation of warmth and temperateness with the distributional limits of zonal forests in eastern North America. Bull Torrey Botan Club 116:145–163
57. Greller AM (2003) A review of the temperate evergreen forest zone of southeastern North America: floristic affinities and arborescent vegetation types. Bot Rev 69(3):269–299
58. Grisebach ARH (1866) Die Vegetationsgebiete der Erde, übersichtlich zusammengestellt. Petermanns Mitteilungen 12:45–53
59. Harper RM (1914) Geography and vegetation of northern Florida. Fla. State Geol. Survey, 6th Ann. Rept., pp 163–437
60. Hart JF (1968) Loss and abandonment of cleared farm land in the eastern United States. Ann Assoc Amer Geographers 58(3):417–440

61. Hartnett DC, Krofta DM (1989) Fifty-five years of post-fire succession in a southern mixed hardwood forest. Bull Torrey Bot Club 116:107–113
62. Hildebrand-Vogel R (2002) Structure and dynamics of southern Chilean natural forests with special reference to the Relation of evergreen versus deciduous elements. Folia Geobotanica 37:107–128
63. Hocker HW (1956) Certain aspects of climate as related to the distribution of Loblolly Pine (Pinus taeda). Ecology 37:824–834
64. Holdridge LR (1959) A simple method for determining potential evapotranspiration from temperature data. Science 130:572
65. Hsieh C-F, Chao W-C, Liao C-C, Yang K-C, Hsieh T-H (1997) Floristic composition of the evergreen broad-leaved forests of Taiwan. Nat Hist Res special issue no. 4:1–16.
66. Hübl E (1988) Lorbeerwälder und Hartlaubwälder (Ostasien, Mediterraneis und Makaronesien). Düsseldorfer Geobot Kolloq 5:3–26
67. Hunt C (1974) Natural Regions of the United States and Canada. W. H., Freeman, San Francisco
68. Hunt DM (1994) Morphology and Ecology of Quercus Series Laurifoliae, Marilandicae and Nigrae. In: Miyawaki A, et al. (eds) Vegetation in Eastern North America, pp. 99–120. Tokyo University Press
69. Isogai T, Okutomi K (1991) Dynamics of an evergreen oak forest in the Kannami virgin forest, central Japan. Jap J Ecol 41:209–223 (in Japanese)
70. Karl TR, Melillo JM, Peterson TC (2009) Global Climate Change Impacts in the United States. Cambridge University Press
71. Kira T (1945) 農業地理学の基礎としての東亜の新気候区分 (Nōgyō Chirigaku-no Kiso to shite no Tō-A no Shin Kikō Kubun) [A new classification of climate in eastern Asia, as a basis for agricultural geography]. Horticult. Inst., Kyoto University, 23 pp (in Japanese)
72. Kira T (1949) 日本の森林帯 (Nippon-no shinrin-tai)[Forest zones in Japan]. Ringyō Kaisetsu, 17:105–141. Nippon Ringyō Gijutsu Kyōkai, Tōkyō (in Japanese)
73. Kira T (1969) 照葉樹林とは何か[What is a laurel forest?] In: 照葉樹林文化 (Shōyō jurin bunka) [Culture in the Laurel Forest Zone] (Sh. Ueyama, ed.), pp 43–84. Chūōkōron-Shinsha, Tokyo. (In Japanese)
74. Kira T (1977) A Climatological Interpretation of Japanese Vegetation Zones. In: Miyawaki A (ed) Vegetation Science and Environmental Protection. Maruzen, Tokyo, pp 21–30
75. Kira T (1991) Forest ecosystems of East and Southeast Asia in a global perspective. Ecol Research 6:185–200
76. Köppen W (1931) Grundriss der Klimakunde. Walter de Gruyter, Berlin
77. Kolstad EW, Breiteig T, Scaife AA (2010) The association between stratospheric weak polar vortex events and cold-air outbreaks in the Northern Hemisphere. Q. J. Royal Meteorol. Soc. 136:886–893
78. Komarek EV (1968) Lightning and lightning fires as ecological forces. In: Procs. 9th annual tall timbers fire ecology conference, vol 9, pp 169–197
79. Komarek EV (1972) Ancient fires. In: Procs. 12th annual tall timbers fire ecology conference, pp 219–240
80. Küchler AW (1964) Manual for map "The Potential Natural Vegetation of the Conterminous United States." American Geographical Society, special research publ. no. 36, New York
81. Kurz H (1944) Secondary forest succession in the Tallahassee Red Hills. Proc Fla Acad Sci 7(2–3):59–100
82. Kurz H, Godfrey RK (1982) Trees of Northern Florida. Univ, Florida Press, Gainesville, p 311
83. Kwit C, Schwartz MW, Platt WJ, Geaghan JP (1998) The distribution of tree species in steepheads of the Apalachicola River Bluffs. Florida J Torrey Bot Soc 125:309–318
84. Laessle AM (1942) The plant communities of the Welaka Area. Biol Sci Ser (Univ. Fla. Presses) 4(1):1–141

85. Laessle AM, Monk CD (1961) Some live oak forests of northeastern Florida. Quart J Fla Acad Sci 234:39–55
86. Lafon ChW (2010) Fire in the American South: vegetation impacts, history, and climatic relations. Geogr Compass 4(8):919–944
87. Lakela O, Wunderlin RP (1980) Trees of Central Florida. Banyan Books, Miami, p 208
88. Larcher W (1976) (1973). Ökologie der Pflanzen. Eugen Ulmer, Stuttgart, 320 pp (2nd ed.: 322pp)
89. Lenihan JM, Neilson RP (1993) A rule-based vegetation formation model for Canada. J Biogeogr 20:615–628
90. Li WH, Li LF, Fu R, Deng Y, Wang H (2011) Changes to the North Atlantic subtropical high and its role in the intensification of summer rainfall variability in the southeastern United States. J Climate 24:1499–1506
91. Lieth H (1975) Primary production of the major vegetation units of the world. In: Lieth H, Whittaker RH (eds) Primary productivity of the biosphere, pp 203–215
92. Loehle C (1998) Height growth-rate tradeoffs determine northern and southern range limits for trees. J Biogeography 25:735–742
93. Long RW, Lakela O (1971) Flora of Tropical Florida. Univ, Miami Press, p 962
94. Loveless AR (1961) A nutritional interpretation of sclerophylly based on differences in the chemical composition of sclerophyllous and mesophytic leaves. Ann Bot 25:168–184
95. Loveless AR (1962) Further evidence to support a nutritional interpretation of sclerophylly. Ann Bot 26:551–561
96. Martin WH, Boyce SG (1993) Introduction: the Southeastern Setting. In: Martin WH et al (eds) Biodiversity of the Southeastern United States. Wiley, New York, pp 1–46
97. Meeker DO, Merkel DL (1984) Climax theories and a recommendation for vegetation classification – a viewpoint. J Range Manag 37(5):427–430
98. Meentemeyer V (1978) Macroclimate and lignin control of litter decomposition rates. Ecology 59:465–472
99. Meentemeyer V (1984) The geography of organic decomposition rates. Annals Assn Amer Geogr 74(4):551–560
100. Miyawaki A (1968) 関東地方の潜在自然植生と代償植生との考察 [The current potential natural vegetation and its substitute communities on the Kanto Plain (central Japan): First Report]. In: Yoshioka (ed) Annual Report JIBP-CT(P) for Fiscal Year 1967, pp 89–95 (in Japanese). Sendai
101. Miyawaki A (1984) A vegetation-ecological view of the Japanese Archipelago. Bull Inst Env Sci Technol. Yokohama Nat. Univ. 11:85–101
102. Miyawaki A (ed) (1989) 日本植生誌 (Nihon Shokusei-Shi) [Vegetation of Japan], vol. 10: Okinawa and Ogasawara. 676pp, + 123 tables (separate volume), and 5 maps. Shibundo, Tokyo (in Japanese, with English summary)
103. Miyawaki A, Fujiwara K (1983) Evergreen broad-leaved forests and secondary forests in the Boso Peninsula (SE parts of Kanto region). Bull Inst Envir Sci Technol. Yokohama Natl. Univ. 9:63–76
104. Miyawaki A, Fujiwara K, Box EO (1987) Toward harmonious green urban environments in Japan and other countries. Bull Inst Environm Sci Technol. Yokohama Nat. Univ. 14:67–82
105. Miyawaki A, Iwatsuki K, Grandtner MM (eds) (1994) Vegetation in Eastern North America. Univ. of Tokyo Press, Tokyo, p 515
106. Miyawaki A, Suzuki K, Fujiwara K (1977) Human impact on forest vegetation in Japan. Naturaliste Canad 104:97–107
107. Miyawaki A, Suzuki K (1980) Natürliche und ihre sommergrünen Sekundär-Wälder in Mitteljapan. Phytocoenologia 7:492–506
108. Monk CD (1965) Southern mixed hardwood forest of north-central Florida. Ecol Monogr 35:335–354
109. Monk CD (1966) An ecological significance of evergreenness. Ecology 47:504–505
110. Monk CD (1968) Successional and environmental relationships of the forest vegetation of north-central Florida. Amer Midland Naturalist 79:441–457

111. Monk CD (1987) Sclerophylly in Quercus virginiana Mill. Castanea 52:256–261
112. Mueller-Dombois D, Ellenberg H (1974) Aims and methods of vegetation ecology. Wiley, New York, p 547
113. Muller CH (1961) The live oaks of the series Virentes. Amer Midland Naturalist 65:17–39
114. Myers RL (1985) Fire and dynamic relationship between Florida sandhill and sand-pine scrub vegetation. Bull Torrey Bot Club 112:241–252
115. Myers RL, Ewel JJ (1990) Ecosystems of Florida. Univ, Central Florida Press, Orlando, p 765
116. Nitta I, Ohsawa M (1997) Leaf dynamics and shoot phenology of eleven warm-temperate evergreen broad-leaved trees near their northern limit in central Japan. Plant Ecol 130(1):71–88
117. Noss RF (2018) Fire Ecology of Florida. University Presses of Florida, 336pp
118. Numata M (ed) (1974) The Flora and Vegetation of Japan. Elsevier, Kodansha (Tokyo), p 294
119. Numata M, Miyawaki A, Itow D (1972) Natural and semi-natural vegetation in Japan. Blumea 20:435–481 (with 26 photos, veg. maps)
120. Orme AR (2002) The Physical Geography of North America. Oxford Univ. Press, New York, p 551
121. Ovington JD (ed) (1983) Temperate Broad-Leaved Evergreen Forests. "Ecosystems of the World", vol 10. Elsevier, Amsterdam, p 241
122. Penfound WT (1952) Southern Swamps and Marshes. Botan. Review 18:413–446
123. Pessin LJ (1933) Forest associations in the uplands of the lower Gulf Coastal Plain (longleaf pine belt). Ecol. 14:1–14
124. Poorter L (2009) Leaf traits show different relationships with shade tolerance in moist versus dry tropical forests. New Phytol 181:890–900
125. Prentice IC, Cramer W, Harrison SP, Leemans R, Monserud RA, Solomon AM (1992) A global biome model based on plant physiology and dominance, soil properties, and climate. J Biogeogr 19:117–134
126. Quarterman E, Keever Ch (1962) Southern mixed hardwood forest: climax in the southeastern coastal plain. USA. Ecol Monogr 32:167–185
127. Ramankutty N, Heller E, Rhemtulla J (2010) Prevailing myths about agricultural abandonment and forest regrowth in the United States. Annals Assoc Amer Geographers 100(3):502–512
128. Rübel EF (1930) Pflanzengesellschaften der Erde. Huber, Berlin
129. Sakai A (1968a) Frost damage on basal stems in young trees. Contrib Inst Low Temp Sci Hokkaido Univ, series B, 15:1
130. Sakai A (1968b) Mechanism of desiccation damage of forest trees in winter. Contrib Inst Low Temp Sci Hokkaido Univ Ser B, 15:15
131. Sakai A (1971) Freezing resistance of relicts from the Arcto-Tertiary flora. New Phytol 70:1199–1205
132. Sakai A, Weiser CJ (1973) Freezing resistance of trees in North America with reference to tree regions. Ecology 54(1):118–126
133. Sakai A, Paton DM, Wardle P (1981) Freezing resistance of trees of the South Temperate zone, especially subalpine species of Australasia. Ecology 62:563–570
134. Sakai A, Larcher W (1987) Frost Survival of Plants: Responses and Adaptation to Freezing Stress. Springer, Berlin, p 321
135. Sargent ChS (1880) Catalogue of the Forest Trees of North America. US Govt. Printing Ofc, Washington
136. Schimper AFW (1898) Pflanzengeographie auf physiologischer Grundlage. Gustav-Fischer-Verlag, Jena (English transl. 1903, by W. R. Fisher, Oxford Press; 3rd edition with F. C. von Faber 1935).
137. Schmithüsen J (ed) (1976) Atlas zur Biogeographie. Meyers Grosser Physischer Weltatlas, vol 3. Bibliographisches Institut, Mannheim/Vienna/ Zürich, p 80

138. Schulze E-D (1982) Plant life forms and their carbon, water and nutrient relations. In: Lange OL, et al. (eds) Encyclopedia of plant Physiology, vol. 12B, pp 616–676
139. Shantz HL, Zon R (1924) Natural Vegetation. Part 1, section E, in Atlas of American Agriculture. US Dept. Agriculture, Washington
140. Sharitz RR, Gibbons JW (1982) The Ecology of Southeastern Shrubs Bogs (Pocosins) and Carolina Bays: A Community Profile. Publ. no. FWS/OBS-82/04. US Fish and Wildlife Service, Div. of Biological Services, Washington, 93 pp.
141. Smith WB, Miles PD, Vissage, JS, Pugh SA (2004) Forest Resources of the United States. In: General Technical Report NC-241, 146. US Forest Service, North Central Research Station, St. Paul
142. Song Y-C (1988) [Evergreen Broad-Leaved Forest Ecological Research]. Southwest Normal University Press, China, 696 pp (in Chinese)
143. Song Y-C (1995) On the global position of the evergreen broad-leaved forests of China. In: Box EO, et al (eds) Vegetation Science in Forestry. Handbook of Vegetation Science, 12/1, pp. 69–84. Kluwer, Dordrecht
144. Song Y-C (2013) [Evergreen Broad-Leaved Forests in China], vol 2, 788pp (in Chinese, with English summary) + 558 pp (tables). Science Press, Beijing
145. Song Y-C, Xu G-S (2003) A scheme of vegetation classification of Taiwan. Acta Botanica Sinica 45(8):883–895
146. Spicer BE (1969) Characteristics of the loess deposits and soils in East and West Feliciana Parishes, Louisiana. MS thesis, Louisiana State University, Baton Rouge, 69 pp
147. Stalter R, Dial S, Laessle A (1981) Some ecological observations of the arborescent vegetation in Highlands Hammock State Park, Florida. Castanea 46:30–35
148. Svoma BM, et al (2013) Expansion of the northern-hemispheresubtropical high-pressure belt: trends and linkages to precipitation and drought. Phys Geogr 34(3):174–187
149. Tagawa H (1997) World-wide distribution of evergreen lucidophyll oak-laurel forests. Tropics (Kyōto) 6(4):295–316
150. Taneda H, Tateno M (2005) Hydraulic conductivity, photosynthesis and leaf water balance in six evergreen woody species from fall to winter. Tree Physiol 25:299–306
151. Tharp BC (1939) The Vegetation of Texas. Anson Jones Press, Houston
152. Thyroff EC, Burney OW, Mickelbart MV, Jacobs DF (2019) Unraveling shade tolerance and plasticity of semi-evergreen oaks: insights from maritime-forest live oak restoration. Frontiers Plant Sci 10:1–11
153. Trelease W (1917) Naming American hybrid oaks. Procs Amer Phil Soc 56:44–52
154. Trelease W (1924) The American oaks. Mem Natl Acad Sci 20:1–255
155. US Geological Survey 1970. National Atlas of the United States of America. Washington
156. Valladares F, Niinemets Ü (2008) Shade tolerance, a key plant feature of complex nature and consequences. Ann Rev Ecol Syst 39:237–257
157. Van Auken OW, Ford AL, Dougherty SL, Stein AG (1980) Woody vegetation of upland plant communities in the southern Edwards Plateau. Texas J Sci 32:23–35
158. Veno PA (1976) Successional relationships in five Florida plant communities. Ecology 57:498–508
159. Walter H (1954) Klimax Und Zonale Vegetation. Angewandte Pflanzensoziologie 1:144–150
160. Walter H (1968) Die Vegetation der Erde in öko-physiologischer Betrachtung. Vol. II: Die gemäβigten und arktischen Zonen. VEB Gustav-Fischer-Verlag, Jena, 1002 pp
161. Walter H (1977) Vegetationszonen und Klima, 3rd edn. Eugen-Ulmer-Verlag, Stuttgart, p 309
162. Walter H (1985) Vegetation of the Earth, 3rd edn. Springer, Berlin/New York, p 318
163. Wang H, Fu R, Kumar A, Li WH (2010) Intensification of summer rainfall variability in the southeastern United States during recent decades. J Hydrometeorol 11(4):1007–1018. 1010.1175/2010
164. Wardle P (1971) An explanation for alpine timberline. NZ J Botany 9:371–402

165. Watts WA (1980) The late Quaternary vegetation history of the southeastern United States. Ann Rev Ecol Syst 11:387–409
166. Watts WA (1983) Vegetational history of the southeastern United States. In: H. E. Wright jr., (ed) Late Quaternary Environments of the United States. University of Minnesota Press, Minneapolis, pp 294–310
167. Wells BW (1928) Plant communities of the coastal plain of North Carolina and their successional relations. Ecology 9:230–242
168. Wells BW (1942) Ecological problems of the southeastern United States coastal plain. Bot Review 8:533–561
169. Wharton CH (1978) The Natural Environments of Georgia. Georgia Dept. Nat, Resources, Atlanta, p 228
170. White DA (1983) Plant communities of the lower Pearl River basin. Louisiana Amer Midl Naturalist 110:381–396
171. White DA (1987) An American beech-dominated original growth forest in southeast Louisiana. Bull Torrey Botan Club 114:127–133
172. Wood VS (1981) Live Oaking. Northeastern University Press, Boston, Southern Timber for Tall Ships
173. Woodward FI (1987) Climate and Plant Distribution. Cambridge University Press, 174 pp
174. Wu Z-Y, Committee (eds.) (1980, 1995) Zhongguo Zhibei [Vegetation of China]. Science Press, Beijing, 1375 pp + 339 B/W photos (in Chinese, Latin-Chinese species lists; no index)

Role of Riparian Zones in Reducing Pollution of Surface and Ground Water, Increase Agricultural Production and Nutrient Acquisition and Storage in River Catchments

Iuliana Florentina Gheorghe, Daniela Strat, and Simona Mihailescu

Abstract Romania is a country relatively poor in water resources: almost 75 000 millions m^3 from which 67 000 millions m^3 surface water and 8000 millions m^3 ground water. Half of the surface water (68%) of Romania is of the first quality, while degraded water represents only about 11% of the whole volume. The average amount of water used in Romania annually is of ca. 9.051 billions m^3, of which: industry 4.823, domestic 2.887, agriculture 1.299 and others 0.042. Irrigation in Romania is fully controlled. Nutrients input of from agriculture into the surface waters by percolation from the soil in river basins is high, majority of domestically wastewater (74%) are not collected and treated [1]. Riparian zones are an important role in nutrient acquisition and storage reducing pollution of surface water, ground water and increase agricultural production. In river catchment a green infrastructure with lakes and rivers, wetland, different types of forest, pastures, shrubs including different types of crops, it represent the ideal structure to harmonize the development and nature conservation. Plants have limited ability in uptake and storing nutrients, and storage time is different, finally the nutrients reach the litter that is decomposed. Decomposition and the nutrient cycles are fundamental to ecosystem biomass production. Most natural ecosystems are nitrogen (N) limited and biomass production is closely correlated with N turnover [2, 3]. In natural ecosystems, external input of nutrients is very low and efficient recycling of nutrients maintains productivity [4]. This chapter presents the effectiveness of different types of riparian zone in nutrient acquisition and storage with a role in reducing pollution of surface water, ground water and increase agricultural production.

I. F. Gheorghe (✉)
Faculty of Natural Science and Ecology, Ecological University of Bucharest,
Blvd. VasileMilea 1G, Bucharest, Romania

D. Strat
Faculty of Geography, University of Bucharest, Blvd. N. Bălcescu 1, Sector 1,
010041 Bucharest, Romania

S. Mihailescu
Institute of Biology - Bucharest - Romanian Academy, 296 Splaiul Independentei,
060031 Bucharest, POBox 56-53, Romania

F. Pedrotti and E. O. Box (eds.), *Tools for Landscape-Scale Geobotany and Conservation*, Geobotany Studies, https://doi.org/10.1007/978-3-030-74950-7_18

Keywords Riparian vegetation · Villages · Small towns · Water pollution · Nutrient storage

1 Introduction

Romania is relatively poor in water resources: almost 75 000 millions m^3 from which 67 000 millions m^3 surface water and 8000 millions m^3 ground water. The specific water resource is 3246 m^3/inhabitant/year, of which 1650 m^3/inhabitant/year come from Danube River. At national level the diffuse and point pollution sources are uniformly distributed and polluted water increases the costs of treatment of its. More than half (68%) of the surface water of Romania is of the first quality class, while degraded water represents only about one tenth (11%) of the whole volume. The average amount of water used in Romania annually is of 9.051 billion m^3, of which: the industry is using 4.823, the housing 2.887, the agriculture 1.299 and others 0.042 [1]. After 1989, the trend, in agricultural irrigation water using, is a decreasing one due to both reductions of the irrigated area and to dismantling of the huge industrial complexes of livestock breeding. Waste water use in irrigation is still very limited, on one side due to the reduced facilities for treatment, but mostly because surface water and ground water resources are enough to satisfy the needs of the areas that are prone to irrigation. Keeping a good water quality and an increase of agricultural production in Romania is an important need for political, economical, moral and scientific reasons. The political reason is that Romania is a member of European Union and it must to respect European legislation, meaning the Water Framework Directive objectives in the field of water conservation policy. The aim of directive is to establish a framework for the protection of intern surface waters, transitional waters, coastal waters and groundwater. The most important moral reason is the water quality. A good water quality in sufficient quantity influences the health condition of the people. The preservation of riparian zone (the interface between land and a river or a stream), and aquatic ecosystems is also a very important scientific and economical reason due to a lot of natural resources produced by these complex ecosystems, including a good water quality. Agriculture and housing wastewater represent two type of pollution source with nutrients. Riparian areas can help as buffer zones and reduce (retention, remove) the effects of non-point source pollution, riparian areas play a key-role in the nutrients flux. Unfortunately, when the buffer zone capacity has been exceeded a very large amount of nutrients is discharged in the river water, for this reason we must to know the buffer zones capacity to stock the nutrients. Nutrient availability in water and soil is highly heterogeneous in space and time. Consequently, efficient growth up of vegetation and uptake of nutrient can be strongly influenced by the ability of the roots system that constantly it does develop, and it's most absorptive elements in the most favorable soil, temperature and humidity [5]. One of the major global issues particularly at European level is the use of fertilizers in agriculture. On the one hand the need for food requires the use of a surplus of nutrients with role in increase in agricultural production, on another part, the excess use of these nutrients leads to soil degradation and pollution of ground water and surface water. Agriculture, previously dominated by productivity, now has multiple objectives. The crops low environmental impact, the quality of crop products, the low cost of production and hence increased nutrients use efficiency, are among these objectives. Understanding the processes that govern nutrient fluxes, particularly nutrients uptake and distribution in crops, is of major importance with respect to both environmental concerns and the quality of crop products [6].

2 Water Resources of Romania

Romania occupies approximately 29% of the Danube basin area (Fig. 1, Table 1) the total length of rivers is 78,905 km, total volume - 40 billions m^3, and almost 1,840 m^3/inhabitant.

Almost all-surface water resources originate from the inland rivers and from the Danube. The average multiannual volume of inland waters amount to an of 42,293 millions m^3, where the largest share is held by the Siret (17%), Mures (13.8%) and Olt (13%) rivers.

Fig. 1 Maps of the Danube River Basin District [8]

Table 1 The hydrologic network capacity in Romania

Source	Total capacity (millions m^3)	Potential used capacity (millions m^3)	Capacity used (millions m^3)	Percent of total use %
Interiors rivers	40.000.000	13.059.071	3.940.724	9.85
Danube river	85.000.000	20.000.000	4.737.664	5.57
Ground waters	9.600.000	6.677.150	758.628	7.9

At the inlet into the country, the Danube's multiannual input is of 175,598 millions m^3. The water resources of Danube are also available to the neighboring countries (Yugoslavia, Bulgaria and Ukraine). The Romanian share from the Danube waters is estimated to 85 000 millions m^3 [1, 7].

In years with normal droughty, the total flow by the inland rivers is almost 29 658 millions m^3 and represent just 70% of the mean multiannual volume that is 42 293 millions $m^{3.}$ During a year with accentuated droughts the total flow is 22 309 millions m^3 and represent just 52.7% of the multiannual one. The Danube displays a different regime. The contribution of the tributaries from the Alps region preserves the flow along the river at a high level even during years with high droughty this being between 84 and 72% of the multiannual flow [7].

On the surface of Romania there are 11 hydrographic basins of relatively equal size, the largest being of the Siret river (42,890 km^2) and the smallest Dobrogea - Litoral (5,480 km^2) (Table 2).

The most populated basin is the basin of Argeş - Vedea River with a relatively small surface (21,847 Km^2). It is one of the most vulnerable the river basin regarding nutrient pollution and represents the ideal case study for following study.

3 Geography Description of River Basin

The Argeş-Vedea catchment lies on central part of the Romanian Plain. This major landform unit is located on the left side of the Danube River where it is the border between Romania and Bulgaria. Tectonically, the Romanian Plain has developed on the northern part of Moesian Platform, over which overlaps a thick sedimentary cover [9].

The most recent deposits consist of loess and loess-like deposits, dominantly composed by silty clay [10–12]. Beneath them are gravels and sands of lower Pleistocene ages [9] that represent a phreatic stratum which host valuable groundwater resources. Along the river valleys, the alluvial deposits are composed by fine gravels, sands and clay. The most distinctive landforms of the Argeş-Vedea catchment are the tabular interfluves, which have been shaped by the rivers network. The altitudes slightly decrease from north to south, ranging from 180 to 50 m. Slopes are nearly level and very gentle, with a mean value of slopes of 1°. These morphologic and morphometric features of this lowland area are key drivers for rivers meandering. Different

Table 2 Human activities, population density in all 11 river basins present at national level

Name of catchments	Surface area (km^2)	Number of inhabitants	Number of city and villages	Industry	Houses
1. Someş-Tisa	22,380	2 090,000	243	722	37,200
2. Crişuri	14,860	1 282,800	195	175	80,477
3. Mureş	27,890	2 190,000	238	381	60,167
4. Banat	18,320	3 640,000	349	415	69,213
5. Jiu	18,975	1 638,900	179	122	5,937
6. Olt	24,050	2 676,000	245	154	4,376
7. Argeş-Vedea	21,847	3 942,500	180	457	24,228
8. Buzău-Ialomiţa	26,205	2 604,000	209	220	22,457
9. Dobrogea-Litoral	5,480	680,000	13	-	-
10. Siret	42,890	2 792,400	143	328	17,793
11. Prut	20,680	1 821,000	193	210	14,306

generations of old meanders of Argeş-Vedea River can be identified and oxbow lakes occur in the flood plain. The wide of flood plain ranges from about 100 to 500 m.

Based on geographical location and altitudinal values, in the area of Argeş-Vedea catchment there is a temperate continental climate, with cold winters hot summers, huge range of extreme temperatures and quite equal distribution of rainfall per seasons. With a mean annual sunshine duration of 2254 h, the mean annual air temperature is 11.1 °C and the annual amount of precipitation is about 532 mm for the period 1961–2013. The mean air temperature of January, the coldest month, is below 0 °C but the annual number of *air frost days fluctuates from 100 to 120.* During the summer, the mean air temperature in July is around 23 °C; the mean maximum is 32 °C and de annual number of days with mean air temperature over 30 °C is around [13]. The degree of continentality of the climate in the study area is depicted by value of the Johansson *Continentality Index* that is 44.6, which means a continental climate [14, 15]. For the same time period, in terms of the annual De Martonne aridity Index (25.2), the climate can be described as a semi-humid one. Annually, potential evapo-transpiration (PET) varies between 598 and 718 mm, surpassing the yearly precipitation, but during the winter (December - February) the PET value is 0 mm, while the highest values are calculated during the summer months, June–August, with an average that range between 143/month and 147 mm/month, much higher than precipitation amount (Fig. 2).

As a result, during the growing season and especially during the summer months there is water deficit. Regarding nebulosity, during the whole year, on average there are 135 clear sky days, with those of 117 cloudy sky and the sky covered with those of 123. During the study period (2008–2013), the average annual air temperature in the sampled catchment area was higher (11.9 °C) that in 1961–2013 period and the amount of precipitations was 529.6 mm. Rainfalls were distributed almost equally, but in terms of quantity, the summer is the rainiest season (117.7 mm) with the peak precipitations in June. Figure 3 depicts the temperature-precipitation diagram

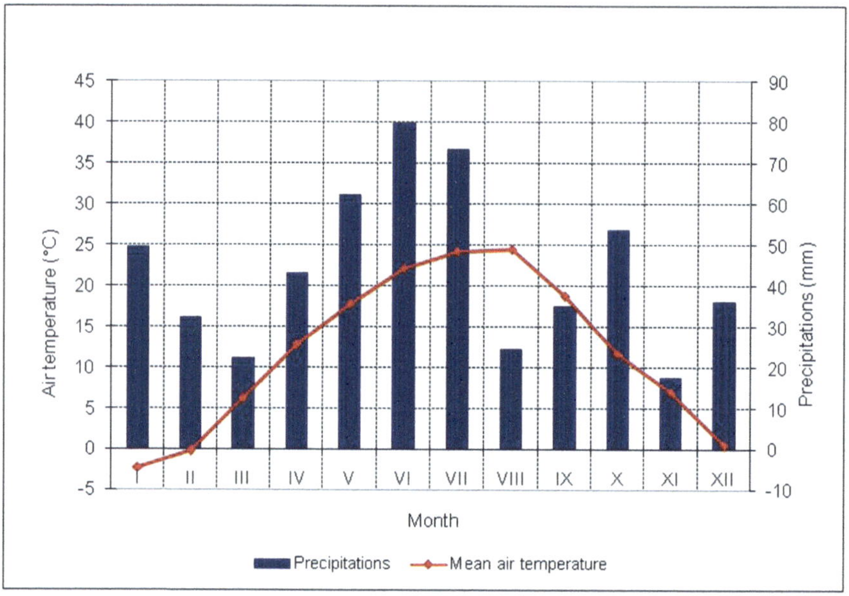

Fig. 2 The monthly air temperature –precipitation diagram during 2008–2013 in the Argeş-Vedea catchment

at the meteorological station that is located at the central-western part of the Argeş-Vedea catchment. The climate diagram shows that only two months—August and September are characterized as climatic arid according to Gaussen classification.

However, the value of the summer drought index, calculated as the ratio between summer precipitation amount and mean maximum air temperature of the hottest month [16], that is 5.3, emphasizes that in the study area the summers are droughty. The annual De Martonne index was 24.1, which depicts a semi-humid climate but compared to the multiannual value (1961–2013) it is a threshold change, from forest steppe, previously, to steppe grassland during the study period time. The mean distribution of the monthly De Martonne aridity index shows the maximum climate diversity—all seven climate types (Fig. 3), according to De Martonne classification [17, 18], from extremely humid, in January, to arid, in August. The greatest diversity of climate types is during summer because each month has different characteristics, humid, semi-humid and arid respectively. Overall, De Martonne aridity index value was 21, which correspond to the Mediterranean climate.

According to climate conditions and natural vegetation, in the area are developed fertile soils that belong tochernozem types and red brown main types respectively with the following particular subtypes: typical red brown, luvicverticred brown, luvicpseudogleyed red brown, gleyed red brown, gleycpodzolicchernozem-like soils, and leached chernozem.

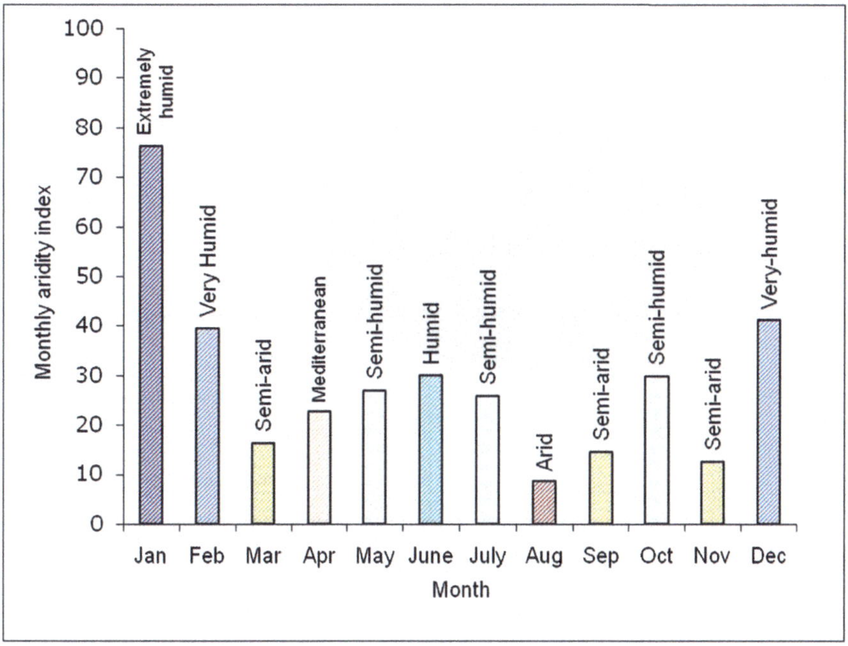

Fig. 3 The monthly distribution of the mean of De Martonne aridity index, from 2008 to 2013 in the Argeş-Vedea catchment

4 Land Use

About About ¾ of the Argeş-Vedea catchment (as the other river basins present in the Romanian Plain) area is covered with agricultural land (72.5%), this is followed by the land covered with various constructions, (13%), it represent the rural areas and consists of villages. The forests occupy 12% of the territory, wetland 2%, rivers and lakes 0.5% (Fig. 4). Although the forest area is not very large there is a great diversity of temperate continental types of forest. Dominant tree species are of the genera Quercus, Salix, Populus, Alnus, Acer, Ulmus, Fraxinus, etc. [19, 20] https://www.ijese.org/download/volume-2-issue-5/.

22.4% of this field is present in the riparian zone and covered with these types of vegetation: wetland with Carex sp. *Lythrum* sp. *Scyrpus* sp.; wetland with *Salix* sp. *Phragmites* and *Typha* sp., *Scyrpus* sp.; pastures; meadows, grasslands, orchards; forest (24 typologies); crops: wheat (81%), sunflower (3%), maize (11%), rape (5%) (Table 3).

For maps of land use were used digitized maps "Corine land cover 2006", being processed in the ArcView GIS 3.2 a, example Fig. 5,

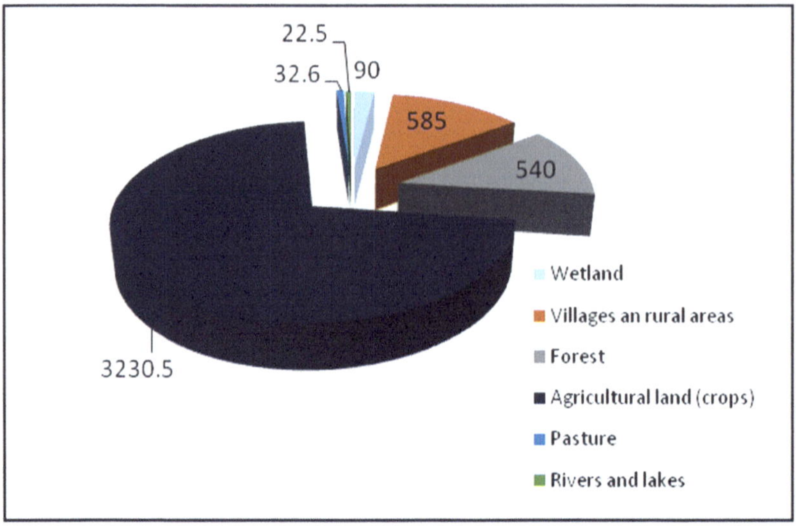

Fig. 4 Type of land cover at catchment level (ha)

Table 3 Types of riparian zones in rivers catchments

Structure of riparian zone	Dominant species plants	Longer (km)	Surface (ha)	% of different land type present in riparian zone
Wetland	*Carex sp., Lythrum sp., Scyrpus sp.*	6.3	69.3	77%
Wetland with *Salix sp.* individuals	*Typha sp., Scyrpus sp.* and *Salix sp.*	1.3	7.8	8.7%
Village		1.8	19.8	3.4%
Forest	*Robinia pseudaccacia, Quercus cerris, Q. frainetto, Q. robur, Q. petraea, Fracxinus excelsior, Populus nigra, P. alba, P. tremuloides x canadensis, Acer campestre, Ulmus laevis, Carpinus betulus, Alnus glutinosa, Tilia cordata, Salix fragilis, S. nigra*	6.9	350.9	65%
Crops		10.4	560	17.2%
Pasture		0.85	5.9	1.12%
Rivers and lakes		26.7	26.7	100%

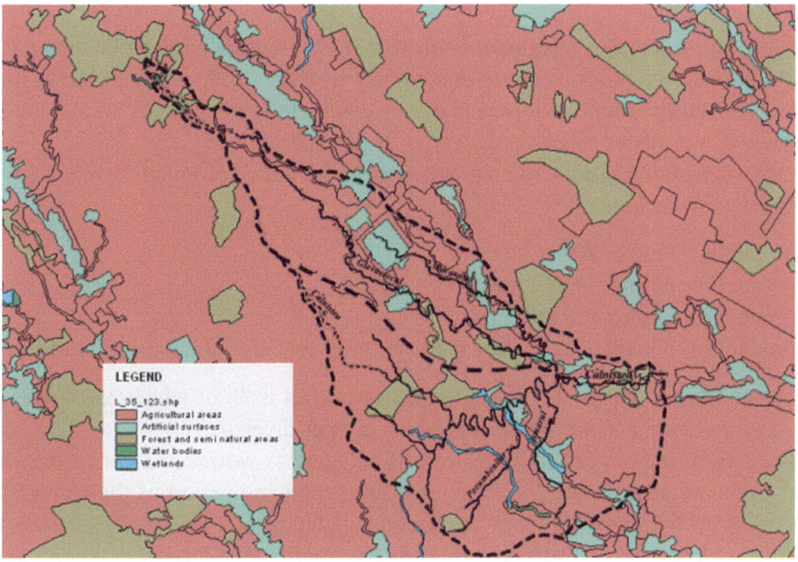

Fig. 5 Land use in river basin (Corine land cover map 2006) [19]

5 System of Agricultural Crops

Using the information from National Statistical Yearbook for 2007 generated by Agency of Payments and Interventions in Agriculture and Land Register Book, has been evaluated the crops system in Romania considering the following terms:

Agricultural area - the lands with agricultural destination, owned by natural or legal persons, classified as follows: arable land, natural pastures and hayfields, vineyards and vine nurseries, orchards and tree nurseries.

Arable land represents area which is ploughed each year or at several years, cultivated with annual or perennial plants.

Pastures represent lands covered with herbal vegetation, grown in natural way, or regenerated by sowing, for animal pasturing.

Hayfields represent lands covered with herbal vegetation, grown in natural way, or regenerated by sowing, for hay harvesting.

Vineyards and nurseries represent areas with vineyards, vine nurseries and land prepared for vineyards.

Orchards and tree nurseries represent areas with tree plantations, fruit younglings, tree nurseries and land prepared for orchards [7].

Cultivated area represents the sown area in agricultural year (October 1 - September 30) and includes:

- sown area in the previous autumn excluding re-sown area in spring with other crops;
- sown area in spring of current year.

Agricultural holding represents economic unit of agricultural production carrying out its activity under a current unique management and includes all the animals owned and all the land area used partly or completely, to perform an agricultural production, no matter of ownership type, legal type or size.

Crop agricultural production represents gross harvested production, minus losses at harvest and includes:

- production in own field;
- production of combined crops;
- production of successive crops;
- production obtained in kitchen gardens (only for vegetables, fruit, grapes) [7].

The number of types of crops at the national level is 36 of which 25(69.4%) are found in the river basins, which represents a great diversity.

The largest area is occupied by cereals for grains—wheat and corn—(62.7%), most crops are annual, and for the most of it, the sowing period, is the spring. The area occupied by industrial crops has declined lately from 24.92% to 16 0.41%, the rapeseed culture was introduced, replacing the sunflower (Fig. 6).

6 Nutrient Input

The point sources of pollution in surface water, at national level are uniformly distributed.

In Romania the diffuse and point sources of pollution are uniform distributed except the hilly and mountain areas where the agriculture activities are reduced and the human population is very low. The most vulnerable zones in pollution by nitrates are: North and North-East of Moldova, and South-Eastern Wallachia, the largest areas are in proximity of Bucharest.

In basin of most rivers are two sources of diffuse pollution: untreated sewage and fertilizers used in agriculture. Both sources of diffuse pollution create an additional intake of nutrients. At national level the population connected to sewerage is between 6.36% and 67.64% with mean value 34.9%.

In the whole area of the country, the annually agriculture input with quantities of nutrients in excess varies in different basins in range 0.39–8.7 kg/ha P; 6.91–23.6 kg/ha N; with the most large quantity in the area (Argeş-Vedea's Basin) 23.6 kg/ha N. The quantities of nitrogen and phosphorus in excess coming from agriculture lands and accumulated in the period 1998–2000 was between 60 and 87 kg/ha/year P; 12–91 kg/ha/year N, with the large quantity in Argeş-Vedea's Basin (91 kg/ha/year N).

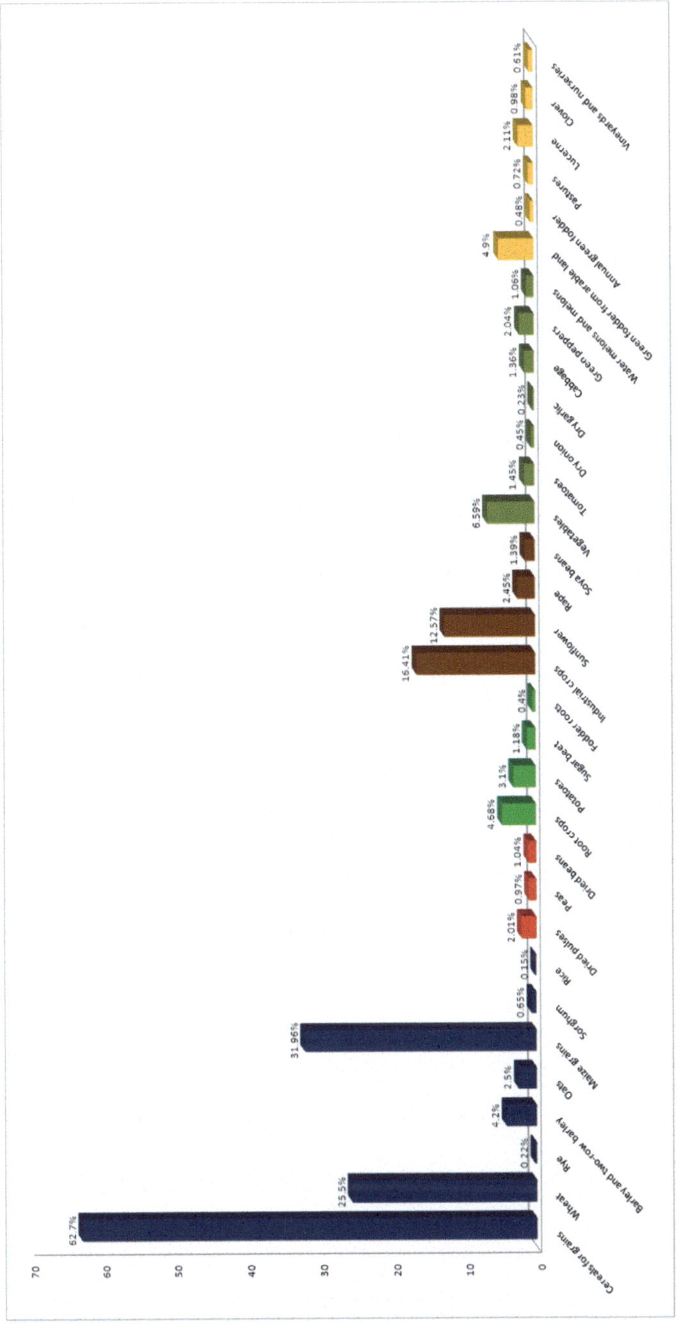

Fig. 6 Type of crops, surface (%)

6.1 Input of Nutrient by Fertilizers

In Argeş-Vedea basin not all agricultural land is fertilized the fertilizers are used only for three main crops (maize, wheat and sunflower). 3% of the cultivated area of grain maize is fertilized with mineral NPK (75 kg N/t, 130 ka P/t and 175 kg K/t) and 10% with organic fertilizer (manure). Surface of the wheat crop fertilized with NPK is about 20% and 5% with the manure is. For fertilization of sunflower crops are used only NPK and only 30% of cultivated area being fertilized. Manure comes from raising cows and has a nitrogen content of 5 kg N/t and 0.49 kg P/t (mean values in dry substance). For fertilization with NPK the farmers use between 250 and 300 kg/ha, and amount of manure used as fertilizer is between 10 and 15 t/ha. Table 4 [19, 20] https://www.ijese.org/download/volume-2-issue-5/.

6.2 Input of Nutrient by Untreated Sewage Water

In most polluted Argeş-Vedea basin, are 6 villages (Negrişoara, Glavacioc, Şelaru, Cătunu, Buteşti, Purani) and a small town (Ştefan cel Mare) situated along the river course. The population is supplied with water from the river Arges and groundwater aquifers. The N content in both sources have an average of 15 mg/L and the P content is 2.5 mg/L. After water use in the house hold, N and P contents increase at values of 25 mg/lN and 3.5 mg/lP (Table 5) [19, 20] https://www.ijese.org/download/volume-2-issue-5/.

Comparing the two sources of input of nutrient it can be noted that most is owed by the fertilization of crops. Total nutrients input introduced by fertilization of crops was 15 483 ka N and 13 206 kg P. Water used by the population and untreated made an annual intake of 8577 kg N and 1299 kg P, the contribution of population is 3169 kg N and 632 kg P. Although a small area of agricultural crops is fertilized and the amount of fertilizers used per area is relatively small these fertilizers produce a significant increase in soil nutrients. Intake of nutrients coming from

Table 4 Quantity of nutrients input in river catchment by fertilizers

Crop type	Fertilizer type	Fertilized surface (ha)	Quantity of fertilizer (t/ha)	Quantity of TN (kg/ha)	Quantity of PT (kg/ha)	Total quantity of fertilizer (t)	Total quantity of TN (kg)	Total quantity of TP (kg)
Maize	NPK	31.29	0.275	20.62	35.75	8.6	645	1119
	Manure	104.3	10	50	4.9	1043	5215	511
Wheat	NPK	166	0.300	22.5	39	49.8	3735	6474
	Manure	41.6	15	75	7.35	624	3120	306
Sunflower	NPK	123	0.300	22.5	39	36.9	2768	4797
Total		466.19	25.875	190.62	126	1762.3	**15,483**	**13,206**

Table 5 Quantity of nutrients input in river catchment by untreated sewage water

Locality	No. of inhabitants	Water volume used/ inhabitant (mean value)/ month (m^3)	Water total volume used /year (m^3)	Content of nutrients in wastewater and difference between the nutrient content of sewage and water supply (mg/L) mean values				Total input (kg/year) of nutrients, increase of nutrients added to water using			
				mg/l N		mg/l P		Kg/year N		Kg/year P	
Negrişoara	796	1.7	16,238.4	24	9	3.05	1.2	390	146	50	19
Glavacioc	814	2.1	20,512.8	26	11	3.35	1.5	533	226	69	31
Şelaru	2140	3.1	79,608	28	13	3.65	1.8	2229	1035	291	143
Cătunu	927	2.1	23,360.4	24	9	3.45	1.6	561	210	81	37
Buteşti	885	2	21,240	21	6	3.15	1.3	446	127	67	28
Purani	1685	2.8	56,616	25	10	4.15	2.3	1415	566	235	130
Ştefan cel Mare	3405	3.5	143,010	21	6	3.55	1.7	3003	858	508	243
Total	10,652	17.3	272,638.8					**8577**	**3169**	**1299**	**632**

fertilizers compared with the intake of nutrient from untreated domestic water is 5 times higher in case of N and 20 for P. Therefore the policy of protection of surface water should be focusing on the sources of diffuse pollution from agriculture and not on the wastewater. In Romania's strategy to reduce pollution of surface water that is focusing on the requirements of European Water Framework Directive the diffuse sources in agriculture are not taken into account [19, 20] https://www.ijese.org/download/volume-2-issue-5/.

7 Vegetation Description of River Basin

Taking into account physico-geographical conditions, the catchment area is located in forest-steppe biome but because of the agricultural practice the natural vegetation is replaced by agricultural lands. The remained patches of forest are dominantly composed by species of the genus *Quercus*, in particular thermophilic species like *Quercus cerris, Q. frainetto*. In in composition of this forest are present also *Fraxinus ornus Acer campestre, Ulmus laevis, Carpinus betulus, Fraxinus excelsior*, and *Tilia cordata*.

Additionally, the floodplain forests are composed by *Populus nigra, P.alba, Alnus glutinosa, Salix fragilis, S. alba.* and *S. nigra*.

The vegetation is part of subunits: U22 - Southern European forests of willow, black and white poplar (*Salix alba, Populus alba, P. Nigra);*

U14 - Panonic - Carpathian Meadow Forests *(Quercus robur, Fraxinus angustifolia)* in complex with poplar plume and willow *(Populus alba, P.nigra, Salix alba);* and U15 - Danube-Pontic forests of meadow *(Fraxinus angustifolia, F. pallisae, Quercus robur, Q. pedunculiflora)*

In subunit U14 and U15 the herbaceous layer consists of hydrophilic species such as *Rubus caesius, Glechoma hederacea, Lysimachia nummularia, Galium mollugo, mesophilous* species such as *Asparagus tenuifolius, Veronica chamaedrys, Potentilla reptans, Geum urbanum*, ruderal species such as *Amaranthus retroflexus, Arctium lappa, Bromus sterilis, Capsella bursa pastoris, Cirsium vulgare, Conyza canadensis, Daucus carota, Setaria pumila, Sonchus asper, Xanthium strumarium, Cichorium intybus, Cirsium arvense, Cirsium canum, Sonchus arvensis*, the edifying species encountered in this area are characteristic of both vegetation subunits.

The arboricol layer is represented by populations belonging to the species *Quercus robur, Q. cerris, Q. pedunculiflora, Fraxinus excelsior, F. ornus, Populus nigra, Acer campestre, A. tataricum, Ulmus laevis (Quercus robur, Fraxinus excelsior, F. ornus, Populus nigra, Acer campestre)*, with the exception of species of the genus Fraxinus, the rest are typical of U14, U15 vegetation subunits.

The shrubs layer consist in *Cornus sanguinea, C. mas, Crataegus monogyna, Evonymus europaeus, Ligustrum vulgare, Rosa canina, Rubus caesius, Salix caprea, Salix triandra, Viburnum opulus, Cornus mas, Corylus avellana, Prunus spinosa*, and contains species of shrubs characteristic of both U14 and U15 subunits.

In subunit 22 the grassy layer consists of hydrophilic species such as: *Bidens tripartitus, Polygonum hydropiper, P. mite, Galium palustre, Stachys palustris, Methha aquatica, Lycopus europaeus, Scutellaria hastifolia, Iris pseudacorus, Lythrum salicaria, Solanum dulcamara; besides these edifying species, there are also species such as: Rorippa silvestris, Gratiola officinalis, Lysimachia nummularia, Eleocharis palustris, Juncus effusus, Veronica anagallis-aquatica*, and due to the anthropic activities that occur in this area (grazing, grubbing, introduced into the vegetal and ruderal communities such as *Amaranthus retroflexus, Arctium lappa, Atriplex patula, Bromus sterilis, Capsella bursa-pastoris, Carduus nutans, Chenopodium album, Conyza canadensis, Datura stramonium, Daucus carota, Galium aparine, Lamium purpureum, Setaria pumila, Solanum nigrum, Sonchus asper, S. oleraceus, Verbascum blattaria, Xanthium strumarium, Aristolochia clematitis, Ballota nigra, Cichorium intybus, Cirsium arvense, Elymus repens, etc.*

The trees layer consists of three species: *Salix alba, Populus alba, P. nigra,* and the grassy layer is dominated by *Rubus caesius.*

Arges- Vedea basin belongs to the boundary between subunits U14, U15 and U22. Seven types of plant communities have been identified: *Molinio-Arrhenatheretea, Fraxino oxycarpe-Ulmetum, Fraxino-pallisae-angustifoliae-quercetum roboris, Populeto-Salicetum, Salicetum albae-fragilis, Phragmition, Populion albae.*

7.1 Types and Structure of Vegetation

Riparian vegetation is important for the health of waterways, contributing to the balance of oxygen, nutrients and sediment, and providing habitat and food for fauna. It grows along banks of a waterway extending to the edge of the floodplain (also known as fringing vegetation). This includes the emergent aquatic plants growing at the edge of the waterway channel and the ground cover plants, shrubs and trees within the riparian zone [19, 20] https://www.ijese.org/download/volume-2-issue-5/. Depending on the degree of representation, the most important are five riparian vegetation types: wetland (W) with *Carex sp. Lythrum sp., Scyrpus sp.*, pasture (P), mixed forest (F1), forest with Quercus species (Querceta) (F2) and agriculture land (A) (wheat, sunflower, corn crops, etc.). For each zone were estimated: structure of vegetation, dominant species, biomass, primary productivity, C, N stocks and C, N uptake.

The vegetation of wetland zone (W) is homogenous (the SD of the cover degree and height not vary more than ±5% and ±7 cm), the species richness value is low (13 taxa). In plain forested area, at catchment river level, are present 24 typologies of forest with nine types. Because the largest surface of forested area is covered with two forest types (mixed and querceta), the study of biomass and storage capacity was carried out in these two forest types. The mixed forest (F1) is a natural forest, plurien, with a great vitality, a complex structure (16 trees species) and an average productivity. The querceta forest is (F2) a young and natural forest, with a great vitality, with a structure a medium complexity (12 trees species) and a high productivity. Concerning the structure of pasture vegetation, the species richness value is highest compared with other herbaceous layer present in the other four vegetation type (32 taxa), the dominate specie is *Elymus repens*. Except for the dominant species individuals of other species are equitable distributed. The herbaceous layer has a low heterogeneity (cover degree and hight of vegetation varies slightly around the average and SD has following values of: ±5% and respectively ±5 cm) Table 6 [19, 20] https://www.ijese.org/download/volume-2-issue-5/. The most low values of species richness is in agricultural land, here, compared with the herbaceous layer present in the other four vegetation types (7 taxa), the dominate species is *Triticum aestivum*. Except cultivated plants, ruderal plants are extremely underdeveloped. The herbaceous layer is very homogeneous (cover degree in all quadrates was 100% and height of vegetation varies slightly around the average, SD of is ±5 cm). In terms of plants specific composition, the species present in common, in all type of vegetation represents less than 17%. Although mixed forest (F1) and querceta forest (F2) belong to the same vegetation unit, specific composition is completely different (less 20% taxa is in common). Vegetation on agricultural land (A) has a very low number of taxa and is similar with the pasture (P). The wetland vegetation (W) has no species in common with the other areas (A, F1, F2 and P) (Fig. 7).

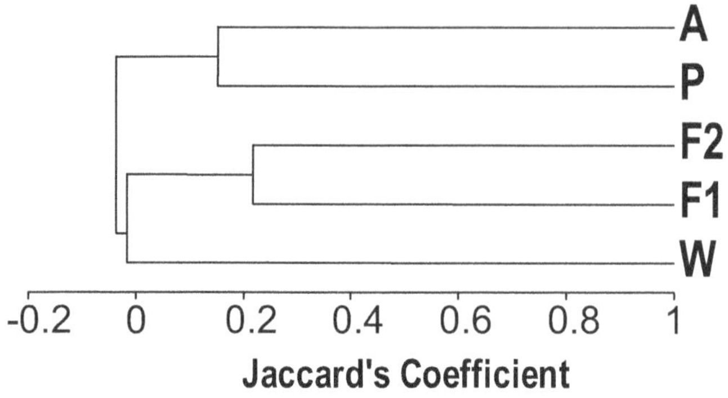

Fig. 7 Jaccard similarity between the species composition in agriculture land (A), pasture (P), querceta forest (F2), mixed forest (F1) and wetland (W) [20] https://www.ijese.org/download/volume-2-issue-5/

7.2 *Biomass–Primary Production and Productivity C, N Stocks and Nutrient Uptake*

The largest quantities of biomass are produce by mixed forest following by querceta forest, the layer of trees has significant contribution. The herbaceous layer present in mixed forest is least productive followed by the one from the querceta forest; because there it is increased competition for space, light and nutrients.

Nitrogen content in soil of riparian zone varies inversely proportional with altitude of land. In agriculture land where the altitude is high the content is low and in wetland where the altitude is low the content is very high. The low content of nitrogen from agricultural land, pasture and forest with low slope is due to takeover by plants, here the oxygenation and humidity conditions of the soil not favor the removed of nitrogen by denitrification. In wetland where the content of nitrogen is high the denitrification conditions are favorable because there is enough substrate for the mineralization processes. In wetland and forest with low slope the fertilization is inefficiently because there not produce an increase of content of nitrogen in soil like, in crops land, pasture and forest with high slope. In all 5 zones the nitrogen content in belowground biomass is higher compared to aboveground biomass. Except wetland, the growth trend of nitrogen content in the other zones is similar. In both type of the forest the nitrogen content in aboveground biomass, in July, is high because the plants present in herbaceous layer are at maturity, at end of biological cycle. The plants present in herbaceous layer in the forest he ends his biological cycle early to avoid competition for light and nutrients with trees leaves. Nutrient availability in soil is highly heterogeneous in space and time (Fig. 8). Soil resources are unevenly distributed in space and time. Water availability can dramatically affect soil nutrient availability, root physiology, and plant nutrient acquisition. If respiration does not decrease at a rate similar to those of water or nutrient uptake, then the roots become more costly to the plant. Moreover, dry surface soil is often associated with high soil temperature, potentially further increasing root costs [21]. Typically, if soil resources such as nitrogen or phosphorus are limiting plant growth, plants increase the amount of root biomass allocation, and thus help maintain a "functional equilibrium" between shoot acquisition of C and root acquisition of mineral nutrients [22]. In an extensive review of plant is specified the responses are non-uniform supplies of nutrients, [23]. Production of root hairs or extrametrical mycorrhizal hyphae can be a very efficient way by which a plant can increase absorptive surface for the same biomass allocation. Higher efficiency does not always lead to higher plant fitness. In a competitive environment where resources are available only for short periods, rapid resource acquisition, rather than high efficiency, may be a key to plant success [24]. High expenditures for rapid resource acquisition may be an ecologically effective strategy if fitness of neighbors is diminished to a greater extent than is the fitness of the individual exhibiting rapid growth. Plants may also overproduce tissues as a means of coping with herbivore, or as insurance against extreme events. The notion that plants might not be efficient in resource use was underscored in a review by Thomas and Sadras (2001) [25].They argue that there may be many instances where plants may support large numbers of "unproductive" tissues that may provide secondary benefits for N storage, as a buffer against herbivore, and as a way of offloading excess C and other nutrients. For example, Thomas and Sadras (2001) [21, 25] speculate that plant species in fertile environments may exhibit high rates of leaf and root turnover, not in response to a reduced need for nutrient conservation [21, 26, 27] but rather because of a greater need to offload excess resources associated with overproduction of carbohydrates. Consistent with the excess tissue hypothesis of Thomas and Sadras (2001) [21, 25] are the arguments that plants may use the alternative respiratory path as an "energy overflow" pathway [21, 28], and the evidence from the elevated CO_2 literature that shows an average of 42% enhanced soil respiration (root plus microbial respiration) in response to elevated CO_2 with-out an increase in shoot growth [29]. Our view is that while there may be times when plants appear "wasteful", especially over short time spans, maintaining redundant absorptive tissues to offset the risks of herbivory or extreme weather events still follows general concepts of optimization in an uncertain environment. Consequently, broad economic analogies [30] of plant resource acquisition and allocation may be useful tools as a first approximation for interpreting plant responses to multiple resource limitations and strategies for tissue deployment. One of the major global issues particularly at European level is the use of fertilizers in agriculture. On the one hand the need for food requires the use of a surplus of nutrients with a role in increasing agricultural production, on another part of the excess use of these nutrients leads to soil degradation and pollution of groundwater and surface water. Nitrogen uptake and accumulation in crops represent two major components of the N cycle in the agro-system. Nitrate ions that not taken up by a crop, in most of them are infiltrate in underground water. Modeling N uptake together with soil water transfers is, therefore, key in quantifying and preventing nitrate leaching [31]. 20% of the phosphorus that coming from fertilizations and

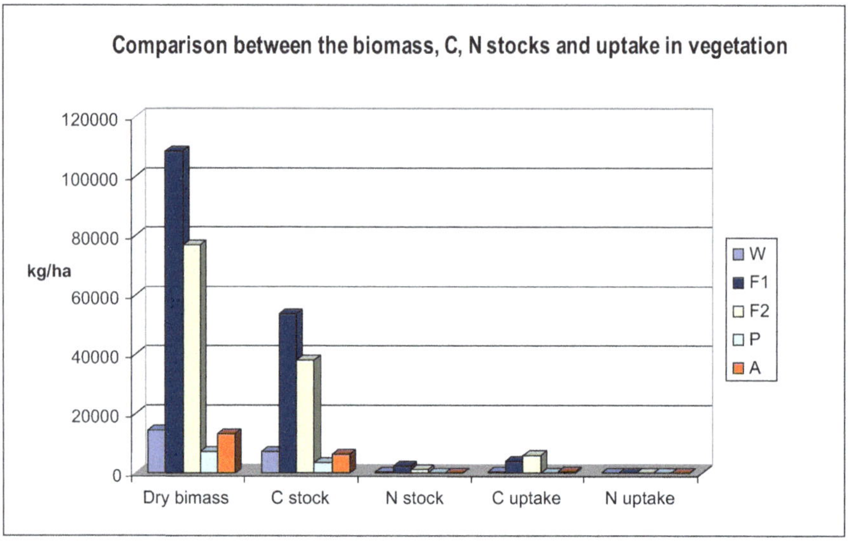

Fig. 8 The biomasses, C, N stocks and uptake in vegetation [39] https://www-pub.iaea.org/MTCD/Publications/PDF/TE_1784_web.pdf

50% of applied nitrogen to land reach receiving waters [3]. So far studies have been conducted only in the field of crop plants, but the wild plant communities that uptake part in excess of N are not been studied. Also is not been studied the role of wild plant communities, that function as buffer zones leading to reduction of nutrient pollution of ground and surface water. Plants have limited ability in retention and storing nutrients, and storage time is different, finally they reach the litter, litter that is decomposed.

7.3 *Decomposition and Nutrient Cycling*

The litter represents the dead plant parts that are in decay, which will decompose. The amount of litter is strictly related f biomass that is transformed into necro-mass. In ecosystems where no are trees and shrubs, the litter is represented by the parties dead of grass. In forests and shrubs areas the litter is represented largely by leafs. Decomposition and nutrient cycling are fundamental to ecosystem biomass production. The productivity of most natural ecosystems, are limited by nitrogen (N) input and biomass production is closely correlated with N turnover [2, 3]. Typically external input of nutrients is very low and efficient recycling of nutrients maintains productivity [4]. Decomposition of plant litter ensure for the majority of nutrients recycled through ecosystems. Rates of plant litter decomposition are highly dependent on litter quality,high concentration of phenolic compounds and

Table 6. Species composition and dominance in riparian vegetation types

Species		Frequency %	Braun-Blanquet index		Dominance
Plant species composition and dominance in wetland vegetation (W)					
Typha latifolia			2	+	
Stachys palustris			5	1	
Scirpus lacustris			100	5	First dominant
S. sylvaticus			15	2	
Lycopus europaeus			0.5	R	
Phragmites australis			2	+	
Lythrum salicaria			69	4	Second dominant
Ranunculus acris			1	+	
Galium palustre			1	+	
Epilobium hirsutum			35	3	
Juncus glomeratus			2	+	
Carex pseudocyperus			56	4	Second dominant
Acorus calamus			0.75	R	
Plant species composition and dominance in mixed forest(F1)					
Trees	Shrubs	Herbaceous			
Fraxinus excelsior	*Cornus mas*	*Bromus sterilis*	5	+	
Fraxinus ornus	*Cornus sanguinea*	*Buglossoides purpurocaerulea*	28	3	
Fraxinus pennsylvanica	*Corylus avellana*	*Galium schultesii*	34	3	
Acer campestre	*Crataegus monogyna*	*Glechoma hederacea*	67	4	Second dominant
Prunus cerasifera	*Prunus spinosa*	*Lolium perenne*	2	+	
Pyrus pyraster	*Rosa canina*	*Lysimachia nummularia*	14	2	
Acer tataricum	*Rubus caesius*	*Plantago major*	7	1	
Malus sylvestris	*Viburnum opulus*	*Plantago media*	9	1	
Quercus cerris	*Ligustrum vulgare*	*Ranunculus acris*	38	3	
Q. pedunculiflora	*Evonymus europaeus*	*Taraxacum officinale*	11	2	
Q. frainetto	*Salix triandra*	*Erigeron canadensis*	78	5	First dominant
Q. robur		*Geranium phaeum*	4	+	
Ulmus laevis		*Asperula glauca*	6	1	
Robinia pseudaccacia		*Alliaria officinalis*	23	2	
P. tremuloides x P. canadensis		*Stellaria aquatic*	4	+	
Populus nigra		*Mercurialis perennis*	0.75	r	
Plant species composition and dominance in Querceta forest (F2)					
Trees	Shrubs	Herbaceous			
Fraxinus excelsior	*Cornus mas*	*Anemone nemorosa*	17	2	
Fraxinus ornus	*Cornus sanguinea*	*Buglossoides purpurocaerulea*	6	1	
Fraxinus pennsylvanica	*Corylus avellana*	*Asparagus tenuifolius*	3	+	
Acer campestre	*Crataegus monogyna*	*Corydalis cava*	21	2	
Populus alba	*Rosa canina*	*Corydalis solida*	24	2	
Prunus cerasifera	*Rubus caesius*	*Circaea lutetiana*	0.5	R	
Pyrus pyraster	*Viburnum opulus*	*Galium mollugo*	11	2	
Quercus cerris	*Ligustrum vulgare*	*Galium schultesii*	9	1	
Q. pedunculiflora		*Geranium phaeum*	7	1	
Q. frainetto		*Geranium pretense*	4	+	
Q. robur		*Geum urbanum*	29	3	Second dominant
Tilia cordata		*Glechoma hederacea*	58	4	First co-dominant
		Heracleum sphondylium	1	+	
		Lamium album	2	+	
		Lysimachia nummularia	14	2	
		Plantago major	9	1	
		Polygonatum latifolium	3	+	
		Potentilla reptans	2	+	
		Ranunculus acris	28	3	
		Ficaria verna	65	4	First co-dominant
		Rumex crispus	0.7	R	
		Salvia nemorosa	2	+	
		Scilla bifolia	18	2	
		Silene vulgaris	12	2	

	Veronica chamaedrys	8	1	
	Viola odorata	46	3	Second dominant
Plant species composition and dominance in pasture (P)				
Amaranthus retroflexus		4	+	
Arctium lappa		2	+	
Bromus sterilis		3	+	
Capsella bursa-pastoris		3	+	
Daucus carota		1	+	
Echinochloa crus-galli		3	+	
Erigeron canadensis		11	2	
Erodium cicutarium		0.5	R	
Setaria pumila		0.7	R	
Achillea millefolium		1	+	
Cichorium intybus		2	+	
Galega officinalis		0.5	R	
Hypericum perforatum		1	+	
Lolium perenne		23	2	
Lotus corniculatus		9	1	
Mentha longifolia		3	+	
Plantago major		4	+	
Potentilla reptans		1	+	
Prunella vulgaris		16	2	
Ranunculus acris		2	+	
Senecio jacobaea		3	+	
Taraxacum officinale		5	+	
Trifolium hybridum		7	1	
Trifolium pratense		8	1	
Trifolium repens		11	2	
Dactilis glomerata		4	+	
Elymus repens		76	5	Dominant
Vicia cracca		11	1	
Cirsium vulgare		0.4	R	
Inula britanica		1	+	
Ranunculus sardous		1	+	
Sonchus asper		0.7	R	
Plant species composition and dominance in agriculture land				
Triticum aestivum or *(Helianthus annuus)*		100	5	Dominant
Cirsium vulgare		1	+	
Setaria pumila		0.5	R	
Sonchus asper		1	+	
Bromus sterilis		2	+	
Viola tricolor		0.5	R	
Vicia cracca		6	1	

especially the lignin, in plant litter has a retarding effect on litter decomposition [32, 33]. Globally, rates of decomposition are mediated by litter quality and climate [37]. Ecosystems dominated by plants with low-lignin concentration often have rapid rates of decomposition and fast nutrient cycles [38].

The decomposition of the litter is a process that follow an exponential law and whose rate can be appreciated by calculating the constant K.

Values of decomposition rate constant (k) are comparables with values present in literature. The highest value of the constant decomposition rate (k) was recorded in forest with low slope—F1—mixed forest; here the decomposition process is most intense. In forest with high slope—F2—querceta forest the k value is close to that of F1. High speed of decomposition in F1 is due a saturated soil in water (here is sufficient water like humidity necessary for bacterial exo-enzymes activity) and the nature of litter (the quantity of lignin and cellulose in trees and shrubs leafs is low to compare the wheat stems and *Scyrpus* sp. *Typha* sp.) (Table 7).

Table 7 Comparison between average values of k with average values present in literature

Zone	K ($days^{-1}$) mean values	K ($days^{-1}$) mean values present in literature
W	2.281×10^{-2}	2.464×10^{-2} Gessner et al. [34]
F1	5.327×10^{-2}	2.354×10^{-2} Nelson et al. [35]
P	4.283×10^{-2}	1.044×10^{-2} Nelson et al. [35]
A	2.578×10^{-2}	0.332×10^{-3} Salamanca et al. [36]
F2	5.061×10^{-2}	0.367×10^{-3} Aerts [27]

The lower value of k was calculated in wetland; plant species present here are adapted to high soil moisture conditions; in tissues structure of these plant the cell wall are impregnated with silica salts that is difficult to break down in small fragments. The wheat straws have also in the structure the tissues impregnated with silica salts; therefore the value of k in agriculture land is low and similarly with wetland. In both agricultural land and in wetland the decomposition process takes place slowly. More complex compounds of C are decomposed more slowly and may take many years to completely breakdown.

8 Conclusions

Riparian zones are the transitional areas between land and water, including the margins of streams, rivers, lakes, and wetlands. They are rich in biodiversity and play an important role in protecting water quality and stream ecosystem health. Riparian vegetation functions as a large sponge that reduces surface flow and absorbs the nutrients (C, N, P) in excess and pollutants from storm water runoff During the development process, riparian areas are degraded when vegetation is removed, the terrain is cultivated or plowed, are installed the utilities, are built the different structures and river borders are regularized. These changes to the landscape and subsequent human activity in the riparian zone have consequences on ecosystem health from impact of nutrients in excess that came from fertilizers, wastes, atmospheric pollutants generated by cars the roads, and soils degradation [40]. The temperate forests play a very important role in terms of the amount of C and N stored, storage period of them and fertility of the soil.

In trees layer the amount of nutrients (C, N) stored in the wood, is 10 times greater than that stored in the leaves. The nutrients stock that accumulate in wood as productivity, grow from year to year and only the leaves supplies the litter, which decomposes. In the two forest types (F1 and F2), 4/5 the amount of litter is decomposed, and 1/5 accumulates at the soil surface and supplies the horizon 0 of the soil with organic matter. This organic matter plays a fundamental role in soil processes; this is an energy source of microorganisms and precursor in soil humic acids [19, 20, 39] https://www-pub.iaea.org/MTCD/Publications/PDF/TE_1784_web.pdf. Although part of N is lost in the process of denitrification,one of the final stage of process of decomposition; a large amount of N returns to soil as nutrients from which is taken by plants. Making a unilateral analysis regarding takeover efficiency and nutrient use by plants in crops, imply the risk of ignore the role of other types of ecosystem in nutrients cycle.

A holistic approach, those make a simultaneous analysis for all function of ecosystems (reducing pollution, creating local microclimates, etc.) outside the production of them can give an overview and help making the best decisions concerning the use of different types of land. Natural and semi-natural ecosystems are the main sources in the production of resources and energy generation and play an important role in reducing of pollution. In modern society the required of resources and energy to developed is greater, also the human pressures exerted on ecosystems and biodiversity is in increase, which implies the need for preservation of these. Keeping an ecosystem mosaic structure is an ideal solution to harmonize the development of society with nature conservation. A green infrastructure with lakes and rivers, wetland, different types of forest, pastures, shrubs including different types of crops, it represent the ideal structure to meet both goals [19, 20, 39].

Acknowledgements Many thanks to Lee Heng officer of the funded program by International Atomic Energy Agency - Soil and Water Management & Crop Nutrition Section, Joint FAO/IAEA Division of Nuclear Techniques in Food and Agriculture, in the research program called Investigation of Nitrogen pathways in typical entropic-modified riparian zones of the Arges river catchment's area, Romania, Contract no. 15352.

References

1. National institute of statistics, National Statistical Yearbook for 2007 [Internet] (2007). https://www.insse.ro/cms/files/Anuar%20arhive/serii%20de%20date/2007/ASR_2007.pdf
2. Vitousek PM, Howarth RW (1991) Nitrogen limitation on land and in the sea: how can it occur? Biogeochemistry 13:87–115
3. Reich PB, Grigal DF, Aber JD, Gower ST (1997) Nitrogen mineralization and productivity in 50 hardwood and conifer stands on diverse soils. Ecology 78:335–347
4. Likens GE, Bormann FH, Johnson NM, Fisher DW, Pierce RS (1970) Effects of forest cutting and herbicide treatment on nutrient budgets in the Hubbard Brook watershed-ecosystem. Ecol Monogr 40:23–47
5. Eissenstat DM (2005) Nutrient acquisition by plants: an ecological perspective. In: The Efficiency of Nutrient Acquisition over the Life of a Root, pp. 185–220. https://doi.org/10.1007/3-540-27675-0_8
6. Tosens T, Laanisto L (2018) Mesophyll conductance and accurate photosynthetic carbon gain calculations. J Experimental Botany 69(22):5315–5318. https://academic.oup.com/jxb
7. Institutul Național de Statistic (2015) România în cifre – ISNN 2066–4079. https://www.insse.ro/cms/files/publicatii/Romania_in_cifre_2015final.pdf
8. Maps of the Danube River Basin District Management Plan 2015, [Internet] (2015). https://www.icpdr.org/flowpaper/viewer/default/files/nodes/documents/drbmp-update2015-maps.pdf
9. Ionesi L (1994) Geologia unităților de platformă și a orogenului Nord-Dobrogean. Editura Tehnică, București, p 278
10. Conea A, Ghițulescu N, Vasilescu P (1963) Considerații asupra depozitelor de suprafață din Câmpia Română de Est,- Studii tehnice și economice. Seria C 11:61–85

11. Conea A (1970) Head deposits in the Romanian Danube Plain. Studii Tehnice Şi Economice Seria C 18:251–268
12. Jipa DC (2014) The loess-Like deposits in the Lower Danube Basin. Genetic significance, Geo-Eco-Marina, vol 20, pp 7–18
13. Călinescu G (2008) Temperaturi extreme caracteristice. In: Clima R, Ion S, Pescaru VI, Poiană II, Geicu A, Cândea I, Tâştea D (eds) Bucureşti, Editura Academiei Române, pp 184–196
14. Gorczynski W (1920) Sur le calcul du degré de continentalisme et son application dans la climatologie. Geographie Annaler 2:324–331
15. Deniz A, Toros H, Incecik S (2011) Spatial variations of climate indices in Turkey. Int J Climatol 31:394–403
16. Lebourgeois F (2010) Cours de bioclimatologie a l'usage des forestiers. Nancy: Departement SIAFEE. UFR Forets, Abres et Milieux Naturels
17. Baltas EA (2010) Surface representation of climate variables and indices in Greece using GIS Methods. Int J Meteorol 35(348):123–136
18. Croitoru A-E, Piticar A, Imbroane AM, Burada DC (2013) Spatiotemporal distribution of aridity indices based on temperature and precipitation in the extra-Carpathian regions of Romania. Theoret. Appl. Climatol. 112(3–4):597–607
19. Gheorghe IF (2014) Comparative efficiency in nutrient retention between the vegetation of the natural ecosystems and of the agro-systems. Int J Innov Technol Exploring Eng (IJITEE) 3 (11):98–103. ISSN 2278–3075. https://www.ijitee.org/download/volume-3-issue-11/
20. Gheorghe IF, Barbu I (2014) Green infrastructure - an important factor in the preservation and use of biodiversity to reduce diffuse pollution and production of biomass. Int J Emerg Sci Eng (IJESE) 2(5):88–94. ISSN 2319–6378. https://www.ijese.org/download/volume-2-issue-5/
21. Eissenstat DM, Volder A (2005) The Efficiency of Nutrient Acquisition over the Life of a Root, Chapter from book Nutrient Acquisition by Plants: An Ecological Perspective Ecological Studies. H. Bassiri Rad (Ed.) Nutrient Acquisition by Plants an Ecological Perspective. Springer, Heidelberg, vol 181 pp 185–220
22. Thornley JHM (1972) A balanced quantitative model for root: shoot ratios in vegetative plants. Ann Botany 36(2): 431–441
23. Robinson SL, Rousseau DM (1994) Violating the psychological contract: not the exception but the norm. J Organ Behav 15(3):245–259
24. Jackson RB, Caldwell MM (1989) The timing and degree of root proliferation in fertile-soil microsites for three cold-desert perennials. Oecologia 81(2):149–153
25. Thomas H, Sadras VO (2001) The capture and gratuitous disposal of resources by plants. Funct Ecol 15:3–12
26. Berendse F, Aerts R (1987) Nitrogen-use-efficiency: a biologically meaningful definition? Funct Ecol 1(3):293–296
27. Aerts R (1997) Climate, leaf litter chemistry and leaf litter decomposition in terrestrial ecosystem: a triangular relationship. Oikos 79:439–449
28. Lambers H (1987) Growth, respiration, exudation, and symbiotic associations: the fate of carbon trans-located to roots. In: Gregory PJ, Lake JV, Rose DA (eds) Root Development and Function. Cambridge Univ. Press, Cambridge
29. Zak DR, Pregitzer KS, Curtis PS, Vogel CS, Holmes WE, Lussenhop J (2000) Atmospheric CO_2, soil-N availability, and allocation of biomass and nitrogen by *Populus tremuloides*. Ecol Appl 10:34–46
30. Bloom A, Chapin III SF, Mooney HA (1985) Resource limitation in plants – an economic analogy. Ann Rev Ecol Syst 16:363–392
31. Gastal F, Lemaire G (2002) N uptake and distribution in crops: an agronomical and ecophysiological perspective. J Exp Bot 53(370):789–799. https://doi.org/10.1093/jexbot/53.370.789
32. Melillo JM, Aber JD, Muratore JF (1982) Nitrogen and lignin control of hardwood leaf litter decomposition dynamics. Ecology 63:621–626

33. Hättenschwiler S, Vitousek PM (2000) The role of polyphenols in terrestrial ecosystem nutrient cycling. Trends Ecol Evol 15:238–243
34. Gessner MO, Meyer E, Schwoerbel J (1991) Rapid processing of fresh leaf litter in an upland stream. Verh. Int. Verein. Limnol. 24:1846–1850
35. Nelson WJ, Kadlec JA, Murkin HR (1990) Seasonal comparison of weight for two types of Typha glauca (Godr.) leaf litter. Aquat. Bot. 37:299–314
36. Salamanca FE, Nobuhiro K, Shigeo K (1998) Effect of leaf litter mixture on the decomposition of *Quercus serrata* and *Pinus densiflora* using field and laboratory microcosm methods. Ecol Eng 10:53–73
37. Meentemeyer V (1978) Macroclimate and lignin control of litter decomposition rates. Ecology 59:465–472
38. Chapin FS, Stuart III (1980) The mineral nutrition of wild plants. Ann Rev Ecol Syst 11:233–260. https://doi.org/10.1146/annurev.es.11.110180.001313.JSTOR2096908
39. Gheorghe IF (2016) Investigation of nitrogen pathways in typical entropic-modified riparian zones of the Arges river catchment's area, Romania, IAEA TECDOC - 1784:135–149. https://www-pub.iaea.org/MTCD/Publications/PDF/TE_1784_web.pdf
40. Harpet Conservancy (2018) Riparian Zones: What are they and why protect them? https://www.harpethconservancy.org/programs/restoration/riparian

Conservation

Vegetation and Flora of Sacred Natural Sites in North-Western Morocco—Landscape Context and Conservation Value

Ulrich Deil, Birgit Frosch, Holger Jäckle, and Allal Mhamdi

Abstract Sacred sites came into the focus of UNESCO, IUCN and conservationists recently. In many societies the surroundings of such sites shelter remnants of vegetation. They are called sacred natural sites (= SNS) and are often considered as primeval forests. While numerous geobotanical studies are available from Southeast Asia and sub-Saharan Africa, such biocultural elements were underexplored until recent years in Muslim societies, especially in the Maghreb countries (NW Africa).

In the last decade (2007 to 2017) we studied and analysed the vegetation around the tombs of local saints ("Marabout") and on graveyards on the Tangier Peninsula (NW Morocco). Total number and size of the SNS in the study area were analysed from topographic maps and verified by field studies in 8 landscape Sects. (36 km^2 plots), representing the mayor ecoregions on a bioclimatic gradient from thermo-mediterranean subhumid coastal sandy terraces to humid mountain ridges of mid altitude with sandstone and schist as bedrock material. On 86 SNS, the vegetation mosaic was recorded: Frequency, spatial extension and naturalness of the habitats were analysed. The habitats were classified, based on physiognomic characters of the plant cover and dominating woody species, and grouped according to naturalness. 203 phytosociological relevés have been sampled in the forested parts of the SNS and were clustered according to floristic similarity. The resulting forest associations are discussed in a broader geographical context (SW Mediterranean) and compared with data from non-sacred forests. Vascular plants were classified according to pre-defined risk factors like stenochory or grazing sensitivity. Naturalness index and refugial value were calculated for each site.

SNS are common phenomenon of the Moroccan cultural landscape. On the Tangier Peninsula, 100 km^2 plots hold on average 29 sacred sites. All over

---

U. Deil (✉) · B. Frosch · H. Jäckle
Faculty of Biology, Department of Geobotany, University of Freiburg, Schänzlestraße 1, 79104 Freiburg, Germany
e-mail: ulrich.deil@biologie.uni-freiburg.de

A. Mhamdi
Département d' Écologie Végétale, Institut Agronomique et Vétérinaire Hassan II, BP 6202, Rabat, Morocco

F. Pedrotti and E. O. Box (eds.), *Tools for Landscape-Scale Geobotany and Conservation*, Geobotany Studies, https://doi.org/10.1007/978-3-030-74950-7_19

Morocco, more than 10 000 SNS can be estimated. The mean spatial extent is less than 1 ha. Only 19% of the SNS area was closed forest. These forests are dominated by *Quercus coccifera*, *Q. suber* or *Olea europaea* var. *sylvestris*, depending on the prevailing substrate and bioclimatic conditions. Only a fraction of these forest patches are near-natural and steno-endemics in these forests are rare. Numerous sacred sites and graveyards are grazed. Under grazing impact, stenoecious perennial forest floor species are replaced by euryoecious annuals. Hence the processes of therophytization and ruderalisation also occur on such sites. Constant overgrazing results in an over-aging of the tree layer ("sub-fossilisation") and ultimately results—in spite of the usage taboo of the tree layer—in the disappearance of the sacred grove, due to missing regeneration of woody plants. Nevertheless, very well preserved holy forests such as the Rusco hypophylli-Quercetum cocciferae and the Teucrio baetici-Quercetum suberis still persist. They have a high conservation value, because a) they are rare examples of intact Mediterranean sclerophyllous forests, b) represent close-to-climax vegetation, c) can serve as stepping stones in a biotope network and d) are aesthetic elements in otherwise intensively used cultural landscapes.

Keywords Cultural ecology · Sacred grove · Holy forest · Marabout · Phytosociology · Quercetea ilicis · Landscape approach · Conservation biology · Naturalness

1 Introduction

In many countries and societies, sacred places are an integral part of the religious self-conception. Most of these sites are designated for cultural and religious reasons, providing a specific space for ceremonies like burials, adoration of local saints, veneration of ancestors, initiation rites etc. [40, 41]. Generally, the idea of nature conservation did not play a role in their foundation. However, taboos like the interdiction of hunting, collecting of plants, deforestation, grazing etc. are indirectly conservation-related and can have positive effects: these taboos can ensure the persistence of near-natural habitats like forests in landscapes otherwise strongly modified by man, and can make sacred sites refuges for threatened species (see for example [8, 22, 62, 70, 72, 118, 132]. They are thus a traditional mode of conserving biodiversity [41, 140].

In the last two decades, conservationists all over the world have become more and more interested in forest patches considered by local societies as "sacred". A symposium [113], organized by UNESCO, IUCN and other international organisations was dedicated to this topic and resulted in the Yamato-declaration, formulating guidelines for the conservation management of sacred places. Such places (including their surrounding vegetation) are called sacred natural sites (= SNS).

A review of the vast body of conservation-related literature about SNS world-wide reveals that the vegetation of many sites is more or less strongly influenced by man. Indicators are, e.g., opening of the tree layer by logging, pasturing or burning, the existence of a mosaic of different degradation stages of the climax vegetation and the invasion or planting of non-native tree species (e.g. [5, 25, 48, 56, 63, 82, 83, 86, 102, 126]). The statements on rare species, habitats or vegetation types on SNS diverge in the scientific literature: There are studies concluding that sacred sites hold a high number of rare and endemic plant species [59, 77, 118]. Other studies report that larger non-sacred forests contain more endemic species than small sacred sites [56]. It can be concluded that "more information about the conservation value of sacred lands is needed", as Dudley et al. [41: 568] pointed out.

However, profound studies about the plant cover, biodiversity and conservation status of SNS are unevenly distributed around the world. The synoptic volume of Ramakrishnan et al. [101] (see first of all the contributions by Hughes & Chandran and Hay-Edie & Hadley therein) documents that holy forests have been intensively analysed in India (e.g. [92, 93]), and, to a lesser extent, in China (e.g. [55]), Korea [28], Indonesia and Japan. A synopsis of SNS studies in Africa and Asia is presented by Deil et al. [34] and Dudley et al. [42]. In Europe, the role of SNS for the conservation of organisms or habitats recently gained some interest (e.g. [6, 7, 49, 50, 78, 108]). Concerning SNS studies in different cultures and religions, the Muslim world has been more or less ignored: In the IUCN-volume, edited by Schaaf and Lee [113], only one contribution is from an Islamic country. Recently, the role of Muslim graveyards in Turkey for protecting orchids from tuber harvest (for salep production) was studied by Molnár et al. [84].

In Africa, sacred site studies concentrate on West and East Africa (see [42, 68, 115, 116]). Inventories of the flora and an analysis of the vegetation and stand structure are for example presented by Guinko [60] and Neumann and Müller-Haude [88] from Burkina Faso, by Tchamié [122], Kokou and Kokutse [71] and Kokou et al. [72] from Togo, by Campbell [25] from Ghana, by Mgumia and Oba [81] from Tanzania, by Virtanen [134] from Mozambique, and by a number of studies in the afro-montane zone of Ethiopia [5, 137, 136, 21, 138, 139].

In contrast to sub-Saharan Africa, the African countries north of the Sahara have been poorly investigated, despite their numerous sacred sites and their biotic relevance. Morocco is part of the Mediterranean Biodiversity Hotspot [87]. About 22% of the known 4200 Moroccan phanerogams are considered endemic [103, 131]. Morocco is faced with the challenge of taking conservation measures for an estimated total number of 1700 endangered plant species (including mushrooms, algae, lichens etc.), among them many narrow endemics (SEEE [114]). The main threats to the survival of endangered species in the southern Mediterranean include overgrazing, deforestation, development (tourism), agricultural intensification [100], urbanisation [129], and, although direct effects are difficult to detect, climate change, above all in mountainous regions [124]. Similar threats affect near-natural habitats like dense sclerophyllous forests. Overgrazing, timber and firewood exploitation, replacement by *Eucalyptus* plantations, wild fires, and intentional burning with subsequent cultivation mainly with Kif (*Cannabis sativa*) result in an ongoing degradation and decline of low and mid altitude thermomediterranean dense sclerophyllous forests in the country [10, 27, 85, 105].

The contribution of sacred sites in Morocco for the conservation of near-natural habitats and the protection of rare and endemic plants has not been extensively and quantitatively studied up to now. Preliminary observations about the flora and the naturalness of the vegetation of SNS in NW Morocco were presented by Deil et al. [32–34]; preliminary data about number, size and locations of SNS were given by Demdan et al. [37] and Taïqui et al. [120].

In Morocco and other Northern African countries, the appreciation of the spiritual authority of patron saints is a common and vivid phenomenon [74], and is expressed by hundreds of collective pilgrimages ("moussem") to the saints' tomb ("koubba" in Arab) every year ([18]; Fig. 1 in [32]). Initially, faithful people assembled around the living saints, called "Marabout", forming a religious brotherhood. In most cases, SNS in Morocco are burial places of Muslim saints. Their foundation lies in the religious belief that the surroundings of the burial place are blessed through the saint's presence, thus rendered sacred themselves [74]. Public rural graveyards lacking a saint's tomb are often equally protected from overexploitation or habitat destruction. Therefore, they have been included into the concept of SNS in this study.

In northern Morocco, some rare thermomediterranean forest types and very well-preserved stands of formerly wide-spread forest types are conserved almost exclusively at sacred sites [12, 52, 96, 98, 110]. However, the exemplary study of two sacred sites in northern Morocco by Deil et al. [33] revealed a mosaic character of the vegetation due to pronounced human impact. As a result of overgrazing and nitrification, sacred groves seem to be often degraded, overaged or reduced to tree groups or a single tree, and processes like the replacement of woody underbrush and perennial forest herbs by ruderals and annual species (therophytization, ruderalisation) take place [51, 52].

Concerning the conservation of rare and endemic plants, Deil et al. [33] speculate that sacred sites in Morocco may provide protection to only few endemic (stenochorous) or stenoecious vascular plant species as sacred sites are generally found where 'normal' site conditions prevail (climatic and edaphic climax). The majority of the known rare and endemic plant species in Morocco however is associated with extreme soil conditions (rocks, dunes, saline habitats) or occurs in high mountain areas [47, 80].

In the current study we hypothesise that sacred sites function to some extent as informally protected 'nature reserves' and can contribute to the protection of endangered plant species and habitats in Morocco. The following main questions are addressed:

1) How many sacred sites exist? What is their cultural context (saints' tomb, cemetery, both functions, destination of pilgrimage)? Where are they located? Is the number and density of SNS different according to ecoregion or land use intensities surrounding them? What is the size of the protected area? Are

topographic maps a probable data source for studying abundance and spatial patterns of SNS? Are SNS restricted to certain relief features or does the variety of their relief positions allow the existence of a broad spectrum of climax communities?

2) Do the sacred sites shelter forests? What percentage of the sites does the forested area have? Are these groves virgin forests or are they subjected to pasturing and other human impacts (beyond burial activities!)? Are they better preserved than non-sacred forests? Does the floristic composition of the groves reflect the abiotic differentiation of the study area and are the well-preserved sacred groves benchmarks of the potential natural vegetation? Can we arrange the holy forests of NW Morocco into associations which are known and described from non-sacred plots?
3) Which other habitats occur on SNS? How natural is the overall vegetation mosaic of SNS in NW Morocco? 'Naturalness' in this context is defined by species combinations, vegetation structure and life-forms. For the site conditions of the study area, a closed sclerophyllous forest is presumed as most natural, while bare soil without plant cover is considered least natural.
4) Is the present land use sustainable and does it guarantee the long-term persistence of the sacred forest?
5) Are the sacred sites refugial areas for rare, vulnerable or threatened plants, especially stenoecious and stenochorous species? Are 'naturalness' of the plant cover and 'refugial value' for plants correlated attributes? Is the extension of the sacred site correlated with the refugial value?

A first starting hypothesis is that the vegetation of sacred sites is closer to the climax vegetation than the surrounding landscape and that the floristic composition of the SNS does mirror the abiotic patterns of the study area. Concerning naturalness and phytodiversity, the hypothesis is that disturbance by men and his livestock can create a vegetation mosaic on the SNS and will hereby increase the species and the habitat diversity (medium disturbance hypothesis). In accordance with the "therophytization" model of the degradation of Mediterranean forests, proposed by Barbero et al. [10], we expect a shift from perennial herbaceous species to annuals in the forest floor vegetation and an increase of ruderals, over-compensating the reduction of stenoecious forest taxa. The opening of the forest canopy and the nitrification of the forest soil should result in a floristic trivialization of the forest floor vegetation.

The study was carried out on the Tangier Peninsula in NW Morocco, an area known for its Marabutism [16]. Finally, the situation on SNS in north-western Morocco will be compared with other parts of the world.

2 Study Area

The study area is located on the Tangier Peninsula in north-western Morocco (Fig. 1). It is delimitated to the west by the Atlantic coast and to the east by the Outer Rif Ranges or Pre-Rif Mountains. The northern limit is 35° 30′ N, the southern limit 34° 45′ N. It features a surface of ca. 7300 km^2 and an altitudinal range from sea level to about 1700 m a.s.l. The climate is of Mediterranean type with winter rains and summer drought. The location close to the Atlantic ocean results in an Atlantic variant of the Mediterranean climate with a relatively high precipitation (mean annual precipitation 600 mm near the coast and more than 1000 mm on the highest peaks) and mild winters (extension of the thermo-mediterranean belt up to 900 m a.s.l.; thermomediterranean sensu [94], mean of the minimum temperature of the coldest month [m] > 3 °C). Figure 1 documents the differentiation of the hygro-climate from sub-humid in the lower parts to per-humid conditions on the Mountain ridges [12].

Geologically the western part of the study area near to the coast is dominated by Pliocene and Quaternary sands, the hilly area in the centre by marls. The Outer Rif Ranges mainly consist of a series of parallel Mountain ridges of Oligocene

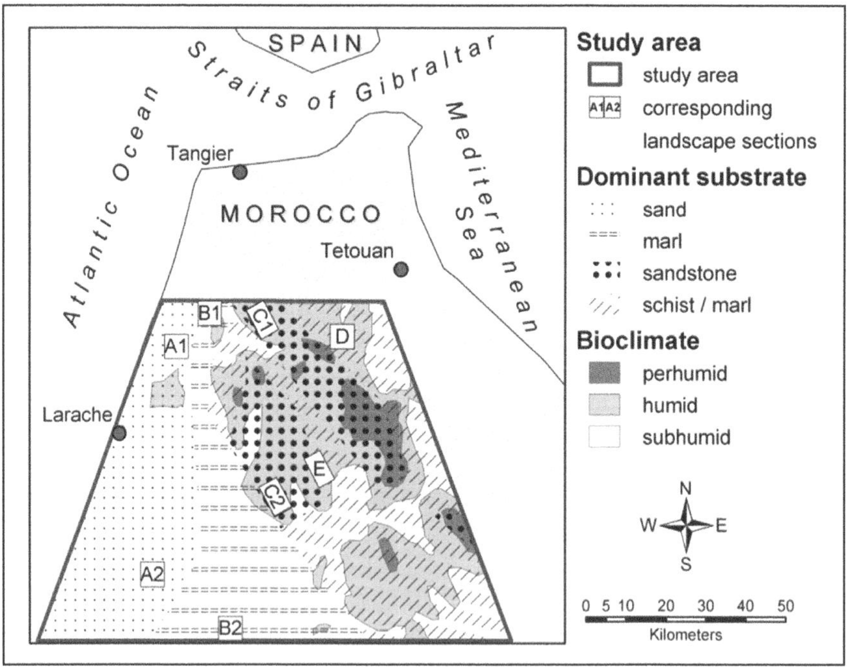

Fig. 1 Study area: delimitation, bedrock strata, climatic differentiation and location of the selected landscape sections (from [66])

sandstone. In between these ranges and further to the east, Cretaceous flysch or schist prevail [58, 119].

The study area falls into the phytochorological units “Tangier Area” and “Loukkos” [131]. The potential natural vegetation throughout the thermomediterranean belt in the study area is Mediterranean evergreen forest with dominating Cork Oak (*Quercus suber)*, Holm Oak (*Q. rotundifolia)*, Kermes Oak (*Q. coccifera)* or Wild Olive (*Olea europaea* var. *sylvestris)* [12, 44, 131]. Details of the bioclimatic niches of the dominating tree species are presented by Frosch and Deil ([52], Fig. 3). The actual vegetation in the lower parts is mainly composed of cropland and pastures. Remaining forests and maquis are intensely used as wood pasture, as well as for firewood collection, timber exploitation and cork extraction. On the sandstone ridges in mid altitude, Cork Oak forests and their first degradation stages (maquis and heathland) are still widespread. Clearing for Kif (*Cannabis sativa*) cultivation, intentional burning and overgrazing are ongoing degradation processes [27, 85, 105].

3 Methods

Stratification and Random Selection of Subplots: The study was performed at 4 spatial scales and with different intensities of field data sampling (Fig. 2). In order to obtain representative data for the Tangier Peninsula, a pre-stratified random sampling procedure was applied. Urban areas were excluded. The study area was divided into 4 eco-regions (= strata), which are homogenous in relief type and dominating bedrock material: A) Littoral sandy plains and hills, B) Prerifean marly hills, C) sandstone ridges, D) flysch and schist mountains (Fig. 2).

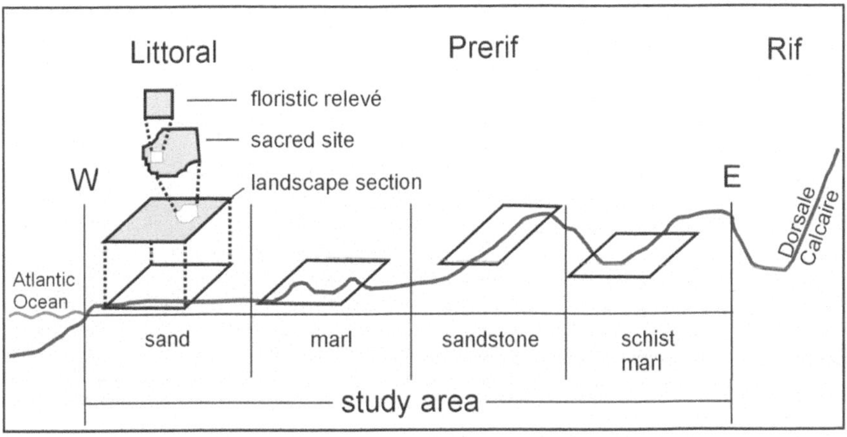

Fig. 2 Schematic topographical and geological profile across the study area, illustrating the 4 spatial scales of data sampling (from [34])

Within each ecoregion, two corresponding landscape sections of 36 km^2 size were selected randomly. They had to fulfil the following additional criteria: 1) a minimum of five SNS according to records on topographic maps, 2) a minimal distance of 20 km to avoid spatial auto-correlation, 3) the two corresponding sections must be situated in the same bioclimate according to Benabid [12].

The sections span an altitudinal range from 40 m a.s.l. near the coast to about 900 m a.s.l. in the Pre-Rif Mountains. Each section has homogeneous geological and climatic conditions. The sampling locations are exclusively situated within the thermomediterranean belt. The hygroclimatic conditions of the studied sacred sites range from subhumid to humid according to the classification in Benabid [11], modified from [111]. Spatial geological information was taken from Suter [119] and generalised to four main strata: littoral sands (sections A), marls (B), sandstone (C), and schist/flysch (D + E). In these eight landscape sections all sacred sites were mapped, apart from landscape section D, where a random sub-sample was chosen due to the unusual high number of sacred sites. All in all, 86 sacred sites with a minimal spatial extension 100 m^2 were investigated. Localities and size (boundaries) of the sacred sites were determined after consultation with local authorities.

Figures 3 and 4 give an impression of two landscape sections and they illustrate the spatial relations of sacred forest patches to the settlements and the surrounding land use pattern.

Fig. 3 Sacred grove Sidi el Mokhfi near the village Er Rhorba. The landscape structure is typical for landscape section D, situated in the tribal area of Beni-Ider in the north-western Rif Mountains. The sacred grove is a mixed stand with *Quercus coccifera* and *Olea europaea* var. *sylvestris*

Fig. 4 Two sacred groves near Mokrisset (Province of Chaouen) in the western Rif Mountains. The landscape structure is typical for landscape section E, situated in the tribal area of Beni-Chaïb with valleys deeply incised into schistose mountain ridges. The holy forests are dominated by *Quercus coccifera*, *Q. suber* and *Olea europaea* var. *sylvestris*

Density and Spatial Pattern of SNS: In a first step, the number and topographic position of SNS (= all the notations "saints tomb" and "cemetery") were extracted from topographic maps of different scale (see [65] for exploited sheets and scales). To verify the information on the maps, a ground check was carried out in all the eight landscape sections. The total number of all existing SNS was identified through field surveys. The area covered by the ground check corresponds to around 4.5% of the surface of the study area. The total number of SNS in the study area was estimated by a linear regression, regressing the number of SNS proved by ground check against the number of sites on the topographic maps 1:50 000. Spatial pattern, relief position and possible explanatory variables (e.g. slope, elevation or dominating land cover in the sections) were analysed by statistical tests described by Jäckle et al. [65] in detail.

Vegetation Analysis Using the Phytosociological Approach: On each sacred site, one to several vegetation relevés were made. A partial species list for each sacred site and a total species list for all investigated sacred sites were derived from 203 relevés and additionally recorded notable plant species. Sampling took place in spring and early summer 2008 and 2009. Nomenclature and taxonomic concept of plant species follow Valdés et al. [131]. Collected specimens are stored in Herbarium Freiburg (FB). Plant communities are named according to Fennane [45] and Rivas-Martínez et al. [106]. Accompanying site parameters like bedrock, pH, and content of $CaCO_3$ were noted or measured. Taking soil samples is a delicate task on sacred sites. We therefore sampled only one per cemetery and only from the upper soil layer (1–10 cm).

On 68 SNS, phytosociological relevés in plots with a tree layer could be sampled, in total 140 relevés. Site names, spatial extension, geographical coordinates of the sites and the complete phytosociological table are given in Frosch and Deil [52]. Data sampling (cover-abundance scale) and data analysis (classification according to floristic similarity) follow the method of Braun-Blanquet [23]. The relevés were stored in a TURBOVEG database and exported into the JUICE program for sorting. Floristic composition as well as vertical structure of the stands was taken into account for classification. A flow chart in Frosch and Deil ([52], Fig. 5) gives an overview of the methodological steps of sampling and analysis of the phytosociological data.

In order to discuss the forest communities on sacred sites in NW Morocco in a broader geographical context and to evaluate their naturalness and their role as documents of potential natural vegetation (PNV), the best preserved holy forests were compared with reference data of similar forest communities, recorded mainly on non-sacred sites in southern Spain and Portugal, and northern Morocco and Algeria. Multivariate analysis was applied to determine the floristic affinity within the data set [123]. The synoptic tables and the ordination diagrams are documented in Frosch and Deil [52].

Physiognomic Approach and Classification of Naturalness: The vegetation of the 86 SNS was mapped based on vertical structure of the vegetation, the combination of growth forms and the dominant species in each stratum. The application of physiognomic and structural parameters follows Küchler [73]. Cover of growth forms was estimated in seven categories and seven height classes were distinguished (Tab. 2 and 3 in [66]). The minimum size for recording of a habitat was 100 m^2.

In a first approach, focussing on the highest vegetation layer, the mapped vegetation patches were grouped to five major habitat types (forest, shrubland, herbaceous formations, open and sparse formations, formations dominated by planted non-native tree or shrub species) (for details see Tab. 1 in [53]). To separate forest from shrubland, a 5 m height threshold (often used for the classification of Mediterranean maquis) was applied (e.g. [46]). To each habitat type, a level of naturalness was assigned on an ordinal scale of 1–8 (Tab. 1 in [53]). Closed forest was presumed to represent the most natural habitat type (highest level 8). Bare soil and formations with more than 50% cover of introduced species represent the least natural habitat types (lowest level 1). Naturalness combined with the respective surface of the habitats on a given locality was used to calculate an average level of naturalness (see formula in [53]) and this value was used to analyse the influence of the naturalness on the refugial value of SNS (see below).

In a more detailed second approach a system of 20 habitat types was developed using the vertical structure of the vegetation, the cover value of the strata, the compositions of growth forms and the dominant species as parameters for the classification (for details see Fig. 2 in [66]). The naturalness values attributed to the habitat types follow the naturalness classes proposed by Ellenberg [43].

Conservation Value of the SNS and Vulnerable Plant Species: 460 autochthonous vascular plant species were recorded on the 86 investigated sites. They were ranked concerning their vulnerability. Ranking was based on five pre-defined risk factors concerning limited geographic range, habitat restriction and sensitivity to grazing (A-E, for definitions see Tab. 2 in [53]). Information on distribution, rarity, and habitat restriction was mainly derived from Fennane and Ibn Tattou [47], complemented by information from Quézel and Santa [97], Castroviejo [26], Valdés et al. [130] and African Plants Database [1]. Risk factors A and B are limited geographic range (narrow endemics respectively distribution restricted to the SW Mediterranean). Risk factor C encompasses rare Mediterranean species which are stenoecious throughout their distribution area like *Davallia canariensis,* as well as species at the margins of their geographic range, and thus stenoecious, like *Cephalanthera longifolia.* Risk factor D (habitat restriction: mesophilous) comprises species more or less restricted to thermomediterranean forests (e.g. *Ruscus hypophyllum*) and shade-tolerant mesophilous herbaceous taxa not restricted to forests (e.g. *Oenanthe pimpinelloides*). Namely diagnostic species of Mediterranean sclerophyllous forests and western Mediterranean thermomediterranean evergreen forests (Querco rotundifoliae-Oleion sylvestris Barbero et al.), according to Rivas-Martínez et al. [107], Barbero et al. [9] and Benabid and Fennane [14] were included in this category. Risk factor E comprises potential grazing sensitive plants.

As more than one of the five risk factors could be attributed to one species, 10 exclusive vulnerability categories were defined through the combination of risk factors. Vulnerability categories were ranked as to their assumed degree of vulnerability from 5 = highest vulnerability to 1 = lowest vulnerability (for definitions and ranking see Tab. 2 in [53]). For each sacred site, vulnerability ranks of the vulnerable species present were added up to obtain a sacred site-based value concerning the vulnerable species representation. We call this value 'refugial value'.

To evaluate the possible influence of different explanatory variables on the refugial value, a multiple linear regression model was calculated with the size of the sacred sites (log-transformed), their calculated naturalness value, and the number of habitat subtypes of sacred sites as independent variables. Since the refugial value showed a non-normal distribution, square-root-transformation was applied to reduce the influence of the few sites with a high refugial value. Statistical analyses were carried out using the statistical environment R (version 2.15.3, [99]). Quantitative analyses were performed using a vector GIS (ArcGIS 9.3, ESRI).

4 Results

4.1 Number, Density and Size of SNS

1450 SNS were documented on the topographic maps 1:50 000 (= TM50) within the study area (Fig. 5). The ground check within the eight landscape sections showed that out of 120 existing SNS only 90 were mapped on TM50. The number of existing SNS varied from 4 to 25 between the landscape sections. The estimation of the total number of SNS using the calculated linear regression model [65] resulted in a total number of 1548 existing SNS and a density of 29.4 ± 6.9 SNS/ $100km^2$. Similar densities were recorded for other regions of the Tangier Peninsula by Demdan et al. [37] and Taïqui et al. [121]. Projection of the average density of 29.4 SNS/100 km^2 on the whole study area resulted in a total number of 2127 SNS. Thus, the 1.450 SNS mapped on the TM50 represent around 67% of the real

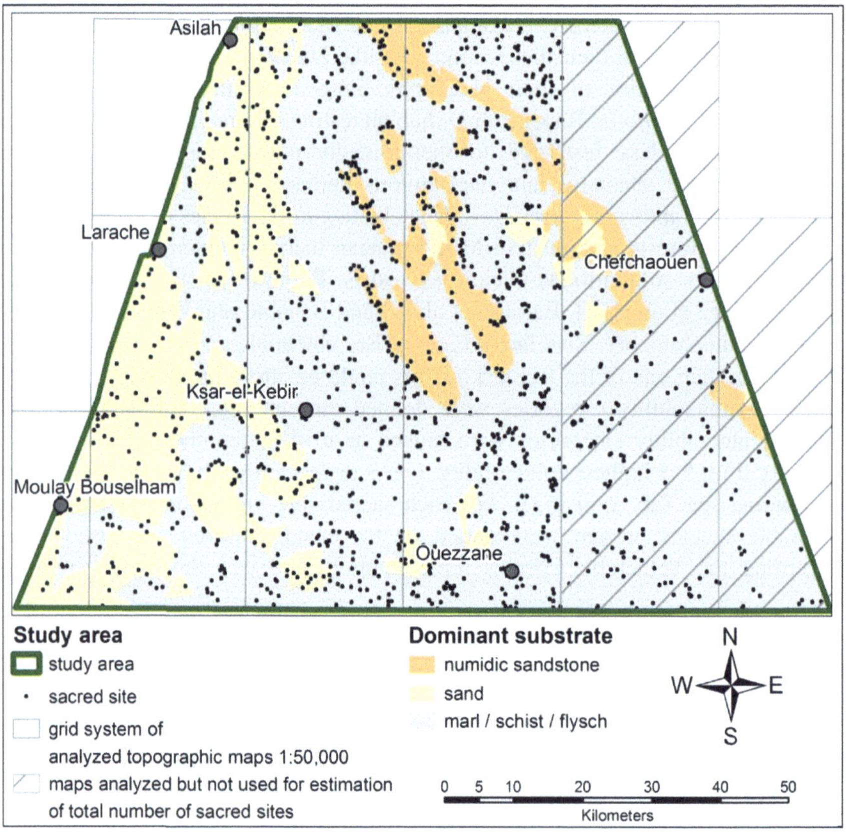

Fig. 5 Number and locations of SNS in the study area, analysed from topographic maps (from [65])

existing SNS. There is a generalization effect, depending from the map scale (67% shown on TP50, 20% on TM100; 2% on TM250; 0% on TM500). For the whole country we can estimate more than 10 000 SNS, distributed over all ecoregions [36]. Because of their abundance and distribution pattern they could serve as elements of a biotope network in degraded landscapes of Morocco.

The 86 investigated sites ranged in size from 85 to 48,950 m^2. The average size was 0.59 ha (median). Only 5 sacred sites were larger than 2 ha (see Fig. 2 in [53] for the frequency of size classes). The total area of the 86 sacred sites amounted to 0.68 km^2. Within the seven completely inventoried landscape sections of 36 km^2 each, the sacred site area varied between 0.09% and 0.36% of the total area. The point pattern analysis shows a non-random distribution (Fig. 5). Sacred sites occurr mainly at elevated topographic positions (hilltops, mountain ridges), but can be found in almost all relief positions, thus offering the potential of supporting different types of climatic climax vegetation. However, depressions and river channels are unlikely positions as well as very steep slopes (<15°) with shallow soils. Such relief positions—representing a special pedoclimax (narrow groundwater level respectively very shallow soils)—are not suitable for graveyards.

By their function as cemeteries, the SNS occur with preference close to the villages [65]. The spatial pattern of settlements in northern Morocco strongly depends on the occurrence of springs and surface water. Especially the borderline between sandstone and underlying marls and flysch is a spring horizon and thus a preferred location for settlements—and in consequence—of cemeteries. Most of the sacred sites consist of a saint's tomb, surrounded by a cemetery. People want to be buried near a sacred tomb to profit from the "Baraka", the spiritual blessing of all objects in spatial context to the Marabout [74].

4.2 Vegetation Cover and Degree of Naturalness of the SNS

Almost half of the investigated sacred site area was covered by forests (42%). The subtype closed forest, attributed with the highest level of naturalness, occurred on 31 sacred sites, covering 19% of the total area. Another 40% of the sacred site surface was shrubland, 30% with and 10% without trees. Herbaceous vegetation with (9%) and without (7%) trees or shrubs occupied less space. Open herbaceous formations and sparse vegetation made up only 2% of the total sacred site area. Habitats dominated by non-native woody plants were found on 9 of the 86 sacred sites, covering less than 0.5% of the total sacred site surface. Per sacred site, one to five habitat subtypes were distinguishable, on average 2.6. At most of the localities (60%), a vegetation mosaic occurred with two or three habitat subtypes. 20% of the sites had a rather uniform vegetation cover with only one habitat subtype. Another 20% had a fine-grained vegetation mosaic with four or five different habitat subtypes. The average degree of naturalness per sacred site was 5.9 (range 2.0–8.0) (see naturalness classes in [53], Tab. 1). Only two sites were completely covered with closed forest. Most of the SNS in the study area underlie a moderate to strong anthropogenic influence.

The classification process applied by Jäckle [66], including all vegetation strata, the combination of growth forms and the dominant species as criteria resulted in 20 habitat types (for details see [66], Tab. 5 and Appendices 2 and 3). These 20 habitat types represent a degradation series from near-natural forest with a more or less closed canopy and typical forest species in the shrub- and herb-layer to sparse vegetation with less than 30% total vegetation cover or patches dominated by planted or sub-spontaneous non-native tree or shrub species. The most natural categories "primeval" and "natural" were not found on the investigated SNS as well as the least natural category "artificial".

The cover value (in ha) of the 20 habitat types on the 86 mapped SNS is illustrated in Jäckle ([66], Fig. 3), the habitat types, their naturalness value and their composition of layer-types is documented in Jäckle ([66], Tab. 5), the vegetation mosaic present on each SNS is shown in Jäckle ([66], Tab 7).

The overall naturalness of the vegetation on the SNS in the different landscape sections is shown in Fig. 6 here, differences in the composition of habitat types among the landscape sections are given in Jäckle ([66], Tab 6). Forest patches occur mainly in the eastern part of the study area in the landscape sections C1, C2, D and E. Very well conserved forest patches were found exclusively in the landscape sections D and E. In sections B1 and B2 forests are rare and, if present, they are overaged and have a nitrified herb layer. In sections A1 and A2 the few forests are also overaged and without tree regeneration in most cases. Maquis with Pistacio-Rhamnetalia shrubs occur in all landscape sections. Vegetation dominated by *Asphodelus ramosus* is restricted to SNS in the marly lowland.

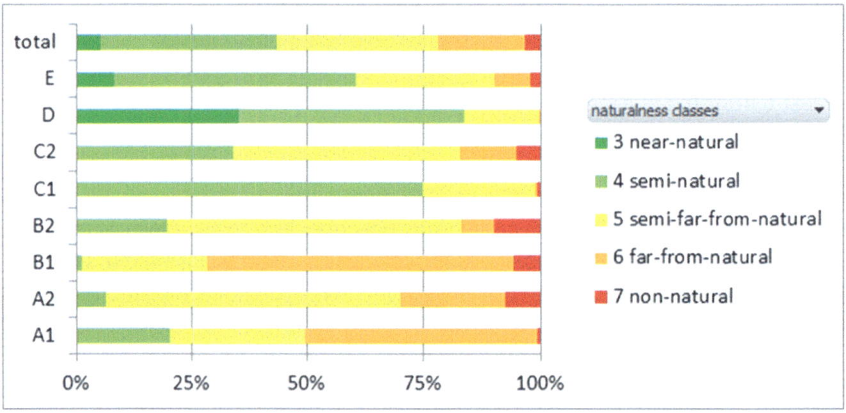

Fig. 6 Pattern of naturalness of the SNS vegetation in the landscape sections A to E (from [66])

4.3 Forest Plant Communities on SNS in NW Morocco

The preliminary typology of sacred forests on the Tangier Peninsula presented by Deil et al. [34] was based on 66 relevés, sampled in holy forests. A bigger data set (140 relevés from 68 SNS) was used by Frosch [51] to compare the vegetation of trees stands on SNS with well-preserved sacred and non-sacred forests in north-western Morocco. The same data set was used to determine the forest communities on the forested parts of SNS and to compare them with well-preserved non-sacred forests in the West-Mediterranean countries. Details and a complete documentation with constancy tables and with tables of the single relevés are given in Frosch and Deil ([52], Tab. 1 and 2, Appendix 2).

The forested vegetation of sacred sites shows a great variety. Abiotic factors, such as substrate and bioclimate, and human interventions (grazing, fire, etc.) were found to be important differentiating factors. Besides well-conserved forests, a broad spectrum of degradation stages was documented. Most stands still shelter the original tree species combination, but vertical structure and floristic composition of the herb layer are strongly modified. Processes related to anthropo-zoogenic pressure, like therophytization, ruderalisation and over-aging of the tree layer occur despite religious taboos.

The sacred sites' forest stands were classified into three clusters, the dominant tree species being either *Quercus suber*, *Q. coccifera* or *Olea europaea*. Overall, 17 forest types could be distinguished. A number of well-preserved stands belong to the associations Teucrio baetici-Quercetum suberis Rivas-Martínez and Rusco hypophylli-Quercetum cocciferae [12] and their subassociations and variants. In comparison to forests on non-sacred sites, these holy forests are very well preserved, as is indicated by the presence of strict forest species of the Quercetalia ilicis and the Querco-Oleion sylvestris, by transgressives of the Quercetalia pubescentis, and by a high constancy and abundance of ombro- and mesophilous taxa.

Cork Oak Forests: Well preserved stands of the Teucrio-Quercetum suberis occur in the thermomediterranean humid bioclimate, on sandstone and sandstone screes, with a subvariant of *Davallia canariensis* on north facing slopes in section C1 and a subvariant with *Quercus coccifera* on shallow sandstone screes in sections D and E. Open and stronger grazed Cork Oak forests are marked by a monospecific tree layer of *Quercus suber* and heliophilous species in the shrub and herb layer. Figure 7 illustrates a variant with *Cistus crispus* on the forest floor, a result of repetitive ground fires.

In the broader SW-Mediterranean context, the well preserved sacred Cork Oak groves on the Tangier Peninsula take an intermediate position between subhumid thermomediterranean and humid thermo- to mesomediterranean Cork Oak forests ([52], Tab. 2, Fig. 11). In general, Teucrio-Quercetum suberis stands from SNS give the impression of very well preserved forests in comparison to non-sacred *Quercus suber* forests. Heliophilous species of the classes Cisto-Lavanduletea and

Fig. 7 Sacred grove with overaged Cork Oak (*Quercus suber*) on the cemetery 'Rija Ibled' in landscape section C2. *Cistus crispus* dwarf shrub vegetation dominates the left foreground, ruderals and annuals to the right, inconspicuous tombs (scattered stones) in the background

Calluno-Ulicetea, indicating fire and opening of the tree canopy by overexploitation, are less frequent and less abundant in their shrub layer. On the other hand, the forests on SNS shelter more ombrophilous and mesophyllous species such as *Davallia canariensis*, *Polypodium cambricum*, *Ruscus hypophyllum*, *Asplenium onopteris* etc., indicating a more humid micro-climate inside the well protected forests.

Kermes Oak Forests: The Rusco-Quercetum cocciferae also occurs in the thermomediterranean humid bioclimatic belt, but grows on schists, marls and shallow sandstone screes. The forests are marked by a species-rich stand structure of several tree- and shrub-layers and a distinct liana layer (Fig. 8). *Quercus coccifera* and *Phillyrea latifolia* are co-dominant in the middle tree-layer (5–15 m); *Pistacia lentiscus* forms a third tree-layer of up to five meters height. Cork Oak contributes to the upper tree stratum (subassociation brachypodietosum sylvatici) on sandstone screes in section D. On calcareous soils, Holm Oak (*Quercus rotundifolia*) and Wild Olive (*Olea europaea* var. *sylvestris*) occur in the middle tree layer (subassociation quercetosum rotundifoliae and variant with *Olea europaea*).

Fig. 8 Near natural Rusco hypophylli-Quercetum cocciferae around Marabout 'Sidi Arabi' in the Western Rif Mountains (landscape section D). The site is dominated by arborescent individuals of *Quercus coccifera* and *Phillyrea latifolia*. Due to opening of the canopy around the qubba, grasses dominate the foreground

The ground floor of the sacred *Quercus coccifera* forests of the Tangier Peninsula is composed of a number of ombrophilous species, e.g. *Ruscus hypophyllum*, *Galium scabrum*, *Brachypodium sylvaticum* and *Luzula forsteri*. These species are restricted to non-grazed sites or to very humid sites with moderate grazing. Remarkable is the high constancy of *Cephalanthera longifolia* in sacred groves of the Rusco-Quercetum brachypodietosum. This orchid is at its southernmost limit of distribution in NW Morocco and is linked to intact meridional brown forest soil with a thick humus layer.

Wild Olive Forests: The Wild Olive groves occur in the thermomediterranean sub-humid bioclimate, on Quaternary and Pleistocene coastal sandy terraces (sections A1, A2 and B1) as well as on sandstone and schists in lower altitudes (up to 490 m a.s.l.). None of the *Olea europaea* groves could be assigned to a syntaxon at association level. The stands must be treated as rankless communities. These units obviously comprise sacred groves which are submitted to strong anthropo-zoogenic impact (grazing, fire, burial activities, and exploitation of wood). Facies with *Hordeum murinum* ssp. *leporinum* or *Brachypodium distachyon* indicate the strong therophytization and nitrification of the ground layer. Figure 9 illustrates such a stand: Under a still quite closed tree canopy with over-aged *Olea europaea* var.

Fig. 9 Sacred site 'Sidi Amour Al Hadi' in Douar Swahal with *Olea europaea* var. *sylvestris*. The forest floor under the overaged Wild Olive trees is ruderalised by heavy grazing, by defaecation of resting animals and by seasonal pilgrimage activities of humans (grove used as campsite)

sylvestris, the forest floor vegetation is strongly ruderalised by heavy grazing, defaecation of resting animals and seasonal pilgrimage activities of humans. In a long run, these sacred groves will disappear despite people respecting the taboo of logging and clearing, because regeneration of the tree layer is missing since decades. We must consider these holy groves as "sub-fossile".

In conclusion, sacred groves often seem to represent the potential natural vegetation. Exceptions are sacred groves with Wild Olive, which occur in an ecoregion (sections A1 and A2 on coastal sandy terraces) where a thermomediterranean *Quercus suber* forest is considered to be the climax (see discussion in [52]).

4.4 Vulnerable Plant Species and Refugial Value of SNS

Of the 460 autochthonous species growing on the investigated sites, 103 (22%) featured one or several of the predefined risk factors indicating vulnerability. Only four species (*Davallia canariensis, Echinops spinosus* ssp. *fontqueri, Scilla monophyllos*, *Silene lagrangei*) were narrow endemics with at least one additional risk factor and thus ranked in the highest vulnerability category. The most common risk factors were "Grazing sensitivity" and "Habitat restriction to forest". 10.9% and 7.8% respectively of all plants on the sacred sites were categorised to these risk factors. Less species (5.2%) were endemic with a southwest Mediterranean distribution. 2% of the 460 species were stenoecious only and 2.6% were more narrow endemics. A complete list of the 103 vulnerable vascular plant species, their risk factors and their frequency on the studied 86 sites is documented in Frosch et al. ([53], Tab. 3, Appendix A).

The number of vulnerable plant species per sacred site varied between 0 and 25. The refugial value, calculated by summarising the vulnerability ranks of the co-occurring species, ranged from 0 to 51. The value shows a non-normal distribution with many sacred sites with a low and only few sacred sites with a high refugial value. A multiple linear regression revealed that the spatial extent of the sacred sites was not significantly associated with their refugial value. However, a higher degree of naturalness and a higher number of habitat subtypes resulted in a higher refugial value (see Fig. 4 and Tab. 4 in [53]). The total number of vulnerable species and the number of species with risk factor "Habitat restriction to forests" was highest on sacred sites with patches of forest and shrubland.

5 Discussion

5.1 *Number, Density and Size of SNS in Morocco in Comparison to Other Countries*

SNS in the study area in NW-Morocco are very numerous (about 29.4 sites per 100 km^2) and in reality more frequent than the locations mapped as "saint tombs" or "cemetery" on TM50. We can expect more than 10 000 SNS all over the country. In most cases they consist of a venerated tomb, surrounded by a cemetery, and they are small in size (on average ca. 0.7 ha). Despite the high number, they cover less than 1% of the landscape. High density and small size of SNS in Morocco is also recorded by other authors [37, 74, 121]. High numbers and small size of SNS are also mentioned from other parts of the world. The number of holy forests all over the Indian subcontinent is estimated to be several hundred thousand. Campbell [25] calculated at least 1900 for Ghana, Mwihomeke et al. [86] counted several hundred in a small district of Tanzania. The groves there are mostly only a few hectares in size.

Often the human impact results in a fine-grained mosaic of different habitats, where forests constitute only one element among others. The mean size of the forest patches in NW Morocco is therefore still smaller (0.5 ha on average per site) than the total mean size of the SNS. We can state a tendency to less conserved sacred sites in lowland areas, where they are embedded into an intensely used agricultural landscape.

5.2 *Naturalness of the Vegetation on Sacred Sites in Northern Morocco*

As climax and close-to-climax forest communities have vanished in vast areas of thermomediterranean Morocco, and the remaining forests on non-sacred sites are more or less degraded [14], sacred groves can be of high conservation value. Our detailed studies in NW Morocco confirm the presumptions of Emberger [44] Sauvage [110] and Benabid [12] that certain sclerophyllous lowland forests types such as Rusco hypophylli-Quercetum cocciferae and Teucrio baetici-Quercetum suberis nowadays exist almost exclusively around Marabout tombs and on cemeteries. On the other hand, Frosch and Deil [52] described for the first time several stages of degradation of these forests. Until recently the percentage of little disturbed forests on sacred sites in Morocco was unknown.

In the studies realised by us on a randomly chosen sample of 86 representative sacred sites on the Tangier Peninsula, a pronounced anthropo-zoogenic impact on many sacred sites is documented: Only about one fifth of the total investigated sacred site area was actually closed forest with a tree canopy cover of $\geq 75\%$. On most of the sites, a vegetation mosaic consisting of several habitat types including degradation stages of forests and maquis was found. This specific vegetation pattern results from different human activities, especially burning, pasturing, and burial activities [33], but also clearing, cork harvesting or logging. This is congruent with the findings of Taïqui et al. [121] who stated that less than 50% of the sites in Larache Province are forests and with the observations of Frosch [51], focussing on the forested part of SNS: Even in some holy forests with a more or less closed tree canopy (especially in groves with Wild Olive and with Cork Oak), a reduced cover of the litter layer and a dominance of annual plants in the ground floor indicate intense grazing and browsing.

The habitats "shrubland" and "herbaceous formations" cover a considerable area on the sacred sites. They represent main stages of a degradation series model for Mediterranean forests, with matorralisation and therophytization as the dominating processes [10, 13, 95, 127]. The term matorralisation refers to the increase in cover and diversity of shrub species in the undergrowth of Mediterranean forests. Therophytization describes the replacement of woody underbrush and perennial herbs through annual species. The conversion of forests to shrublands or herbaceous formations is a gradual process, caused by lack of tree regeneration, often due to intense browsing, and the reduction of the tree cover through logging or natural mortality.

In a global context, the focus of many studies has been on at least partly forested sacred sites, the so-called 'sacred groves'. This was also the case in most research in Morocco in the period 1960 to 2005. Accordingly, results in these earlier studies may be biased in favour of the best preserved sacred sites, and these selective studies might have favoured the un-reflected impression of sacred groves as "virgin forests". Nevertheless, in many non-exhaustive studies from other parts of the world, a human influence on vegetation is at least mentioned (e.g. [25, 67, 76, 82,

83, 102]), and sometimes considered to be the most important factor shaping vegetation [70]. Furthermore, quantitative vegetation studies suggest that human influence/degradation may be an important factor on sacred sites and sacred groves worldwide [5, 54, 63, 86, 126]. In a few studies, less and more disturbed patches, classified according to canopy cover level or dominant life form were described from the same sacred site [25, 83, 126], indicating vegetation mosaics similar to those often found in northern Morocco.

Reasons for the degradation of SNS can be manifold. Weakening of the religious system (secularisation processes) and growing resource pressure by increasing population have been observed by many authors from sub-Saharan Africa (see [25] and contributions in [116] and are here confirmed by us for Morocco. In a few cases, the tree layer of a sacred grove was cut down to finance the renovation of the Mosque or the Qubba (the tomb of the Saint) through the purchase of the timber. On the other hand, the high veneration a certain Saint is held in can also increase the anthropogenic pressure on the vegetation via frequent visits and/or regularly held pilgrimages which may temporarily transform a SNS into a campsite. From an ecocentric viewpoint, the usage of the term "degradation" for this process may be justified. From an ethnocentric viewpoint, taking into account the spiritual value as an ecosystem service of the sacred sites, this process cannot be evaluated as degradation, even if it leads in the long run to the extinction of some organisms [48, 76]. Degradation in the spiritual sense is the transformation of a pilgrimage into a mass tourism event by an urban society with tendencies of secularisation, as it occurs in some Marabout places along the Moroccan coast [17], at Moulay Abdessalam in the western Rif Mountains [141] and in the Western Ghats in India for example [133].

5.3 Sacred Sites as Refuges for Rare or Threatened Plant Species

It is often argued that sacred sites serve as refuges for rare and endemic species [69, 112]. The results presented here only partially confirm this assumption for northern Morocco, as stenochorous and stenoecious plants occur only on a few of the studied sites. An exception are stenoecious forest floor herbs and epiphytes (e.g. *Davallia canariensis*) which need a forest micro-climate and an intact soil profile for their existence. The 'service' of the SNS in north-western Morocco as refuges for threatened plants seems to be limited and our study confirms the statement of Fournier [48] for Burkina Faso: "Although they are religions sanctuaries, wooded shrines are not necessarily biological sanctuaries in the sense of strictly protected areas sheltering exceptional species".

This result is not too surprising, given that many endemics and plants with narrow niches are restricted to mountainous regions or a specific pedoclimax (cliffs, rocks, screes) or saline conditions (coastal cliffs and dunes). The investigated sacred sites were usually not situated in extreme locations concerning inclination, altitude, soil depth, and water availability [65]. In most cases, SNS occur in a climatic climax situation. Nonetheless, places with special environmental features or objects of nature like mountains, sources, ponds, coastal dunes, or caves can be sacred in Morocco [15, 38, 57]. They are often well-known, highly frequented pilgrimage places (e.g. [104, 141]). These examples are, however, the exception rather than the rule (see [35] for two sites in coastal environments).

The bulk of potentially threatened plant species found on the sacred sites consisted of mesophilous forest species and perennial potential grazing decreasers. Overgrazing has been identified as a main threat to terrestrial ecosystems in the Mediterranean [79, 100]. Mesophilous forest floor species in the southern and southwestern Mediterranean biome might become locally endangered due to the ongoing decline and degradation of relatively intact close-canopy forests (see e.g. [2, 105]). The survival of viable populations of mesophilous forest species and potential grazing decreasers on sacred sites probably contributes to the conservation of genetic diversity in the respective species, especially in heavily exploited agrarian landscapes.

As suggested by Bhagwat and Rutte [19] for India and for regions with fertile soils and high population densities, a potential nature conservation role of sacred sites in north-western Morocco could be the protection of 'normal' habitats. Such 'normal' sacred sites with a relatively low anthropogenic impact could fill a considerable void in the existing conservation network. In Morocco, as elsewhere in the Mediterranean Basin, densely populated lowland and mid altitude regions are underrepresented in the network of protected areas [39, 128].

It is a general rule in phytogeography that species, distributed over a broad macroclimatic gradient, become more and more stenoecious at the margin of their range and exist there in fragmented and small populations. Such locally rare plants could possibly be protected on 'normal' sacred sites. The Tangier Peninsula is the southernmost range limit of a number of Euro-Siberian plants, some of them endangered in NW Morocco due to increasing land use pressure and spreading urbanisation [79]. The long-term protection of locally rare taxa is important because peripheral populations of widespread species can feature a genetic distinctiveness, potentially providing flexibility in times of global change [75, 125].

Surprisingly, the size of the sacred site had no significant effect on the refugial value. This outcome is probably related to the relatively small size variance. Most of the 86 sites were smaller than 1 ha. In India, small moist evergreen sacred forest remnants had less endemic tree species in comparison to adjacent large state forest reserves [56, 91]. Wassie [135] and Abiyu et al. [3], on the contrary, found that size had no significant effect on species composition of sacred dry evergreen forests in Ethiopia and concluded that even small forest patches could be useful tools for biodiversity conservation and serve as source populations for native tree species and recolonization of deforested areas. To investigate this hypothesis for sacred sites in Morocco however, more data on the dispersal strategies of the species and the effects of fragmentation on extinction risk and in-breeding depression are needed.

6 Conclusions

The study of SNS in northern Morocco in the last decade provided new insights into the role SNS can play for the conservation of near-natural habitats and vulnerable plant species. The importance of sacred sites for the protection of near-natural forests in northern Morocco has already been known and studied since the 1930s (e.g. [12, 44, 98]) and could be confirmed by us [52, 53]. It seemed promising that almost half of the total investigated sacred site area was composed of forest, but large parts of the forests were degraded and/or subjected to intense browsing, preventing successful regeneration [53, 66]. Figure 10 illustrates such a sacred forest on the mountain top of Jbal Lhabib, characterized by an old-growth *Quercus suber* forest with senescent Cork Oaks and missing tree regeneration since decades. A similar stand structure has been reported by Ajbilou et al. [4] for sacred forests in NW Morocco, an over-aged tree layer in Tibetian SNS by Salick et al. [109]. Frascaroli et al. [50] report a clustering of giant trees and patches of old-growth forests in the vicinity of SNS in Central Italy, with higher total basal area than in non-sacred forests nearby. Gao et al. [55] stated a higher tree layer and larger DBH in temple and cemetery forests in SE China in comparison to no-sacred forests.

Fig. 10 Over-aged Cork Oak forest of Sidi Habib (Western Rif ranges). Grazing, trampling and camping during the annual pilgrimage period prevented any regeneration of *Quercus suber* in the last few decades

Fig. 11 Sacred grove with Dwarf Palm (*Chamaerops humilis*) in arborescent form on a cemetery near Basra, Province Ouezzane, NW Morocco

The ecological role of small forest patches has recently been reviewed for Central Europe. A number of ecosystem services listed by Decocq et al. [30] for such forest fragments (lifeboat habitat for forest species, buffering of macroclimatic changes by a closed tree canopy, aesthetic value, construction of a local identity etc.) can be assigned to vegetation patches surrounding Marabout tombs in Morocco. Their main conservation potential can be seen in their role as refuges for ombrophilous forest-floor specialists and grazing sensitive plants. Furthermore, their ecosystem values "aesthetic landscape elements" and "contribution to the tribal identity" (called "materialization of the noosphere in the biosphere" by [31]) are obvious.

The founders of the sacred sites did not intend to conserve organisms or habitats. This was just a side effect in the protection of a place due to spiritual reasons (see [117] and [68] for West Africa, [133] for India). Site selection did not follow any spatial conservation prioritisation strategies. Many sacred sites in Morocco are situated near settlements in non-extreme topographic locations and have zonal soils. This circumstance may be seen as a disadvantage, concerning the protection of stenoecious and endemic species, but it may also be seen as an advantage: for the required protection of 'normal' habitats and their biocoenoses (e.g. the avifauna, see [22]) in the light of agricultural intensification processes and ongoing degradation of forests in the southern Mediterranean. Due to their high number in the rural landscape, sacred sites may serve as stepping stones in a habitat network [140].

Concrete measures should be taken to support local communities and caretakers in the protection and management of their sacred sites. The exclusion of livestock—at least for a number of years or on parts of the site—would be a suitable management to guarantee the regeneration of woody species and to conserve the forest patches in the long term. A second strategy, to be applied to SNS with semi-natural vegetation, is moderate browsing of the shrubland and seasonal grazing of the herbaceous patches. This will keep these secondary habitats open, ensure the persistence of a fine-grained vegetation mosaic with a high beta-diversity and support, nevertheless, grazing-sensitive non-forest species. To reduce the considerable impact on some sacred sites caused by the defecation of livestock resting under tree shade (see Fig. 9), the planting of tree groups nearby the sacred sites could be promoted.

Highest priority of conservation (no grazing, fire or logging) however should be given to the SNS with near-natural forests sensu Jäckle [66]: These lasts remnants of close-to-climax forests are the only sites to study ecosystem qualities such as maximum standing crop, maximum stand density and height, regeneration behaviour of the matrix species, nutrient cycles, intact soils profiles and natural regeneration of tree species. The potential growth rates of woody plants, existing outside of SNS only in degraded shrubby form (e.g. *Quercus coccifera*, *Phillyrea latifolia*, *Pistacia lentiscus*, *Arbutus unedo*, *Chamaerops humilis*), can be studied here. An example of such ecosystem and forest regeneration studies in Indian sacred groves is summarized by Pandey [93]. These characters of sacred groves might become even more important in the future, because degradation of forests and reduction of the forested area is still going on in the southern Mediterranean countries [20, 29, 89, 105].

At the moment, only very few sacred sites in Morocco are nature reserves in a legal sense like Sidi Bourhaba near Rabat. The proposed classification system of naturalness, the risk factors and the vulnerability categories for plant species developed here may serve as a starting point for a screening of Moroccan SNS with regard to their conservation value. Due to the large number of sacred sites a prioritisation according to attributes like size, level of naturalness, occurrence of vulnerable species and location in intensely used agricultural landscapes seems inevitable. Local societies and caretakers should be supported in the protection and management of their sacred sites by the nature conservancy administration and/or by NGOs. However, imposing the value system of an urban Moroccan society or a western secular NGO on a traditional rural group is not without problems [24, 61, 64, 90, 115]. It should always be kept in mind that such a top-down approach of conservation policy may conflict with the spiritual value. Imposing restrictions and regulations from outside could undermine the intrinsic motivations of the local societies to protect "their" sacred site.

Acknowledgements Sincere thanks are given to the people living in the study area for their great hospitality and the local authorities for allowing us to enter their sacred sites and cemeteries. We are indebted to the Moroccan administration for research permissions. We thank our colleagues and co-workers Achmed Achhal El Kadmiri (deceased), Khadija Fateh, Lahcen Jahouer and Vlad Tataru for the fruitful cooperation. Financial support by the German Research Foundation (DFG) and by the German Federal Ministry of Economic Cooperation and Development (BMZ) is greatly acknowledged (Az. DE 402/6-1, 2).

References

1. African Plants Database (version 3.3.4) (2011) Conservatoire et jardin botaniques de la ville de Genève and South African National Biodiversity Institute Pretoria. https://www.ville-ge.ch/musinfo/bd/cjb/africa/. Accessed June 2011
2. Aafi A (2007) Étude de la diversité floristique de l'écosystème de chêne-liège de la forêt de la Mamora. Dissertation, Institut Agronomique et Vétérinaire Hassan II, Rabat
3. Abiyu A, Tekelay D, Glatzel G, Gratzer G (2016) Seed production, seed dispersal and seedling establishment of two Afromontane tree species in and around a church forests: implications for practice. For Ecosyst. https://doi.org/10.1186/s40663-016-0076-5
4. Ajbilou R, Marañon T, Arroyo J (2006) Ecological and biogeographical analyses of Mediterranean forests of northern Morocco. Acta Oecol 29:104–113
5. Aerts RK, Van Overtveld K, Haile M, Hermy M, Deckers J, Muys B (2006) Species composition and diversity of small afromontane forest fragments in northern Ethiopia. Plant Ecol 187:127–142
6. Agnoletti M, Rotherham ID (2015) Landscape and biocultural diversity: an introduction. Biodivers Conserv 24:3155–3165
7. Agnoletti M, Emanueli F (eds) (2016) Biocultural diversity in Europe. Environmental history, vol 5
8. Anderson DM, Salick J, Moseley RK, Ou X (2005) Conserving the sacred medicine mountains: a vegetation analysis of Tibetan sacred sites in Northwest Yunnan. Biodivers Conserv 14:3065–3091
9. Barbero M, Quézel P, Rivas-Martínez S (1981) Contribution à l'étude des groupements forestiers et préforestiers du Maroc. Phytocoenologia 9:311–412
10. Barbero M, Bonin G, Loisel R, Quézel P (1990) Changes and disturbances of forest ecosystems caused by human activities in the western part of the Mediterranean basin. Vegetatio 87:151–173
11. Benabid A (1983) Études biogéographique et dynamique des peuplements forestiers du Rif (Maroc). Annales De La Recherche Forestière Au Maroc 23:49–129
12. Benabid A (1984) Étude phytoécologique des peuplements forestiers et préforestiers du Rif centro-occidental (Maroc). Travaux de l'Institut Scientifique, Série Botanique 34:1–64
13. Benabid A (1995) Les problèmes de préservation des écosystèmes forestiers marocains en rapport avec le développement socio-économique. In: Bencherifa A, Swearingen WD (eds) L'Afrique du nord face aux menaces écologiques. Université Mohammed V., Rabat. Publications de la Faculté des Lettres et Sciences Humaines 50:109–124
14. Benabid A, Fennane M (1994) Connaissances sur la végétation du Maroc: Phytogéographie, phytosociologie et séries de végétation. Lazaroa 14:21–97
15. Ben-Ami I (1998) Saint veneration among the Jews of Morocco, Jerusalem
16. Berriane M (1989) Tourisme national et migrations de loisir au Maroc. Étude Géographique. Thesis, Tours
17. Berriane M (1990) Tourisme national et migrations de loisir au Maroc: acculturation ou évolution interne? In: Bencherifa A, Popp H (eds) Le Maroc—espace et société, Passau. Passauer Mittelmeerstudien 1:195–214
18. Berriane M (1992) Tourisme national et migrations de loisir au Maroc (Étude géographique). Publ. Fac. Lettres Université Mohammed V, Sér. Thèses et Mém. 16, Rabat

19. Bhagwat SA, Rutte C (2006) Sacred groves: potential for biodiversity management. Front Ecol Environ 4:519–524
20. Bijaber N, Ahlafi Z (2005) Projet FORMA: Etudes détaillées pour la cartographie des changements dans le cadre de l'Inventaire Forestier National. Géo Observateur 14:13–21
21. Bongers F, Wassie A, Sterck FJ, Bekde T, Tekelay D (2006) Ecological restoration and church forests in northern Ethiopia. J Dryland 1:35–45
22. Brandt JS, Wood EM, Pidgeon AM, Han L, Fang Z, Radeloff VC (2013) Sacred forests are keystone structures for forest bird conservation in southwest China's Himalayan Mountains. Biol Conserv 166:34–42
23. Braun-Blanquet J (1964) Pflanzensoziologie. Grundzüge der Vegetationskunde, 3rd edn. Wien, New York, 865 p
24. Brown K (2003) Three challenges for a real people-centered conservation. Glob Ecol Biogeogr 12:89–92
25. Campbell MON (2005) Sacred groves for forest conservation in Ghana's coastal savannas: assessing ecological and social dimensions. Singap J Trop Geogr 26:151–169
26. Castroviejo S (Coord.) (1986–2010) Flora Ibérica. Plantas vasculares de la Península Ibérica e Islas Baleares, vol 21. Real Jardin Bot. CSIC, Madrid
27. Chabbi A (1994) Conditions écologiques et dynamiques de la végétation des basses montagnes du Rif occidental au Maroc. Dissertation Université Michel de Montaigne, Bordeaux
28. Choi JK, Kim DY (2014) Characteristics and management of old and sacred Dangsan forests in Korea. In: Hong SK, Bogaert J, Min Ch (eds) Biocultural landscapes—diversity, functions and values, pp 129–142
29. Dahmani-Megrerouche M (2002) The holm oak (Quercus rotundifolia) in Algeria: climate-vegetation relations. Phytocoenologia 32:129–141
30. Decocq G, Andrieu E, Brunet J, Chabrerie O, De Frenne P, Wulf M et al (2016) Ecosystem services from small forest patches in agricultural landscapes. Curr For Rep 2:30–44
31. Deil U (2003) Holy forests in Northern Morocco. a materialization of the noosphere in the biosphere. Bocconea 16:879–904
32. Deil U, Culmsee H, Berriane M (2005) Sacred groves in Morocco—a society's conservation of nature for spiritual reasons. Silva Carelica 49:185–201
33. Deil U, Culmsee H, Berriane M (2008) Sacred groves in Morocco: vegetation mosaics & biological values. In: Sheridan MJ, Nyamweru C (eds) African sacred groves. Ecological dynamics and social change, Oxford, pp 87–102
34. Deil U, Frosch B, Jäckle H, Mhamdi A, Achhal El Kadmiri A (2009) A geobotanical analysis of forest patches on sacred sites in Northern Morocco. Ber. d. Reinh.-Tüxen-Ges. 21:174–195
35. Deil U, Mhamdi A (2015) Les sites maraboutiques—un patrimoine naturel et culturel du Maroc. In: El Fasskaoui B, Kagermeier A (eds) Patrimoine et tourisme culturel au Maroc. Actes du 9ème colloque maroco-allemand de Meknes 2014 = Publ. Fac. Lettres et Sc. Humaines, Universitè Moulay Ismail, Sér. Actes de Colloques 43:171–178
36. Deil U, Mhamdi A (in Press) Flore et végétation des sites sacrés au Maroc—cas de marabouts et cimetières, 185 p
37. Demdam H, Taiqui L, Seva E (2008) Vers une base de données spatiales des sites sacrés de la province de Tétouan (N du Maroc). Apport de la cartographie officielle. Mediterranea 19 (2):9–68
38. Dermenghem E (1954) Le culte des saints dans l'Islam maghrébin, Paris
39. Derneği D (2010) Ecosystem profile: mediterranean basin biodiversity hotspot. Final report for submission to the CEPF Donor Council, 27 July 2010
40. Dudley N, Higgins-Zogib L, Mansourian S (2006) Beyond belief: linking faith and protected areas for biodiversity conservation. WWF, Gland, 143 p
41. Dudley N, Higgins-Zogib L, Mansourian S (2009) The links between protected areas, faiths, and sacred natural sites. Conserv Biol 23:568–577

42. Dudley N, Bhagwat S, Higgins-Zogib L, Lassen B, Verschuuren B, Wild R (2010) Conservation or biodiversity in sacred natural sites in Asia and Africa: a review of the scientific literature. In: Verschuuren B, Wild R, McNeely J, Oviedo G (eds) Sacred natural sites, pp 19–32
43. Ellenberg H (1963) Vegetation Mitteleuropas mit den Alpen in kausaler, dynamischer und historischer Sicht, Stuttgart, 943 p
44. Emberger L (1939) Aperçu général sur la végétation du Maroc. Commentaire de la carte phytogéographique du Maroc 1:500.000. In: Rübel E, Lüdi W (eds) Ergebnisse der Internationalen Pflanzengeographischen Exkursion durch Marokko und Westalgerien, Bern, pp 40–158
45. Fennane M (2003) Inventaire des communautés végétales à l'aide du phytosociologie, au Maroc. Ecologia Mediterr 29:87–106
46. Fennane M (2006) Esquisse des bases d'une typologie des habitats au Maroc. Bull de l'Institut Scientifique, Séction Sci de la Vie 28:1–5
47. Fennane M, Ibn Tattou M (1998) Catalogue des plantes vasculaires rares, menacées ou endémiques du Maroc, vol 8. Bocconea, pp 1–243
48. Fournier A (2011) Consequences of wooded shrine rituals on vegetation conservation in West Africa: a case study from the Bwaba cultural area (West Burkina Faso). Biodivers Conserv 20:1895–1910
49. Frascaroli F (2013) Sacred sites for the conservation of biodiversity. Dissert. University of Zurich, 195 p
50. Frascaroli F, Bhagwat S, Guarino R, Chiarucci A, Schmid B (2016) Shrines in Central Italy conserve plant diversity and large trees. Ambio 45:468–479
51. Frosch B (2010) Characteristics of the vegetation of tree stands on sacred sites in comparison to well-preserved forests in northwestern Morocco. Ecologia Mediterr 36:83–95
52. Frosch B, Deil U (2011) Forest vegetation on sacred sites of the Tangier Peninsula (NW Morocco) - discussed in a SW-Mediterranean context. Phytocoenologia 41:153–181
53. Frosch B, Jäckle H, Mhamdi A, El Kadmiri A, Rudner M, Deil U (2016) Sacred sites in north-western Morocco—naturalness of their vegetation and conservation value for vulnerable plant species. Feddes Repertorium 127:83–103
54. Gadgil M, Vartak VD (1976) The sacred groves of Western Ghats in India. Econ Bot 30:152–160
55. Gao H, Ouyang Z, Chen S, van Koppen CSA (2013) Role of culturally protected forests in biodiversity conservation in Southeast China. Biodivers Conserv 22:31–544
56. Garcia CA, Pascal JP, Kushalappa CG (2006) Les forêts sacrées du Kodagu en Inde: écologie et religion. Bois et Forêts des Tropiques 288:5–13
57. Gellner E (1969) Saints of the Atlas, Chicago
58. Gharbaoui A (1981) La terre et l'homme dans la péninsule Tingitane. Trav. Institut Sci., Série Géologie et Géographie Physique, vol 15, pp 1–439
59. Githitho AN (2003) The sacred Mijikenda Kaya forests of coastal Kenya and biodiversity conservation. In: Lee C, Schaaf T (eds) The importance of sacred natural sites for biodiversity conservation. Proceedings of the international workshop held in Kunming and Xishuangbanna biosphere reserve, P. R. China, 17–20 February 2003. UNESCO, Paris, pp 27–35
60. Guinko S (1985) Contribution à l'étude de la végétation et de la flore du Burkina Faso. 1. Les reliques boisées ou bois sacrés. Rev. Bois Forêts Trop 208:29–36
61. Hay-Edie T, Hadley M (1998) Natural sacred sites - a comparative approach to their cultural and biological significance. In: Ramakrishnan PS, Saxena KG, Chandrashekara UM (eds) Conserving the sacred for biodiversity management, Oxford, New Delhi, pp 47–67
62. Hong SK, Bogaert J, Min Ch (eds) (2014) Biocultural landscapes—diversity, functions and values, 218 p

63. Huabin H (2003) Sacred natural sites in Xishuangbanna, south-western China. In: Lee C, Schaaf T (eds) The importance of sacred natural sites for biodiversity conservation. Proceedings of the international workshop held in Kunming and Xishuangbanna biosphere reserve, P. R. China, 17–20 February 2003. UNESCO. Paris, pp 127–133
64. Hughes JD, Chandran MDS (1998) Sacred groves around the earth: an overview. In: Ramakrishnan PS, Saxena KG, Chandrashekara UM (eds) Conserving the sacred for biodiversity management, Oxford, New Delhi, pp 71–86
65. Jäckle H, Rudner M, Deil U (2013) Density, spatial pattern and relief features of sacred sites in northern Morocco—an analysis based upon topographic maps. Landscape Online 32:1–16
66. Jäckle H (2017) Conservation status of the vegetation of sacred natural sites in north-western Morocco. Ecologia Mediterr 43:65–83
67. Juhé-Beaulaton D (2008) Sacred forests and the global challenge of biodiversity conservation: The case of Benin and Togo. J Study Relig Nat Cult 2:351–372
68. Juhé-Beaulathon D (ed) (2010) Forêts sacrées et sanctuaires boisés—des creations culturelles et biologiques (Burkina Faso, Togo, Bénin), Paris
69. Khumbongmayum AD, Khan ML, Tripathi RS (2005) Sacred groves of Manipur, northeast India: biodiversity value, status and strategies for their conservation. Biodivers Conserv 14:1541–1582
70. Kibet S (2011) Plant communities, species diversity, richness, and regeneration of a traditionally managed coastal forest, Kenya. For Ecol Manag 261:949–957
71. Kokou K, Kokutse AD (2006) Rôle de la régénération naturelle dans la dynamique actuelle des forêts sacrées littorales du Togo. Phytocoenologia 36:403–419
72. Kokou K, Adjossou K, Kokutse AD (2008) Conserving sacred and riverside forests and criteria and indicators of forest management in low wood producing communities: the case of Togo. Ecol Ind 8:58–169
73. Küchler AW (1988) A physiognomic and structural analysis of vegetation. In: Küchler AW, Zonneveld IS (eds) Vegetation mapping. Handbook of vegetation science, Dordrecht, vol 10, pp 37–50
74. Lang H (1992) Der Heiligenkult in Marokko. Formen und Funktionen der Wallfahrten, Passau
75. Lesica P, Allendorf FW (1995) When are peripheral populations valuable for conservation? Conserv Biol 9:753–760
76. Liberski-Bagnoud D, Fournier A, Nignan S (2010) Les "bois sacrés", faits et illusions. A propos des sanctuaires boisées des Kasena (Burkina Faso). In: Juhé-Beaulaton D (ed) Forêts sacrées et sanctuaires boisées: Des creations culturelles et biologiques (Burkina Faso, Togo, Benin), Paris, pp 59–90
77. Malhotra KC, Gokhle Y, Chatterjee S, Srivastava S (2001) Cultural and ecological dimensions of sacred groves in India. INSA, New Delhi
78. Mallarach JM, Papayannis T, Vaisanen R (eds) (2012) The diversity of sacred lands in Europe. Proceedings of the third workshop of the Delos initiative—Inari/Ananaar 2010. IUCN, Gland, 292 p
79. Mathez J, Quézel P, Raynaud C (1985) The Maghreb countries. In: Gomez-Campo C (ed) Plant conservation in the Mediterranean area, The Hague, pp 141–157
80. Médail F, Quézel P (1997) Hot-spots analysis for conservation of plant biodiversity in the Mediterranean Basin. Ann Mo Bot Gard 84:112–127
81. Mgumia FH, Oba G (2003) Potential role of sacred groves in biodiversity conservation in Tanzania. Environ Conserv 30(3):59–265
82. Miehe G, Miehe S, Will M, Opgenoorth L, Duo L, Dorgeh T, Liu J (2008) An inventory of forest relicts in the pastures of Southern Tibet (Xizang A.R., China). Plant Ecol 194:157–177
83. Mishra BP, Tripathi OP, Tripathi RS, Pandey HN (2004) Effects of anthropogenic disturbance on plant diversity and community structure of a sacred grove in Meghalaya, northeast India. Biodivers Conserv 13:421–436
84. Molnár AV, Nagy T, Löki V, Tökölyi J, et al (2017) Turkish graveyards as refuges for orchids against tuber harvest. Ecol Evol https://doi.org/10.1002/ece3.3562

85. Moore HM, Fox HR, Harrouni MC, El Alami A (1998) Environmental challenges in the Rif mountains, northern Morocco. Environ Conserv 25:354–365
86. Mwihomeke ST, Msangi TH, Mabula K, Ylhäisi J, Mndeme KCH (1998) Traditionally protected forests and nature conservation in the North Pare Mountains and Handeni District, Tanzania. J East Afr Nat Hist 87:279–290
87. Myers N, Mittermeier RA, Mittermeier CG, da Fonseca GAB, Kent J (2000) Biodiversity hotspots for conservation priorities. Nature 403:853–858
88. Neumann K, Müller-Haude P (1999) Forêts sèches au sud-ouest du Burkina Faso: végétation - sols - action de l'homme. Phytocoenologia 29:53–85
89. Nsibi R, Souayah N, Khouja LM, Khaldi A, Bouzid S (2006) Impacts des facteurs biotiques et abiotiques sur la dégradation de subéraie tunisienne. Geo-Eco-Trop 30:25–34
90. Ormsby AA (2011) The impacts of global and national policy on the management and conservation of sacred groves of India. Hum Ecol 39:783–793
91. Page NV, Qureshi Q, Rawat GS, Kushalappa CG (2010) Plant diversity in sacred forest fragments of Western Ghats: a comparative study of four life forms. Plant Ecol 206:237–250
92. Pala NA, Negri AK, Gokhale Y, Todaria NP (2013) Three regeneration status of sacred and protected landscapes in Garhwal Himalaya, India. J Sustain For 32:39–246
93. Pandey HN (2010) Sacred forests: their ecology and diversity, New Delhi, 197 p
94. Quézel P (1974) Les forêts du pourtour méditerranéen. UNESCO, Notes Techniques du MAB, vol 2, pp 1–53
95. Quézel P, Barbero M (1982) Definition and characterization of Mediterranean-type ecosystems. Ecologia Mediterr 8:15–29
96. Quézel P, Barbero M (1990) Les forêts méditerranéennes - problèmes posés par leur signification historique, écologique et leur conservation. Acta Bot Malacitana 15:145–178
97. Quézel P, Santa S (1962–63) Nouvelle Flore de l'Algérie et des régions désertiques méridionales, Paris, vol 2
98. Quézel P, Médail F (2003) Écologie et biogéographie des forêts du bassin méditerranéen, Paris
99. R Core Team (2013) R: a language and environment for statistical computing. R Foundation for Statistical Computing, Vienna. https://www.R-project.org/
100. Radford EA, Catullo G, de Montmollin B (eds) (2011) Important plant areas of the south and east Mediterranean region: priority sites for conservation. IUCN, Malaga
101. Ramakrishnan PS, Saxena KG, Chandrashekara UM (eds) (1998) Conserving the sacred for biodiversity management, Oxford, New Delhi, 480 p
102. Ramanujam MP, Kadamban D (2001) Plant biodiversity of two tropical dry evergreen forests in the Pondicherry region of South India and the role of belief systems in their conservation. Biodivers Conserv 10:1203–1217
103. Rankou H, Culham A, Jury SL, Christenhusz MJM (2013) The endemic flora of Morocco. Phytotaxa 78:1–69
104. Regato P, Salman R (2008) Mediterranean mountains in a changing world: Guidelines for developing action plans. IUCN Centre for Mediterranean Cooperation, Malaga
105. Rejdali M (2005) Forest cover changes in the Maghreb countries with special reference to Morocco. In: Mazzoleni S, di Pasquale D, Mulligan M, di Martino P, Rego F (eds) Recent dynamics of the Mediterranean vegetation and landscape, Chichester, pp 23–31
106. Rivas-Martínez S, Fernández-González F, Loidi J, Lousã M, Penas A (2001) Syntaxonomical checklist of vascular plant communities of Spain and Portugal to association level. Itinera Geobotanica, vol 14, pp 5–341
107. Rivas-Martínez S, Díaz TE, Fernández-González F, Izco J, Loidi J, Lousã M, Penas A (2002) Vascular plant communities of Spain and Portugal. Addenda to the syntaxonomical checklist of 2001. Itinera Geobotanica, vol 15, pp 5–922
108. Rotherham ID (2015) Bio-cultural heritage and biodiversity: emerging paradigms in conservation and planning. Biodivers Conserv 24:3405–3429

109. Salick J, Amend A, Anderson D, Hoffmeister K, Gunn B, Fang Z (2007) Tibetan sacred sites conserve old growth trees and cover in the eastern Himalayas. Biodivers Conserv 16(3):693–706
110. Sauvage C (1961) Recherches géobotaniques sur les subéraies marocaines. Trav. Institut Sci. Chérifien, Série Botanique, Rabat, vol 21, pp 1–462
111. Sauvage C (1963) Étages bioclimatiques. Notices explicatives [in Atlas du Maroc]. Comité National de Géographie du Maroc, Rabat
112. Schaaf T (2006) Linking cultural and biological diversity: the UNESCO-MAB approach. In: Schaaf T, Lee C (eds) Conserving cultural and biological diversity: the role of sacred natural sites and cultural landscapes. Proceedings of the international workshop held in Tokyo (Japan), 30 May–2 June 2005, Paris, pp 12–14
113. Schaaf T, Lee C (eds) (2006) Conserving cultural and biological diversity: the role of sacred natural sites and cultural landscapes. UNESCO, Paris, 341 p
114. SEEE (Secrétariat d'Etat chargé de l'Eau et de l'Environnement - Département de l'Environnement) (2014) Cinquième rapport national sur la biodiversité. Rabat : Secrétariat d'Etat auprès du Ministère de l'Energie, des Mines, de l'Eau et de l'Environnement, chargé de l'Eau et de l'Environnement, Dép. Environnement, Rabat. https://ma.chm-cbd.net/implementation/rap_nat/ma-nr-05-fr.pdf
115. Sheridan MJ (2008) The dynamics of African sacred groves. Ecological, social and symbolic processes. In: Sheridan MJ, Nyamweru C (eds) African sacred groves, Oxford, pp 9–41
116. Sheridan MJ, Nyamweru C (eds) (2008) African sacred groves. Ecological dynamics and social change, Oxford
117. Siebert U (2008) Are sacred forests in Northern Benin "traditional conservation areas"? In: Sheridan MJ, Nyamweru C (eds) African sacred groves, Oxford, pp 164–177
118. Sukumaran S, Jeeva S, Raj ADS, Kannan D (2008) Floristic diversity, conservation status and economic value of miniature sacred groves in Kanyakumari District, Tamil Nadu, Southern Peninsular India. Turk J Bot 32:185–199
119. Suter G (1980) Carte géologique de la chaîne rifaine 1:500.000. Notes et mémoires du service géologique du Maroc 245a, Rabat
120. Taïqui L, Seva E, Roman JL, R'ha A (2005) Los bosquetes de los khaloa (morabitos) del Rif, Atlas Medio y región de Sus de Marruecos. Ecosistemas 14:31–41
121. Taïqui L, Bensalah H, Seva E (2009) La conservation des sites naturels sacrés an Maroc: est-elle incompatible avec le développement socio-économique? Mediterr Serie de Estudios Biológicos 20:106–142
122. Tchamié TTK (2000) Évolution de la flore et de la végétation des bois sacrés des Massifs Kabyè et des régions environnantes (Togo). Lejeunia 164:1–36
123. Ter Braak CJF, Šmilauer P (2002) CANOCO reference manual and CanoDraw for Windows user's guide: software for canonical community ordination (version 4.5). Microcomputer Power, Ithaca, 500 p
124. Thuiller W, Lavorel S, Araújo MB, Sykes MT, Prentice IC (2005) Climate change threats to plant diversity in Europe. Proc Natl Acad Sci USA 102:8245–8250
125. Thuiller W, Albert C, Araújo M, Berry PM, Cabeza M, Guisan A, Hickler T, Midgley GF, Patterson J, Schurr FM, Sykes MT, Zimmerman N (2008) Predicting global change impacts on plant species' distributions: future challenges. Perspect Plant Ecol Evol Syst 9:137–152
126. Tiwari BK, Barik SK, Tripathi RS (1998) Biodiversity value, status, and strategies for conservation of sacred groves of Meghalaya, India. Ecosyst Health 4:20–32
127. Tomaselli R (1976) La dégradation du maquis méditerranéen. In: Forêts et maquis méditerranéens: écologie, conservation et aménagement. Notes techniques du MAB, vol 2, pp 5–76
128. Underwood EC (2009) Expanding the global network of protected areas to save the imperiled Mediterranean biome. Conserv Biol 23:43–52
129. Underwood EC, Viers JH, Klausmeyer KR, Cox RL, Shaw MR (2009) Threats and biodiversity in the Mediterranean biome. Divers Distrib 15:188–197

130. Valdés B, Talavera S, Fernández-Galiano E (eds) (1987) Flora vascular de Andalucía occidental, Barcelona, vol 3
131. Valdés B, Rejdali M, Achhal El Kadmiri A, Jury SL, Montserrat JM (eds) (2002) Catalogue des plantes vasculaires du Nord du Maroc, incluant des clés d'identification, Madrid, vol 2
132. Verschuuren B, Wild R, McNeely J, Oviedo G (eds) (2010) Sacred natural sites. Conserving nature and culture, London, 300 p
133. Vipat A, Bharucha E (2014) Sacred groves: the consequences of traditional management. J. Anthropol https://doi.org/10.1155/2014/595314
134. Virtanen P (2002) The role of customary institutions in the conservation of biodiversity: sacred forests in Mozambique. Environ Values 11:227–241
135. Wassie A (2007) Ethiopian church forests. Opportunities and challenges for restoration. Dissertation, University of Wageningen
136. Wassie A, Tekelay D (2006) Soil seed banks in church forests of northern Ethiopia: implications for the conservation of woody plants. Flora 201:32–43
137. Wassie A, Tekelay D, Powell N (2005) Church forest in northern Gondar. Administrative zone, northern Ethiopia. For Trees Livelihoods 15:329–374
138. Wassie A, Sterck FJ, Tekelay D, Bongers F (2009) Tree regeneration in church forests of Ethiopia: effects of microsites and management. Biotropica 41:110–119
139. Wassie A, Sterck FJ, Bongers F (2010) Species and structural diversity of church forest in a fragmented Ethiopian Highland landscape. J. Veget. Sci. 21:938–948
140. Wild R, McLeod C (eds) (2008) Sacred natural sites: guidelines for protected area managers. Best practice protected area guidelines series, no 16. IUCN, Gland
141. Zouanat Z (2009) 'Abd al-Salâm ibn Mashîsh and Jabal La'lâm site, Northern Morocco. In: Papayannis T, Mallarach JM (eds) The sacred dimension of protected areas. Proceedings of the second workshop of the Delos initiative ouranoupolis, Gland, Athens, vol 2007, pp 66–75

The Kaga Coast (Japan): A Natural Ecosystem and Cultural Landscape Ensuring Biodiversity

Kazue Fujiwara, Elgene O. Box, Jin Ōsaka, Hideki Sakai, and Youji Kitaguchi

Abstract The Kaga Coast, along the Sea of Japan, has highly diverse ecosystems involving natural and anthropogenic landscapes. The most natural area is old Kashima island, with a Shintō shrine; it is connected to the main island and is a National Natural Monument of Japan. *Persea thunbergii* forest remains under protection and is the habitat of a terrestrial crab (akate-gani: *Chiromantes haematocheir*). The crabs live in the forest, eat evergreen leaves, and hibernate in burrows. The scene of akate-gani egg-laying is a good example of a food chain, and the *Persea* forest seems like a terrestrial mangrove forest holding the habitat of crabs. The long (16.4 km), high (46 m) sand dune along the Kaga Coast has been managed since 1766, by plantation of black pine (*Pinus thunbergii*) on the back dune and *Vitex rotundifolia* on the foredune. Nowadays the planted *Pinus thunbergii* forests and sand-dune vegetation protect human lives and also restore and support biodiversity. Even though people planted *Pinus thunbergii,* it still continues nowadays and creates a unique landscape. First, as a planting for defense against mobile sand, the local people combined traditional Japanese technology with European (e.g. French) technology to make the planting succeed. Actually, the method of making artificial foredunes and planting coastal plants was invented independently in Edo-era Japan and in Europe. In Japan, they used dune herbs like *Carex kobomugi*, *Calystegia soldanella* etc. and shrubs like *Vitex rotundifolia.* On the back dune they planted mainly black-pine trees to protect farms and villages from being buried by sand blown by winter winds. They gained partial, temporary results, but could not get a final solution. On the Kaga Coast, it is confirmed that

K. Fujiwara (✉)
Yokohama City University, Yokaohama, Japan
e-mail: fujiwara-kazue-sh@ynu.ac.jp

E. O. Box
Geography Department, University of Georgia, Athens, GA 30602, USA

J. Ōsaka
Daishoji Honcho 56, Kaga City, Ishikawa Prefecture 922-0041, Japan

H. Sakai · Y. Kitaguchi
Culture Promotion Division, Kaga City Hall, Daishoji Minamicho 41, Kaga City, Ishikawa Prefecture 922-0041, Japan

F. Pedrotti and E. O. Box (eds.), *Tools for Landscape-Scale Geobotany and Conservation*, Geobotany Studies, https://doi.org/10.1007/978-3-030-74950-7_20

there are 17 kinds of animals and over 290 kinds of birds (including *Nipponia nippon*). There are 18 kinds of "important species," as designated in the Red Data Book of Japan and of Ishikawa prefecture. Insects utilize both shrubs and herbs, which also represented technological progress in the planting project. This technology and experience was highly significant for sand-defense forests in Japan. The Kaga Coast can thus be described as an area where the human landscape and biodiversity are in harmony.

Keywords Dune vegetation · Harmony of nature and civil engineering · Man-made coastal landscape · Shintō shrine forest · Terrestrial mangrove forest

1 Introduction

The "Kaga Coast" is the coastline of Kaga city, which extends for 16.4 km along the Sea of Japan, from Shioya to Ikiri, in Ishikawa Prefecture. The Kaga coast consists of three distinct areas: sandy beach from Shioya to Katano; reef-bound beach from Katano to Cape Amagozen-misaki (this coastal area is designated as a special protected area of Echizen-Kaga Kaigan Quasi-National Park); and gravel beach from Amagozen to Ikiri [6, 13]. The Daishōji dune (4 km long and 1 km

Fig. 1 During the Heian Era (794-1185/1192) there was forest along the Kaga coast (photo by Kaga City)

wide, from Shioya to Katano) is in the sandy-beach portion of the Kaga Coast and is a single line of dunes [2]. This dune is not only an important element that signifies the natural landscape and ecology of the Kaga Coast; it also has had a great historical influence on industry, social activities, and culture (Fig. 1) [1].

Sand blown by seasonal (winter) winds and actively moving sand dunes were great threats to people's lives and occupations. It was not rare that both their farms and houses were threatened with burial and that people were forced to leave. In one instance, winter winds buried the mouth of the *Daishōji-gawa* (river) under sand, and the people had to remove the sand. The domain of the *Daishōji* clan set up the position of the *Matsu-bugyō* (Pine Magistrate) and made sand-defense and wind-break forests in cooperation with people living around the dunes.

From *Katano-hama* to the *Uwagi-hama* coast, there was Japanese cypress (*Chamaecyparis obtusa*) forest, with a pond called *Ō-ike* (large pond) that was made by people for agricultural use. Ō-ike was buried by blowing sand and separated into five ponds including *Katano Ō-ike* (the present *Katano-kamoike*, or 'duck pond'). The outlet of Katano ō-ike was also buried by sand, and the volume of water became larger year after year. Because the water of Katano ō-ike was about to flood, the Daishōji domain made a tunnel in 1678 to let water flow to the *Kanjo-dani* valley at Katano village [7, 17]. Today, the previous Ō-ike (Katano-kamoike) is the smallest Ramsar-listed wetland in the world (see Fig. 2).

Halfway between *Katano* and *Shioya* there was, from about 1645 to 1844, a village called *Nakahama*. The residents of *Nakahama* had come from *Sanri-hama* village (Echizen province) and made a living by fishing or by making salt. They cut many cypress and other trees as fuel for salt manufacture, so more sand was blown

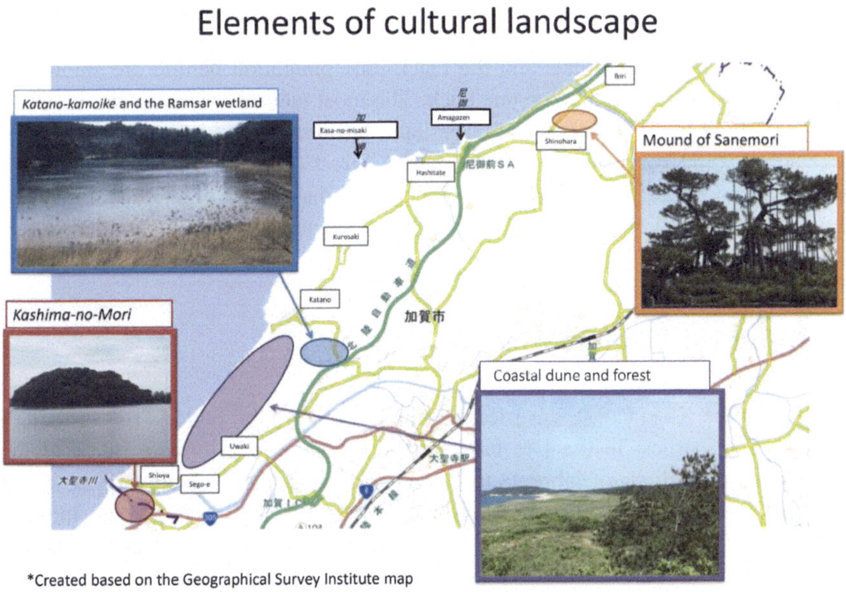

Fig. 2 Elements of the natural and cultural landscape around Kaga City

Fig. 3 Kashima Island, the most natural area along the Kaga coast, with evergreen broad-leaved forest dominated by *Machilus (Persea) thunbergii*

and the sand dunes became larger. Finally, in the Tempō period (1830 to 1844), Nakahama was abandoned. In Uwagi village, blowing sand became violent and buried about ten thousand square meters of farmland in 1744; in 1774 it buried more farms and 21 houses. Especially in 1774, sand was deposited along about 450 m, 24 m high, and some residents were forced to move, creating a new *Uwagi-demura* village [12].

Nowadays these foredunes and the black-pine plantation behind are stabilized, and more natural species have come back. The most natural area is Kashima island (Fig. 3), which has a Shintō shrine. The natural forest around the shrine, the dune vegetation, some parts of the recovered pine forest (plantation), and the wetland have high biodiversity. In 2004–2006, Kaga City made more plantations to restore natural forests on the coastal hills, using potted seedlings raised by students.

2 Historical Planting Activities

Planting on the Kaga Coast was tried beginning in the Empō period (1673 to 1680) but started in earnest in the late Edo era. In 1766 the local branch office of the Daishōji sent hundreds of laborers to Uwagi-hama and made a sand-defense fence about 1600 m long; they then planted black pines (*Pinus thunbergii*) and mimosa (*Albizia julibrissin*) along the sand defense fence. The office continued planting black pines, mimosa, willows (*Salix*), and miscanthus grass (*Miscanthus sinensis*), while also building fences along the coastline in Katano, Uwagi, Segoe, Shioya,

Table 1 Sand-defense plantings by the domain of Daishōji

Period	Area	Kind of trees	Extension
1751 to 1763	Katano	Black pine, *Celtis*, etc.	About 360 m
From 1766	Uwagi	Black pine, mimosa, *Celtis*, etc.	About 1620 m
In 1786	Shioya	Black pine, *Celtis*, etc.	
In 1790	Uwagi	Black pine, willow, *Celtis*, etc.	
From 1833	Shioya	Black pine, mimosa, *Celtis*, etc.	About 324 m
From 1833	Segoe	Black pine, mimosa, *Celtis*, etc.	About 864 m
From 1836	Shinohara	Black pine, mimosa, *Celtis*, etc.	About 648 m
From 1851	Uwagi	Black pine, mimosa, *Celtis*, etc.	About 324 m

Shinohara and Ikiri. Daishōji's retainer, Tojuro Koduka, planted hundreds of thousands of black-pine trees in the sand-dune area. (According to *Sozan Iko*, he planted about 20 million trees and other plants.) Also, the local leader at *Oshiotsuji* village, Koshiro Kano 9th (generation), cooperated with Tojuro, inventing a way of planting pines using straw bags. This sand-defense planting made the basis of today's coastal forest in the National Park of the Kaga Coast (Table 1) [7, 10, 14].

After the Meiji Restoration, the sand-defense plantings lost the support of the feudal clan and were neglected for a long time, because of political and social changes. In consequence, from 1876 to 1883, sand was deposited along about 130 m of coastline, 110 m high, and buried large areas of farmland. In that situation, Chōhei Nishino, son of the previous local leader Chobe, expended his own funds on planting trees. After that, an important national project of sand-defense planting was started in 1911 and continued until 1925, hiring young laborers at Uwagi village (cf Fig. 4). The workers made two lines of fencing in the sand-dune area, 90 m inland from the sea. Then they planted over 4 million black-pine trees, along with alder (*Alnus japonica*), robinia (*Robinia pseudoacacia*), and mimosa [10].

During World War II, people cut pine branches for fuel, as well as whole pine trees to create farmland. They also damaged the trunks of pine trees in order to gather turpentine, which the Japanese Navy used as a substitute for gasoline. As a result, the forest went wild again. Still, people such as Nihei Nishino, son of Chōhei, continued maintaining the sand-defense plantings by putting straw in the fences and planting shrubs, which fertilize the land [5, 11]. The sand-defense forest was designated "the National Park of Echizen-Kaga Coast" in 1968, and the Forestry Agency designated the forest as a "natural recreation forest."

The coast from Shioya to Katano is a typical cultural landscape of the Kaga Coast, and has special significance. For sand-defense plantings, the people combined traditional Japanese technology with European (especially French) technology to make the planting succeed. At the Daishōji domain they planted mainly black-pine trees to protect farms and villages from being buried by sand blown by winter winds. As a result, they gained partial, temporary success, but could not gain a radical solution.

After the Meiji Restoration, based on the demands of the people, interested parties and Ishikawa prefecture, the government began a big project called "the

Sand dune in the coast (in 1910).

National project of planting started in 1911 (photo in 1912).

Sand defense fence (in 1913)

Planting project continued until 1951 (photo in 1917).

Fig. 4 Kaga pine plantation during the Meiji Era (photo by Kaga City)

coastal sand-defense project of beach national forest" and drastically solved the sand problem. This project took 13 years, from 1911 to 1924, and was carried out by the Kanazawa Forestry Office (the present Ishikawa Forest Management Office) (cf Fig. 4). Excellent forest engineers working there made the project progress, investigating the environment (e.g. the seasonal winds) and studying previous experiences, including foreign cases, systematically and in detail.

One important method of planting involved utilizing wind power to make 10 meter-high sand dunes called *mae-kyū* (foredunes) on the seaward side of planting areas; this protects the trees from blowing sand. On the foredunes the project planted coastal plants such as *Vitex rotundifolia*, *Elaeagnus pungens* (shrub), sea bells (*Calystegia soldanella*), *Ixeris repens*, *Salsola komarovii*, *Carex kobomugi*, *Wedelia prostrata*, *Linaria japonica*, *Artemisia capillaris*, *Elymus mollis*, *Imperata cylindrica* var. *koenigii*, and *Arabis stelleri* var. *japonica*. To prevent these plants from being buried by sand, they made bamboo-mat fences. They also made *ko-kyū* dunes (back dunes) behind the planting area [11].

Actually the method of making artificial sand dunes and planting coastal plants was invented independently in Edo-era Japan and in Europe. In Japan people used shrubs like *Vitex rotundifolia*, while in Europe people planted herbaceous plants such as sea bells. On the Kaga Coast, people utilized both shrubs and herbs, which represented a technological advance in the planting project. This technology and experience had great significance for the creation of sand-defense forests in Japan.

3 Methodology

The vegetation of Kaga City was surveyed in 1999–2001 by phytosociological methods [3, 15] and was reported in 2002 (Kaga City 2002) [9]. The vegetation of the 320 relevés collected was classified into 117 associations and communities. Types of forest and dune vegetation on the Kaga Coast have been selected from this report and are compared here for planted trees and herbs. Analysis of the data identified those species that remained and their vegetation types, and identified newcomers as endangered species, such as *Viola* spp. Species of natural forests that could come back in planted pine forests were also identified.

Community Types

1. Natural Vegetation

a) Dune Grasslands (Table 2)
The species planted since 1911, on the 10 m sand dune, remain and have adapted to each habitat. Sand-dune herbs planted at that time include *Calystegia soldanella*, *Ixeris repens*, *Salsola komarovii*, *Carex kobomugi*, *Wedelia prostrata*, *Linaria japonica*, *Artemisia capillaris*, *Elymus mollis*, *Imperata cylindrica* var. *koenigii*, and *Arabis stelleri* var. *japonica*. These remain as the phytosociological units Calystegio soldanellae-Salsoletum komarovi, Caricetum kobomugi association group, Ichaemetum anthephoroidis association group, and an *Elymus mollis* community.

1) *Salsola komarovi* occurs along the front line of the sand-dune vegetation. It was planted as a dune-fixing herb (in 1911–1924) and remains on the front dunes, growing mainly at about 2 m above the sea level. *S. komarovi* is nitrophilic and, after plantation of the herbs on the dunes, has remained and created its habitat on sites created by accumulation of seaweed and other drifting organic materials. *S. komarovi* vegetation is classified, based on its characteristic species *S. komarovi*, in the Calystegio soldanellae-Salsoletum komarovi, belonging to the higher units Salsonion komarovii, Salsoletalia komarovii and Salsoletea komarovii.
2) The other main dune-fixing herbs (from 1911–1924) are *Carex kobomugi*, *Wedelia prostrata*, *Ixeris repens*, *Salsola komarovii*, *Linaria japonica*, and *Artemisia capillaris*; these continue to form the main dune vegetation. *Carex kobomugi* forms a community with *Calystegia soldanella*, *Ixeris repens*, and *Glehnia littoralis* on unstable dunes where wind-blown sand is still accumulating. A Caricetum kobomugi association group occurs on the Kaga coast. One member is the Wedelio-Caricetum kobomugi, which occurs in warm-temperate coastal areas of Japan. Another association is the Elymo-Caricetum kobomugi, which occurs in typical southern-temperate coastal areas. The Kaga coast has both of these along its transitional warm-temperate to typical-temperate coast. This can also be said for the Ichaemetum anthephoroides association group. The Caricetum kobomugi association group is summarized as the Wedelio prostratae-Caricetum

Table 2 Salsoletea komarovii and Glehnietea littoralis

Running number:	1	2	3	4	5	6	7	8	9	10	11
Mean altitude(m):	2	4	3	4	5	5	13	10	5	8	14
Mean height of vegetation (cm):	10	8	11	20	21	30	28	10	13	80	70
Mean number of species:	6	6	3	5	6	7	8	5	5	6	7
Number of relevés:	2	2	5	5	4	3	6	2	4	1	1
Chr. species of association:											
Salsola komarovii	2(3)	·	·	·	·	·	·	·	·	·	·
Chr. species of ass.-group:											
Carex kobomugi	·	2(+)	V(2-3)	V(2-3)	·	·	·	·	·	·	1(+)
Diff. species of association:											
Wedelia prostrata	·	2(4)	·	I(+)	4(+-4)	1(+)	·	·	3(+)	·	·
Diff. species of association:											
Linaria japonica	·	1(+)	·	V(+-1)	1(+)	3(+)	V(+)	·	·	·	1(+)
Artemisia capillaris	·	·	·	·	1(+)	3(+-1)	V(+-3)	·	·	1(+)	1(+)
Chr. species of ass.-group:											
Ischaemum anthephoroides	·	·	I(+)	I(1)	4(+-3)	3(3-4)	V(+-2)	2(1)	2(+)	1(+)	·
Diff. species of subassociation:											
Vitex rotundifolia	·	·	·	·	·	3(1)	·	·	·	1(+)	·
Diff. species of subassociation:											
Viola senamiensis	·	·	·	·	·	·	V(+-5)	·	·	·	·
Chr. & diff species of association:											
Zoysia macrostachya	·	·	·	·	1(+)	·	·	2(3)	·	·	1(+)
Carex pumila	2(+)	·	·	·	·	·	·	2(+-1)	·	·	·
Chr. species of association:											
Fimbristylis sericea	·	1(+)	·	·	·	·	·	·	4(3-4)	·	·
Diff. species of community:											
Elymus mollis	·	·	·	I(2)	·	·	·	·	·	1(5)	·
Diff. species of community:											
Imperata cylindrica var. *koenigii*	·	·	·	·	·	·	·	·	·	·	1(4)
Chr. species of Glehnietea littoralis:											
Calystegia soldanella	2(+)	1(+)	IV(1-2)	V(+-2)	2(+-2)	3(+-1)	V(+)	2(+-2)	3(+)	1(+)	·
Ixeris repens	1(+)	2(+-1)	III(+)	IV(+-2)	1(+)	1(+)	I(+)	·	3(+)	·	·
Glehnia littoralis	·	1(2)	III(+-1)	IV(+-1)	3(+-1)	2(+)	V(+)	·	4(+)	1(+)	1(+)
Companions:											
Heteropappus hispidus var. *arenarius*	·	·	·	·	2(+)	·	V(+)	1(+)	1(+)	·	·
Cuscuta pentagona	·	1(+)	·	·	1(+)	·	I(+)	·	·	·	·
Viola mandshurica var. *crassa*	·	·	·	·	1(+)	1(+)	I(+)	·	·	·	·
Cocculus orbiculatus	·	·	·	·	·	·	III(+-1)	·	·	·	·
Digitaria adscendens	1(1)	·	·	·	·	·	·	·	1(+)	·	·
etc.											

Location: Shioya-kaigan: 1-7; Katano-Kaigan: 1-7, 9-11; Daishōji-Uwagi-machi: 7; Hamayama-misaki: 8.

kobomugi, with *Wedelia prostrata* as character species, and the Elymo-Caricetum kobomugi, with *Elymus mollis* and *Linaria japonica* as differential species. Both of these species occur on northern warm-temperate and typical-temperate coastlines. The Caricetum kobomugi association group and Ischaemetum anthephoroides association group (see next sub-section) belong to the higher units Caricion kobomugi, Glehnetalia littoralis, and Glehnietea littoralis. *Calystegia soldanella*, *Ixeris repens*, and *Glehnia littoralis* occur as characteristic species of higher units. *Glehnia littoralis* is a near-threatened species in Ishikawa Prefecture (for more information see https://www.pref.ishikawa.lg.jp/sizen/reddata/rdb_2010/data/documents/hamabouhuu521.pdf) [20].

3) The Ischaemetum anthephoroides makes back-dune vegetation, where sand movement is reduced. It occurs in two sub-types within the so-called Ischaemetum anthephoroides association group. One association is the Wedelio-Ischaemetum anthephoroides, with *Wedelia prostrata* as differential species. It occurs on warm-temperate coasts. The other sub-type is the Linario japonicae-Ischaemetum anthephoroides, with *Linaria japonica* and *Artemisia capillaris* as differential species. This occurs on temperate coasts. Within this Linario japonicae-Ischaemetum anthephoroides there occur several clumps of *Viola senamiensis* (=*V. grayi*) (see https://www.pref.ishikawa.lg.jp/sizen/reddata/rdb_2010/data/documents/isosumire306.pdf) [21], which is a species of endangerment class II. This *Viola senamiensis* (see Fig. 5) is a perennial plant, flowering in April-May. Its rhizomes are very long and branch upward so that the plant can recover from sand burial. When the winter monsoon is strong, old leaves and stalks stop and gather the moving sand, under which *Viola senamiensis* spends the winter. The Kaga coast, especially the Shioya coast, is famous for clumps of *Viola senamiensis* and *V. mandshurica* var. *crassa* also occurs in the Ischaemetum anthephoroides association group on the Kaga coast. Here it is a coastal phenotype of *V. mandshurica*, with thick, shiny, curved leaves. These *Viola* species are newcomers after the plantation of dune herbs.
4) *Elymus mollis* is dominant on the Kaga coast in plantations for sand stabilization. It also occurs naturally in the Elymo-Caricetum kobomugi. *E. mollis* was one of the species planted in 1911 and has remained on the Kaga coast, albeit in smaller populations. Recently it is still planted for controlling moving sand, but it does not occur as a sand-dune herb otherwise.

Fig. 5 *Viola senamiensis*, part of the largest such community in the world (photo by I. Honda, on 23 April 2008)

5) The other communities and associations are the Ixerido-Fimbristyldetum sericeae, the Wedelio prostratae-Zoysietum macrostachyae, and an *Imperata cylindrica* var. *koenigii* community. The Ixerido-Fimbristyldetum sericeae occurs in slightly concave areas of unstable dunes. The Wedelio prostratae-Zoysietum macrostachyae occurs on small terraces of coastal cliffs of accumulated sand. Grassy areas of *Imperata cylindrica* var. *koenigii* occur on back dunes. Usually, *I. cylindrica* var. *koenigii* occurs in *Miscanthus sinensis* grasslands. It came onto the Kaga coast after the sand became stable, next to the areas of pine plantation.

b) Dune Shrub Communities (Table 3)

Dune shrub species were planted with herb species in 1911, including mainly *Vitex rotundifolia* and *Elaeagnus pungens*. *V. rotundifolia* adapted especially well to the sand dunes and made back-dune scrub vegetation. This is the so-called Linario-Viticetum rotundifoliae, with *V. rotundifolia* as characteristic species, and *Linaria japonica* and *Artemisia capillaris* as differential species. In this association *Viola* species make sub-associations. Violetosum crassae is differentiated by *V. mandshurica* var. *crassa*, *Cuscuta pentagona*, *C. japonica* and *C. chinensis*. It occurs less (10–50% dominant) than the area of *Vitex rotundifolia* and has been disturbed less by humans. Other subassociations are typicum, with no differential species; and violetosum senamiensis, with *V. senamiensis*, *Arabis stelleri* var. *japonica*, and *Cocculus orbiculatus* as differential species. Typicum has dominant *Vitex rotundifolia* with more than 60% cover on relatively stable dunes. Violetosum senamiensis occurs on stable back dunes, where *Vitex rotundifolia* is not dominant (only 5–20% cover). *V. senamiensis* (=*V. grayi*) makes relatively large colonies on the tops of the main dunes and on back dunes. *Vitex rotundifolia* sites on the Kaga coast can hold characteristic *Viola* colonies. *Arabis stelleri* var. *japonica* also occurs here.

c) Evergreen Broad-Leaved Forests (Tables 4, 5)

Evergreen broad-leaved forests (Camellietea japonicae) were mostly destroyed and remained only around Shintō shrines. Forests of *Machilus* (*Persea*) *thunbergii* on the Kaga coast are especially remarkable around the Kashima Shintō shrine and Daishōji temple (e.g. Fig. 6). These forests are summarized as the Polysticho-Perseetum thunbergii, with truly laurophyllous *P. thunbergii* dominant in the forest canopy and with several understoreys of evergreen tree and herb species. The Kashima shrine forest is especially famous for its *akate-gani* (*Chiromantes haematocheir*) crab population and habitat (Fig. 7). The canopy of the Kashima forest is dominated by *M. thunbergii*, with a few trees of *Zelkova serrata* and some *Castanopsis cuspidata* var. *sieboldii*. Forest height is about 25 m. Of the evergreen forests in Kaga City, only the Kashima forest has *Cinnamomum japonicum* and *Dryopteris erythrosora* var. *dilatata*. The Kashima forest was an island (elevation 30 m, perimeter 600 m) in ancient times, but nowadays it is connected to the mainland. Kashima forest is constituted mostly of evergreen broad-leaved species. Sub-canopy trees include *Camellia japonica*, *Aucuba japonica* var. *borealis*, and *Neolitsea sericea* in the tree and shrub layers; *Dryopteris erythrosora*, *Kadsura japonica* and *Trachelospermum asiaticum* f. *intermedium* occur in the herb layer. Vine species such as

Table 3 Viticetea rotundifoliae

Running number:	1	2	3	4
Mean altitude(m):	7	4	12	15
Mean height of vegetation (cm):	18	25	33	18
Mean number of species:	7	7	13	7
Number of relevés:	8	3	12	2
Chr. & diff. species of association and higher units:				
Vitex rotundifolia	V(2-4)	3(4-5)	V(+-3)	·
Linaria japonica	V(+)	2(+-1)	9(+)	2(+)
Artemisia capillaris	III(+-2)	2(+-1)	V(+-3)	2(+)
Diff. species of subassociation:				
Viola mandshurica var. *crassa*	V(1-2)	·	III(+)	·
Cascuta pentagona	III(+)	·	·	·
Cascuta japonica	II(+)	·	·	·
Cascuta chinensis	I(1)	·	+(+)	·
Diff. species of subassociation:				
Viola senamiensis	·	·	V(+-3)	·
Arabis stelleri var.japonica	·	·	IV(+-1)	·
Cocculus orbiculatus	·	·	IV(+-1)	1(+)
Chr. species of association:				
Juniperus conferta	·	·	·	2(5)
Chr. & diss. species of Glehnietea littoralis:				
Glehnia littoralis	IV(+-1)	3(+)	IV(+)	·
Ischaemum anthephoroides	II(+-1)	3(2-3)	IV(+-1)	2(+-2)
Calystegia soldanella	II(+)	3(+-2)	III(+-2)	·
Ixeris repens	III(+)	·	III(+)	1(+)
Carex kobomugi	·	1(+)	·	·
Zoysia macrostachya	I(+)	·	·	·
Carex pumila	·	·	+(+)	·
Companions:				
Heteropappus hispidus var. *arenarlus*	V(+-2)	1(+)	IV(+-1)	1(+)
Scutellaria strigillosa	I(+)	1(+)	IV(+)	2(+)
Elaeagnus umbellata	·	1(2)	III(+-3)	·
Paederia scandens var. *mairei*	·	1(+)	I(+)	·
Oenothera biennis	·	·	II(+)	·
Albizia julibrissin	·	·	II(+)	·
Luzula capitata	·	·	I(+-1)	·
Carex sp.	·	·	I(+)	·
etc.				

Location: 1, 3, 4: Shioya-kaigan; 2: Katano-kaigan.

Table 4 Summarized table of Ardisio-Castanopsietum sieboldii and Polysticho-Persetum thunbergii

Running number:		1	2	3	4	5	6
Mean altitude(m):		18	40	18	70	120	50
Mean height of tree layer 1 (m):		18	20	22	18	18	20
Mean number of species:		22	28	28	31	37	20
Number of relevés:		3	2	3	3	2	1
Characteristic and differential species of association:							
Machilus thunbergii	T1,T2,S,H	3(4–5)	2(3–4)	3(4–5)	3(1)	1(2)	1(+)
Zanthoxylum ailanthoides	T1	1(2)	1(3)	1(1)	·	·	·
Polystichum tagawanum	H	1(+)	2(+–1)	·	·	·	·
Polystichum polyblepharum	H	1(+)	·	2(+)	·	·	·
Acer palmatum	T2,S,H	1(+)	2(+–2)	·	·	·	·
Differential species of lower units:							
Celtis sinensis var. *japonica*	T1,S	3(+–2)	·	·	·	·	·
Thea sinensis	H	2(+)	·	·	·	·	·
Cyrtomium fortunei	H	·	2(+–1)	·	·	·	·
Iris japonica	H	·	2(+)	·	·	·	·
Cinnamomum japonicum	T2,S	·	·	2(1)	·	·	·
Dryopteris erythrosora var. *dilatata*	H	·	·	2(+)	·	·	·
Ophiopogon planiscapus	H	·	·	2(+)	·	·	·
Rhus succedanea	T2	·	·	2(+)	·	·	·
Characteristic and differential species of association:							
Castanopsis cuspidata var. *sieboldii*	T1,T2,S,H	·	2(+)	3(2–3)	3(4–5)	2(4–5)	1(4)
Quercus salicina	T1,S,H	·	·	·	·	1(+)	1(2)
Differential species of lower units:							
Ainsliaea apiculata	H	·	·	·	1(+)	2(+)	·
Pyrola japonica	H	·	·	·	2(+)	1(+)	·
Cymbidium goeringii	H	·	·	·	1(+)	2(+)	·
Rhus ambigua	H	·	·	·	2(+)	·	·
Rohdea japonica	H	·	·	·	2(+)	·	·
Dioscorea tokoro	H	·	·	·	·	2(+)	·
Disporum smilacinum	H	·	·	·	·	2(1)	·
Arachniodes aristata	H	·	·	·	·	·	1(3)
Cleyera japonica	T2	·	·	·	·	·	1(1)
Characteristic and differential species of higher units:							
Camellia japonica	T2,S,H	2(+–4)	2(3)	3(2)	2(+–3)	2(1–2)	1(1)
Dryopteris erythrosora	H	2(+–1)	2(1)	3(1–2)	3(+–1)	2(+–1)	1(1)
Eurya japonica	S,H	3(+–2)	1(+)	1(+)	3(+–1)	1(+)	1(+)
Kadsura japonica	H	2(+–1)	1(+)	3(+)	3(+)	1(+)	1(+)
Aucuba japonica var. *borealis*	S,H	2(3)	2(2)	3(+–2)	3(1–2)	2(+)	1(1)
Neolitsea sericea	T2,S,H	3(+)	2(+–1)	3(+–1)	3(+–1)	1(+)	1(1)
Trachelospermum asiaticum f. *intermediu*	T2,H	2(+–1)	2(+)	3(+–2)	1(2)	2(+–2)	1(+)
Ophiopogon ohwii	H	3(+–1)	2(+–1)	1(+)	3(+–1)	·	1(+)
Elaeagnus glabra	S,H	1(+)	·	1(+)	1(+)	2(+)	1(+)
Ligustrum japonicum	S,H	2(+)	1(1)	2(+)	1(+)	1(+)	·
Hedera rhombea	T2,S,H	2(+–1)	2(+–1)	3(+)	3(+–1)	2(+)	·
Ilex integra	T1,T2,S	2(1–2)	2(+–1)	3(+–1)	3(1–2)	1(1)	·
Ficus oxyphylla	H	1(+)	2(+–1)	1(+)	1(+)	1(+)	1(+)
Ophiopogon japonicus	H	·	·	2(+)	1(+)	2(+)	·
Ardisia japonica	H	2(+)	·	2(+)	3(+–1)	2(1–2)	·

Table 4 (continued)

Epimedium sempervirens	H	•	1(+)	1(+)	•	2(+–1)	•
Fatsia japonica	S,H	2(+)	1(+)	2(+)	3(+)	•	•
Liriope muscari	H	1(+)	1(+)	2(+)	3(+)	•	•
Ardisia crispa	H	•	2(+)	2(+–1)	2(+)	•	•
Dryopteris varia var. *setosa*	H	1(1)	2(+)	•	3(+)	•	•
Ternstroemia gymnanthera	T2,S	1(+)	•	1(+)	•	•	•
Dryopteris lacera	H	1(+)	•	1(+)	•	•	•
Ilex pedunculosa	H	•	•	•	1(+)	1(+)	•
Stauntonia hexaphylla	H	•	•	•	•	1(+)	1(+)
Companions:							
Zelkova serrata	T1,T2	•	2(2–5)	3(1)	2(+–1)	1(1)	1(1)
Callicarpa japonica	S	1(+)	2(1)	2(+)	2(+)	•	•
Parthenocissus tricuspidata	S,H	•	•	3(+)	1(+)	1(+)	•
Pseudosasa japonica	S,H	1(+)	1(3)	1(2)	1(+)	•	•
Wisteria floribunda	T2,S,H	3(+–1)	•	•	2(+)	•	•
Lepisorus thunbergianus	S,H	•	•	•	2(+)	1(+)	•
Zanthoxylum piperitum	H	•	•	•	1(+)	1(+)	•
Acanthopanax sciadophylloides	S	•	•	•	1(+)	1(+)	•
Pleioblastus chino var. *viridis*	H	•	•	•	1(+)	1(+)	•
Rhus trichocarpa	H	•	•	•	1(+)	1(+)	•
Smilax china	H	•	•	•	1(+)	2(+)	•
Cephalanthera erecta	H	•	•	•	1(+)	1(+)	•
Cornus controversa	T1,T2	1(+)	1(1)	•	•	•	•
Ampelopsis glandulosa var. *heterophylla*	S,H	•	1(+)	1(+)	•	•	•
etc.							

Location: 1-4: Kaga coast:(1: Hashitate-machi, Kamoike, Hachiman-jinja; 2: Daishōji,; 3: Kashima-no-mori, Inari-jinja in Daishōji-Ogyu-machi; 4: Hashitate-machi and Daishōji

Trachelospermum and *Kadsura* climb to the tree layers. Kashima forest is protected as a National Natural Monument, especially since it is the important habitat for the akate-gani crab (see below).

The evergreen broad-leaved forests in Kaga City include tall forests dominated by *Machilus thunbergii* (Polysticho-Perseetum thunbergii) but also forests dominated by *Castanopsis cuspidata* var. *sieboldii* (Ardisio-Castanopsietum sieboldii). These two kinds of tall evergreen forest do not have obvious differential species in Kaga City. Polysticho-Perseetum thunbergii occurs on humid sites such as around the Kashima shrine, alluvial areas, and lower hill slopes. The Ardisio-Castanopsietum occurs on slightly drier sites, such as ridges or upper hill slopes. Evergreen, wind-exposed shrub thickets also occur on the Kaga coast, especially the Euonymo-Pittosporetum tobira. These thickets, usually 5–10 m tall, have *Pittosporum tobira*, *Camellia japonica* and *Ilex integra* as dominant species. Sometimes *Pinus thunbergii* also occurs and the height becomes 10 m. The combined cover of tree and shrub layers is 80–100%, so these thickets work well as

Table 5 Euonymo-Pittosporetum tobira

Mean altitude(m):		18
Mean height of tree layer (m):		9
Mean number of species:		17
Number of relevés:		3
Ch. & diff. spp. of association:		
Euonymus japonicus	S,H	2(1)
Rosa multiflora	S,H	2(+)
Paederia scandens	H	2(+)
Pittosporum tobira	S,H	2(2-4)
Farfugium japonicum	H	2(+-1)
Cocculus orbiculatus	S	2(+)
Pinus thunbergii	T1,S	1(3)
Neolitsea sericea	S,H	1(1)
Rhaphiolepis umbellata	S	1(+)
Cyrtomium falcatum	H	1(+)
Ch. & diff. spp. of Camellietea japonicae:		
Camellia japonica	T1,S,H	3(2-4)
Machilus thunbergii	S	2(1-2)
Ilex integra	T1,S	3(+-3)
Castanopsis cuspidata var. *sieboldii*	T1,T2,S,H	2(1-2)
Ligustrum japonicum	S,H	2(+)
Kadsura japonica	S,H	2(+)
Companions:		
Wisteria floribunda	T1,S	3(+)
Celtis sinensis var. *japonica*	S,H	2(+-1)
Smila china	S,H	2(+)
Aucuba japonica var. *borealis*	S,H	2(+-1)

etc.

Location: Hashitate-machi, Kashima-no-mori

windbreaks. Most evergreen broad-leaved forests were recorded on the Kaga coast areas (Table 4, running numbers 1–4, Kaga City 2002). These evergreen broad-leaved forest species recover in old pine plantations over the years.

2) *Pinus thunbergii-Pinus densiflora* Plantations (Table 6)

Historical treatments describe the planting of *Pinus thunbergii* and *P. densiflora* on back dunes, for protection against moving sand, since 1911. Nowadays these plantations have developed into somewhat more natural pine forests and have become stable. Pine plantations are normally 5 m to 20 m tall, depending on plantation year and site conditions.

The pine plantations have developed into two types of forest. One is already the potential vegetation (evergreen broad-leaved forest: Camellietea japonicae), with many naturally recovering species, e.g. *M. thunbergii*, *I. integra*, *I. pedunculosa*,

Fig. 6 Old *Machilus* (*Persea*) *thunbergii* in remnant of the natural forest around the Kashima Shintō shrine

Fig. 7 Native *akate-gani* crabs (*Chiromantes haematocheir*) climbing in arborescent evergreen *Aucuba japonica* var. *borealis*, on the Kaga coast. (Photo by Prof. Seiji Yanai, Ishikawa Prefectural University)

Table 6 *Pinus thunbergii* and *Pinus densiflora* plantation

Running number:		1	2	3	4
Mean altitude(m):		30	30	25	63
Mean height of tree layer 1 (m):		8	20	17	14
Mean number of species:		34	36	44	14
Number of relevés:		1	1	4	3
Planted trees:					
Pinus thunbergii	T1,T2,S,H	1(4)	1(4)	4(1-4)	3(1-4)
Pinus densiflora	T1,T2,S,H	·	·	3(1-4)	2(+-4)
Diff. spp. of lower units:					
Euonymus alatus f. *ciliato-dentatus*	S,H	1(+)	1(+)	4(+-2)	·
* *Machilus thunbergii*	T2,S,H	·	1(+)	4(+-3)	·
Prunus verecunda	T2,S,H	1(+)	1(+)	4(+-1)	·
Ligustrum obtusifolium	S,H	1(+)	1(+)	3(+-1)	·
Liriope minor	H	1(2)	1(1)	3(+-1)	·
Patrinia villosa	H	1(+)	1(1)	3(+)	·
* *Ilex integra*	T2,S,H	1(+)	·	3(+-1)	·
Smilax china	S,H	1(1)	1(2)	3(+-1)	·
* *Ardisia japonica*	H	1(+)	1(+)	2(1-4)	·
* *Ilex pedunculosa*	T2,S,H	1(+)	·	2(+-1)	·
Polygonatum odoratum var. *pluriflorum*	H	1(+)	1(1)	2(+)	·
Alnus sieboldiana	T1,T2,S	1(+)	1(+)	2(2)	·
Paederia scandens	S,H	1(1)	1(1)	2(+-1)	·
* *Hedera rhombea*	H	1(+)	1(+)	2(+)	·
Celtis sinensis var.japonica	S,H	·	1(+)	3(+)	·
* *Eurya japonica*	S,H	·	2(2)	2(+-2)	·
Rhus ambigua	S,H	·	1(+)	2(+-1)	·
Solidago virgaurea var.*asiatica*	H	·	1(+)	3(+)	·
Zanthoxylum piperitum	S,H	1(+)	1(+)	1(+)	·
* *Neolitsea sericea*	T2,S,H	1(+)	·	2(+-1)	·
Celastrus orbiculatus	H	1(+)	·	2(+)	·
Luzula capitata	H	1(+)	·	2(+)	·
Diff. spp. of lower units:					
Juniperus taxifolia var. *lutehuensis*	H	1(5)	·	·	·
Arabis stelleri var. *japonica*	H	1(+)	·	·	·
Linaria japonica	H	1(+)	·	·	·
Pleioblastus chino var. *viridis*	H	·	1(3)	·	·
Miscanthus sinensis	H	·	1(2)	·	·
Iris ensata var. *spontanea*	H	·	1(+)	·	·
Hemerocallia fluva var. *littorea*	H	·	1(+)	·	·
Campanula punctata	H	·	1(+)	·	·
Thalictrum minus var. *hypoleucum*	H	·	1(+)	·	·
Mallotus japonicus	S,H	·	·	4(+-1)	·
Sorbus alnifolia	S,H	·	·	4(+)	·
Cocculus orbiculatus	S,H	·	·	3(+)	·
Akebia trifoliata	H	·	·	3(+-2)	·
Aralia elata	S,H	·	·	3(+)	·
Lonicera japonica	H	·	·	3(+)	·

Table 6 (continued)

* *Aucuba japonica* var. *borealis*	H	·	·	2(+-1)	·
Carex lenta	H	·	·	2(1)	·
Callicarpa japonica	S	·	·	2(+-1)	·
Euscaphis japonica	T2,S,H	·	·	2(1)	·
* *Ligustrum japonicum*	S,H	·	·	2(+)	·
Mitchella undulata	H	·	·	2(+)	·
Viburnum dilatatum	H	·	·	2(+)	·
Acanthopanax sciadophylloides	T2,S	·	·	2(+)	·
Rhus sylvestris	S,H	·	·	2(+)	·
Zanthoxylum schinifolium	S,H	·	·	2(+)	·
* *Quercus glauca*	S	·	·	1(1)	·
* *Quercus salicina*	S	·	·	1(1)	·
* *Camellia japonica*	S	·	·	1(+)	·
* *Ohiopogon ohwii*	H	·	·	1(+)	·
* *Cinnamomum japonicum*	H	·	·	1(+)	·
Diff. spp. of lower units:					
Hypnum plumaeforme	M	·	·	2(2-3)	2(4)
Carex tristachya var. *pocilliformis*	H	·	·	1(+)	2(+)
Carex fibrillosa	H	·	·	·	2(+-1)
Cladonia sp.	M	·	·	·	2(1-2)
Dicranum scoparium	M	·	·	·	1(1)
Campylopus japonicus	M	·	·	·	1(1)
Companions:					
Parthenocissus tricuspidata	T2,S,H	1(3)	1(+)	4(+)	2(+)
Quercus serrata	T2,S,H	1(+)	1(+)	2(1-2)	2(+-1)
Rhus trichocarpa	S,H	1(+)	·	3(+-1)	2(+)
Vaccinium oldhamii	S,H	1(+)	·	3(+-1)	1(+)
Castanea crenata	T1,S	·	·	3(+)	2(+)
Prunus grayana	S,H	·	·	3(+)	1(+)
Carpinus laxiflora	S	·	·	1(+)	1(+)
Carpinus tschonoskii	S,H	·	·	1(+)	1(+)
Oplismenus undulatifolius var. *japonicus*	H	·	·	1(+)	1(+)
Cymbidium goeringii	H	·	·	1(+)	1(+)
Rosa multiflora	H	1(+)	·	1(1)	·
Pyrola japonica	H	1(+)	·	1(+)	1(+)
Albizia julibrissin	H	1(+)	·	1(+)	·
Rhus javanica var. *roxburgii*	S	1(+)	·	1(+)	·
Vitis ficifolia var. *lobata*	H	·	1(+)	1(+)	·
Styrax japonicus	S,H	·	·	1(+)	1(+)
Leucobryum sp.	M	·	·	1(+)	1(+)
Rhamnella franguloides	S,H	·	·	1(+)	1(+)

etc.

Location: Kaga coast: 1: Katano-Kaigan; 2: Hamayama-Misaki; 3: Amagozen-Misaki, Katano-machi, Shioya-machi; 4: Katano-machi.

Eurya japonica, *Neolitsea sericea*, *Ardisia japonica*, and *Hedera rhombea*. At the front of the back dunes and along the contact with the foredunes, these forests are relatively short and may have *Juniperus taxifolia* var. *lutchuensis* as dominant, along with other dune species such as *Arabis stelleri* var. *japonica* and *Linaria japonica*. Less built-up areas have *Miscanthus sinensis* grassland species, such as *Pleioblastus chino* var. *viridis*, *M. sinensis*, *Iris ensata* var. *spontanea*, *Hemerocallis flava* var. *littorea*, *Campanula punctata* and *Thalictrum minus* var. *hypoleucum*. The most stable areas have four-layer forests, with evergreen broad-leaved forest species (* mark in Table 6) and secondary-forest species. Sometimes *P. thunbergii* covers more than 30% in the sub-canopy. These forests have a higher mean number of species, perhaps 44 species. On the other hand, areas of undeveloped soil support only short pine forests, and mosses and lichens dominate on the forest floor; the mean number of species is around 24.

3) Crab Communities and Biology

Although Akate-gani is highly adapted to land life, it must live temporarily in the sea during its growth process. Akate-gani is omnivorous. It can climb trees to eat fresh leaves, such as of *Aucuba japonica* var. *borealis* (Fig. 7), and can eat the dropped leaves of *Machilus thunbergii*, etc. [18]. It can also use tree trunk flow to get down to the seacoast [19]. Kashima Shintō shrine forest thus seems somewhat like a terrestrial mangrove forest.

4 Current Protection Activities and Results

1. Collaboration with traditional and European technology for stability of sand dune

For sand-defense plantings, the Japanese combined traditional technology with European (such as French) technology to make the planting succeed.

After the Meiji Restoration, the government carried out a big project called "the coastal sand defense project of beach national forest" from 1911. This project took 13 years, from 1911 to 1924, and drastically solved the sand problem (cf Fig. 4). Following traditional methods (see above), they made 10 meter-high foredunes and back dunes behind. On the foredunes they planted coastal plants such as *Vitex rotundifolia* and sea bells (*Calystegia soldanella*), stabilized by bamboo mats. On the Kaga Coast, they utilized both shrubs and herbs. These survived in each habitat, and even endangered species came into these habitats (see next section).

2. Pine and sand-dune herb plantations, and natural forests made large ecosystems

Black pine (*Pinus thunbergii*) plantations and coastal plant communities made by the project have become "nature created by humans" reflecting the characteristics of this region. This is the result of interaction between the human approach, which considers natural laws, and natural activity such as succession.

Plantation in 2006 (The 6th plantation ceremony for creating homeland forest) at Kasano-Misaki cape.

Current view of pine plantation on manmade back sand-dune

Fig. 8 Plantation activity on the Kaga coast, with current view of pine plantation on the back dune

The area of black-pine plantation did not only come into existence as a defense against sand and wind. It has also become a beautiful forest with an excellent landscape that reflects the characteristics of this area. This forest has been created over about one century by both human engineering and plant successional processes. Today the forest is also a place for recreation of children and adults (Fig. 8).

The tree overstorey of the black-pine area is black pine, but depending on the location, red pines (*P. densiflora*) were mixed with black pines. There are also a few individuals of exotic *Pinus pinaster* (*P. mariana*), which remain from the early experimental planting. Adjacent to black pines there are plantations of *Robinia pseudoacacia* and *Alnus sieboldiana*, which have the same origin. Euonymo-Pittosporetum tobirae is the natural windbreak shrub community on the coast of Ishikawa prefecture. *Euonymus japonica* shrubs were planted by the Daishōji domain. Recently some black pines have been damaged by pine-wood nematodes, where *Machilus thunbergii* occurred and succession to Polysticho-Perseetum thunbergii has been in progress. Citizen planting activities based on the potential natural vegetation have pushed this succession forward.

Coastal plants other than *Vitex rotundifolia* and *Elaeagnus umbellata*, originally planted by humans, have also grown up on the foredune. Today the community is evaluated as the richest habitat for rare species, such as *Viola senamiensis*, *V. mandshurica* var. *crassa*, and *Iris ensata*. The Kaga community of *Viola senamiensis* in particular is the largest such colony in the world (Fig. 5). The plants in this community take turns blooming over the whole period from March to October. Each plant species is integrated into the community by succession. In the sandy-beach shrub community, there are some other communities, such as the *Linaria-Vitex* community. Of herbaceous communities, about 20 are differentiated (Kaga City 2002). The planted herbaceous species have developed their own communities and occur on appropriate habitats (Fig. 9) [9].

Fig. 9 Restored Sand-Dune Vegetation on the Kaga Coast

It is confirmed that there are 17 species of larger animals, such as serow. There are also over 290 bird species that come to or live all year on the Kaga Coast. Wild ducks and wild geese migrate seasonally to Katano-Kamoike and its Ramsar-listed surrounding wetlands. It is also confirmed that *Nipponia nippon*, the Japanese national bird, fly along the Kaga Coast.

There are also about 450 insect species recorded just in the coastal plant communities. There are 18 "important species," which are listed in the Red Data Book of Japan or of Ishikawa prefecture. Insects on the Kaga Coast mainly live in three areas: the coastal plant community, the Kashima-no-mori forest, and the wetlands around Katano-Kamoike. It is ascertained that over 900 species of insects are living in the whole Kaga-Coast area. The most important species are *Agrotis ripae* and *Cicindela laetescripta*, which live in only a few places. The Kaga Coast is also a stable habitat of *Bembix niponica* (a moth of the family Noctuidae) and *Herpetogramma albipennis* (a moth in the family Crambidae), both of which were found only recently. Spiders *Araneida* and many *Lycosa ishikariana* also live in the coastal plant communities [8].

The Kashima-no-mori, a National Natural Monument of Japan, is an evergreen broad-leaved forest dominated by *Machilus thunbergii*. The forest is the largest habitat of the Akate-gani crabs (*Chiromantes haematocheir*). Akate-gani has a unique mode of living—it lives in the forest and hibernates in burrows. The scene of Akate-gani laying eggs shows a good example of a food chain.

Why are there so many animals and plants living on the Kaga Coast? Because there is a large area of coastal dune vegetation, and this forest and foredune support large, diverse ecosystems.

3. The Kaga Coast was the last remaining undeveloped coastal area in Japan

The dunes along the Kaga Coast represent a rare case in coastal Japan, since the Kaga dunes had not been diverted to farmland, factory or residential area, amusement park, sports facility, or land for public use such as park and school. Such diversion is generally seen in sand-defense or windbreak forests in Japan. In addition, even if such forest remains, the landscape often gets destroyed, as by highway construction, installation of tetrapods or other breakwaters, or development of ports or other waterfront areas. Destruction of the sand-defense forest is now accepted by society because sand disasters now occur less frequently, since less sand is now delivered by rivers where dams are constructed [16]. For these reasons, from a national viewpoint, the value of the "Kaga Coast" as a cultural landscape is increasing year after year.

References

1. Anonymous (1794) A summary memorandum of local office
2. Anonymous (1816) The memorandum of official maps of Kaga, Ecchu, Noto
3. Braun-Blanquet J (1964) Pflanzensoziologie, Grundzüge der Vegetationskunde, 3rd edn
4. Editorial Committee for Flora of Ishikawa Prefecture (ed) (1997) Flora of Ishikawa prefecture. Ishikawa Prefecture, Kanazawa
5. Editorial Committee for The History of Forestry in Ishikawa Prefecture (ed) (1997) The history of forestry in Ishikawa prefecture. Forest association of Ishikawa prefecture
6. Editorial Office of Urban Kubota (1992) Feature Article: Hills and plains of Hokuriku, Urban Kubota, Issue 31, Kubota Corp., Tokyo
7. Enuma Country Office of Ishikawa Prefecture (1925) The topography of Enuma country
8. Executive Committee of Present State Survey of Hashitate Hills of Kaga City (ed) (2007) The present-state survey report on the preparation project of Hashitate Natural Park, Kaga City
9. Institute for Global Environmental Strategies-Japanese Center for International Studies in Ecology (ed) The vegetation of Kaga City, Kaga City, Kaga (in Japanese with English summery)
10. Ishikawa Prefecture (1925) The history of Ishikawa prefecture, 3rd volume, Kanazawa. Ishikawa Prefecture, Kanazawa
11. Kanazawa Forestry Office (1952) The outline of the coastal sand defense project of beech national (revised edition), Kanazawa Forestry Office
12. Kano T, year unknown, Sozan Iko
13. Konno Y (ed.) (1993) The geological History of Ishikawa prefecture, Ishikawa Prefecture and Hokuriku Research Institute of Geology, Kanazawa
14. Kozuka H (1844) The topography of Kaga and Enuma
15. Mueller-Dombois D, Ellenberg H (1974) Aims and methods of vegetation ecology. Wiley, p 547
16. Murai H et al (ed) Coastal forests in Japan, Soft Science Corp., Tokyo
17. Tsukaya S (1803) Bakkei Kibun

18. Yamagishi A, Yanai S (2012) Defoliation by terrestrial crabs in a laurel forest on the coast of the Sea of Japan. ESJ59/EAFES5 Abstract. P3–305J (in Japanese)
19. Yanai S (2017) The ecology of Aka-te-gani (Chiromantes haematocheir) inhabiting coastal forests. Inochi-Shiki-Saiji (Life, seasonal color) D & U, vol 24 (in Japanese). http://www.sizenken.jp/info/inochi_sikisaiji/vol_24.pdf
20. https://www.pref.ishikawa.lg.jp/sizen/reddata/rdb_2010/data/documents/hamabouhuu521.pdf
21. https://www.pref.ishikawa.lg.jp/sizen/reddata/rdb_2010/data/documents/isosumire306.pdf

Measurement and Conservation of Species and Habitat Biodiversity at Different Scales on Small Mediterranean Islands

Gianniantonio Domina, Anna Maria Mannino, Francesco M. Raimondo, and Patrizia Campisi

Abstract In this paper we discuss current methodologies for measuring biodiversity, analyse pros and cons at different scales, and finally propose distribution mapping as the most suitable method for measuring floristic biodiversity on small Mediterranean islands. These territories often host endangered flora and habitats, and have peculiarities that distinguish them from the mainland and the larger islands.

Keywords Remote sensing · Distribution mapping · Floristic lists · Population size

1 Introduction

Small islands are geographically and ecologically well-defined areas in which biological processes are easier to schematize than on the mainland [15]. Size is the best geographical feature for characterizing the different Mediterranean islands biologically. Larger islands show biological traits similar to other coastal areas of the Mediterranean [7, 8]. Other features, such as distance from the mainland and shape, have been shown to be less influential for the study of biological colonization [8].

G. Domina (✉)
Department SAAF, University of Palermo, viale delle Scienze Bldg. 4, 90128 Palermo, Italy
e-mail: gianniantonio.domina@unipa.it

A. M. Mannino · F. M. Raimondo · P. Campisi
Department STEBICEF, University of Palermo, via Archirafi 38, 90123 Palermo, Italy
e-mail: annamaria.mannino@unipa.it

F. M. Raimondo
e-mail: francesco.raimondo@unipa.it

P. Campisi
e-mail: patrizia.campisi@unipa.it

F. Pedrotti and E. O. Box (eds.), *Tools for Landscape-Scale Geobotany and Conservation*, Geobotany Studies, https://doi.org/10.1007/978-3-030-74950-7_21

The size limit for distinguishing small versus large Mediterranean islands may be about 80 km^2, approximately the size of Formentera, Pantelleria, and Santorini.

According to Gaston [12], biodiversity can be considered from the point of view of ecology, of organisms and of genetics. It is therefore clear that several methods exist to measure biodiversity but all agree on measuring the number of elements and their variability. The range of spatial, temporal and taxonomic scales used increases further the variables to be measured. Many existing methods are well designed and informative. But, very often they are inadequate for purposes beyond those for which they were specifically designed. Species diversity indexes, for example, are based on counting species, and do not take into account their total or relative abundance distribution. Knowledge on distribution of species and their population trends are essential for reliable plans of conservation and management of the resources. Knowledge on trends in biodiversity loss is hindered by the absence of reliable basic data for most groups of organisms as well as habitats. Plants are primary producers and key structural elements for most ecosystems and islands are among the best floristically known territories in the Mediterranean. Greuter [14] reports that in Mediterranean islands the percentage of records of taxonomic doubtful taxa per total number of records is about half of that recorded in mainland, whereas the percentage of records of doubtful occurrence per total number of records is about doubled. These data can be explained by a higher attention of researcher for taxa occurring in the islands and with a greatest difficulty in ascertaining historical records. From these considerations comes the need to choose a suitable method for measuring biodiversity in small islands. In this contribution, we review the current methods for the measurement of biodiversity and propose distribution mapping as the most suitable method for small Mediterranean islands.

2 Materials and Methods

A preliminary survey of current methods used to assess biodiversity was made from available literature. The approaches used concern: habitat measurement, population size measurement, taxon lists, and distribution mapping. Pros and cons of each method, at different scales, are summarized in Table 1.

3 Discussion

Habitat Measurement. Advances in remote sensing and GIS software have greatly aided large-scale habitat measurement. For example, the extent of forest fires in Southeast Asia and losses of primary forest have been monitored with satellite images [18]. Nowadays, it is possible to obtain satellite images with ground resolution of 40 cm. This resolution can be sufficient for trees or habitats well differentiated from their surroundings, but it is not yet adequate for many other

Table 1 Methods to measure biodiversity, their pros and cons and their scale of applicability

Method	Pros	Cons	Most suitable for
Remote habitat measurement	– low cost and high speed – automatic recognition – there is no need to visit all sampling sites – satellite data available for the whole earth – high interest in current scientific literature	– mapped phenomena must be recognizable on available images	Broad scale
Measures of population size	– highly informative – adopted as international standard	– need to visit all sampling sites – long time requirement – few historical data available	Fine scale
Floristic lists	– low cost – large historical data-sets available	– no reliable information on the phenomenon size – little interest in current scientific literature	Both broad and fine scale
Distribution mapping	– gives information about the phenomenon size – large data-sets available	– geo-referenced data are needed – need to visit all sampling sites	Both broad and fine scale

purposes, such as monitoring many habitats and plants that occur on islands. Typical examples are Mediterranean temporary ponds, a highly threatened ecosystem [24], and ephemeral therophytic grasslands [6], which occur on surfaces so small that they cannot be recognized on satellite or aerial photos (Fig. 1).

Population size for threatened taxa, together with analysis of population trends [11], are the bases for assessments according to IUCN methodologies (IUCN 2012). Unfortunately, historical data on population size are not available for most species, and new measures are expensive and not practical in many cases. The problem on

Fig. 1 Temporary ponds in the island of Lampedusa from a satellite image (a) and a live photo (b)

Fig. 2 Habitat of *Silene hicesiae* Brullo & Signor. in the island of Alicudi. The instability of steep rocky slopes makes this area inaccessible for measurements

islands is accentuated by the occurrence of sea cliffs and other areas of difficult access that make impracticable the counting of single individuals or the measurement of surfaces occupied by populations (Fig. 2).

Floristic Lists. Information on the presence of higher plants is available for many Mediterranean islands and often from different periods. This is because islands have always attracted naturalists who have historically produced reliable records that can be found dispersed within national and supranational floras. For example, several references to the flora of the East Aegean Islands can be found in Boissier's *Flora Orientalis* (Boissier 1867–1884) or in targeted contributions (e.g. [5]. Nowadays, visits by individual researchers have been joined by formal surveys undertaken by governments and NGOs. The PIM (Petites Iles de Méditerranée) initiative was set up in 2005 by the Conservatoire du Littoral in Aix-en-Provence (France), in order to promote the conservation of biodiversity on small Mediterranean islands. It is a virtuous example of how it is possible to carry on naturalistic exploration of the small Mediterranean islands [17], involving local administrations and recording all the results into a large virtual library (https://www.initiative-pim.org) (Fig. 3).

In addition, on islands the lack of records for a species is more easily correlated to its disappearance from that territory than it is on the mainland. Particular attention has to be paid to measurement of rarity and extinction risk. For example, *Arbutus unedo* L. and *Erica multiflora* L. ssp. *multiflora* occupied large areas on Lampedusa until the XIX Century, before the systematic deforestation of the island. At present, only few, isolated individuals are present [20]. Most extinctions, after a

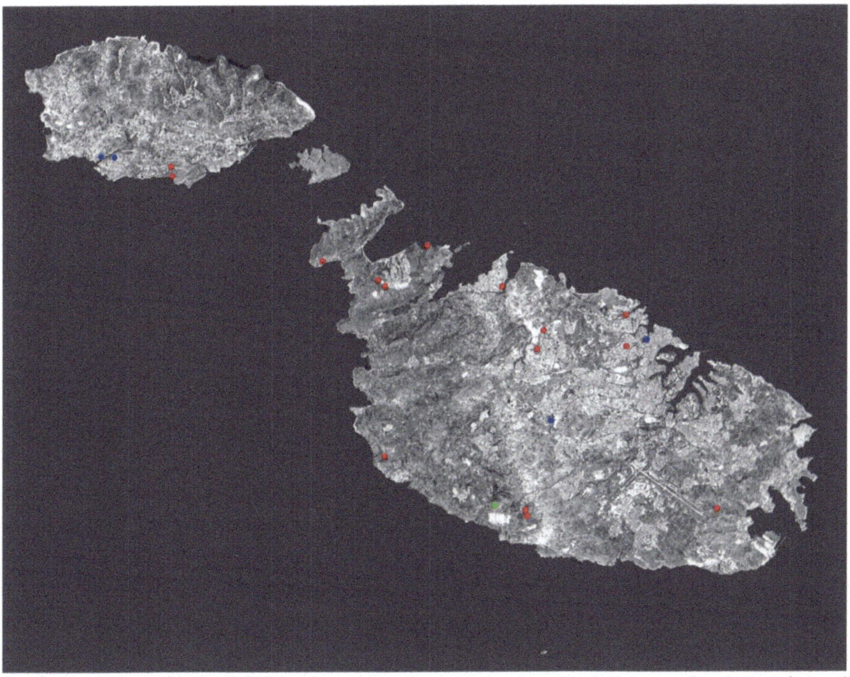

Distribution of *Elatin gussonei* in the Maltese Islands (red-existing populations, blue-recorded but not confirmed, green-destroyed by vandals from [19].

Fig. 3 An example of freely distributes dot mapping for *Elatine gussonei* (Sommier) Brullo et al., an endangered species living on Mediterranean temporary ponds [19], Fig. 7), in the Maltese islands

first peak due to some specific phenomenon, have long-term after-effects, whereby the species may persist at low numbers with a negligible chance of recovery and a severely diminished role in the ecosystem. In addition, actual presence is no longer comparable to that of an earlier time in terms of genetic resources [4].

Distribution mapping of taxa is probably the most commonly used surrogate for overall biodiversity, at both local and broader scales. The first well known European example of an atlas covering an entire regional flora was carried out for the British Isles [21]. Several projects have been undertaken in the last years to map flora in Europe and in the Mediterranean area, both at large and regional scale. Coastal biodiversity of Sicily was assessed by dividing the territory into elementary areas. These cells were used to map the presence of the biogeographically most important taxa (fauna and vascular, bryophytic and algal flora), in order to evaluate punctual and general levels of biodiversity conservation diachronically [10]. The WIKIPLANTBASE system [2] was set up in four Italian administrative regions (Toscana, Sardinia. Liguria, and Sicily) and allows free entry of historical and modern records as dots on a base map, based on field observations, bibliographic references or herbarium specimens. This system has the great advantage of being open to collaborative work greatly enriched by amateur contributions [22].

The species level can be an accepted standard for Mediterranean island monitoring and conservation plans. This is because the concept of a species is well understood also by the public, stakeholders and policy makers. Varieties or subspecies, although basic to understanding ongoing speciation, risk creating confusion and giving an unclear idea of the phenomena, diverting important resources from the entities that need conservation most. It must be kept in mind, though, that there are several species, such as *Pancratium maritimum* L., that show great interpopulation variability, so they deserve protection at as many sites as possible [13].

4 Conclusion

From the above discussion it is intuitive that remote habitat measurement is the best solution for broad-scale phenomena that can be recognized on photos. Distribution mapping of taxa is probably the most commonly used surrogate for overall biodiversity, at both local and broader scales. Freely distributed dot mapping can be used when highly detailed distribution data are available. This allows identification of the geographical position of a taxon on a cartographic support with the highest possible precision (e.g. [1]. For further statistical elaboration with commensurable units of investigation free-dot conversion in a fixed matrix is advisable [13].

References

1. Barina Z (2017) Distribution atlas of vascular plants in Albania. Hungarian Natural History Museum, Budapest
2. Bedini G, Pierini B, Roma-Marzio F, Caparelli KF, Bonari G, Dolci D, Gestri G, D'Antraccoli M, Peruzzi L (2016) Wikiplantbase #Toscana, breaking the dormancy of floristic data. Pl. Biosyst. 150(3):601–610
3. Boissier PE (1867–1884) Flora Orientalis, 1–5. H. Georg, Basileae
4. Boivin NL, Zeder MA, Fuller DQ, Crowther A, Larson G, Erlandson JM, Denham T, Petraglia MD (2016) Ecological consequences of human niche construction: examining long-term anthropogenic shaping of global species distributions. PNAS 113(23):6388–6396
5. Candargy CA (1889) Flore de l'ile de Lesbos. Plantes sauvages et cultivées, Diggelmann, Zürich
6. Čarni A, Matevski V, Šilc U, Ćušterevska R (2014) Early spring ephemeral therophytic non-nitrophilous grasslands as a habitat of various species of Romulea in the southern Balkans. Acta Bot Croat 73(1):107–129
7. Celesti-Grapow L, Bassi L, Brundu G, Camarda I, Carli E, D'Auria G, Del Guacchio E, Domina G, Ferretti G, Foggi B, Lazzaro L, Mazzola P, Peccenini S, Pretto F, Stinca A, Blasi C (2016) Plant invasions on small Mediterranean islands: an overview. Pl Biosyst 150 (5):1119–1133
8. Domina G, Mazzola P (2011) Sulla presenza di specie aliene nella flora vascolare del Mediterraneo. Biogeographia 30:269–276

9. Domina G, Venturella G, Gargano ML (2018a) Synthetic cartography for mapping biodiversity in the Mediterranean region: Sicily as a case study. PhytoKeys 109:77–92
10. Domina G, Campisi P, Mannino AM, Sparacio I, Raimondo FM (2018b) Environmental quality assessment of the Sicilian coast using a multi-disciplinary approach. Acta Zool Bulg Suppl 11:11–18
11. Ercole S, Fenu G, Giacanelli V, Pinna MS, Abeli T, Aleffi M, Bartolucci F, Cogoni D, Conti F, Croce A, Domina G, Foggi B, Forte T, Gargano D, Gennai M, Montagnani C, Oriolo G, Orsenigo S, Ravera S, Rossi G, Santangelo A, Siniscalco C, Stinca A, Sulis E, Troia A, Vena M, Genovesi P, Bacchetta G (2017) The species-specific monitoring protocols for plant species of Community interest in Italy. Pl Sociol 2:77–83
12. Gaston KJ (2010) Biodivesity. In: Sodhi NS, Ehrlich PR (eds) Conservation Biology for All. Chapter 2, Oxford University Press, Oxford, pp 27–44
13. Giovino G, Domina G, Bazan G, Campisi P, Scibetta S (2015) Taxonomy and conservation of *Pancratium maritimum* (Amaryllidaceae) and relatives in the Central Mediterranean. Acta Bot Gallica BotLett 162:289–299
14. Greuter W (1995) Origin and peculiarities of Mediterranean island floras. Ecol Medit 21:1–10
15. Greuter W (2001) Diversity of Mediterranean island floras. Bocconea 13:55–64
16. IUCN (International Union for Conservation of Nature) (2012) IUCN Red List Categories and Criteria. Version 3.2. Second Edition. IUCN Species Survival Commission. IUCN, Gland, Switzerland. https://jr.iucnredlist.org/documents/redlist_cats_crit_en.pdf. Accessed 08 Aug 2018
17. Médail F, Charrier L, Charrier M, Kashta L (2013) Vegetation map of Sazani Island (Albania). Conservatoire du littoral, Aix-en-Provence
18. Miettinen J, Stibig H-J, Achard F (2014) Remote sensing of forest degradation in Southeast Asia - aiming for a regional view through 5–30 m satellite data. Global Ecol Conserv 2:24–36
19. Mifsud S (2006) A comparative study between *Elatine gussonei* (from Malta) and *Elatine macropoda* (from Majorca), Aquatic Gardeners. https://www.aquatic-gardeners.org/Articles/elatine/Comparitive_study.pdf. Accessed 16 Oct 2018
20. Pasta S (2001) Contributi alla conoscenza botanica delle isole minori circumsiciliane. I. Sintesi aggiornata delle conoscenze botaniche sull'isola di Lampedusa finalizzata alla conservazione delle sue emergenze floristico-vegetazionali. Naturalista Sicil., s. 4, 25 (Suppl.):19–70
21. Perring F, Walters SM (1962) Atlas of the British Flora. Botanical Society of the British Isles, London
22. Peruzzi L, Bagella S, Filigheddu S, Pierini B, Sini M, Roma-Marzio F, Caparelli KF, Bonari G, Gestri G, Dolci D, Consagra A, Sassu P, Caria MC, Rivieccio G, Marrosu M, D'Antraccoli M, Pacifico G, Piu V, Bedini G (2017) The Wikiplantbase project: the role of amateur botanists in building up large online floristic databases. Fl. Medit. 27:117–129
23. Whittaker RH (1972) Evolution and measurement of species diversity. Taxon 21:213–251
24. Zaccharias I, Zamparas M (2010) Mediterranean temporary ponds. a disappearing ecosystem. Biodiv Conserv 19(14):3827–3834

The manufacturer's authorised representative in the EU is Springer Nature Customer Service Centre GmbH, Europaplatz 3, 69115 Heidelberg, Germany. If you have any concerns regarding our products, please contact ProductSafety@springernature.com

Printed and bound by CPI Group (UK) Ltd, Croydon, CR0 4YY
30/09/2026
02225691-0010